AF799147

Lecture Notes in Intelligent Transportation and Infrastructure

Series Editors

Janusz Kacprzyk, Systems Research Institute, Polish Academy of Sciences, Warsaw, Poland

Olegas Prentkovskis, Vilnius Gediminas Technical University, Vilnius, Lithuania

The series "Lecture Notes in Intelligent Transportation and Infrastructure" (LNITI) publishes new developments and advances in the various areas of intelligent transportation and infrastructure. Merging theoretical foundations, practical applications, and forward-looking insights, LNITI provides a comprehensive understanding of both the state-of-the-art and the future prospects within this dynamic field.

LNITI is designed to be an inclusive platform that covers an extensive array of topics including, but not limited to intelligent transportation systems, smart mobility, intelligent logistics, critical infrastructure, smart architecture, smart cities, intelligent governance, construction design, data security, operational analysis, optimal route planning, digitalization, autonomous vehicles, the evolution of transport systems as well as green and sustainable urban structures. The series contains monographs, conference proceedings, edited volumes, lecture notes and textbooks. Of particular value to both the contributors and the readership are the short publication timeframe and the world-wide distribution, which enable wide and rapid dissemination of high-quality research output.

Proceedings published in the series are indexed by INSPEC.

All books, including proceedings, published in the series are evaluated by Web of Science

HongSheng Qi

Stochastic Two-Dimensional Microscopic Traffic Model

Theory and Applications

Springer

HongSheng Qi
College of Civil Engineering
and Architecture
Zhejiang University
Hangzhou, China

ISSN 2523-3440 ISSN 2523-3459 (electronic)
Lecture Notes in Intelligent Transportation and Infrastructure
ISBN 978-981-97-3596-9 ISBN 978-981-97-3597-6 (eBook)
https://doi.org/10.1007/978-981-97-3597-6

© The Editor(s) (if applicable) and The Author(s), under exclusive license to Springer Nature Singapore Pte Ltd. 2024

This work is subject to copyright. All rights are solely and exclusively licensed by the Publisher, whether the whole or part of the material is concerned, specifically the rights of translation, reprinting, reuse of illustrations, recitation, broadcasting, reproduction on microfilms or in any other physical way, and transmission or information storage and retrieval, electronic adaptation, computer software, or by similar or dissimilar methodology now known or hereafter developed.
The use of general descriptive names, registered names, trademarks, service marks, etc. in this publication does not imply, even in the absence of a specific statement, that such names are exempt from the relevant protective laws and regulations and therefore free for general use.
The publisher, the authors and the editors are safe to assume that the advice and information in this book are believed to be true and accurate at the date of publication. Neither the publisher nor the authors or the editors give a warranty, expressed or implied, with respect to the material contained herein or for any errors or omissions that may have been made. The publisher remains neutral with regard to jurisdictional claims in published maps and institutional affiliations.

This Springer imprint is published by the registered company Springer Nature Singapore Pte Ltd.
The registered company address is: 152 Beach Road, #21-01/04 Gateway East, Singapore 189721, Singapore

If disposing of this product, please recycle the paper.

Preface

This book is devoted to the development of two-dimensional microscopic traffic flow models and applies them to specific real-world tasks. The application of the models and methods are specifically designed for the mixed autonomous vehicles scenario, although the main body of the book focuses on the modeling of human-driven vehicles. The modeling scenarios include both the roads (urban roads, highway roads, and on-off ramps) and intersections. However, as the main topic of the book is microscopic level, the road network scale is not involved. Nonetheless, the road network traffic flow can be simulated via the microscopic models by appropriate customization.

Modeling the traffic flow has long been the focus of traffic flow theory, which can be traced back to the thirties of the last century. In the beginning era of traffic flow, the microscopic traffic models play an important role. Just like any academic discipline, any research field begins with the definition of concept, and then the theories are developed around these concepts. Finally, the methods that apply the concepts and theory to real-world problems come into paly. Microscopic traffic model is no exception. Throughout the history, various microscopic traffic models are developed based on different assumptions. The microscopic traffic models have been successfully applied to a variety of tasks. We list some important tasks: (1) traffic simulation. Many commercial or open-source microscopic traffic simulation packages are developed. Microscopic traffic models serve as the theoretic foundations for these packages. (2) Capacity evaluation and explanation at various facilities, such as intersection, highway basic segment, on or off ramps, etc. With the help of microscopic traffic model, the improvement can be made. (3) Interpretation of macroscopic traffic flow phenomena. These macroscopic-level phenomena include stop-and-go wave, oscillation, fundamental diagram, capacity drop, etc. Many of these phenomena can be explained from microscopic traffic perspective, which involves the behavioral characteristics of drivers.

With the development of the autonomous vehicles, the transportation is reshaped a lot. Not only the traffic flow gradually shifts from pure human-driven vehicles to mixed autonomous vehicles and finally to fully autonomous vehicle flow. This process would take a long time. Thus, for a long period, the road traffic would be a mixture of human-driven vehicles and autonomous vehicles. As a result, road traffic flow operational efficiency, safety, etc., are both influenced by the HDVs and AVs. Besides, the HDVs, when driving along with AVs either as followers or as neighboring environment vehicles, may change their behaviors. These factors implicate that the study of the principles that govern the movement of HDVs is more necessary in the mixed AV flow era than the pure HDV flow or AV flow.

As the title of this book suggests, the main argument of the book is that the microscopic traffic model, although well developed, lacks two important ingredients: two-dimensional nature and stochasticity. These two attributes are basics of vehicular movement: the vehicle drives along the lane, and its planar coordinate can be expressed by (x, y), which is basically a two-dimensional variable; and needless to say, the vehicular movement is influenced by many factors, many of which are stochastic: for instance, the driver behavior is random and uncertain, the road pavement is random from the viewpoint of the vehicle, etc. Most existing microscopic traffic models are one-dimensional and thus cannot capture the operational characteristics of vehicular movement.

What Does the Book Cover?

Roughly speaking, the book can be divided into the following four components, including behavioral analysis of real-world vehicular trajectory data, the theoretic modeling of two-dimensional movement, the application, and the two-dimensional movement within intersections. They are briefly summarized as follows:

(1) The first component is the study of the movement from real-world data. Nowadays, there is a plethora of vehicle trajectory data. These data cover different countries and contain data from highway, on or off-ramp, and urban streets. With these data, the basic exploratory analysis can be implemented. The results show the two-dimensional and stochastic nature of the vehicular movement. Further, we show that the behavioral difference between different countries can be found using the predictability index, i.e., the permutation entropy. Besides, we established the behavioral inference methods from these vehicular trajectory data;

(2) The second component may be the most important part of the book. It established the two-dimensional stochastic microscopic traffic models from various perspectives. These perspectives include (i), we establish the two-dimensional microscopic traffic model considering the vehicle kinetic characteristics. Although the model is deterministic, the results show that the model can be applied well to the off-ramp trajectory prediction. (ii) The above deterministic type model is extended to the stochastic setting. The two-dimensional noise is considered. The resulting two-dimensional stochastic microscopic traffic model is analyzed regarding the stability attribute, using the Lyapunov function technique. The results show that the model can describe the vehicular movement in

stochastic configuration. (iii) Regarding the heterogeneous road (i.e., the on or off-ramp), the vehicle movement is influenced by the infrastructure, which is lacking in current literatures. We assume that there is a virtual boundary defined by the infrastructure, and this virtual boundary generates a virtual field, which influences the vehicular movement. From the observations of vehicle movements, the virtual boundary can be calibrated. The results show that the model can be used to predict the trajectories under the configuration of heterogeneous road. (iv) Lateral dimensional modeling using SDE theory. In this perspective, we focus on the lateral movement of the vehicle, and Newtonian mechanics is applied to model the lateral movement. The stochastic stability is analyzed. This stochastic lateral movement model contains parameters that have clear physical meaning interpretations, including the level of noise, the sensitivity of driver to lateral deviation, and the sensitivity of driver to the lateral noise. (v) Most researches model the two-dimensional speed and acceleration. As an example, the stimulus-reaction framework of CF models treats the acceleration as reaction and models the relationship between acceleration and its influencing factors. However, the jerk, which is the differential of acceleration, plays an important role in traffic safety. Real-world traffic data shows that jerks exist in both longitudinal and lateral dimensions. We also prove that current stochastic models, either longitudinal models (i.e. CF models) or lateral models cannot reproduce the jerk profiles existing in the real-world data. This motivates us to model the two-dimensional movement considering the jerk attributes. The jump-diffusion approach is applied and the resulting trajectory shows that the model can reproduce the speed-dependent jerk distribution.

(3) The third component mainly applies the model developed in the above models into real-world problems. These problems include (i) the capacity with two-dimensional noise for mixed autonomous vehicle flow. The famous HCM is purely devoted to the systematic measurement of capacity in various facilities. However, conventional capacity is evaluated for HDV flow. In the era of progressively mixed AV flow, the behavioral change of vehicular movement necessitates the re-evaluation of the capacity. We study the capacity adjustment factor for the multi-lane highway considering two-dimensional noise when the MPR changes from pure HDVs to pure AV flow. The results show that current methodologies overestimate the capacity when the lateral noise is overlooked. (ii) We apply the stochastic lateral movement model to the evaluation of lateral wandering when the flow is mixed AV flow, which is one of the major factors that influences the pavement life. While there are empirical observations of lateral wandering, there is a lack of theoretical investigation into this phenomenon in scenarios involving mixed automated vehicle flow. To address this, we developed a theoretical model that represents stochastic lateral movement, considering influences from road boundaries, lane markings, neighboring vehicles, and stochastic noise. The findings demonstrate that the model effectively replicates empirical

observations of the wandering pattern. Moreover, the study examines the influence of AV market penetration rates on both one-lane and multi-lane roads, providing insights into the analysis of pavement sustainability under mixed AV flow conditions, including situations with dedicated lanes. (iii) Quick and accurate inference of lane-changing events is crucial for autonomous vehicles. The lane-changing events are characterized by many attributes, including change lane or not, lane-changing duration, and the moment when vehicle traverses the lane mark. However, most practices treat these sub-tasks of inference separately. Besides, the data-driven approach requires traffic flow data as a prerequisite, and the resulting methods cannot be applied to a new place where the data may not be applicable. To address this drawback, we propose a logistic diffusion approach based on a stochastic logistic diffusion type of stochastic differential equation. Our proposed framework incorporates parameter calibration, LCIM inference, CLMM estimation, and LCD estimation within a unified framework, and successfully provides real-time output for these desired information.

(4) Intersection traffic flow modeling. The role of intersections is to separate the flows from different approaches, such that they can traverse the intersection in an efficient and safe way. However, during rush hours, the vehicles stuck within the intersection, and the intersection deadlock forms. This issue becomes more severe in mixed autonomous vehicle flow, due to the fact that, not like HDVs which can cooperate and check the deadlock visually, nowadays AVs cannot achieve this. In this regard, we developed three models. (i) A concept called outlet saturation flow rate (OSFR) is defined, which reflects the impact of game behavior among drivers at the outlet. A model is proposed that divides vehicle movements into three parts and includes a game model for outlet lane-choosing. Based on empirical data, the model is calibrated with a stop-line headway distribution model, speed model, and payoff function. Results suggest that the proposed model can successfully generate real-world OSFR distributions and describe the interaction between vehicles, making it a useful tool for predicting outlet headway distribution in scenarios with mismatched lanes. (ii) We employ PN to investigate both the probability and duration of deadlock formation. We define three classes of system states: deadlock state, trap state, and live state. The occurrence of deadlock is formalized as an integer programming problem, while the probability and duration of deadlock formation are investigated through the construction of a reachability graph that leads to the formation of a Markov chain for the system. (iii) We develop a protocol to detect, prevent, and recover from traffic deadlock through the utilization of connected and autonomous vehicle technology. The formation and characteristics of deadlock are initially explored, with the introduction of two distinct forms: weak deadlock and strong deadlock. Subsequently, two detection algorithms are proposed for identifying weak and strong deadlock scenarios, using evasion distance propagation as the underlying basis. Following this, a cooperative control method is presented to avert deadlock, employing strategies that are both chain-spillover-free and loop-free. The

results indicate that, on average, the algorithm can anticipate the occurrence of deadlock approximately 13 seconds in advance.

Who Is the Book For?

This book deals with the microscopic traffic models, which falls in the field of transportation engineering. However, given that the microscopic traffic model plays an important role in autonomous vehicle R&D development, this book also offers valuable resources for autonomous vehicle industry practitioners.

For traffic engineers, this book extends the current microscopic traffic model frameworks, by incorporating both stochastic nature and two-dimensional movement. As a result, the provided models are more realistic for vehicular movement description. Such improvements can be embedded into the traffic simulation package or procedures. The resulting conclusions would be more realistic.

For autonomous vehicles industry practitioners, the books contain behavioral modeling of the driver's motivations and the resulting stochastic two-dimensional trajectories. The stochasticity existing in both dimensions is universal. Yet current microscopic traffic models cannot describe them very well. A typical example is the lateral description limitations. The materials provided in this book fill this gap by developing a systematic framework considering various factors. The resulting models and methods are beneficial to the autonomous vehicle testing, particularly in the background traffic flow generation.

This book can also be used by entry-level graduate students in transportation. In addition, civil engineering juniors and seniors may find some in-depth information about traffic flow. Further, applied mathematics majors may find concrete examples of mathematical modeling with specific domain knowledge. Advanced readers are referred to other traffic flow theory books for in-depth coverage.

Which Basic Knowledge is Required to Read This Book?

The major methods of models required to understand the book including the following knowledge: ordinary differential equation, stochastic differential equation, basic control theory include state space model, Lyapunov stability theory, jump-diffusion method, partial differential equation, likelihood estimation methods, basic game theory concepts, and vehicle dynamics.

In closing, it is important to acknowledge the inherent limitations of this work. Despite diligent efforts to uphold the standards of quality and accuracy, fallibility is an inherent aspect of the human endeavor, and as such, errors may inadvertently arise. Consequently, it is advised that readers exercise discernment and discretion when utilizing the content provided within this book.

Hangzhou, China
December 2023

Hongsheng Qi

Acknowledgements

This work was supported by the National Natural Science Foundation of China (No. 52272314; 52131202); the Ministry of Education in China Project of Humanities and Social Science (21YJCZH116); and Zhejiang Province Public Welfare Scientific Research Project (LGF22E080007).

During the writing of this book, many people provide invaluable support to the author. There cannot be a list of them as any list is incomplete. Special thanks are given to the master students of the term, including Jiahao Zhang, Yukai Zhang, and Yuyan Ying. They provide enormous help to the author.

Contents

Abbreviations

ACC	Adaptive Cruise Control
AIGC	AI-Generated Content
BG	Blockage Graph
CACC	Cooperative Adaptive Cruise Control
CAF	Capacity Adjustment Factor
CAV	Connected and Autonomous Vehicles
CellularPN	Cellular automate equipped Petri Net
CF	Car Following
CLMM	Cross-lane-mark moment
CoG of CG	Center of gravity
ConflictPN	Conflict solution Petri Net
CPP	Compound Poisson Process
CTM	Cell Transmission Model
DED	Distance Escaping from Deadlock
DemandPN	The input demand Petri Net. It provides the inflow of the intersection
DES	Discrete Event System
DL	Dedicated Lane
Doerf	Derivate of error function
DRL	Deep Reinforcement Learning
DTW	Dynamic Time Wrapping
EDP	Escape Distance Propagation
EKF	Extended Kalman Filtering
FD	Fundamental Diagram
FEFS	First Enter First Service. It illustrates a special priority rule at the confliction point
FPE	Fokker-Planck Equation
GL	General Lane
GPT	Generative Pre-trained Transformer
GRU	Gated Recurrent Units
HCM	Highway capacity mannual

HDV or HV	Human-Driven Vehicles
IDM	Intelligent Driver Model
IntersectionPN	Intersection Petri Net. IntersectionPN = DemandPN+ TrafficLightPN+
LC	Lane-Changing
LCD	Lane-Changing Duration
LCE	Lane-Changing Execution
LCIM	Lane-Changing Initiation Moment
LLM	Large Language Model
LSTM	Long Short-Term Memory
MAE	Mean Absolute Error
MAPE	Mean Absolute Percentage Error
MEPDG	Mechanistic-Empirical Pavement Design Guide
MLP	Multi-layer Perception
MPR	Market Penetration Rate
MSE	Mean Squared Error
NCHRP	National Cooperative Highway Research Program
NDD	Naturalistic Driving Data
NN	Neural Network
OBD	On Board Device
ODE	Ordinary Differential Equation
OSFR	Outlet Saturation Flow Rate
OU process	Ornstein-Uhlenbeck Process
OV	Optimal Velocity
PCU	Passenger Car Unit
PDE	Partial Differential Equation
PE	Permutation Entropy
PID	Proportional-Integral-Derivative Control
PN	Petri Net
PSO	Particle Swarm Optimization
RBF	Radius Basis Function
RMSE	Root Mean Squared Error
RNN	Recurrent Neural Network
SAE	Society of Automotive Engineers
SDE	Stochastic Differential Equation
SFR	Saturation Flow Rate
STD	Standard Deviation
SVM	Support Vector Machine
SVR	Support Vector Regression
TOD	Time of Day
TrafficLightPN	Traffic Signal Petri Net
TTC	Time to Collision

Chapter 1
Introduction

1.1 Before We Begin

The understanding of vehicular movement has been a long-standing focal point in traffic flow theory, representing an intriguing topic that warrants further investigation. The modeling of such movement methodologies not only finds direct application in traffic analysis, but also holds substantial potential within the AVs industry, particularly as CAVs progressively infuse the system. Notably, vehicular movement models have significant roles to play in the development of AVs' technical stack.

Throughout the history of traffic flow development, researchers have developed various models, methods, and tools for describing and analyzing the movement of vehicles. The models and methods presented in this book represent a select few among them, proposed to rectify some of the limitations of current counterparts. These limitations have rendered present models inadequate in capturing critical traffic flow characteristics as well as rendering them impractical for real-world tasks. This introductory chapter is devoted to the discussion of current practices, and paves the way to the model development of the book.

1.2 Microscopic Versus Macroscopic

Generally, any textbook on traffic flow would typically discuss the various scales of traffic flow modeling, namely microscopic and macroscopic. Some additional scales, such as mesoscopic and picoscopic, are also occasionally mentioned in certain materials. The present discussion aims to provide a brief overview of the microscopic and macroscopic approaches, and their implications for two-dimensional modeling in this particular book.

The microscopic traffic flow modeling approach treats each vehicle individually, representing its movement with a single mass point. A stack of microscopic traffic

© The Author(s), under exclusive license to Springer Nature Singapore Pte Ltd. 2024
H. Qi, *Stochastic Two-Dimensional Microscopic Traffic Model*, Lecture Notes in Intelligent Transportation and Infrastructure,
https://doi.org/10.1007/978-981-97-3597-6_1

models is constructed to describe the movement of these points, commonly referring to the midpoint of the front bumper as the representative point. The state variable primarily considers acceleration, which is influenced by speed and distance differences, among other factors. Consequently, the microscopic traffic flow model can be described using a series of ordinary differential equations (ODEs) with or without delay, summarized within a stimulus-reaction framework. It is important to note that, despite its accuracy, this modeling technique has inherent limitations due to the lacking of considerations of vehicular kinetics or dynamics model.

On the contrary, macroscopic traffic flow theory adopts a fluid viewpoint for analyzing traffic flow. Therefore, the target system, or state variables, are defined continuously throughout, encompassing flow rate, density, and velocity. The governing equation primarily takes the form of a partial differential equation (PDE), which describes the evolution of the system's state. These PDEs can be solved using numerical methods, among which the cell transmission model (CTM) stands as one of the most renowned approaches.

This book, as the title implies, centers around microscopic traffic flow modeling. Nevertheless, it employs distinct methodologies with a specific focus on vehicle kinetics and the lateral dimension, which are the subjects of discussion in subsequent sections.

1.3 The Basics of Microscopic Traffic Flow Modeling

As previously discussed, the microscopic traffic model employs a single mass point to represent a vehicle, as shown in Fig. 1.1. In Fig. 1.1a, a leader vehicle and a follower vehicle are traveling along a roadway. Assuming the leader's speed remains constant, the objective of the follower is to position itself at the target location, which can be defined by the equilibrium density-speed relationship. Traditional traffic flow models assume that the vehicle adjusts its state using acceleration, as depicted in Fig. 1.1b. Various models have been developed based on this approach, including the Pipe model, GM model, Gipps model, and IDM model among others.

However, it is important to note that drivers adapt their behavior by manipulating both steering and acceleration. Even when a driver does not change lanes, they must adjust the steering to compensate for deviations from the lane centerline. Additionally, the movement of a vehicle is influenced by neighboring vehicles. If a neighboring vehicle gets too close, the ego vehicle will steer away to avoid potential conflicts. Consequently, when describing vehicular movement, it is necessary to consider both the vehicle's kinetic characteristics and its two-dimensional movement nature. The resulting dynamics of traffic flow can be illustrated as shown in Fig. 1.1c. The ego vehicle continuously fluctuates around the middle line of the lane and influences Vehicle 1 at the lateral dimension, pushing it away from the ego vehicle. The half ellipse represents the area of influence, which depends on the speed and characteristics of the driver. Vehicle 1 is also within the influence of Vehicle 2, resulting in Vehicle 2 being affected by Vehicle 1. This influencing chain can be recursively

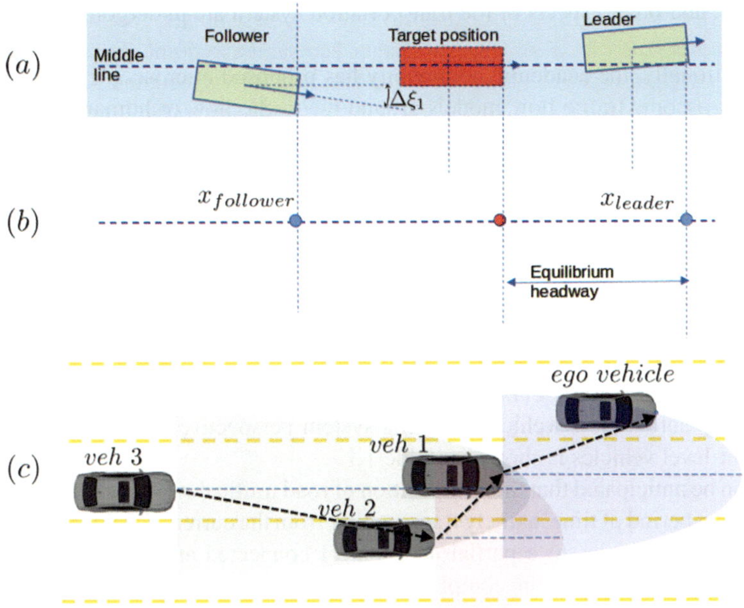

Fig. 1.1 Microscopic description of vehicle

extended infinitely. Furthermore, it is essential to incorporate the vehicle's kinetic model, as depicted in Fig. 1.1a. The control input of the follower, specifically the steering angle, should be included in the model.

The aforementioned extension represents more than just an enhancement in modeling techniques. The requirements for adopting such a modeling approach are driven by practical applications, particularly in the imminent era of autonomous vehicles. It is anticipated that autonomous vehicles will coexist with human-driven vehicles (HDVs) for an extended period of time, posing significant challenges for the development of autonomous vehicles. These challenges will be thoroughly examined and elucidated in the subsequent section.

1.4 Autonomous Vehicles and Challenge

The development of the new generation of information technology and artificial intelligence technology has brought new impetus to the informatization and intelligentization of the transportation system. Under a series of policy guidelines, such as the "Outline of Long-term Development Plan for Science and Technology Innovation in the Field of Transportation (2021–2035)" and the "Fourteenth Five-Year Plan for Science and Technology Innovation Development," the analysis theory, technical methods, and other aspects of the transportation system are undergoing a paradigm shift.

Traditionally, the academic community has proposed microscopic, mesoscopic, and macroscopic traffic flow models around the traffic flow of human-driven vehicles. By relying on multi-source traffic data to analyze and estimate traffic states, multiple levels of traffic management and optimization control methods have been developed, playing an important role in ensuring the efficient operation of the transportation system. However, the above-mentioned traditional theories and methods are difficult to adapt to the increasingly intelligent and networked heterogeneous traffic participants and infrastructure. According to the Chinese national standard GB/T 40429-2021, intelligent vehicles are classified into six levels from 0 to 5. The Society of Automotive Engineers (SAE) also divides autonomous driving into six levels: L0–L5. L0 represents conventional human-driven vehicles, and L5 represents fully autonomous vehicles. From a system perspective, the controllability of different-level vehicles is shown in Table 1.1.

It can be anticipated that the composition of road traffic flow will gradually evolve over a long period of time, namely transitioning from the current dominance of manually operated vehicles to a partial adoption of connected and automated vehicles (CAVs) and eventually to the complete dominance of CAVs. This evolution presents challenges to the theories and methods of traffic system analysis, as well as perception and decision-making for intelligent connected vehicles. These challenges are particularly manifested in the following aspects:

(1) The subject of traffic control is the traffic flow as a whole, rather than individual vehicles. The rich source of real-time microscopic trajectory information provided by intelligent connected vehicles introduces the issue of consistency between macroscopic and mesoscopic spatiotemporal traffic flow analysis models and microscopic trajectory states.

Table 1.1 Classification of autonomous vehicles from a controllability perspective

Control hierarchy Automatic driving level (SAE level)	Travel end level	Travel path level	Section class Lane level	Trajectory level	Control level (throttle/brake, steering)
Levels 0 and 1 (L0 and L1)	○	○	○	○	◐
2 (L2 level)	○	○	○	◐	◐
3 (Level L3)	○	○	◐	◐	◐
4 (Level 4)	◐	◐	◐	◐	●
5 (L5 level)	●	●	●	●	●

Note "●" indicates that the vehicle is fully controllable at the level of autonomous driving; "◐" indicates that it is controllable or partially controllable only in certain scenarios, and "○" indicates that it is not controllable

(2) From the perspective of intelligent connected vehicles, different levels of control and decision-making require varying information. At the level of vehicle control and trajectory planning, information on neighboring vehicles' short-term motion states is needed, while decisions at the level of road segments, routes, and trip destinations require long-term information on distant traffic flow states. Therefore, the microscopic decision-making of intelligent connected vehicles is a result of incorporating multiscale spatiotemporal information.

The multiscale characteristics of intelligent connected transportation systems mentioned above are illustrated in Fig. 1.2. With respect to the temporal granularity of traffic flow modeling, the scales include real-time (intelligent connected vehicles can provide two-dimensional coordinates, acceleration, etc., at any given moment), seconds, minutes (for path decisions, signal cycles), and even time-of-day (TOD) levels. Regarding the spatial granularity, the scales range from the entire road network to routes, road segments, and lanes, all the way to the continuous two-dimensional trajectories of intelligent connected vehicles. In terms of control granularity, this includes the coordination of trajectories among intelligent connected vehicles, lane-level traffic flow (lane selection by intelligent connected vehicles), route-level traffic flow (route selection by intelligent connected vehicles), signal control, and multimodal sharing.

The interweaving of multiple temporal and spatial scales and optimization methods presents challenges for the modeling of traffic flows and raises new issues for traffic management. In terms of traffic flow modeling, the traditional classification system of micro, meso, and macro scales is no longer applicable. Next-generation

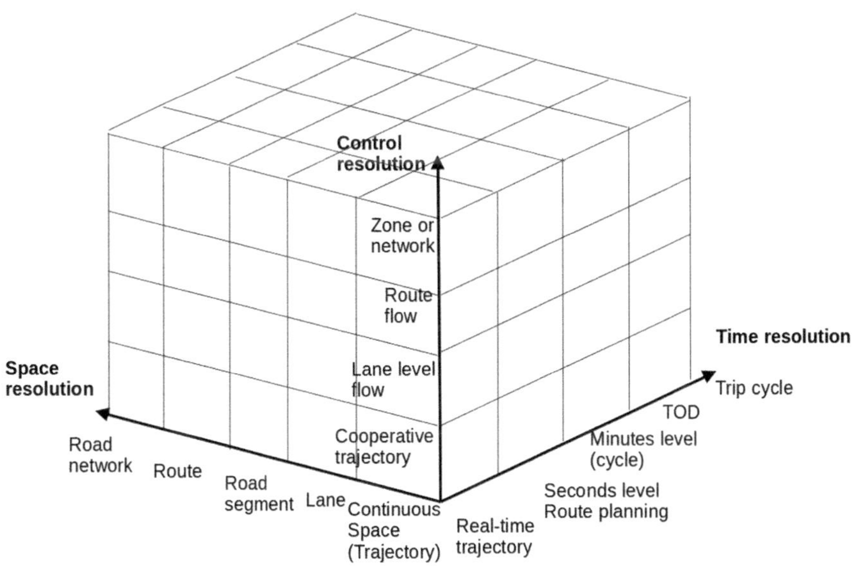

Fig. 1.2 Multi scale of mixed autonomous vehicles system

traffic flow models must combine micro-vehicle modeling with macro-traffic flow description to construct multi-scale traffic flow models. With respect to different control methods, their temporal and spatial scales and influence vary. For instance, multi-vehicle trajectory coordination planning targets both intelligent connected vehicles and surrounding vehicles within a time window, whereas signal cycle optimization targets the arrival and departure flows of vehicles at intersections within a time window. Therefore, optimization methods need to consider the multi-scale temporal and spatial domains of control measures in the context of heterogeneous traffic flows.

1.5 A Review of Literature on Microscopic Traffic Flow Modeling

1.5.1 CF Models: 1D and 2D

1.5.1.1 Deterministic 1D and 2D CF Models

Car following may be the most studied theoretical microscopic traffic flow behavior model. Typically, it is formulated by one or several ordinary differential equations. In the CF model, each vehicle is regarded as a single point, which is a reference point on the vehicle body. This reference point is typically defined as being at the middle point of the front bump. Traditional CF is constructed, based on the stimulus-reaction framework, i.e., the driver reacts to a change in the state of the leading vehicles, either in the current lane or in a neighboring lane, by adjusting the ego vehicle's speed (or equivalent, through acceleration). Different models are developed, based on different assumptions under this stimulus-reaction framework, including Pipes model, GM model, Gipps model, IDM model, etc. In the above models, the stimulus includes the gap, speed difference, space headway, etc., while the reaction is mainly acceleration. An appropriate microscopic traffic model should replicate the characteristics of macroscopic traffic flow. For instance, the flow-density relationship, generated by the model, should be consistent with empirical observations.

The basic CF model only describes longitudinal dimension movement, while the lateral counterpart is totally neglected. On the other hand, with the development of CAV technology, the demand for detailed two-dimensional vehicular modeling is increasing. To deal with this issue, the basic CF framework is extended toward a two-dimensional model, or is enhanced with weak-lane discipline. The former one equips the traditional CF model with lateral dimensional dynamics (lateral acceleration or speed), while the latter counterpart mainly refers to the non-lane based model. Regarding the former type, Delpiano et al. (2020) proposed a two-dimensional car model that considers the lateral friction effect; In Sharath and Velaga (2020), the one dimensional IDM model was extended to a two-dimensional case. The effective lateral gap is defined. Basically, the input–output structure is the same as IDM: the outputs of the model are accelerations, although the lateral and longitudinal components are decomposed. For the weak-lane based models, the common car-following framework is also inherited. The lateral gap is defined. These models partly solve the problem of traditional car-following models. However, although there are a plethora of models available that can describe vehicle movement, these models do not respect vehicular dynamics and, thus, the results from the model may not be drivable for the vehicle. Besides, the stochasticity is still not very well represented in the model framework. The simulation platform, such as SUMO or Carla, cannot replicate traffic-flow characteristics. It is concluded that, to better understand traffic phenomena, a microscopic traffic-flow model, with vehicle dynamics and stochasticity representation, is necessary.

1.5.1.2 Stochastic CF Model

It is obvious that road traffic flow is stochastic and major cars that follow models are developed as a deterministic system. However, some macroscopic level phenomena are closely related to stochasticity, such as oscillation. Some researchers extend traditional models with a different randomness consideration. These extensions include white acceleration noise, Brownian motion, geometric Brownian motion, etc. The base models are either a simple model with different regimes or a complex IDM model. With these stochasticity considerations, some phenomena, including speed-density relationship and oscillation can be explained or reproduced. However, it should be noted that noise exits at both lateral and longitudinal dimensions, while a majority of the stochastic models focus on a single dimension noise. On the other hand, just like the traditional microscopic traffic-flow model, a vehicle is represented by a single point, which neglects vehicular dynamics.

1.5.2 Lane Changing Process Modeling

Lane changing modeling has long been a topic of interest in traffic flow theory. Generally, the complete lane changing process can be divided into two stages: lane changing decision and lane changing execution (LCE). There are four types of lane changing decision models: rule-based models, discrete choice-based models, artificial intelligence models, and incentive-based models. A representative rule-based model proposed by Gipps (1981) describes driver lane changing behavior based on predefined rules, such as a favored lane. The discrete choice model employs a typical choice model framework, such as the logit choice model, to quantify the advantages of specific lanes considered as target lanes. Artificial intelligence models encompass fuzzy logic and neural networks. The incentive-based models assume that drivers maximize their benefits when making lane changing decisions. It is important to note that both the lane changing decision process and the lane changing execution are inherently stochastic. Several studies provide empirical observations of the lane change duration (LCD). For instance, Sun and Li (2015) concluded that free lane changes last approximately 1.5–3.5 s, while cooperative lane changes last 2–6.5 s. Observations by Hill et al. (2015) and Yang et al. (2012) also confirm variations in LCD.

1.5.3 Lateral Dynamics

1.5.3.1 Lateral Movement in Lane Keeping Stage

As previously stated, microscopic modeling of traffic behavior can be classified into the CF theory and LC theory. Lateral movement is typically disregarded in most CF models, whereas it is not independently addressed in the LC formulation. To explore lateral movement, it is possible to extend either the conventional CF or LC model (as previously demonstrated by Raju et al. (2021)), or alternatively, formulate a statistical model that is based on observational data. This stochastic model can involve techniques such as distribution fitting, or linear/nonlinear regression (as elucidated by Delpiano et al. (2015)).

When considering the lateral component in theoretical car-following models, the non-lane-based approach may be the first to take into account the vehicle's lateral displacement. This type of model includes the impact of the lateral gap between the pair of the lead and following vehicles. However, the stochastic evolution of the lateral movement is often neglected in these non-lane-based models, as the focus lies in capturing vehicle-following behaviors. Other theoretical models aim to incorporate the lateral movement into account with the development of two-dimensional car-following models. For example, Raju et al. (2021) embedded the lateral factor into the Intelligent Driver Model and the Gipps model, where the acceleration/speed of the above two models were extended by a normalized lateral shift. Mulla et al. (2019) proposed a two-dimensional or lane-less microscopic model that takes into account the lateral movement of the vehicle. In their work, the lateral movement is assumed to place the vehicle at the midpoint of surrounding vehicles. Mullakkal-babu et al. (2021) suggested a sub-microscopic traffic flow model that combines car-following, lane-changing, and vehicle dynamics. In their proposed model, the vehicle's physical model explicitly accounts for the lateral movement. Lin et al. (2016) introduced a lateral movement model that considers regulating the lateral acceleration for lane keeping or moving toward a target lane. In this model, the lateral acceleration is computed based on the lateral distance to the target location. In contrast, Delpiano et al. (2020) developed a two-dimensional car-following model that considers the lateral movement while accounting for the acceleration force, repulsive force, and lane force that influence the vehicle's movement. However, all aforementioned models are deterministic, and may therefore not be capable of uncovering or reproducing the randomness in the lateral dimension.

Literature covering real-world instances of lateral movement can be found in several works, which consider the presence of factors such as mixed traffic flow involving motorcycles or weak lane discipline. Specifically, Chand et al. (2016), Lin et al. (2016), Mulla et al. (2019), and Munigety and Mathew (2016) contain observations of lateral movement. In Chand et al. (2016), the authors employ the Hurst exponent, a measure from chaos theory, to quantify observed lateral movement. They conclude that there is a strong positive correlation between Hurst exponent values greater than 0.5 and lateral movement. It is important to note that the Hurst exponent is influenced by factors such as vehicle type. An investigation by Pal and Chunchu (2019) on the lateral gap maintained between vehicles when passing slow-moving ones results in a wide range of lateral gaps. Further, Kotagi et al. (2020) study lateral movement in India, determining that lateral displacement can reach up to 3 m and establishing a relationship between lateral movement and speed using multiple linear regression analysis. However, it should be highlighted that the majority of real-world observations of lateral movement occur at fixed locations (fixed section or fixed segment) and do not include a temporal factor, except as in Chand et al. (2016).

1.5.3.2 Lateral Movement in Lane Changing Stage

The process of executing lane changes involves lateral movement and can be categorized into discretionary and mandatory types. Broadly, the overall lane change procedure can be divided into distinct stages: lane change intention, lane change decision, and lane change execution (LCE). During LCE, both the longitudinal dimension (as observed by Yang et al. 2016) and the corresponding lateral dimension (including parameters such as lateral speed, acceleration, and position) follow a certain stochastic evolution (as noted by Xia et al. 2021). Researchers have employed various data-driven approaches to analyze the stochastic characteristics of lane changes. These methodologies encompass clustering, regression, time series analysis, survival analysis [17], and deep networks. Yang et al. (2015) established that lateral speed initially increases and subsequently decreases. Lateral trajectories generally exhibit an 'S'-shaped pattern and can be fitted using empirical data. For example, utilized a polynomial function and found that a fifth-order polynomial provided the best fit. In the study by Yang et al. (2015), a parabolic curve was used to model lateral speed, while a third-order polynomial was employed for lateral position. Asaithambi and Joseph (2018) investigated the distribution of lateral shift duration during lane changes. They discovered that this duration ranged from 0.5 to 15 s and followed a Weibull distribution, which was influenced by factors such as vehicle types and the space gap with leading vehicles.

The lane changing event is characterized by a duration known as lane changing duration (LCD). Numerous factors contribute to the LCD, and scholarly literature presents various definitions and observations of this phenomenon [17]. For instance, Corey et al. (2015) and Olsen et al. (2002) define the beginning of the event as the moment when the vehicle initiates lateral movement and the end as when the vehicle is centered in the target lane. Christian et al. (2008) defines the LCD as the difference between two moments based on half the vehicle's width. Li et al. [17] determines the onset and end of the event when the lateral speed reaches zero, while employs lateral speed to segment the lane changing process. Consequently, reported LCD values vary. Corey et al. (2015) reports an average LCD of approximately 5.5 s, Olsen et al. (2002) reports 6.28 s, and [17] reports 5.24 s.

1.5.4 Data Driven Approach Toward CF and LC

1.5.4.1 Data Driven CF Modeling

There are many data-driven methods. As the vehicular trajectories datasets are readily available, various methods are devoted to the modeling of data-driven methods, which include NN, SVM, etc. We briefly discuss the existing modeling approach. The list of these methods are given in Table 1.2.

Neural network

Kehtarnavaz et al. [14] conducted a study where they trained two controllers using time-delay back propagation neural networks for the purpose of speed control and steering control. The training data consisted of measurements of the preceding vehicle's distance and heading angle. Remarkably, these controllers achieved autonomous control without the necessity of characterizing the dynamic properties of the vehicle. Additionally, Xu et al. (2007) developed a car-following model that relied on a hybrid neural network model, employing car-following data obtained from a highly precise on-board GPS system. The hybrid model's weights were fine-tuned using real measurement data, yielding significantly improved accuracy.

SVM (support vector machine)

Wei and Liu [31] proposed a self-learning support vector regression (SVR) method to investigate the asymmetric characteristics in vehicle following behavior and their impact on traffic flow evolution. By employing the self-learning approach, they analyzed asymmetric driving behavior and reproduced different congestion propagation patterns based on the distribution of the "Neutral Line" among heterogeneous drivers. Additionally, they discovered a positive correlation between asymmetric driving behavior and macroscopic hysteresis in the fundamental diagram. Zhang et al. (2018) introduced road conditions into the decision-making mechanism for autonomous driving, known as the DDM. They optimized the DDM by incorporating support vector regression models and particle swarm optimization algorithms, resulting in a DDM suitable for autonomous vehicles. Through comparisons with BP neural network models and SVR models based on the RBF kernel function, they validated the accuracy of the optimized SVR model.

K-Nearest Neighbors

He et al. [10] proposed a non-parametric vehicle following model based on field data, which matches K similar historical driving scenarios to output the most probable driving behavior in the current scenario. Huang and Zheng [11] developed a traffic flow model based on cellular automata and partitioned K-nearest neighbors. This model first provides updating rules for vehicle states and lane-changing rules, establishing a vehicle state prediction model based on partitioned K-nearest neighbors.

Recurrent Neural Network (RNN)

Zhou et al. [34] proposed a micro-level vehicle following model based on recurrent neural networks. This model is capable of effectively simulating the interactions between vehicles and the dynamic changes in traffic flow, thereby enhancing traffic flow stability and safety. The paper compared the proposed model with classical vehicle following models such as the Intelligent Driver Model (IDM) and demonstrated that the proposed model outperforms IDM in predicting traffic fluctuations with different driver characteristics. Das et al. [3] presented a gated recurrent unit neural network model for predicting the response of a following vehicle interacting with two preceding vehicles simultaneously. By learning the interactions and dynamic changes among vehicles based on historical data of vehicle following behavior, the model significantly improves the accuracy of trajectory reconstruction.

Long Short-Term Memory (LSTM)

Wang et al. (2018) utilized the speed, speed difference, and position difference between the preceding and following vehicles within a recent time interval as inputs to construct a long short-term memory network, which mimics the predictive ability and memory effect of human drivers. By testing with real trajectory data, this vehicle following model exhibited superior accuracy compared to existing models. Wang et al. [29] explored the influence of historical vehicle following information at different time scales on the predictive ability of neural networks, confirming the significance of long-term memory for prediction accuracy and stability of the model. Additionally, the paper proposed a deep learning-based vehicle following model that captures long-term features in vehicle following by utilizing long short-term memory networks (LSTMs). Huang et al. [12] established a following model based on long short-term memory networks, focusing on capturing the asymmetric characteristics displayed by drivers during vehicle acceleration and deceleration. They investigated the phenomenon of time delay closely associated with asymmetric driving behavior and the driver's unstable driving behavior. The proposed model successfully captures asymmetric driving behavior, leading to more accurate predictions of vehicle following behavior. The experiment demonstrated high prediction accuracy and stability of the model. Lin et al. [20] analyzed the reasons behind the propagation of temporal and spatial errors in traditional training methods from the perspective of vehicle queue generation. The proposed long short-term memory network overcomes the propagation problems of temporal and spatial errors in the vehicle following model by extracting trajectory features from the perspective of the vehicle queue, thereby improving the prediction accuracy of vehicle following behavior.

Seq2Seq

Ma and Qu [22] proposed a vehicle following model based on sequence-to-sequence learning, which considers memory effect and reaction delay and is able to better learn historical information and accurately complete multi-step predictions. Compared to the IDM and LSTM models, this model performs better in trajectory reconstruction and capturing heterogeneous driving behavior. Lu et al. [21] proposed an improved CAV system following model based on sequence-to-sequence deep learning, which considers the motion information of multiple preceding vehicles and adopts an improved sequence-to-sequence deep learning framework. The model integrates an encoder with bidirectional gated recurrent units (GRUs) neural network and a decoder with attention-based GRUs neural network in an end-to-end manner. The experimental results show that this model outperforms the baseline vehicle-following model in learning heterogeneous driving behavior and has higher accuracy in speed and position prediction.

DRL

Zhu et al. [40] propose an automatic vehicle following planning framework based on deep reinforcement learning. By inputting historical driving data (velocity, relative velocity, and vehicle spacing) into a simulation environment, the optimal strategy for controlling the acceleration of the following vehicle is obtained through trial-and-error learning based on a reward function. This model can continuously update when inputting long-term sequential data. Song et al. [27] propose a vehicle following control strategy that combines reinforcement learning (RL) and supervised learning (SL). RL is used for multi-objective cooperative optimization of vehicle following control, while SL is used to achieve human-like following behavior control.

GAN

Zhou et al. [38] proposed a following model based on generative adversarial imitation learning, which can learn driving strategies from the behaviors of real drivers without specifying rewards. By integrating the Gated Recurrent Neural Network (GRNN) into the reward function, the model is endowed with the ability to explore historical information. Cross-validation results demonstrate that compared to intelligent driver models and RNN-based following models, this model can more accurately reconstruct the vehicle following trajectory and driving style of the driver. Xu et al. [32] introduced a novel vehicle following model based on Transformer-GAN (TransGAN) that considers missing data. Firstly, a multi-head attention Transformer network is used to extract latent features from incomplete vehicle state data. Then, a generative adversarial network is constructed to generate multi-step control states of the target vehicle in the future. Chen et al. [1] proposed a hybrid conditional autoencoder generative adversarial network (CAE-GAN) model based on multiple loss functions. This model adopts an encoder-decoder structure that effectively handles noise and uncertainty in trajectory data. Generative adversarial networks (GANs) are used to extract confident features of the generated trajectories, making the generated trajectories closer to real trajectories. Experimental results demonstrate that the CAE-GAN model exhibits high accuracy and robustness in predicting vehicle trajectories.

Attention mechanism

Fang et al. [5] proposed a novel data-driven dynamic transformation CF model consisting of a long-term prediction sub module based on modified Seq2Seq model and a short-term prediction sub module based on Transformer, aiming to balance the accuracy of simulating traffic flow fluctuations between multiple HVs and prediction efficiency. Wang et al. [28] introduced an attention-based and data fusion vehicle following model (ABF-CF), which constructed a multi-source dataset by fusing field data and simulation data to improve the model's generalization ability. Experimental results demonstrated that compared to the baseline model, the ABF-CF model exhibited higher accuracy and stability in predicting vehicle following trajectories. Ablation experiments indicated that the encoder-decoder structure and attention mechanism could enhance prediction accuracy, while data fusion could improve generalization ability. Zhu et al. [39] developed a vehicle following trajectory prediction model based on Transformer with multi-head attention, focusing on long-term sequences. By using an encoder-decoder architecture, the model took historical speed and distance data as input and formed a blended representation of past driving contexts using multi-head self-attention mechanism. Experimental results showed that the model outperformed IDM, MLP, and LSTM models in terms of accuracy for long sequence trajectory prediction.

1.5.4.2 Data Driven LC Modeling

CNN for lane changing

Zhang et al. (2022) proposed a lane change prediction method based on a time convolutional network to capture the stochastic time series of high-resolution lane change behavior. This method processes vehicle trajectory data to extract key features and models these features using a time convolutional network to predict lane change behavior. Mozaffari et al. [23] introduced a multi-task model that estimates both the probability and time-to-lane-change (TTLC) of vehicles. In both tasks, a convolutional neural network (CNN) based on attention mechanisms is used as a shared feature extractor. The spatial attention focuses on the most relevant regions of the surrounding environment to enhance feature extraction. Evaluation and comparison of the proposed method against existing lane change prediction models on benchmark datasets demonstrates superior long-term prediction performance. Cheng et al. [2] proposed a lane change decision method based on a convolutional neural network. This approach utilizes dynamic image representations to reveal traffic conditions, extracts potential features and learns driving strategies from labeled dynamic image datasets, and adds a safety constraint layer to avoid vehicle collisions.

Table 1.2 The data-driven CF modeling

Category	Machine learning method	References	Dataset	Verification method
Machine learning	Neural network	Kehtarnavaz et al. [14]	Micro data such as vehicle speed, workshop distance, steering wheel Angle and front Angle	Compare the trajectories of front and rear vehicles (model output)
Machine learning	Neural network	Xu [33]	Micro-tracking data is collected by vehicle high precision GPS	The back propagation neural network, radial basis neural network and hybrid neural network were trained respectively. Calculate the acceleration prediction error
Machine learning	Support vector machine	Wei and Liu [31]	NGSIM(US-101)	(1) Model test and comparison with real speed; (2) Simulation test, using real data to build simulation scenarios and compare real values with predicted values
Machine learning	Support vector machine	Zhang et al. (2018)	The UC-win/Road driving simulator platform collects the position coordinates, speed and acceleration of the tested vehicle and surrounding vehicles	The data sets 75 and 25% were used for training and validation, respectively, to calculate the prediction accuracy of the weighted mixed kernel function based support vector regression model, radial basis kernel function based support vector regression model and backpropagation neural network model for driving decisions (free driving, following, lane change) and the fitness function error value in the optimization process
Machine learning	K nearest neighbor	He et al. [10]	NGSIM(US-101)	(1) No collision, calculate the relative error of workshop time interval; (2) Portability, applying the US-101-trained model to the I-80 dataset: calculating the shop floor interval estimation error: (3) the minimum model, removing one input variable: (4) the reduction ability, by simulating three typical traffic scenarios, and matching the results presented with real data compare

(continued)

Table 1.2 (continued)

Category	Machine learning method	References	Dataset	Verification method
	RNN	Das et al. [3]	Guwahati, India	The mean square error is calculated by comparing the predicted value with the true value
Deep learning	RNN	Zhou et al. [34]	NGSIM (I-80)	According to different driving styles, the mean square error of trajectory prediction under different driving styles is calculated by comparing recurrent neural network model and intelligent driver model
Deep learning	LSTM	Wang et al. (2018)	NGSIM(US-101)	The mean square error distribution of following position and speed predicted by feedforward neural network intelligent driver model and gated cycle unit is compared
Deep learning	LSTM	Huang et al. [12]	NGSIM(US-101)	The prediction errors of speed and workshop distance were compared between long short-term memory neural network and recurrent neural network asymmetric total velocity difference model
Deep learning	LSTM	Lin et al. [19, 20]	NGSIM(I-80)	The average absolute error and average maximum absolute error of trajectory prediction are calculated by comparing the proposed one-way correlation long short-term memory neural network model with intelligent driver model, long short-term memory neural network model, sampling model based on long short-term memory neural network, and fleet level model based on long short-term memory neural network
Deep learning	Seq2Seq	Ma and Qu [22]	NGSIM(I-80)	The mean square error is calculated by comparing the predicted value with the true value
Deep learning	Seq2Seq	Lu et al. [21]	NGSIM(US-101)	The mean square error is calculated by comparing the predicted value with the true value
Deep learning	DRF	Song et al. [27]	HighD	The mean square error is calculated by comparing the predicted value with the true value

(continued)

Table 1.2 (continued)

Category	Machine learning method	References	Dataset	Verification method
Deep learning	DRF	Zhu et al. [40]	Shanghai Nature driving data	(1) Compare the RMS percentage errors of 7 models in shop distance and speed prediction; (2) The following scenarios outside the spinning data set were randomly selected from the empirical data, and the prediction errors of the model, intelligent driver model and recurrent neural network model in position, speed and workshop distance were compared
Deep learning	GAN	Zhou et al. [38]	NGSIM(I-80)	The mean square root percentage error of the speed and shop interval of the training set and the test set is calculated by cross-over test, and the results are compared with the intelligent driver model and recurrent neural network model
Deep learning	GAN	Chen et al. [1]	NGSIM(I-80), NGSIM(US-101)	Compared with the most advanced existing models, the average accuracy of the proposed model is improved by 16.34%
Deep learning	GAN	Xu et al. [32]	NGSIM(I-80)	Compared with IDM, FVD, LSTM, Seq2Seq, Int-LSTM
Deep learning	Attention	Fang et al. [5]	NGSIM	Compared with K-nearest neighbor, IDM and Seq2seq CF models
Deep learning	Attention	Wang et al. [28]	NGSIM(US-101)	Ablation experiments and comparison with advanced models
Deep learning	Attention	Zhu et al. [39]	Shanghai Nature driving data	Compared with IDM, MLP and LSTM
Mix model with data-driven approach	Gipps	Ding [4]	NGSIM	The mean square error is calculated by comparing the predicted value with the true value
Mix model with data-driven approach	IDM	Qin and Ying [24]	NGSIM(I-80)	The mean square error is calculated by comparing the predicted value with the true value
Mix model with data-driven approach	IDM+LSTM	Li et al. [18]	Nanjing city driving data	The mean square error is calculated by comparing the predicted value with the true value

Encoder-Decoder for lane changing

Hui et al. [13] proposed a trajectory prediction model based on a deep encoder-decoder and deep neural network (DNN) that achieves accurate long-term prediction of vehicle lane-changing behaviors for different road types by introducing attention mechanisms and optimizing the deep neural network structure. Wei et al. [30] presented a prediction model based on an attention-assisted encoder-decoder structure and deep neural network. By analyzing the vehicle lane-changing process and dividing it into four stages, the model employs the optimal attention-assisted encoder-decoder architecture for prediction. This approach utilizes fine-grained descriptions and heuristic networks to improve the prediction accuracy of lane changes.

RNN for lane changing

Scheel et al. [25] proposed a attention mechanism based on recurrent neural networks (RNNs) to improve the performance of vehicle lane change prediction tasks. They also introduced a scene-based evaluation scheme to complement traditional prediction metrics. Li et al. [17] developed an intent inference model based on RNNs to solve time series forecasting problems. They considered the dynamic interaction among surrounding vehicles, using the sequential motion information of neighboring vehicles as inputs and combining them with the target vehicle's state to calculate the congestion level in different lanes. Griesbach et al. [8] presented two classifiers, namely Echo State Network and recurrent neural networks with long short-term memory units, for predicting vehicle lane change behaviors in urban areas. The input variables were steering wheel angles and turn signal indicators, and the data was extracted from the Chemnitz natural driving dataset in Germany. Specifically, the Echo State Network achieved higher accuracy in predicting left lane changes, while the recurrent neural networks achieved higher accuracy in predicting right lane changes.

LSTM for lane changing

Zhao et al. [35] proposed a theory-data combination driven vehicle following model that considers the influence of lateral lane-changing. Firstly, a two-lane speed difference following model (FS-MAVD model) is proposed by taking into account the impact of the safety distance and speed difference between the following vehicle and the preceding vehicle as well as the lateral vehicles. The parameters of the model are calibrated using a differential optimization algorithm. Then, a data-driven vehicle following model is constructed using CNN-Bi-LSTM-Attention, which utilizes convolutional neural network (CNN) layers to extract traffic features of the preceding and lateral vehicles, and bidirectional long short-term memory network (Bi-LSTM) to describe driver memory effects. The attention mechanism layer is utilized to allocate model weights. Finally, an optimal weighted combination of the theoretical model and the data-driven model is used for prediction. Gao et al. [6] proposed a data-driven lane-changing behavior detection (DLCD) system based on sequence learning. The system collects vehicle driving data and models and recognizes vehicle lane-changing behavior using sequence learning algorithms. Additionally, an improved focus loss based long short-term memory network (FL-LSTM) is introduced to learn co-occurring features and capture the dependency relationships of lane-changing events. Shi et al. [26] proposed a vehicle trajectory prediction algorithm based on deep neural networks. They designed a model based on attention mechanism, bidirectional long short-term memory network (BiLSTM), and temporal convolutional neural network (TCN) that can simultaneously consider vehicle following and lane-changing behavior.

DRL for lane changing

Zhao et al. [36] proposed a Mask Attention Double Deep Q Network (MAQ) model for lane changing decision-making, which combines deep reinforcement learning with molecular dynamics theory. First, vehicular driving state information both inside and outside of the connected vehicle network was collected in a SUMO simulation environment. Then, the MAQ model was constructed using both the mask and attention mechanisms, and the model input size was fixed to achieve permutation invariance and complete the estimation of lane changing decisions. Zhou et al. [37] modeled the lane changing decisions of multiple autonomous vehicles (AVs) in a highway environment as a multi-agent reinforcement learning (MARL) problem, and proposed a cooperative control-based lane changing decision-making method for autonomous vehicles in mixed traffic. By designing a multi-objective reward function, the decision-making process is optimized to simultaneously consider fuel efficiency, driver comfort, and safety. Li et al. (2023) proposed a Transformer-based lightweight end-to-end model for decision-making in autonomous vehicles. First, the trajectory data's temporal sequence is processed using deep separable convolution and Transformer modules to extract image semantics. Then, a probability model evaluates driving risk, after which the model is combined with deep reinforcement learning to obtain a policy that minimizes expected risk. Li et al. [15, 16] proposed a lane changing decision-making framework based on deep reinforcement learning to obtain a driving strategy with minimal expected risk. This method is based on driver behavior data, and real-time evaluation and prediction of driving risk is achieved by building driver behavior and driving risk assessment models.

Attention mechanism for lane changing

Gao et al. [7] proposed a dual-Transformer model which includes a lane change intention prediction model and a trajectory prediction model. The vehicle lane change intention prediction model extracts the vehicle driving state from the vehicle status, enhances prediction accuracy by designing multi-head attention, and outputs a probability vector for lane change intention. For the trajectory prediction model, performance is optimized by fusing the intention probability vector and the spatio-temporal attention long short-term memory (STA-LSTM) network proposed by Lin et al. [19] to predict lane change probability. The STA-LSTM model achieves the same level of prediction performance as other state-of-the-art models. Furthermore, the model can explain the effects of historical trajectories and surrounding vehicles on the target vehicle's lane change behavior.

The above data-driven lane changing researches are summarized in the following Table 1.3.

1.5.5 *Mixture of Theoretical Model and Data-Driven Model*

There are also efforts that combine the theoretical model developed from physical law and basic assumptions with the data-driven approach. For instance, Ding et al. (2017) proposed a linear combination vehicle-following model (LC-CF) that combined the advantages of the Gipps model and the neural network following model, aiming to establish a data-mixed following model that was both safe and accurate. By adjusting the weights of the neural network and Gipps models, the model achieved the goal of controlling the accuracy and safety of predicted speed. Qin and Ying [24] presented a low-speed following model based on the IDM model and radial basis function neural network, which combined the model-driven following model with the data-driven following model by optimizing weights. Li et al. [18] combined the adaptive Kalman filtering algorithm with a data-driven model based on long short-term memory neural network (LSTM) and IDM to establish a vehicle following model. The model was tested with real driving data from a single driver and demonstrated higher accuracy than the individual models. Furthermore, the model optimized the generalization capacity of the data-driven model, and had better fitting effects on specific situations such as acceleration and stationary state when speed was zero.

1.6 Drawbacks of Current Models

A wide range of microscopic traffic models are currently available. The utilization of these models in addressing specific tasks has demonstrated their effectiveness in various domains, such as transportation planning and traffic analysis. However, the development of two-dimensional traffic flow models is necessary due to inherent limitations in the current models. These limitations primarily stem from inadequate considerations of two-dimensional movement, vehicular model, and stochasticity, and can be summarized as follows:

1. Two-dimensional movements: Vehicles move in a planar space where both longitudinal and lateral dimensions play a significant role. Most current microscopic traffic flow models focus only on describing longitudinal movement, neglecting the lateral component. Commonly employed models, such as car-following (CF) models, primarily capture the behavior of vehicles following one another, completely excluding the lateral dimension. As demonstrated in subsequent sections, vehicular movement is fundamentally two-dimensional, with both dimensions being interdependent. Extending the existing one-dimensional

Table 1.3 Data driven lane changing

Category	Machine learning method	References	Dataset	Verification method
Deep learning	Encoder-Decoder	Hui et al. [13]	InD, RounD, HighD	Module cascading verification
Deep learning	Encoder-Decoder	Wei et al. [30]	HgihD	Compared with six mainstream models
Deep learning	CNN	Zhang et al. (2022)	Data collected using a driving simulator	Compare with the benchmark model
Deep learning	CNN	Mozaffari et al. [23]	HgihD	Compare with the benchmark model
Deep learning	RNN	Li et al. [17]	NGSIM	
Deep learning	RNN	Griesbach et al. [8]	I-FAS, "Carai-studie [raw data files]"	
Deep learning	RNN	Scheel et al. [25]	HgihD, NGSIM	Compare with the benchmark model
Deep learning	LSTM	Gao et al. (2022b)	NGSIM(I-80), NGSIM(US-101)	Compare with the benchmark model
Deep learning	LSTM	Zhao et al. [35]	Southeast University drone data	MAE, MSE, RMSE
Deep learning	LSTM	Gao et al. [6]	Use the Transportable Instrumentation Package (TIP) to collect data	MAE, MSE, RMSE
Deep learning	LSTM	Shi et al. [26]	NGSIM(I-80), NGSIM(US-101)	MAE, MSE, RMSE, Compare with the benchmark model
Deep learning	DRF	Zhao et al. [35]	SUMO simulation	MAE, MSE, RMSE
Deep learning	DRF	Li et al. (2023)	CARLA simulation	Compared with Baseline, DSCNN and DSCNN LSTM
Deep learning	DRF	Li et al. (2023)	CARLA simulation	Compared with Baseline, DSCNN and DSCNN LSTM
Deep learning	DRF	Zhou et al. [37]	–	Compare with the benchmark model
Deep learning	CNN	Cheng et al. [2]	NGSIM	SUMO simulation verification

(continued)

Table 1.3 (continued)

Category	Machine learning method	References	Dataset	Verification method
Deep learning	Attention	Guo et al. [9]	SPMD data	The mean square error is calculated by comparing the predicted value with the true value
Deep learning	Attention	Gao et al. [7]	HgihD, NGSIM	The mean square error is calculated by comparing the predicted value with the true value
Deep learning	Attention	Lin et al. [19]	NGSIM(US-101)	The mean square error is calculated by comparing the predicted value with the true value

models to two-dimensional models is essential to tackle the challenges posed by autonomous vehicle (AV) technology.

2. Modeling assumptions: In most microscopic traffic flow models, the trajectories of vehicles are the main modeling object and can be represented by parameters such as location (x), speed (v), acceleration (a), and time (t). From this perspective, vehicles are simplified as mass points. However, vehicles possess physical dimensions. While this simplification suffices in the case of longitudinal descriptions, it is questionable in the context of two-dimensional modeling. Furthermore, vehicle movement is fundamentally driven by driver maneuvers, including acceleration and steering. Incorporating the vehicle model and driver maneuvers into microscopic traffic models would enhance their realism.
3. Stochasticity: It is evident that vehicular movement is stochastic in nature. Consequently, considerable efforts have been dedicated to the development of stochastic traffic flow models. However, these models primarily enhance traditional deterministic microscopic traffic flow models by introducing stochastic elements. When stochasticity interacts with vehicle dynamics and the two-dimensional nature of traffic, the resulting vehicular dynamics become increasingly complex.

In response to the aforementioned limitations present in existing microscopic traffic flow models, this book proposes a set of methodological enhancements that aim to improve modeling accuracy. Specifically, the lateral dimension treatment is modeled and analyzed separately through the utilization of stochastic differential theory. Furthermore, explicit incorporation of vehicle dynamics into microscopic behavioral modeling considers the control inputs of the model, including steer and acceleration/deceleration, among others. Additionally, the stochasticities at both dimensions are captured to ensure modeling realism, as demonstrated through the validation results.

The proposed enhanced models are subsequently employed in several tasks involving mixed autonomous vehicles' flow configurations. These tasks include identifying human driven vehicles/autonomous vehicles using lateral dynamics, as well as inferring lane change-related events such as lane-changing probability and duration. The proposed models also investigate lateral wandering patterns and capacity considerations with lateral dimensions. These represent a non-exhaustive list of the potential advantages that the enhanced models offer. Moreover, two-dimensional movement within intersections is also analyzed in this book using game theory, Petri net theory, and other related methodologies.

1.7 Book Structure

The structure of the book is presented in Fig. 1.3. Basically the book contains four components. The first include the introduction and the analysis based on real world dataset. We will introduce various real world vehicular datasets, and make the following analysis: two-dimensional noise analysis, parameter estimation for several datasets, and behavioral predictability analysis. The second component basically is theoretic focus of this book. It establishes several two-dimensional microscopic traffic flow models. These models include deterministic microscopic traffic flow model considering vehicular kinetic attribute, stochastic microscopic traffic flow model considering vehicular kinetic attribute, two-dimensional microscopic traffic flow model for heterogeneous road, lateral dimension stochastic modeling and jump-diffusion approach for two-dimensional jerk; the third component applies the theory in the second component to real world problems. These problems include the CAF modeling, lateral wandering patterns analysis and the lane changing event inference. The last component focuses on the traffic flow at intersections. Three models are developed to encapsulate the real world operational phenomena which are overlooked. The first is the game behavior within the intersection, which leads to the SFR reducing at the outlet of the intersection. After that the PN theory is employed to study the intersection deadlock formation probability and influencing factors. Lastly we consider the mixed autonomous vehicles flow, and develop an intersection deadlock protocol which embeds the EKF of the HDVs.

Finally, we close the book with discussion of future microscopic traffic theory. We discuss several possible directions where the microscopic traffic model can be improved. These directions include integrated mobility agent modeling, incorporating new technologies such as LLM, modeling irregular driving behavior, developing simulation platforms, and multi scale modeling.

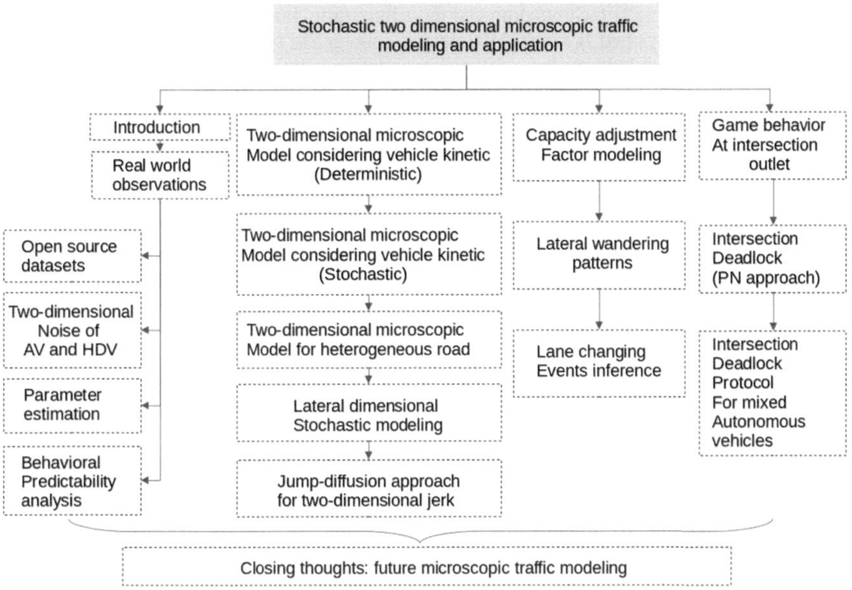

Fig. 1.3 Structure of the book

References

1. Chen L, Zhou Q, Cai Y et al (2022) CAE-GAN: a hybrid model for vehicle trajectory prediction. IET Intell Transp Syst 16:1682–1696. https://doi.org/10.1049/itr2.12174
2. Cheng S, Wang Z, Yang B, Nakano K (2023) Convolutional neural network-based lane-change strategy via motion image representation for automated and connected vehicles. IEEE Trans Neural Netw Learn Syst. https://doi.org/10.1109/TNNLS.2023.3265662
3. Das S, Maurya A, Dey A (2023) A recurrent neural network model for predicting two-leader car-following behavior. Transp Lett, 1–15. https://doi.org/10.1080/19427867.2023.2205190
4. Diandian D, Sun L, Chen S (2017) A car-following model coupling machine learning and dynamic. J Transp Syst Eng Inf Technol 17(6):33–38 [In Chinese]
5. Fang S, Yang L, Zhao X et al (2023) A dynamic transformation car-following model for the prediction of the traffic flow oscillation. IEEE Intell Transp Syst Mag, 2–27. https://doi.org/10.1109/MITS.2023.3317081
6. Gao J, Murphey YL, Yi J, Zhu H (2022) A data-driven lane-changing behavior detection system based on sequence learning. Transp B Transp Dyn 10:831–848. https://doi.org/10.1080/21680566.2020.1782786
7. Gao K, Li X, Chen B et al (2023) dual transformer based prediction for lane change intentions and trajectories in mixed traffic environment. IEEE Trans Intell Transp Syst, 1–14. https://doi.org/10.1109/TITS.2023.3248842
8. Griesbach K, Beggiato M, Hoffmann KH (2022) Lane change prediction with an echo state network and recurrent neural network in the urban area. IEEE Trans Intell Transp Syst 23:6473–6479. https://doi.org/10.1109/TITS.2021.3058035
9. Guo H, Keyvan-Ekbatani M, Xie K (2022) Lane change detection and prediction using real-world connected vehicle data. Transp Res Part C Emerg Technol 142:103785. https://doi.org/10.1016/j.trc.2022.103785
10. He Z, Zheng L, Guan W (2015) A simple nonparametric car-following model driven by field data. Transp Res Part B: Methodol 80:185–201. https://doi.org/10.1016/j.trb.2015.07.010

11. Huang K, Zheng J (2019) Simulation and modeling traffic flow based on Division K Nearest Neighbor. Mod Phys Lett B 33:1950407. https://doi.org/10.1142/S0217984919504074
12. Huang X, Sun J, Sun J (2018) A car-following model considering asymmetric driving behavior based on long short-term memory neural networks. Transp Res Part C Emerg Technol 95:346–362. https://doi.org/10.1016/j.trc.2018.07.022
13. Hui F, Wei C, ShangGuan W et al (2022) Deep encoder–decoder-NN: a deep learning-based autonomous vehicle trajectory prediction and correction model. Phys Stat Mech Its Appl 593:126869. https://doi.org/10.1016/j.physa.2022.126869
14. Kehtarnavaz N, Groswold N, Miller K, Lascoe P (1998) A transportable neural-network approach to autonomous vehicle following. IEEE Trans Veh Technol 47:694–702. https://doi.org/10.1109/25.669106
15. Li G, Qiu Y, Yang Y et al (2022) Lane change strategies for autonomous vehicles: a deep reinforcement learning approach based on transformer. IEEE Trans Intell Veh, 1–15. https://doi.org/10.1109/TIV.2022.3227921
16. Li G, Yang Y, Li S et al (2022) Decision making of autonomous vehicles in lane change scenarios: deep reinforcement learning approaches with risk awareness. Transp Res Part C Emerg Technol 134:103452. https://doi.org/10.1016/j.trc.2021.103452
17. Li L, Zhao W, Xu C et al (2021) Lane-change intention inference based on RNN for autonomous driving on highways. IEEE Trans Veh Technol, 1–1. https://doi.org/10.1109/TVT.2021.3079263
18. Li Y, Lu X, Ren C, Zhao H (2019) Fusion modeling method of car-following characteristics. IEEE JS Mag 7:162778–162785. IEEE Xplore. https://ieeexplore.ieee.org/document/8882217. Accessed 4 Mar 2024
19. Lin L, Li W, Bi H, Qin L (2022) Vehicle trajectory prediction using LSTMs with spatial-temporal attention mechanisms. IEEE Intell Transp Syst Mag 14:197–208. https://doi.org/10.1109/MITS.2021.3049404
20. Lin Y, Wang P, Zhou Y et al (2022) Platoon trajectories generation: a unidirectional interconnected LSTM-based car-following model. IEEE Trans Intell Transp Syst 23:2071–2081. https://doi.org/10.1109/TITS.2020.3031282
21. Lu W, Yi Z, Liang B et al (2023) Learning car-following behaviors for a connected automated vehicle system: an improved sequence-to-sequence deep learning model. IEEE Access, 1–1. https://doi.org/10.1109/ACCESS.2023.3243620
22. Ma L, Qu S (2020) A sequence to sequence learning based car-following model for multi-step predictions considering reaction delay. Transp Res Part C Emerg Technol 120:102785. https://doi.org/10.1016/j.trc.2020.102785
23. Mozaffari S, Arnold E, Dianati M, Fallah S (2022) Early lane change prediction for automated driving systems using multi-task attention-based convolutional neural networks. IEEE Trans Intell Veh 7(3):758–770. https://doi.org/10.1109/TIV.2022.3161785
24. Qin W, Ying L (2020) Combination low-speed car-following model based on IDM and RBFNN. Appl Res Comput 37(8):2354–2357. [In Chinese]
25. Scheel O, Nagaraja N, Schwarz L et al (2022) Recurrent models for lane change prediction and situation assessment. IEEE Trans Intell Transp Syst 23:1–17. https://doi.org/10.1109/TITS.2022.3163353
26. Shi K, Yuankai W, Shi H et al (2022) An integrated car-following and lane changing vehicle trajectory prediction algorithm based on a deep neural network. Phys Stat Mech Its Appl 599:127303. https://doi.org/10.1016/j.physa.2022.127303
27. Song D, Zhu B, Zhao J et al (2023) Personalized car-following control based on a hybrid of reinforcement learning and supervised learning. IEEE Trans Intell Transp Syst, 1–16. https://doi.org/10.1109/TITS.2023.3245362
28. Wang L, Li J, Zhang H et al (2023) An attention-based car-following model based on fused data. IEEE Access, 1–1. https://doi.org/10.1109/ACCESS.2023.3280409
29. Wang X, Jiang R, Li L, Yilun L (2019) Long memory is important: a test study on deep-learning based car-following model. Phys Stat Mech Its Appl 514:786–795. https://doi.org/10.1016/j.physa.2018.09.136

30. Wei C, Hui F, Yang Z et al (2022) Fine-grained highway autonomous vehicle lane-changing trajectory prediction based on a heuristic attention-aided encoder-decoder model. Transp Res Part C Emerg Technol 140:103706. https://doi.org/10.1016/j.trc.2022.103706
31. Wei D, Liu H (2013) Analysis of asymmetric driving behavior using a self-learning approach. Transp Res Part B Methodol 47:1–14. https://doi.org/10.1016/j.trb.2012.09.003
32. Xu D, Gao G, Qiu Q, Li H (2023) A car-following model considering missing data based on TransGAN networks. IEEE Trans Intell Veh, 1–13. https://doi.org/10.1109/TIV.2023.3270336
33. Xueming X, Jian R, Wang L (2007) Development of a car-following model based on combined neural network model. J Highw Transp Res Dev 3(132):130–132. [In Chinese]
34. Zhou M, Qu X, Li X (2017) A recurrent neural network based microscopic car following model to predict traffic oscillation. Transp Res Part C Emerg Technol 84:245–264. https://doi.org/10.1016/j.trc.2017.08.027
35. Zhao J, Xiaoyu H, Zhixin Y, Mingmin H (2023) A combination model for connected and autonomous vehicles lane-changing decision-making under multi connectivity range. J Transp Syst Eng Inf Technol 23(1):77–85. [In Chinese]
36. Zhao J, Lanxin J, Zhimin Z, Yunchao Q, Huijun S (2023) A car-following model driven by combination of theory and data considering effects of lane change of side cars. J South China Univ Technol (Natural Science Edition) 51(6):10–19. [In Chinese]
37. Zhou W, Chen D, Yan J, Li Z, Yin H, Ge W (2022) Multi-agent reinforcement learning for cooperative lane changing of connected and autonomous vehicles in mixed traffic. Auton Intell Syst 2(1):5. https://doi.org/10.1007/s43684-022-00023-5
38. Zhou Y, Fu R, Wang C, Zhang R (2020) Modeling car-following behaviors and driving styles with generative adversarial imitation learning. Sensors 20. https://doi.org/10.3390/s20185034
39. Zhu M, Du S, Wang X et al (2022) TransFollower: long-sequence car-following trajectory prediction through transformer
40. Zhu M, Wang X, Wang Y (2018) Human-like autonomous car-following model with deep reinforcement learning. Transp Res Part C Emerg Technol 97:348–368. https://doi.org/10.1016/j.trc.2018.10.024

Chapter 2
Real World Observations, Maneuver Estimation and Behavioral Predictability

2.1 Introduction of This Chapter

Any theory is developed based on the observation of real world phenomena, or certain assumptions. The microscopic traffic flow model is of no exception. The release of traffic datasets (for instance, NGSIM) greatly boosted the development of traffic models. Currently there are many available vehicular trajectories datasets, which comes from different countries, and collected from different traffic facilities. These data are termed NDD, or naturalistic driving datasets. This chapter is devoted to the following:

– Famous datasets are introduced firstly.
– Maneuver estimation method from these NDD is developed. The motivation of the method is that, the NDD only present the trajectory data, which is the results of the drivers' maneuvers, and the maneuver themselves, cannot be observed directly and hence are not included in the original datasets. The proposed maneuver estimation method is tested against various datasets. The results prove the effectiveness of the proposed method.
– Although many researches have investigated the diversity of driving behavior, few literatures focus on the behavioral differences between countries. However, it should be noted that the AV algorithm may be trained or calibrated using dataset from one country, it is questionable whether it works well when the algorithm is deployed to other countries. In this regard, we use the PE (permutation entropy) to investigate the behavioral predictability of the vehicles. The results show that there exist considerable differences regarding the predictability among different countries.

© The Author(s), under exclusive license to Springer Nature Singapore Pte Ltd. 2024

H. Qi, *Stochastic Two-Dimensional Microscopic Traffic Model*, Lecture Notes in Intelligent Transportation and Infrastructure,
https://doi.org/10.1007/978-981-97-3597-6_2

2.2 Real World Observation Datasets

Currently, there are numerous open-sourced real world observation datasets available [1–7]. These datasets originate from various countries, including the United States, Japan, China, Germany, and Greece, and encompass scenarios such as urban roads, highway segments, on and off ramps. Data collection methods employed consist of cameras [4], drones [2], Radar and LiDAR, GNSS [3], among others. Essentially, these datasets provide trajectories of relevant vehicles, with time resolutions ranging from 0.04 s to 0.1 s. Additionally, most datasets offer supplementary information regarding the road characteristics and collection scenarios.

The availability of traffic flow data has greatly facilitated the development of models, including those presented in this book. The models featured herein are no exception, as many of them have been established based on observations from real world datasets. Moreover, the model parameters have been calibrated using these datasets. In the following, we provide an introduction to some of the datasets employed in this book.

However, datasets are collected using different methods. The teams who collect these data used different techniques to process the raw data. As a result, these datasets have different noise or errors. Besides, majority of these data provide the vehicular trajectories, which is the result of the maneuver of the drivers. However, the maneuvers themselves cannot be observed directly and are not included in the datasets. Hence, a maneuver estimation method is given.

2.2.1 Datasets Introduction

2.2.1.1 Ngsim

NGSIM[1] may be one of the most famous traffic dataset and it is the predecessor of current release of different datasets. The NGSIM dataset was collected by researchers through a project called "Next Generation Simulation". This dataset covers hotspots in areas such as structured road intersections and highway on and off ramps, which are of interest in the field of vehicle–road coordination research. The researchers involved in the project processed the original video information and generated trajectory data for each vehicle in the traffic flow. The dataset was collected in four different regions in the United States: southbound US 101 in California, Lankershim Boulevard in Los Angeles, eastbound I-80 in Emeryville, California, and Peachtree Street in Atlanta, Georgia. As many literatures employ the dataset (Fang et al. 2014; Zhang et al. 2010;

---

[1] https://data.transportation.gov/stories/s/Next-Generation-Simulation-NGSIM-Open-Data/i5zb-xe34.

https://datahub.transportation.gov/Automobiles/Next-Generation-Simulation-NGSIM-Vehicle-Trajector/8ect-6jqj.

Thiemann et al. 2008; Li et al. 2020; Ahmed et al. 2019; Montanino and Punzo 2013), we omit the details of the dataset introduction.

2.2.1.2 OpenACC[2]

OpenACC is an open database that encompasses experiments involving vehicles equipped with Adaptive Cruise Control (ACC) systems. The goal of the OpenACC database is to provide the scientific community with data regarding ACC behavior, in order to enhance understanding of ACC vehicle characteristics and the impact of their widespread deployment on traffic dynamics, as well as predict potential issues associated with them. As new experimental data becomes available, it will be incorporated into OpenACC. The database includes vehicle trajectories from various following experiments, the majority of which involve ACC driving. However, for comparative purposes, some trajectories are also available for manual driving. The database comprises data sets from a total of seven different testing scenarios:

AstaZero. The test was conducted in the second quarter of 2019 and involved a two-day following test on a rural road at the AstaZero test track in Sweden. The rural road had a length of approximately 5.7 km. This activity included five high-end cars from four different brands, representing various models. The system used for trajectory data collection was the RT-Range S multi-target ADAS measurement solution from Oxford Technical Solutions, which provided differential GPS accuracy. The data collection system offered a frequency of over 100 Hz, necessitating under sampling to achieve 10 Hz. In all tests, the leading vehicle was the same and driven with ACC enabled. Generally, the following vehicles applied two different following modes: (a) maintaining a constant speed car platoon and (b) disturbing the platoon from its equilibrium point by decelerating to a new desired speed. Except for the two laps designed to test manual driving scenarios, the followers were driving with ACC engaged.

Casale. The experiment was carried out on public roads in northern Italy between Ispra and Cassale Monferrato and vice versa, involving a one-day following test with two cars driving in a platoon. The order of the platoon was kept the same, and ACC was constantly engaged. In some cases, the drivers needed to brake manually, resulting in ACC being disabled. Such instances were documented and reported in a CSV file. Additionally, there were two test vehicles equipped with both Ublox9 and OBD systems, and data processing involved removing sections related to tolls and tunnels from the data series. Three cubic spline interpolations were implemented to achieve a frequency of 10 Hz.

Cherasco. The experiment took place during the last quarter of 2018 and involved a two-day following test on a public highway in northern Italy, from Ispra (Virginia, USA) to Cherasco (Colorado, USA), with two or three cars forming a platoon. The tests were conducted during off-peak hours to minimize interference from other

[2] https://data.jrc.ec.europa.eu/dataset/9702c950-c80f-4d2f-982f-44d06ea0009f. https://jeodpp.jrc.ec.europa.eu/ftp/jrc-opendata/TransportExpData/JRCDBT0001/LATEST.

road users. The experiment was conducted on-road, with all vehicles equipped with Ublox 8 GNSS data acquisition devices. The acquired data had a sampling frequency of approximately 3–5 Hz, which was interpolated using cubic splines to achieve a frequency of 10 Hz. The lead vehicle was instructed to drive manually and occasionally perform random decelerations and accelerations in a realistic manner at the expected speed. On the day testing manual driving scenarios, the followers used ACC as much as possible. The total number of vehicles involved in each trip was four (involving two or three vehicles). The equipment used included Ublox 8 and OBD systems, which were only utilized to check the status of the brake pedal position. In data processing, sections with significant noise were removed from the data sequence. The interpolation method of cubic splines was implemented to achieve a frequency of 10 Hz.

Other locations can be found on the official document which is omitted here.

2.2.1.3 Zen Traffic Data[3]

Zen Traffic Data was born out of the image sensing technology developed by Hanshin Expressway Company, with the aim of achieving safe, reliable, and comfortable vehicle usage. This dataset consists of comprehensive and long-term trajectory data of all vehicles, as well as any other data that influences traffic events. As a result, truly complex traffic phenomena that have been beyond human grasp until now have been digitized. Hanshin Expressway Company believes that 'Zen Traffic Data' will contribute to making road traffic services safer, more reliable, and more comfortable'. Basic data fields are given shown in Fig. 2.1.

2.2.1.4 HighD[4]

The highD dataset comprises data extracted from 60 recordings. For each recording, a total of four files are provided: the image of the highway section covered by the recording, a csv file describing the location of the recording, another csv file containing an overview of the recorded vehicle trajectories, and a csv file containing the trajectory data of each vehicle. These files are created for each recording to ensure ease of data processing. Details can be found from its official website.

2.2.1.5 Mirror Traffic[5]

In terms of data acquisition, the Mirror dataset relies on visual algorithms to extract real trajectory data of vehicles and pedestrians from traffic surveillance videos and

---

[3] https://zen-traffic-data.net/english/outline/.

[4] https://www.highd-dataset.com/format.

[5] http://www.scenarios.cn//html/dataset.html.

No.	COLUMN	TYPE	FRACTIONAL DIGITS	UNIT	DESCRIPTION	SAMPLE	UNIQUE
1	vehicle_id	int			Vehicle ID	4316	*
2	datetime	string			Datetime HHMMSSFFF	070216100	*
3	vehicle_type	int			1: normal vehicle 2: large vehicle(bus, truck, etc.)	1	
4	velocity	decimal	1	km/h	Velocity	63.1	
5	traffic_lane	int			Traffic lane ID 1: driving lane 2: passing lane 3: entrance lane	1	
6	longitude	decimal	7	degree	Longitude measured as WGS84 system	135.4598866	
7	latitude	decimal	7	degree	Latitude measured as WGS84 system	34.7209920	
8	kilopost	decimal	1	meter	Distance from starting point of the expressway route	5070.7	
9	vehicle_length	decimal	1	meter	Estimated vehicle length from image recognition. Value is rounded to 0.5 [m].	5.5	
10	detected_flag	int			1: detected record by image recognition 0: interpolated record	1	

Fig. 2.1 Basic field name of the Zen dataset

drone videos, effectively addressing the issue of data scarcity and supplementing traditional on-board data collection scenarios, thus enriching the database. By considering different road structures, a trajectory library called Mirror-Traffic is generated, encompassing various types of road scenarios including ramps, straight roads, curves, intersections, etc., formed by the presence of traffic participants. More details can be found from the official website.

2.2.1.6 Ad4che[6]

The Aerial Dataset for China Congested Highway and Expressway (AD4CHE) currently contains 68 segments of data extracted from multiple highways and expressways in five Chinese cities. For each dataset, five files will be provided, including images of the highway/expressway displaying the lane positions and road topology, a CSV file providing an overall description of the data, a summary CSV file of vehicle trajectories, and a detailed CSV file documenting all trajectories. The file format has significant similarities with that of the German HighD dataset, making it relatively convenient to process. In order to improve the usability of the dataset, AD4CHE format references the HighD dataset to a great extent. However, due to the complexity of AD4CHE scenarios, there are still some key differences, such as the definition of the vehicle position in the coordinate system, and additional information unique to AD4CHE datasets, such as the number of buses (numBuses), road angle (angle), vehicle orientation (orientation), yaw rate (yaw_rate), and lane offset (ego_offset).

For each recorded drone video segment, highway segment images are added. These images are obtained by applying a filtering algorithm to remove dynamic vehicles. Each lane is displayed in a different color. Trajectories with lane line numbers exceeding 100 contain less information, such as vehicle state estimation. It is recommended to utilize trajectory information with lane line numbers below 100. The world coordinate system corresponds to the image coordinate system in which the video is recorded, with the origin located at the top-left corner of the image.

The data format for XX_recordingMeta.csv (meta info of the dataset) contains metadata for each drone recording, providing general information such as recording time and total number of trajectories. The data column is as follows (Table 2.1).

Trajectories data summarization is contained in XX_tracksMeta.csv. This file contains an overview of all the trajectories. For each trajectory, there are summary data such as the distance traveled and average vehicle speed. The purpose of this file is to filter trajectories based on certain conditions, such as type or direction of travel. Typical data is as follows (Table 2.2).

Detailed trajectories are given in XX_tracks.csv. This file contains all the temporal information related to each trajectory, including the current speed, field of view, and surrounding vehicles (similar to HighD dataset) (Table 2.3).

[6] https://auto.dji.com/ad4che-dataset?site=auto&from=nav#30259.

Table 2.1 The field of the AD4CHE meta data

Name	Description	Unit
id	The ID of the video, each data has a unique ID	[−]
frameRate	The frame rate at which the video was recorded	[hz]
locationId	The ID of the video location (the location is virtual)	[−]
speedLimit	The speed limit of the driving lane. In all recordings, the speed limit is the same for each lane	[m/s]
...		

Table 2.2 The field of the AD4CHE tracks description

Name	Description	Unit
id	The ID of the trajectory, arranged in ascending order. There may be some gaps in the trajectory IDs	[−]
Width	The width of the vehicle bounding box after post-processing algorithms, corresponding to the length in the direction of vehicle movement. Calculated using L-shape	[m]
Height	The height of the vehicle bounding box after post-processing algorithms, corresponding to the width in the direction of vehicle movement	[m]
initialframe	The initial frame at which the trajectory starts	[−]
...		

Table 2.3 The field of the AD4CHE tracks

Name	Description	Unit
frame	The current frame number	[−]
id	The trajectory ID	[−]
x	The x-coordinate of the center point of the vehicle bounding box in the vehicle coordinate system	[m]
y	The y-coordinate of the center point of the vehicle bounding box in the vehicle coordinate system	[m]
Width	The width of the vehicle bounding box	[m]

2.2.2 *Two Dimensional Noise Observation: Longitudinal*

In this section, we demonstrate the two-dimensional noise existing in the vehicular movement using real world dataset. The two-dimensional noise exist both for AVs and HDVs. Regarding AVs, we use the openACC as the source data. Regarding the HDV, we use NGSIM as demonstration purpose.

2.2.2.1 HDVs

The IDM model used in this study is a typical one-dimensional car-following model proposed by Treiber. The specific calculation formulas are as follows.

$$a_f = a\left[1 - \left(\frac{v}{v_d}\right)^4 - \left(\frac{s^*}{s}\right)^2\right] \tag{2.1}$$

$$s^* = s_0 + max\left(0, vT + \frac{v\Delta v}{2\sqrt{ab}}\right) \tag{2.2}$$

where a represents the maximum vehicle acceleration, b represents the most comfortable vehicle acceleration, v_d represents the desired vehicle speed, s_0 represents the minimum safety distance, and 'T' represents the headway time. The values of these parameters adopted in this study are based on the work of Sharath (2020). He calibrated these parameters using vehicle data from 1050 NGSIM trajectories, resulting in the following values: $a, b, s0, T, vd = 2.95\,\frac{\mathrm{m}}{\mathrm{s}^2}, \frac{2.21\,\mathrm{m}}{\mathrm{s}^2}, 2.58\,\mathrm{m}, 1.3\,\mathrm{s}, 12.1\,\mathrm{m/s}$. Here, the noise refers to the difference between the actual lateral position of the vehicle at time 't' and the calculated value using the IDM model. The specific calculation formula is as follows.

$$\begin{cases} x_{t_i}^{\mathrm{IDM},k} = x_{t_{i-1}}^k + 0.5 * a_{\mathrm{IDM}(\cdot)} * \mathrm{dt}^2 + v_{\mathrm{IDM}(\cdot)} * dt(a) \\ \varepsilon_{t_i}^k = x_{t_i}^k - x_{t_i}^{\mathrm{IDM},k}(b) \end{cases} \tag{2.3}$$

where $x_{t_i}^k$ represents the lateral position of the k-th vehicle in the NGSIM dataset at time t_i, $x_{t_i}^{\mathrm{IDM},k}$ represents the theoretical value of the lateral position of the k-th vehicle at time $\mathrm{t_i}$ calculated based on the IDM model, $\mathrm{x_{t_{i-1}}^k}$ represents the lateral position of the k-th vehicle in the NGSIM dataset at time $\mathrm{t_{i-1}}$, $\mathrm{a_{IDM(\cdot)}}$ and $\mathrm{v_{IDM(\cdot)}}$ represent the acceleration and velocity calculated based on the IDM model. The noise is presented in the following figure. It is shown that there is considerable exist in the real world datasets (Fig. 2.2).

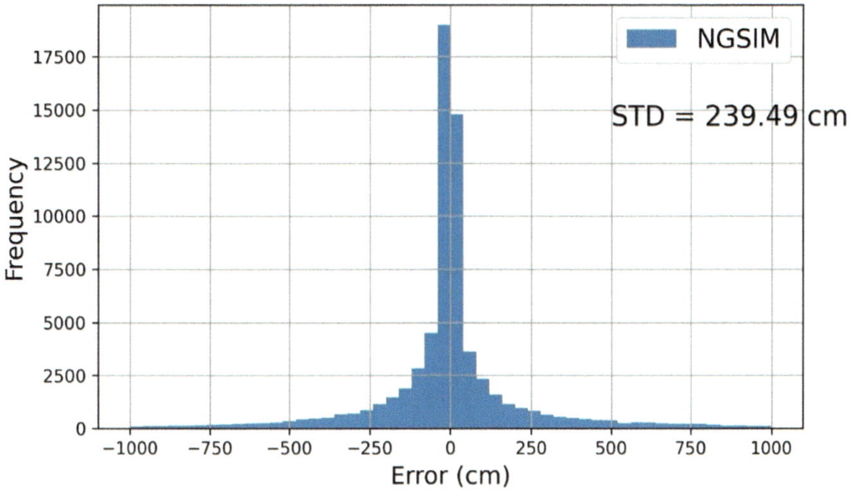

Fig. 2.2 Longitudinal error of HDVs

2.2.2.2 AVs

In this section, we use the real world data source to investigate the noise of the autonomous car following algorithms. The openACC dataset is an open-access database of different car-following experiments involving 28 vehicles, 22 of which equipped with state-of-the-art commercial ACC systems. Experiments are carried out in the framework of four test campaigns, among which two are on public roads. There are 28 vehicles involved. The speed profiles of the leader vehicles are specially defined such that frequent small perturbations in a random yet realistic way are created. Interesting readers may refer to [3] for detailed description. The noises is calculated as follows. Suppose the leader's trajectories are $x_{LD}(t)$ and $v_{LD}(t)$, which represent the longitudinal location and speed respectively. The follower's trajectories are $x_{FL}(t)$ and $v_{FL}(t)$ respectively. The ACC formulation of the follower is expressed as:

$$dv_{FL}(t) = k_1(D(t) - t_{hw}v_{FL} - L) + k_2(v_{LD} - v_{FL}) \tag{2.4}$$

t_{hw} is the time headway; L is the vehicle length of the leader vehicle. D(t) is the real-time gap between the two vehicles. Suppose the time step of the observation of the Δt, then the theoretical speed and location of the follower are calculated as follows.

$$\begin{cases} v_{FL}(t+\Delta t) = v_{FL}(t) + \Delta t \cdot dv_{FL}(t) \ (a) \\ x_{FL}(t+\Delta t) = x_{FL}(t) + \Delta t \cdot v_{FL}(t) \ (b) \end{cases} \tag{2.5}$$

Suppose the location observation is $\hat{x}_{\mathrm{FL}}(t)$. The difference which is calculated as $\xi(t) = x_{\mathrm{FL}}(t) - \hat{x}_{\mathrm{FL}}(t)$, is treated as the stochasticity in the system. Next, we show the characteristic of $\xi(t)$ using available datasets. The model parameters are determined based on calibration researches for openACC. The durations of the observations are different for openACC and Waymo dataset. Typically the data extracted from Waymo is short while openACC contain platoons with long durations over 1 h. Thus we firstly select the noise for the long platoon and then present the distribution of the noise. The longest platoon in openACC data occurs when vehicle driver from Ispra (VA) to Cherasco (CO) in Italy. Noise results are given in Figs. 2.3 and 2.4. Figure 2.3 presents the longitudinal location error. The time horizon of this platoon data is over 1 h (4069 s). The errors of the second and third vehicle in the platoon indicate that: (1) each vehicle is accompanied with error nearly all the time. The error is stochastic and time-variant; (2) different vehicles have different error profiles. During certain moment, the error reaches 4 m (nearly one vehicle length), which is considerably large.

To calculate the level of stochasticity, we use the standard deviation as an index. A time window is chosen. For specific moment t, we calculate the standard deviation within the interval $[t - t_{window}/2, t + t_{window}/2]$. The result can describe the randomness level of the vehicle. The results are given in Fig. 2.4. It can be concluded that: (1) there is a considerable time-varying pattern of the deviation; (2) different vehicle have similar patterns; (3) the standard deviation is relatively stable within certain local time window. For instance, the highest deviation is observed within $[2200\,s, 2500\,s]$, and reaches about 0.4 m. The duration of such deviation is about 5min.

The above empirical observations reveal that: (1) the stochasticity in the ACC vehicle cannot be overlooked. (2) the stochasticity level is time-varying and keep stable in short time window. The above two conclusions imply that, current models developed for deterministic system may not be suitable anymore. Hence a stochastic formula is necessary, with self-adaption mechanism capabilities

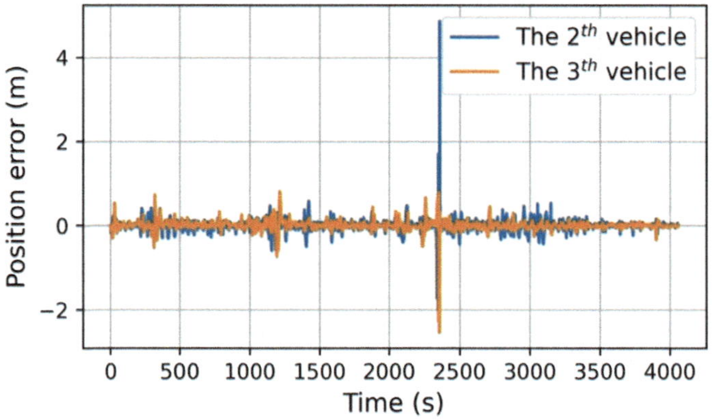

Fig. 2.3 Longitudinal error

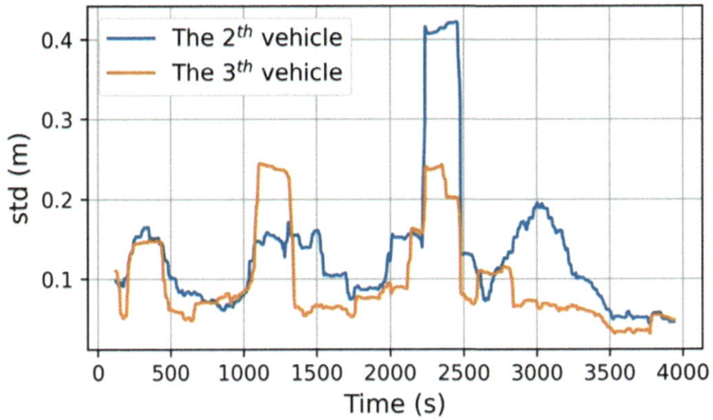

Fig. 2.4 Standard deviation within 4 min

2.2.3 *Two Dimensional Noise: Lateral*

We directly present the lateral noise for both lane keeping and lane changing in Fig. 2.5. The resulting two-dimensional trajectories show considerable stochasticity. Regarding lane keeping, the trajectories wander within the lane, while for lane changing trajectory, the duration, the initiation, and the termination vary for each lane changing instance. Figure 2.6 presents the lateral stochasticity of openACC dataset. The results show that the lateral dimension of the AVs also is random. Modeling details regarding AV lateral stochasticity can be found at the lateral wandering section.

2.3 Maneuver Estimation of NDD: State Space Model Approach

2.3.1 *Introduction*

In recent years, with the rapid development of autonomous driving technology, there has been an increasing focus on developing more accurate and powerful techniques for collecting naturalistic driving data sets (NDDs). It is estimated that the driving behavior parameters of NDDs can be used to gain insights into how drivers interact with different driving scenarios and environments, as well as to assist researchers in identifying hazardous driving situations and devising mitigation strategies through the analysis of these parameters. In this chapter, commonly used open-source data sets from different countries are employed to develop a noise model for NDDs and denoising processing is conducted. Then, a predictive control method based on the

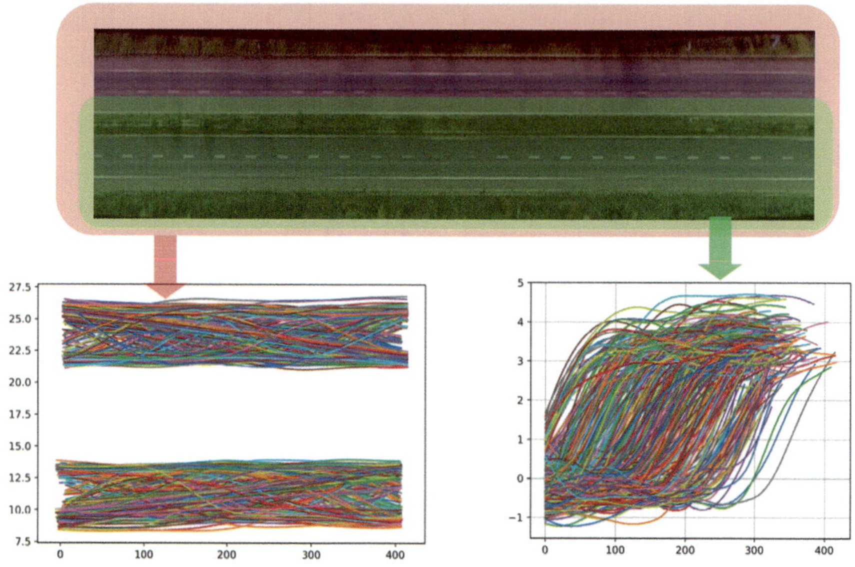

Fig. 2.5 Lateral stochasticity for highD dataset

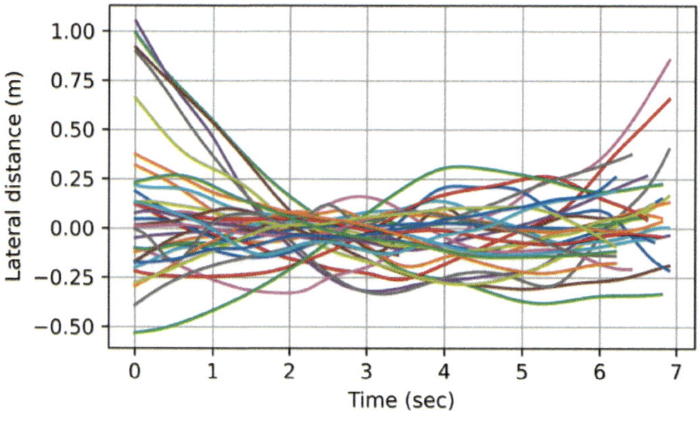

Fig. 2.6 OpenACC lateral movement

state-space model is constructed for estimating and predicting the driving behavior parameters of NDDs, and the validity of the model is verified through error analysis.

2.3.2 Description of Employed Datasets

The state estimation employs the following datasets, namely NGSIM, HighD, Mirror, ZTD, and DJI, which are collected from the United States, Germany, Japan, and China. Autonomous vehicle simulation trajectories from the CARLA environment were also employed. The NGSIM dataset is a collection of detailed driving data on US highways collected by the Federal Highway Administration. Trajectories were recorded at a high sampling rate of 10 Hz, resulting in a detailed and accurate dataset for studying driver behavior.

HighD is an open-source dataset that consists of naturalistic driving data collected on German highways from six different locations near Cologne. The dataset, which can be accessed at https://www.HighD-dataset.com/, is characterized by its large scale and diversity. It includes data from over 20,000 trips, covering a total distance of over 300,000 km. Additionally, the dataset offers high-precision vehicle localization with positioning errors typically smaller than 10 cm, achieved through the use of state-of-the-art computer vision algorithms. This makes it suitable for tasks such as parameterizing driver models, autonomous driving, and traffic pattern analysis.

The Mirror-Traffic dataset was jointly released by Tsinghua University's Suzhou Automotive Research Institute and the Jiangsu Provincial Intelligent Connected Vehicle Innovation Center. It utilizes image recognition and tracking techniques to identify and track real-world traffic participants from actual road traffic data in China. The dataset covers various types of road scenarios (ramps, straight roads, curves, intersections, etc.) and different traffic flow states (light, medium, congested, etc.). With trajectory accuracy reaching centimeter-level, it is highly suitable for tasks such as driving scenario extraction, driving scene generation, driving behavior and intent prediction model development, naturalistic driving behavior imitation learning, autonomous driving prediction and planning algorithm development and validation, traffic flow analysis (speed, density), and micro-traffic model simulation.

ZTD (Zen Traffic Data) is an open-source dataset that can be downloaded from https://zen-traffic-data.net. In this research, the Kitada Route from the ZTD dataset was used, which represents a typical road design with slopes, despite having lane restriction extensions. The restrictions may have minimal impact on lane change decision-making, which is of interest in studying its potential effects on lane-changing. For the traffic conditions on the Kitada Route, it initially starts with no congestion, gradually transitioning to congested conditions. Furthermore, the vehicle behaviors of all vehicles traveling on the targeted road segment over several kilometers were converted into long-term trajectory data at a 0.1-s interval.

CARLA data originated from the CARLA simulation platform. The platform offers open digital assets (urban layouts, buildings, vehicles) and supports flexible specification of sensor suites, environmental conditions, complete control over all static and dynamic actors, map generation, and more. The highway was designed using RoadRunner and consists of a five-lane road with a total length of approximately 2000 m. The dataset includes 1045 trajectories of vehicles such as Jeep, Tesla, Mini, Mercedes-Benz, etc. Additionally, real-time information for each vehicle can

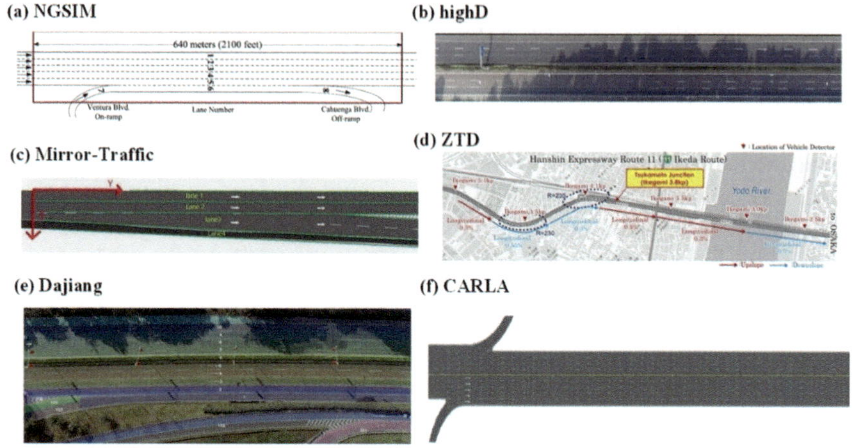

Fig. 2.7 Road topologies

be directly obtained from the CARLA Simulator, including position, speed, acceleration, heading angle, front-wheel angle, and yaw angle. The road topology where the data is collected is given in Fig. 2.7.

2.3.3 *Preprocessing*

The raw data generated from these sources may contain errors, redundancies, or inconsistencies. Additionally, there can be certain attribute differences among different NDDs, thus necessitating data preprocessing. This section demonstrates the processing of NDD data, including data transformation, data imputation, attribute standardization, and extraction of key data, as illustrated in the flowchart (Fig. 2.8).

Data transformation refers to converting the format of different NDDs, such as standardizing date formats or converting data with different units. In this section, data

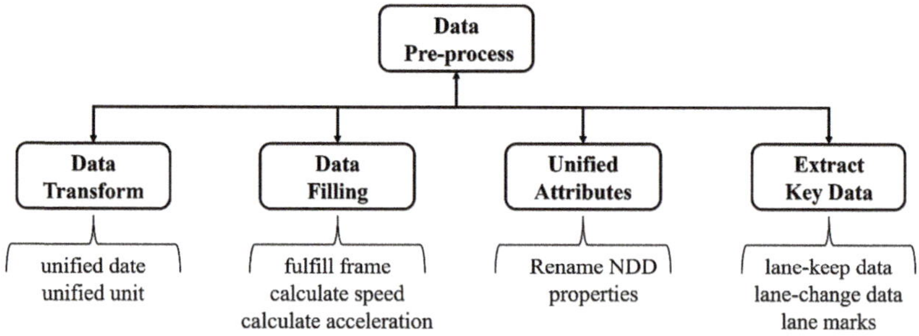

Fig. 2.8 Data process

transformation includes converting the longitude and latitude of ZTD data to standard XY coordinates and unifying time attributes to "frames" to achieve consistency with other NDDs. Additionally, NGSIM data is measured in "feet", and is converted to the international standard unit "meters" by multiplying with the appropriate conversion factor. Data imputation refers to filling in missing data. Despite the high collection frequency of NGSIM, missing data can still occur in the dataset, such as certain vehicles lacking data for certain frames. This section utilizes one-dimensional linear interpolation method based on available data to fill in missing values. Furthermore, as this section examines vehicle two-dimensional motion, NGSIM and ZTD data do not include lateral and longitudinal velocity and acceleration; their values are calculated by taking the first and second derivatives of lateral and longitudinal positions.

To enable subsequent batch processing of data, the attributes of these NDDs need to be renamed. This section adopts the attribute naming convention of HighD data and standardizes the naming of lateral and longitudinal positions, velocities, and accelerations to "x", "y", "xVelocity", "yVelocity", "xAcceleration", and "yAcceleration", respectively. Other attribute names renamed include "frame" for time, "id" for vehicle name, "laneld" for fleet number, and "length" and "width" for vehicle length and width, respectively.

Furthermore, it is necessary to extract behavioral information from NDDs, such as lane-keeping data, lane-changing data, and lane line extraction. Lane-keeping data refers to instances where the total number of lane changes for the current vehicle is "0", while lane-changing data indicates that the total number of lane changes for the current vehicle is greater than or equal to "1", thus allowing for the extraction of NDDs' lane-keeping and lane-changing data, respectively. Lane line extraction is based on lane-changing data. For example, by using vehicle lane-changing data, the coordinates both before and after the lane change can be obtained, and their average can be calculated to determine a coordinate point for a lane line. Subsequently, the least squares method can be used to fit these coordinate points to identify a lane line's coordinates.

2.3.4 *Driving Behavior State Estimation*

2.3.4.1 State Space Model of the Vehicle

In NDDs, for each given vehicle trajectory, it is represented as a multidimensional state vector denoted by $S^{\mathrm{m}\times\mathrm{n}}$. Although different NDDs may have varying attributes, they all include position coordinates, velocity, and acceleration, as given by the following equation:

$$S^{\mathrm{m}\times\mathrm{n}} = \left\{\left[\hat{x}_t, \hat{y}_t, \hat{v}_t, \hat{a}_t\right]^{\mathrm{T}} | t = 1,2,3, \ldots m\right\} \tag{2.6}$$

However, as $\boldsymbol{S}^{\mathrm{m}\times\mathrm{n}}$ does not account for driving behavior parameters, we redefine the state of the vehicle at time t as s_t:

$$\mathbb{s}_t = \left[x_t, y_t, \beta_t, \psi_t, v_t\right]^\top \tag{2.7}$$

where x_t, y_t represent the longitudinal and lateral positions of the vehicle, respectively; v_t denotes the vehicle's velocity; β_t is the lateral slip angle; and ψ_t represents the heading angle. Additionally, the longitudinal and lateral behaviors of the vehicle are controlled by the acceleration a_t and the steering angle δ_t. Therefore, in this paper, we model them as the control input vector $\mathbb{u}_t$, as given by the following equation:

$$\mathbb{u}_t = \begin{bmatrix} a_t \\ \delta_t \end{bmatrix} \tag{2.8}$$

In order to predict the real-time state of the vehicle, this paper proposes a state space model predictive control to estimate $\mathbb{u}_t$ and $\mathbb{s}_t$, as SSMPC allows for continuous stable controller design and system analysis. Firstly, we construct a new state vector $\mathcal{X}_t = [\mathbb{s}_t, \mathbb{u}_t]^\top$, and an output state vector $\mathcal{Y}_t$ with the same dimensions as $\mathcal{X}_t$. SSMPC is given by the following equation:

$$\begin{cases} \mathcal{X}_{t+1} = A\mathcal{X}_t + B\begin{bmatrix} \Delta t \cdot \boldsymbol{\Gamma}_t \\ \boldsymbol{O}_1 \end{bmatrix} + G\begin{bmatrix} \boldsymbol{O}_2 \\ \mathbb{u}_t \end{bmatrix} \\ \mathcal{Y}_{t+1} = D\mathcal{X}_{t+1} \end{cases} \tag{2.9}$$

where Δt represents the time step; $\boldsymbol{O}_1$ and $\boldsymbol{O}_2$ are two zero vectors with the same dimensions as $\mathbb{s}_t$ and $\mathbb{u}_t$ respectively; $\boldsymbol{A}$ and $\boldsymbol{D}$ are identity matrices; and $\boldsymbol{B}$ and $\boldsymbol{G}$ are specific matrices defined as:

$$\boldsymbol{B} = \begin{bmatrix} 1\,0\,0\,0\,0\,0\,0 \\ 0\,1\,0\,0\,0\,0\,0 \\ 0\,0\,1\,0\,0\,0\,0 \\ 0\,0\,0\,1\,0\,0\,0 \\ 0\,0\,0\,0\,1\,0\,0 \\ 0\,0\,0\,0\,0\,0\,0 \\ 0\,0\,0\,0\,0\,0\,0 \end{bmatrix}, \boldsymbol{G} = \begin{bmatrix} 0\,0\,0\,0\,0\,0\,0 \\ 0\,0\,0\,0\,0\,0\,0 \\ 0\,0\,0\,0\,0\,0\,0 \\ 0\,0\,0\,0\,0\,0\,0 \\ 0\,0\,0\,0\,0\,0\,0 \\ 0\,0\,0\,0\,0\,1\,0 \\ 0\,0\,0\,0\,0\,0\,1 \end{bmatrix} \tag{2.10}$$

To derive the state transition matrix $\boldsymbol{\Gamma}_t$ in Eq. (2.9), we adopt the bicycle model proposed by Rajamani, which uses longitudinal and lateral forces as well as the front steering angle to describe the vehicle's kinematics. The configuration of the bicycle model is shown in Fig. 2.9.

According to Eq. (2.7), the detailed mathematical formulas of the bicycle model are as follows.

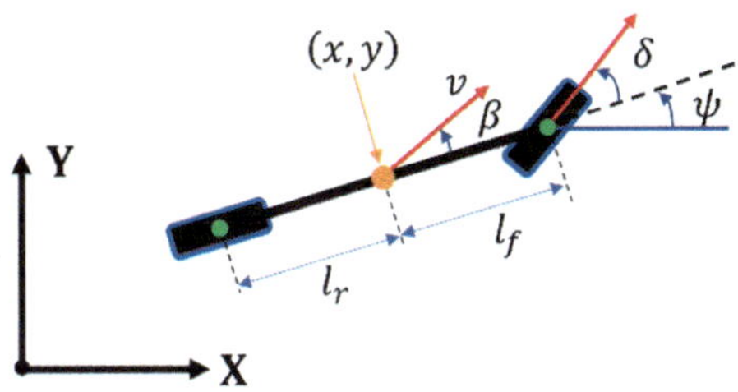

Fig. 2.9 Bicycle model

$$\begin{cases} \dot{x}_t = v_t cos(\psi_t + \beta_t), (i) \\ \dot{y}_t = v_t sin(\psi_t + \beta_t), (ii) \\ \dot{\psi}_t = \frac{v_t \cos(\beta)}{l_r + l_f} \cdot \tan(\delta_t), (iii) \end{cases} \tag{2.11}$$

where l_r, l_f represent the distance from the vehicle's center of mass to the rear and front wheel axles respectively, which are related to the vehicle's dimensions and their sum equals the wheelbase. The term β_t is given by the following equation:

$$\beta_t = \tan^{-1}\left(\frac{l_r \tan(\delta_t)}{l_r + l_f}\right) \tag{2.12}$$

Based on Eqs. (2.11) and (2.12), by discretizing the bicycle model with a time step Δt, the relationship between s_(t + 1) and s_t can be established as:

$$\mathbb{s}_{t+1} = \begin{bmatrix} \dot{x}_t \\ \dot{y}_t \\ \beta_t \\ \dot{\psi}_t \\ v_t \end{bmatrix} + \begin{bmatrix} v_t cos(\psi_t + \beta_t) \\ v_t sin(\psi_t + \beta_t) \\ \tan^{-1}\left(\frac{l_r \tan(\delta_t)}{l_r + l_f}\right) \\ \frac{v_t \cos(\beta_t)}{l_r + l_f} \cdot \tan(\delta_t) \\ a_t \end{bmatrix} \cdot \Delta t =: \mathbb{s}_t + \Delta t \cdot \boldsymbol{\Gamma}_t(\cdot) \tag{2.13}$$

According to Eq. (2.13), $\boldsymbol{\Gamma}_t$ can also be derived accordingly. Given the initial state of the vehicle, the future state of the vehicle can be estimated. Additionally, this paper utilizes model predictive control (MPC) to optimize the control input vector $\mathbb{u}_t$, as it can control the vehicle's speed and steering based on the measured trajectory of the NDD and guide the vehicle through a given path. In Eq. (2.6), the observed vehicle trajectory is denoted as $\left[\hat{x}_t, \hat{y}_t, \hat{v}_t, \hat{a}_t\right]$. Subsequently, the error between the MPC-estimated trajectory and the observed trajectory is defined as $g(\mathbb{s}_t, \mathbb{u}_t)$:

$$g(\mathbb{s}_t, \mathbb{u}_t) = {e_t}^\top \mathcal{K}_t e_t + {\mathbb{u}_t}^\top R_t \mathbb{u}_t \tag{2.14}$$

where e_t is the vector error between the state predicted by MPC and the observed state in the NDD, $\mathcal{K}_t$ represents the state cost matrix of the vehicle, and R_t is the control input matrix of the vehicle. This paper only considers the longitudinal and lateral positions of the vehicle, as they are essential for describing the vehicle information.

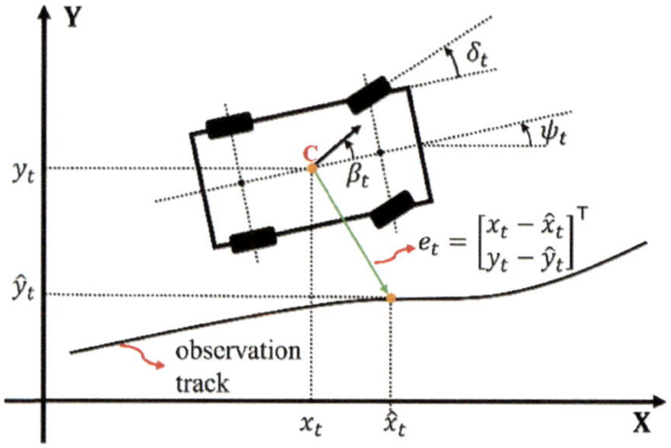

Fig. 2.10 Error between observation and prediction

The specific formula for e_t is as follows.

$$e_t = \begin{bmatrix} x_t \\ y_t \end{bmatrix}^\top - \begin{bmatrix} \hat{x}_t \\ \hat{y}_t \end{bmatrix}^\top = \begin{bmatrix} x_t - \hat{x}_t \\ y_t - \hat{y}_t \end{bmatrix}^\top \tag{2.15}$$

where $[x_t, y_t]$ represents the predicted position of the vehicle at time t, $[\hat{x}_t, \hat{y}_t]$ represents the observed position of the vehicle at time t, and their relationship is shown in Fig. 2.10.

To optimize the control input parameters in order to make the vehicle follow the observed trajectory as closely as possible, this paper minimizes the difference between the predicted trajectory and the observed trajectory. The formula is as follows.

$$\min J(\mathbb{s}_t, \mathbb{u}_t | t \in [T]) = \sum_{t \in [T]} g(\mathbb{s}_t, \mathbb{u}_t) \tag{2.16}$$

In Eq. (2.16), due to limitations of the vehicle's physical properties, the control inputs and state variables are also bounded. The constraints on the steering angle and acceleration are defined as follows.

$$\delta \in \left[\underline{\delta}, \overline{\delta}\right], a \in \left[\underline{a}, \overline{a}\right] \tag{2.17}$$

In Eq. (2.17), their allowable ranges are: $\delta \in [-0.91\text{rad}, 0.91\text{rad}]$ and $a \in \left[-11.5\text{m/s}^2, 11.5\text{m/s}^2\right]$. It is worth noting that the prediction horizon plays an important role in MPC, and in this paper, it is set to 15 sample points. Equations (2.16) and (2.17) can be solved using Python libraries. Therefore, the vehicle's state variable $\mathcal{Y}_t = [\mathbb{s}_t, \mathbb{u}_t]^\top$ can be computed.

2.3.4.2 Error Measurment

This section utilizes the dynamic time warping (DTW) algorithm to calculate the error between the observed trajectory of NDD and the predicted trajectory of SSMPC. The detailed calculation process of DTW is illustrated in Fig. 2.11. In Fig. 2.11, the blue line represents the observed trajectory of NDD, and the red line represents the predicted trajectory of SSMPC. q_i and c_i denote two coordinate points on the two trajectories, respectively. The calculation of dtw_{ij} is as follows.

$$\mathrm{dtw}_{i,j} = |q_i - c_i| + \min\{\mathrm{dtw}_{i,j-1}, \mathrm{dtw}_{i-1,j}, \mathrm{dtw}_{i-1,j-1}\} \tag{2.18}$$

where $|q_i - c_i|$ denotes the distance between two coordinate points. The DTW of Q and C is defined as the accumulated distance between these points from the initial points q_0 and c_0 until reaching the end points q_I and c_J, and the cumulative distance is the similarity between sequences Q and C:

$$\mathrm{DTW}(\boldsymbol{Q}, \boldsymbol{C}) = \sum\nolimits_{i=0}^{I}\sum\nolimits_{j=0}^{J}\mathrm{dtw}_{i,j} \tag{2.19}$$

Thus, the smaller the value of $\mathrm{DTW}(\boldsymbol{Q}, \boldsymbol{C})$, , the closer the predicted trajectory is to the observed trajectory. Other common error metrics, such as mean absolute error (MAE), root mean square error (RMSE), and mean absolute percentage error (MAPE), can also be used to compute their similarity. Their calculation formulas are as follows.

$$\begin{cases} MAE = \frac{1}{N}\sum\nolimits_{n=1}^{N}\left|p_{obs}^{n} - p_{est}^{n}\right|, \text{(i)} \\ MSE = \frac{1}{N}\sum\nolimits_{n=1}^{N}\left(\left|p_{obs}^{n} - p_{est}^{n}\right|\right)^2, (ii) \\ RMSE = \sqrt{\frac{1}{N}\sum\nolimits_{n=1}^{N}\left(\left|p_{obs}^{n} - p_{est}^{n}\right|\right)^2}, (iii) \\ MAPE == \frac{100\%}{N}\sum\nolimits_{n=1}^{N}\left(\frac{\left|p_{obs}^{n}(x) - p_{est}^{n}(x)\right|}{p_{obs}^{n}(x)} + \frac{\left|p_{obs}^{n}(y) - p_{est}^{n}(y)\right|}{p_{obs}^{n}(y)}\right), (iv) \end{cases} \tag{2.20}$$

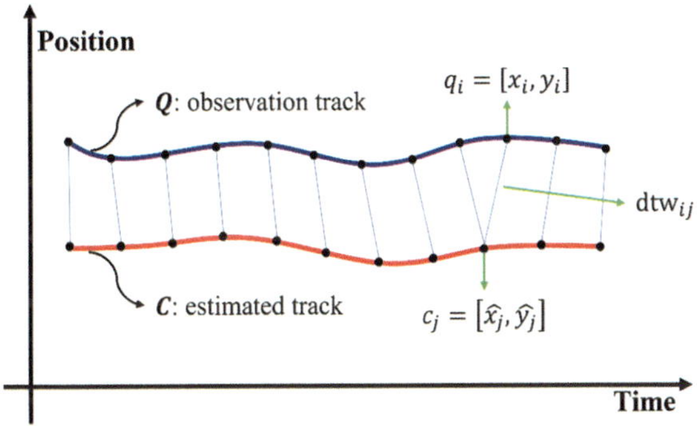

Fig. 2.11 DTW

where p^n_{obs} and p^n_{est} denote the coordinate of the nth point on the observed and estimated trajectories, respectively, and the symbol $|\cdot|$ represents the Euclidean distance.

2.3.4.3 Case Study

Firstly, two trajectories from HighD were randomly chosen: a lane keeping (LK) trajectory and a lane changing (LC) trajectory, to evaluate the effectiveness of the SSMPC model proposed in this article. The results are shown in Fig. 2.12. In (a), the red and pink lines represent the predicted LK and LC trajectories which closely fit the observed trajectories, indicating the good predictive performance of the SSMPC model. Additionally, the predicted acceleration, yaw angle, and heading angle of the LK and LC trajectories are shown in Fig. 2.12b–d. The yaw angle and heading angle of both trajectories fluctuate near zero, and the magnitude of these fluctuations in the LC trajectory is greater than in the LK trajectory, which is consistent with actual driving situations.

Furthermore, tests were also conducted using NGSIM data and the results are shown in Fig. 2.13. The accuracy is lower than in the HighD dataset due to the earlier data collection, but the results show that the SSMPC model is still applicable to lower-quality NDD data.

The error between the predicted and observed trajectories for both HighD and NGSIM were calculated using Eq. (2.20), and the results are shown in Table 2.4. All DTW values are almost less than 1, and the MAE and RMSE values are all

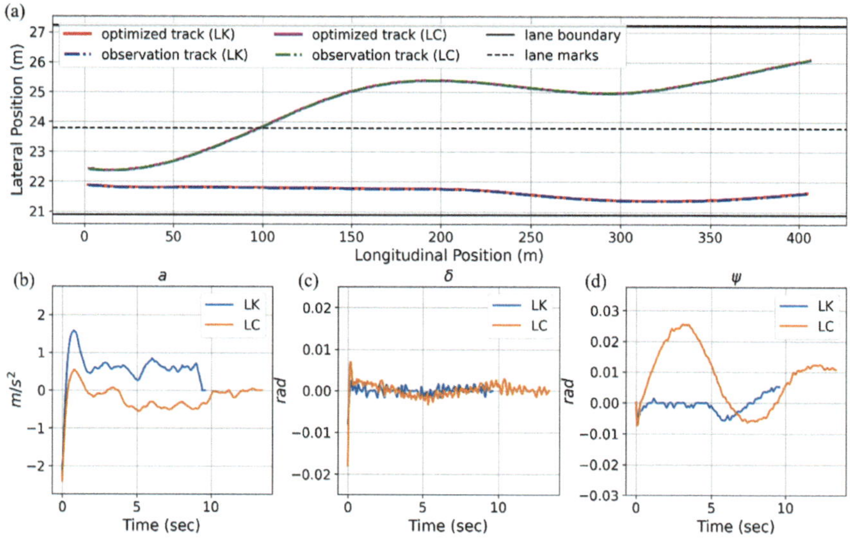

Fig. 2.12 **a** HighD prediction and observation; **b** HighD acceleration prediction; (c) HighD steering angle prediction; **d** HighD heading angle prediction

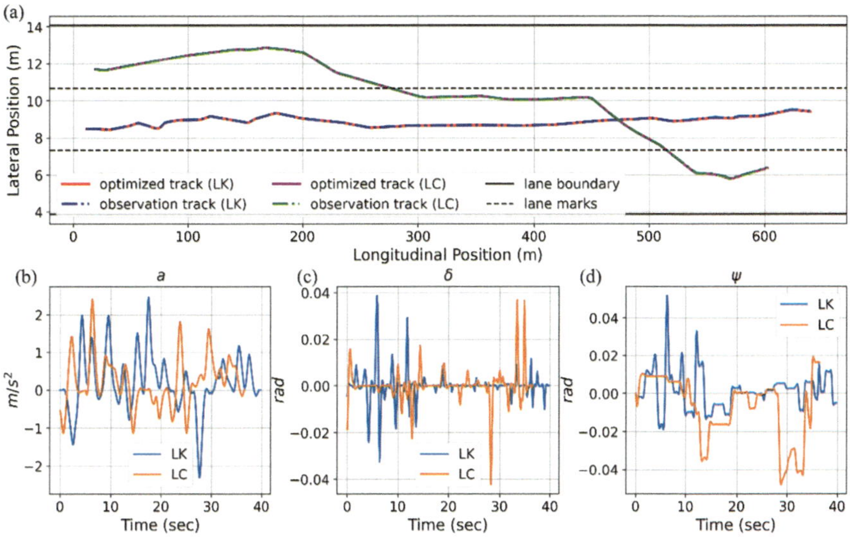

Fig. 2.13 **a** NGSIM prediction and observation; **b** NGSIM acceleration prediction; (c) NGSIM steering angle prediction; **d** NGSIM heading angle prediciton

less than 0.065, while all MAPE values are less than 0.2%. They are all very small, indicating that the predicted and observed trajectories using the SSMPC model are almost identical. Therefore, the SSMPC model proposed in this article is reliable and accurate for predicting driving parameter trajectories.

The SSMPC model was then used to predict the driving behavior parameters of the entire dataset for different NDD, as shown in Fig. 2.14. Figure 2.14a shows the distribution of acceleration which is aggregated, while Fig. 2.14b shows the distribution of yaw angle, displaying a normal distribution. The error between the predicted and observed trajectories for different NDD was then calculated using Eq. (2.20), and their distributions are presented in Fig. 2.15. Figure 2.15a shows the DTW distribution, while Fig. 2.15b–d display the distribution of MAE, RMSE, and MAPE, respectively. The average values of these different error metrics are shown in Table 2.5, indicating a gamma distribution for these errors. The DTW values in ascending order were DJI, NGSIM, HighD, Mirror, and ZTD. Similarly, the MAE,

Table 2.4 Error comparison between highD and NGSIM prediction

NDD	Type	Error index			
		DTW	MAE	RMSE	MAPE
HighD	LK	0.988	0.037	0.064	0.16%
	LC	0.818	0.020	0.045	0.11%
NGSIM	LK	1.020	0.033	0.051	0.05%
	LC	0.909	0.032	0.023	0.11%

RMSE, and MAPE errors in ascending order were NGSIM, HighD, DJI, Mirror, and ZTD. These errors are very small, demonstrating the high precision of the proposed SSMPC.

In addition, the SSMPC model is used to predict the driving behavior parameters of CARLA data, and its observed values can be directly obtained from the simulation platform. The predicted steering angle and heading angle are compared with the observed actual values using Eq. (2.20), and the probability histogram of the error is plotted as shown in Fig. 2.16. The histograms of the Mean Absolute Error (MAE), Mean Squared Error (MSE), and Mean Absolute Deviation (MAD) corresponding to the steering angle and heading angle are shown in Fig. 2.16a–c and Fig. 2.16d–f, respectively. The maximum error values for these metrics in each subplot are also

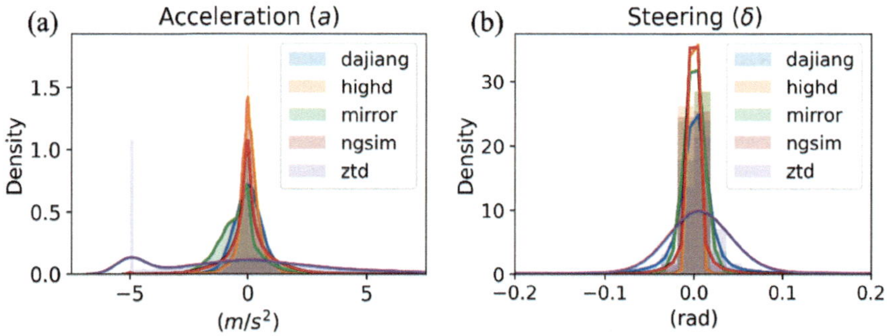

Fig. 2.14 Distribution of acceleration and steering angle

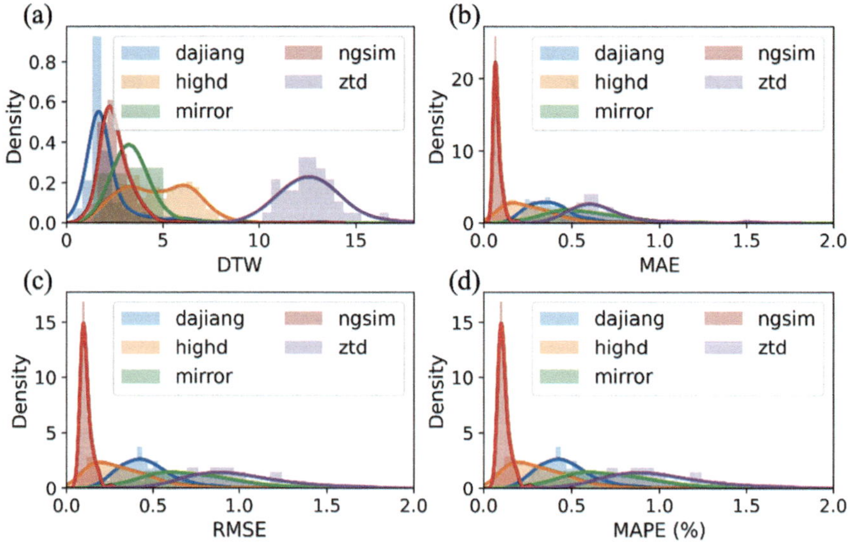

Fig. 2.15 NDD error distribution

Table 2.5 Average of five NDD errors

NDD	Average error			
	DTW	MAE	RMSE	MAPE
NGSIM	2.59	0.08	0.12	0.28%
HighD	4.75	0.27	0.33	0.25%
Mirror	3.60	0.58	0.74	1.19%
Dajiang	2.11	0.38	0.46	1.06%
ZTD	13.1	0.67	1.02	0.23%

marked with red numbers. The maximum MAE, MSE, and MAD for the steering angle are 0.108 rad, 0.087 rad, and 0.109 rad, respectively, while the maximum MAE, MSE, and MAD for the heading angle are 0.08 rad, 0.049 rad, and 0.082 rad, respectively. These values are very small, indicating that the SSMPC model performs well in predicting the driving behavior parameters.

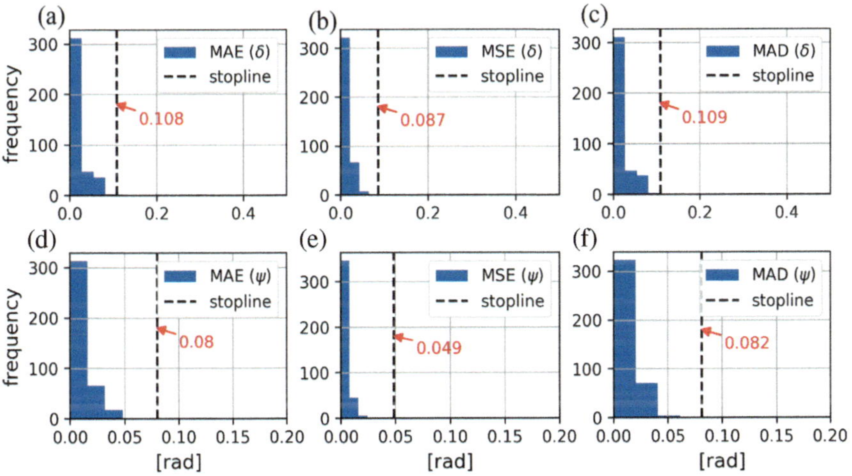

Fig. 2.16 Comparison between CARLA observation and estimation

2.4 Predicability of Driving Behavior: A Comparative Study Among Different Countries

2.4.1 Introduction

Driving behavior modeling, analyzing, and predicting are the core of the traffic flow theory. In the upcoming era of connected and autonomous vehicles (CAV), behavioral inference of driving behavior plays an important role in the autonomous vehicles' workflow (including sensing, planning, decision, and maneuvering). The microscopic driving behavior for a general road consists of car following and lane changing. Empirical observations show that considerable volatility exists among driving behaviors. For instance, the drivers can be classified into timid or aggressive types [8]. Also, the driving volatility can be represented by many indexes using large-scale data (Wen 2022). These indexes include standard deviation, variation quintiles of observation (Yu 2022) such as speed; acceleration, jerk, speed recovery time etc.

To model, analyze, or infer these driving behaviors, generally, two approaches are available: data-driven or model driven. The former one relies on huge amounts of data to train certain models such as deep network, bayesian network etc. [9]. In contrast, the latter develops the model from basic physical principles, such as the car following procedures of the drivers. No matter which approach is used, many parameters or model structures are calibrated/trained using field observations from certain locations/countries. When these models/algorithms are deployed to different locations/countries, a question then may arise: as there exist certain differences in driving behavior, would the performances are getting better or worse?

A typical example is the autonomous vehicles' inference of surrounding human-driven vehicles behaviors. Currently, there are many surrounding vehicle trajectory prediction algorithms available [10]. These models are trained/calibrated using specific sources of data. After deployment, the performance may deteriorate if the target environment is different from the training dataset's environment. It is supposed that if the driving behavior is more predicable, the underlying mechanism can be well encoded into the model structure and parameters. Thus, as an attribute, predictability can be used to describe the level of the chaos of a single driver or even collective drivers group's driving behaviors.

Currently, there are many predictability indexes available. Permutation entropy is proposed by Bandt and Pompe (2002) as an index of the complexity measure of a dynamical system and can be used to describe the predictability attribute of the target system. It has been applied to many fields, including EEG analysis (Ra 2021), financial time series [11], wind speed prediction [12], etc. According to empirical studies, the PE can track the dynamical changes (Li 2007) or analyze the anomaly in the time series [13]. For instance, [14] find that permutation entropy can be used to effectively classify EEG signals into normal vs. epileptic with an accuracy of 85% even for highly noisy EEG. After its introduction in 2022 [15], it is widely accepted as established means of detecting and characterizing complex dynamics within a time series [16]. The advantage of PE is that PE is robust to noise, invariant under nonlinear scale transformations, and can be computed with comparative speed. The above advantage leads to a wide application of PE in electrocardiogram signals [17], financial markets [18] etc.

For the predictability application of the PE, Garland et al. [19] suggest that redundancy is an effective way to measure complexity and predictive structure in an experimental time series and that weighted permutation entropy is an effective way to estimate that redundancy. Their results show that prediction accuracy is indeed correlated with weighted permutation entropy: the higher the WPE, in general, the higher the prediction error.

As PE can be used to describe the predictability of the dynamical system, we employ it to explore the predictability of the driving behaviors, which are extracted from the naturalist driving datasets. We use four popular naturalist driving datasets: NGSIM, highD, mirror, and zen-data. They correspond to four countries respectively. We also use the traffic flow simulation dataset and the autonomous vehicles simulation dataset to compare with the NDD. Using the PE as a predictability index, the difference between these datasets can be identified and ranked.

2.4.2 *Permutation Entropy*

2.4.2.1 Basics of the Permutation Entropy

The PE definition directly applies to arbitrary real-world time series data. Naturalist driving datasets are a typical example of nonlinear system observations. For a general one-dimensional time series $\{x_t, t = 1,2,3,..,T\}$, we reconstruct its phase space and get reconstructed time series $X_t = \left[x_t, x_{t+\tau}, x_{t+2\tau}, \ldots, x_{t+(m-1)\tau}\right]$, where m represents the embedding dimensions (or permutation order) and τ is a time delay. Then we sort these elements in X_t in ascending order by their values, and we have the following relationship:

$$x_{t+(j_1-1)\tau} \leq x_{t+(j_2-1)\tau} \leq \cdots \leq x_{t+(j_m-1)\tau} \tag{2.21}$$

In the above formulation, $j_1, j_2, \ldots, j_m$ represents the column index of each element in the X_t. If there exists equal elements, their positions do not change. After the sorting, we have different sorting patterns $\pi_k = (j_1, j_2, \ldots, j_m)$ of the reconstructed time series X_t, where $k \leq m!$. As there are m elements in X_t, there is a total m! different patterns, and each pattern π_k has a probability $P(\pi_k)$. Then the PE of embedding dimension $m \geq 2$ and time-lag τ is:

$$H(m, \tau) = -\sum_{i=1}^{k} P_i \cdot \ln P_i \tag{2.22}$$

The value of H(m, τ) represents the complexity and randomness of a given time series. The domain of H is [0,1]. A smaller value of H(m, τ) corresponds to a simpler system and more regular law which govern the time series. Otherwise, the system is more complex. The change in the value of H(m, τ) can also reflect the subtle local changes in the time series.

An illustrative example is presented in Fig. 2.17. The upper curve is the original data series. We use the rolling step S and the time window W (in this case, the parameter m is W and τ equals 1) to reconstruct the data. For each sub-series with length W, there are W! permutation patterns. As an example, if W is 3, there would be 6 patterns as shown in Fig. 2.18. For a data series with length T, the occurrences of different modes represent the complexity of the dynamical system. If the occurrences of these modes are nearly the same, prediction of the time series may be difficult, as each mode has the same occurrence possibility.

The above example and definition are for the one-dimensional data series. The definition can be extended to multi-dimensional cases and also the conditional scenario. We consider the state space of a vehicle as $\mathbb{s}(t)=[x(t), y(t), v_x(t), v_y(t), a_x(t), a_y(t)]$, where x and y are two-dimensional coordinates, which indicate longitudinal and lateral components, respectively; v_x and v_y

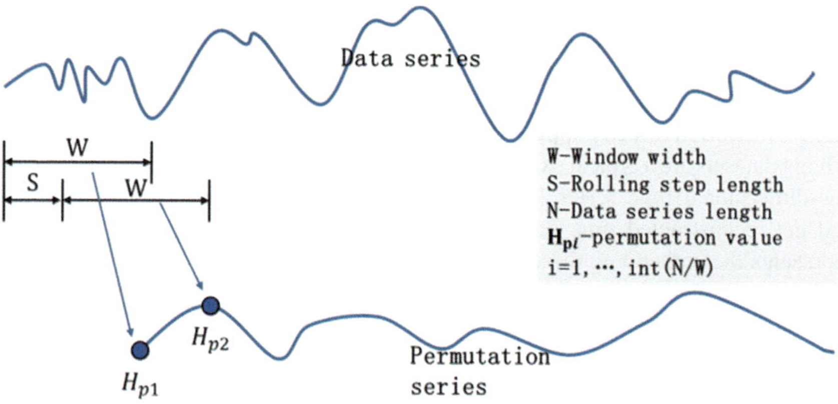

Fig. 2.17 Construction of the permutation entropy

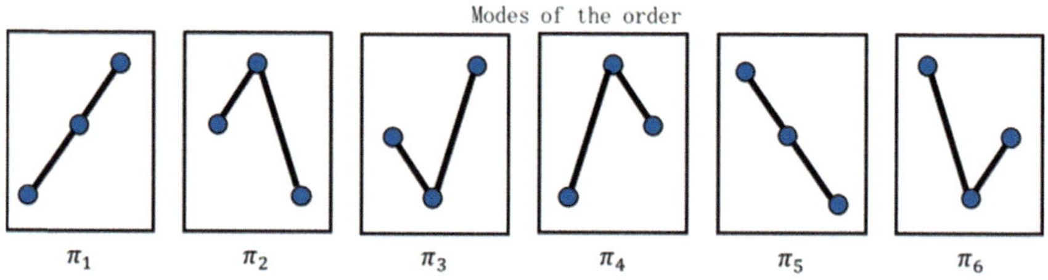

Fig. 2.18 Ordering modes

are the two-dimensional speeds and a_x and a_y are two-dimensional acceleration. As the amplitude of the above variables is different, we first normalize the variable for dimension j as follows.

$$\tilde{\mathbb{s}}_j(t) = \frac{\mathbb{s}_j(t) - \min(\mathbb{s}_j)}{\max(\mathbb{s}_j) - \min(\mathbb{s}_j)} \tag{2.23}$$

$\tilde{\mathbb{s}}_j(t)$ is normalized value for the j th dimension in $\mathbb{s}(t)$; $\min(\mathbb{s}_j)$ is the minimum value in the j th dimensional time series; $\max(\mathbb{s}_j)$ is the maximum value in the j th dimensional time series. After the normalization, we reconstruct the phase space using an embedding dimension parameter m and a time delay τ as follows.

$$\zeta(t) = \left[\tilde{\mathbb{s}}(t), \tilde{\mathbb{s}}(t+\tau), \tilde{\mathbb{s}}(t+2\tau), \ldots, \tilde{\mathbb{s}}(t+(m-1)\tau)\right] \tag{2.24}$$

Similar to the case in one-dimensional PE, the elements of time series ζ(t) should be rearranged in ascending order.

$$\zeta\ \tilde{\mathbb{s}}(t+(j_1-1)\tau) \leq \tilde{\mathbb{s}}(t+(j_2-1)\tau) \leq \cdots \leq \tilde{\mathbb{s}}(t+(j_m-1)\tau) \tag{2.25}$$

where $j_1, j_2, \ldots, j_m$ represents the column index of each element in the ζ(t). There are m! possible patterns. The relative probability of each pattern π_i gives:

$$p(\pi) = \frac{\#\{t | t \leq T-(m-1)\tau, [\mathbb{s}(t), \mathbb{s}(t+\tau), \ldots, \mathbb{s}(t+(m-1)\tau)]\ \textit{has type}\ \pi\}}{T-(m-1)\tau} \tag{2.26}$$

Then the multi-dimensional PE or MPE of embedding dimension $m \geq 2$ and time-lag τ is:

$$H_M(m, \tau) = -\sum_{i=1}^{k} p(\pi_i) \cdot \ln p(\pi_i)/\ln(m!) \tag{2.27}$$

2.4.2.2 Interpretation of the PE Concept

According to the definition, if the time series is strictly monotone (either decreasing or increasing), the resulting H would be zero. On the other hand, if each sorting pattern occurs with the same probability, the PE achieves its maximum (i.e., 1). When it is the case, the accurate prediction of the system is difficult. The above interpretation can be extended to other types of PE, including conditional PE or weighted PE.

For driving behavior analysis, the PE is interpreted similarly. Given the surrounding vehicles states, if the ego vehicle trajectories' PE index is relatively low, the ego vehicle trajectory prediction may be more accurate. On the contrary, if the resulting PE is higher, prediction of the vehicle movement would be more difficult.

2.4.3 *PE Extraction from NDD*

2.4.3.1 PE Extraction

We extract the PE index as follows for a general trajectories dataset (one among the four datasets, i.e. NGSIM, highD, mirror and ztd). For each vehicle in the dataset (which is indicated as ego vehicle in Fig. 2.19), we filter the following five neighboring vehicles: ego lane leader, left lane leader, left lane follower, right lane leader and right lane follower. The distance for the leader (either at left or right lane) and the follower (either at left or right lane) are indicated as Δ_{ll} (Δ_{rl}) and Δ_{lf} (Δ_{rf}) respectively. The distance between the ego vehicle and the leader in the same lane is Δ_{el}. All the five distances are time varying. Clearly, the movement of the ego vehicle is influenced by the surrounding vehicle's states. Hence we use the conditional PE to describe the predictability of the driving behavior of the ego vehicle.

Note that not all the five surrounding vehicles exist. For instance, for some moments, the ego lane leader may be absent for certain duration. Thus we filter all the combinations of the surrounding vehicles' existences. As each of the five surrounding vehicles has two possibilities: exist or not exist, there are total 32 (i.e. 2^5) combinations of the existences. We use σ to denote certain combination of the vehicle's existences. The conditional PE is given by the following formula:

$$H_{CM}((m,\tau)|\sigma) = p(\sigma)\cdot\left(-\sum_{i=1}^{k} p(\pi_i|\sigma)\cdot \ln p(\pi_i|\sigma)\right)/\ln(m!) \tag{2.28}$$

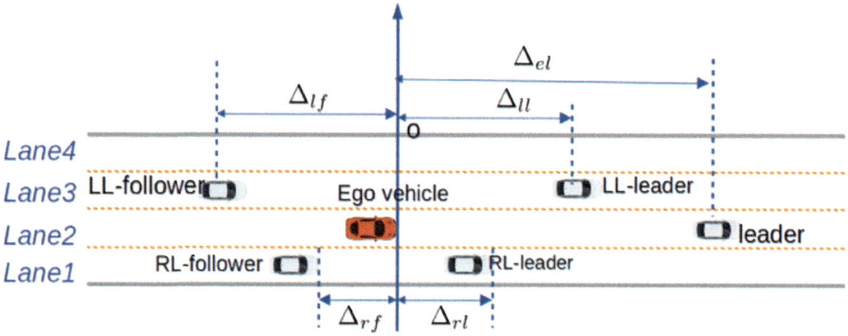

Fig. 2.19 Road example

For a vehicle, the final conditional PE or CPE is given by the average over all the possible combination, i..e.:

$$H_{CM}((m,\tau)) = \sum_{n=0}^{5} p(\sigma = n) \cdot \left(- \sum_{i=1}^{k} p(\pi_i | \sigma = n) \cdot \ln p(\pi_i | \sigma = n) \right) / \ln(m!) \tag{2.29}$$

2.4.3.2 Parameters Determination

When calculating the values of PE, two crucial parameters must be taken into account: the embedding dimension m and the delay time τ. Despite the widespread usage of PE, it is often unclear how these parameters should be chosen, with trial and error being the most typical method. For m, it is obvious that if m is too small, there is not enough data to describe the dynamical information of the time series; on the contrary, if m is too large, PE will contain more redundant information and the calculation becomes more computationally intensive. Bandt and Pompe (2002) recommend $m = 3,4,\ldots,7$. Under the assumption that a data sequence has length N, [20] proposed $5 \times m! \leq K$, where K is ordinal patterns of the sequence. [21] concluded that $m = 5$, 6, or 7 is the best suited value for complex system with the condition of $m! < N$. We chose $m = 5$. And about τ, we set $\tau = 3$.

2.4.4 Case Study

Four real-world vehicle trajectories from US, German, China, and Japan are used in the manuscript: ZTD, highD, NGSIM, and Mirror-Traffic.

The topology of the four datasets are presented in Fig. 2.20.

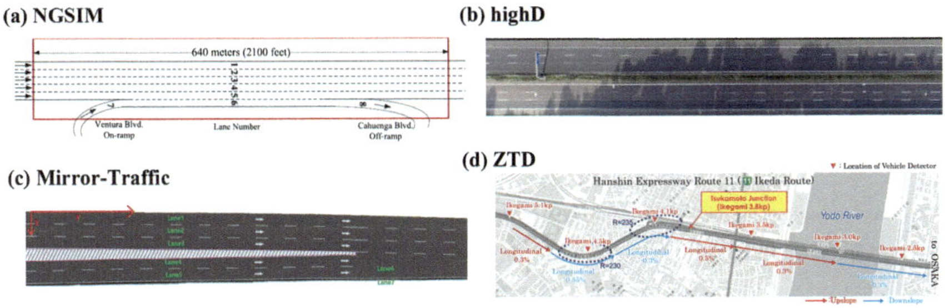

Fig. 2.20 Four datasets used in the manuscript

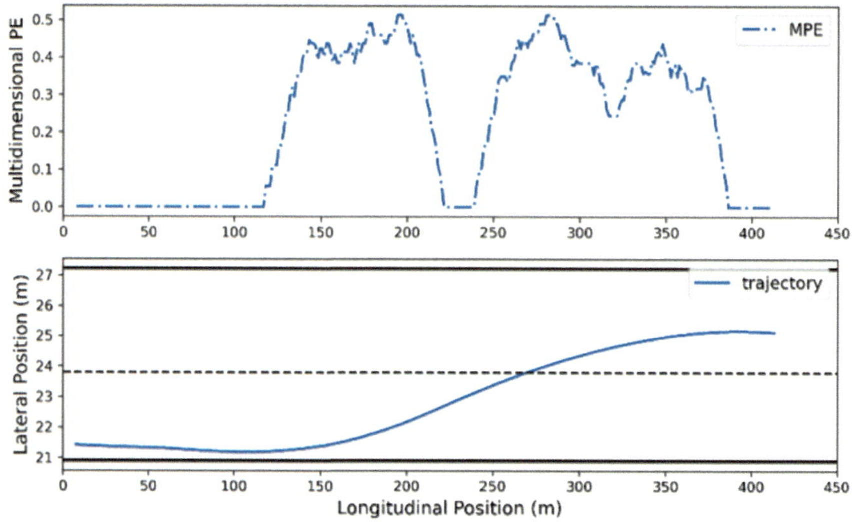

Fig. 2.21 PE and the trajectory of highD dataset

2.4.4.1 Single Vehicle Trajectories PE

Firstly we select one lane-changing trajectory in highD data to describe the evolution of the change of the PE. The lane-changing trajectory and the corresponding PE are presented in Fig. 2.21. In the lower plot, the vehicle changes lane from right to left. In the upper plot, the PE is presented. It is shown that before and after the lane changing process, the PE is nearly zero, indicating that the trajectory's predictability is well reserved. During the lane changing process, the PE increases to about 0.5. While at the middle of the lane changing process, the PE also is small.

The NDD typically include certain noise, and the noise itself would influence the results of the PE. The NGSIM data while contain nosie is presented in the following plot Fig. 2.22. The lane changing trajectory is not so smooth as highD dataset. The resulting PE is stochastic. Therefore, we resort to the aggregation analysis of the PE distribution of the whole datasets in the next sub-section.

2.4.4.2 Collective Analysis

Next we present the results of the whole dataset for each country. We first split the data into two cases: lane changing and lane keeping. After that the PE is calculated for each trajectory. For certain dataset (such as NGSIM), the PEs of all vehicles presents certain distributions. The distribution for the four NDDs are given in Fig. 2.23. Figure 2.23a is for lane keeping data and Fig. 2.23b is for lane changing data. In Fig. 2.23c, the lane keeping and lane changing are aggregated. It is shown that for the lane keeping data, the highD generally has a small PE value while ztd has a high

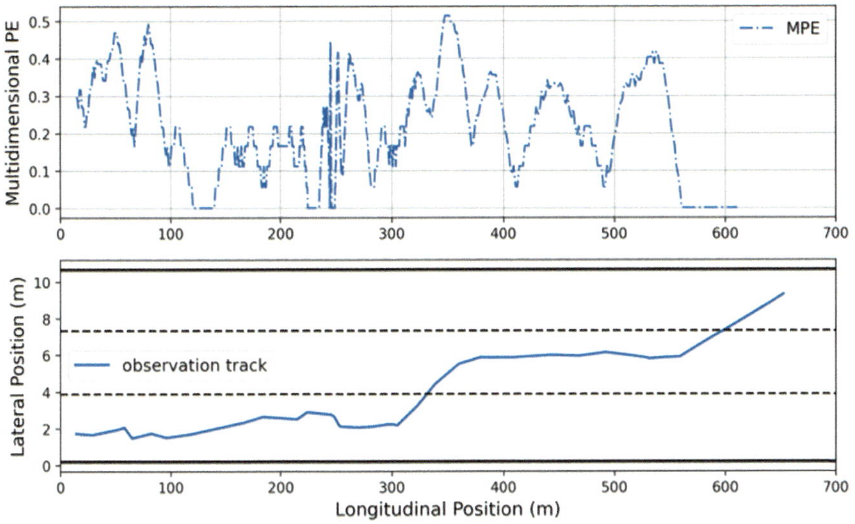

Fig. 2.22 PE and the trajectory of NGSIM dataset

PE value. As the driving behavior is stochastic, PE of part of highD trajectories is aso high. Similar patterns can be observed for the lane changing trajectories. For the aggregation of lane changing and keeping, the ascending order of the PE values are highD, NGSIM, mirror and ztd, respectively. The above order corresponds to the most predicable to the most unpredictable cases. Hence regarding the predictability, the highD performs the best.

Note that the traffic flow consists of different types of vehicles. As the data contains the vehicle size, we split the data into big vehicles and small vehicles (standard passenger car unit, or PCU). Big vehicles are those not PCUs. The PEs for the two types are presented in Fig. 2.24. The rank of the four datasets is the same as in Fig. 2.23. However, the ratio of small vehicles in highD dataset is higher than that of big vehicles. This implies that smaller vehicles are more predictable than their big counterparts.

The average PE for the lane keeping and lane changing process of the four datasets are presented in Table 2.6. The PE for the two vehicles types are presented in Table 2.7.

Currently there are many researches that employ simulation data as partial input. We nex consider traffic flow simulation results and autonomous vehicles simulation results. The former is generated by SUMO package, and the latter is generated by Carla platform. Both simulations are implemented using the US101 highway scenario. The comparison between the six datasets (four NDD plus two simulation results) are presented in Fig. 2.25. It is shown that the PE of the SUMO simulation is very like the mirror dataset, while Carla simulation has the smallest PE value. This is due to the fact that in Carla, the vehicles movement is implemented by exact path following algorithm, which lead to highly predicable behavior.

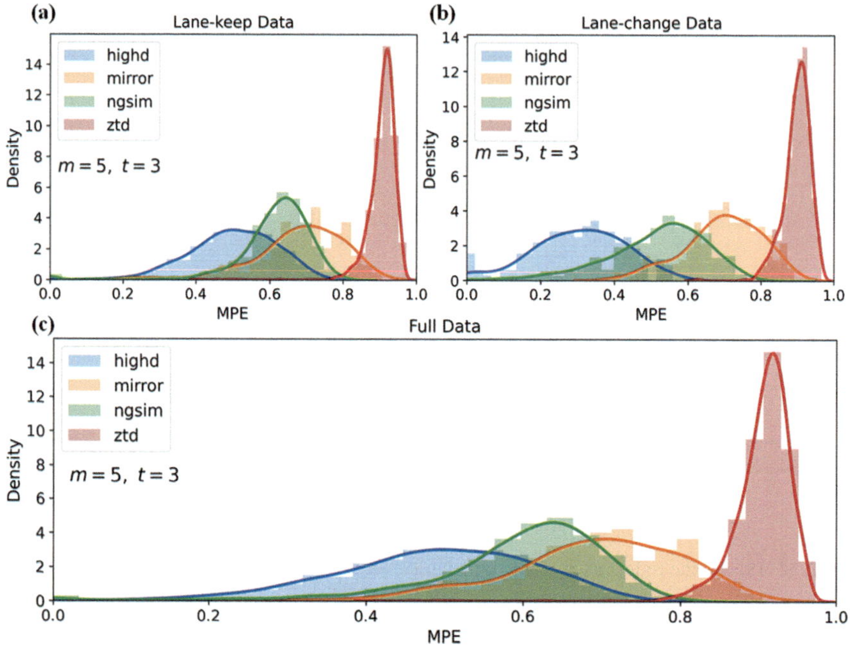

Fig. 2.23 PE for lane keeping, lane changing and both

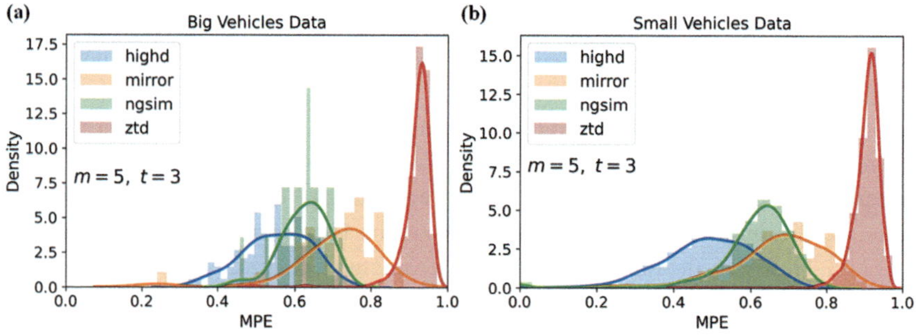

Fig. 2.24 PE for the big vehicle and small vehicle

Table 2.6 The statistics of the PE for all datasets for lane keeping and lane changing case

Mean of MPE	Lane-keep Data	Lane-change Data	Full Data
highD	0.499	0.303	0.483
Mirror-Traffic	0.682	0.697	0.688
NGSIM	0.615	0.516	0.592
ZTD	0.908	0.899	0.907

Table 2.7 The statistics of the PE for all datasets with small vehicles and big vehicles seperated

Mean of MPE	Big Vehicles Data	Small Vehicles Data
highD	0.554	0.492
Mirror-Traffic	0.711	0.678
NGSIM	0.627	0.615
ZTD	0.921	0.906

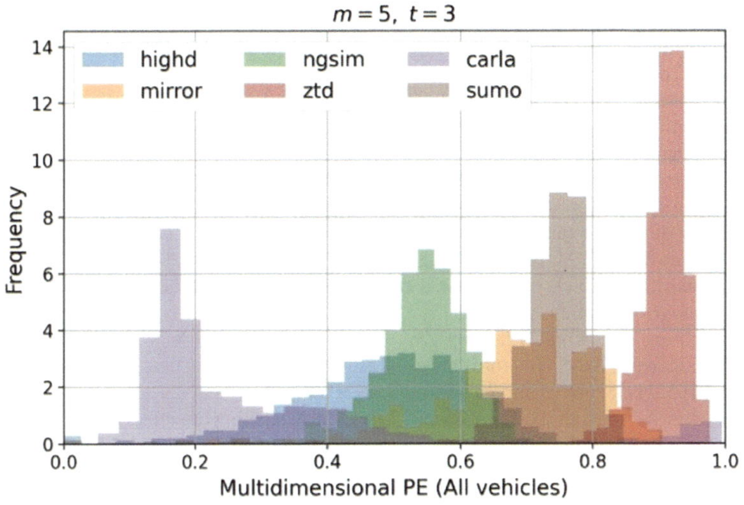

Fig. 2.25 NDD and the simulation

2.4.5 Conclusion and Remarks

The understanding of the heterogeneity in driving behavior is crucial for many real world applications, including traffic flow analysis, autonomous vehicles etc. Empirical observations and theoretical models both prove that there exists considerable volatility in driving behavior. Either physically developed models such as car following or data-driven models such as deep network are calibrated or trained using real world data from certain locations/countries. It is unclear whether the models still perform as expected if deployed to different locations or countries. To answer this question, it is crucial to analyze the behavioral differences among different locations/countries. This research employs the concept of permutation entropy to explore the predictability of the vehicles behaviors. The conditional driving permutation entropy is established and applied to four datasets from four countries. The results show that the descending order regarding predictability is Carla simulation, highD, NGSIM, mirror, SUMO and ztd.

References

1. Barmpounakis E, Nikolas G (2020) On the new era of urban traffic monitoring with massive drone data: The pNEUMA large-scale field experiment. Trans Res Part C Emerg Technol 111:50–71. https://doi.org/10.1016/j.trc.2019.11.023
2. Krajewski R, Bock J, Kloeker L, Eckstein L(2018) The highD dataset: a drone dataset of naturalistic vehicle trajectories on german highways for validation of highly automated driving systems. In: 2018 21st international conference on intelligent transportation systems (ITSC), pp 2118–2125. https://doi.org/10.1109/ITSC.2018.8569552
3. Makridis M, Konstantinos M, Aikaterini, Biagio C (2021) OpenACC. An open database of car-following experiments to study the properties of commercial ACC systems. Trans Res Part C Emerg Technol 125:103047. https://doi.org/10.1016/j.trc.2021.103047
4. U.S. Department Of Transportation Federal Highway Administration (2017) Next generation simulation (NGSIM) vehicle trajectories and supporting data. Not Available. https://data.transportation.gov/d/8ect-6jqj.
5. Ye XQ, Mao S, Li HY, Shi YF, Li YY, Wang GJ, Tan X, Errui D (2022) Rope3D: the roadside perception dataset for autonomous driving and monocular 3D object detection task. arXiv: 2203.13608. arXiv. http://arxiv.org/abs/2203.13608
6. Yu HB, Luo YZ, Shu M, Huo YY, Yang ZB, Shi YF, Guo ZL et al (2022) DAIR-V2X: a large-scale dataset for vehicle-infrastructure cooperative 3D object detection. arXiv:2204.05575. arXiv. http://arxiv.org/abs/2204.05575
7. Yu HB, Yang WX, Ruan HZ, Yang ZW, Tang YJ, Gao X, Hao X et al (2023) V2X-Seq: a large-scale sequential dataset for vehicle-infrastructure cooperative perception and forecasting. arXiv:2305.05938. arXiv. http://arxiv.org/abs/2305.05938
8. Adavikottu A, Nagendra V, Sabyasachee M (2023) Modelling the effect of aggressive driver behavior on longitudinal performance measures during car-following. Trans Res F Traffic Psychol Behav 92:176–200. https://doi.org/10.1016/j.trf.2022.11.006
9. Guo M, Zhao XH, Yao Y, Bi CF, Su YL (2021) Application of risky driving behavior data in crash detection and analysis. Physica A 591(December):126808. https://doi.org/10.1016/j.physa.2021.126808
10. Jia XS, Sun LT, Masayoshi T, Zhan W (2021) IDE-net: interactive driving event and pattern extraction from human data. Ieee Robot Autom Lett 6(2):3065–3072. https://doi.org/10.1109/LRA.2021.3062309
11. Zhao XJ, Ji MF, Zhang N, Shang PJ (2020) Permutation transition entropy: measuring the dynamical complexity of financial time series. Chaos, Solitons Fractals 139:109962
12. Ruiz-Aguilar JJ, Turias I, Gonzalez-Enrique J, Urda D, Elizondo D (2021) A permutation entropy-based EMD-ANN forecasting ensemble approach for wind speed prediction. Neural Comput Appl 33(7):2369–2391. https://doi.org/10.1007/s00521-020-05141-w
13. Wang SG, Lu MY, Kong SY, Ai J (2020) A dynamic anomaly detection approach based on permutation entropy for predicting aging-related failures. Entropy 22(11). https://doi.org/10.3390/e22111225
14. Veisi I, Pariz N, Karimpour A (2007) Fast and Robust detection of epilepsy in noisy EEG signals using permutation entropy. In: Proceedings of the seventh IEEE international conference on bioinformatics and bioengineering, Boston, MA, pp 200–203.
15. Bandt C, Bernd P (2002) Permutation entropy: a natural complexity measure for time series. Phys Rev Lett 88(17):174102
16. Little DJ, Deb MK (2016) Permutation entropy of finite-length white-noise time series. Phys Rev E 94(2):022118
17. Frank B, Pompe B, Schneider U, Hoyer D (2006) Med Biol Eng Comput 44:179
18. Zunino L, Zani M, Tabak BM, Pérez DG, Rosso OA (2009) Physica A 388:2854

19. Garland J, Ryan J, Elizabeth B (2014) Model-free quantification of time-series predictability. Phys Rev E 90(5):052910
20. Amigó JM, Zambrano S, Sanjuán MAF (2008) Combinatorial detection of determinism in noisy time series. Europhys Lett 83:60005
21. Cao Y, Tung W, Gao J et al (2004) Detecting dynamical changes in time series using the permutation entropy. Phys Rev E 70(4):046217

Chapter 3
Two Dimensional Microscopic Traffic Model with Vehicle Model: Deterministic Formulation

3.1 Introduction

The microscopic traffic behavior of vehicles is the core subject of study in traffic flow theory. In general, the micro-behavior of vehicles can be divided into longitudinal behavior and lateral behavior. The former is primarily described by car-following models, while the latter is characterized by lane-changing models. Car-following models investigate the responsive behavior of a following vehicle in response to changes in the motion state of the leading vehicle from a kinematic perspective. Lane-changing behavior, on the other hand, is a complex process that incorporates drivers' characteristics, their evaluation of surrounding vehicles' driving states, and information about the surrounding environment to achieve their own driving goals. In this chapter, we will establish two-dimensional dynamic driving models based on vehicle dynamics, taking into account both car-following and lane-changing behaviors. Furthermore, we will analyze the microscopic driving behavior in different driving scenarios using an intelligent driving model. The established micro-scale models will be validated through simulation using the natural driving dataset discussed previously.

3.2 Intelligent Driver Model Considering Vehicle kinetic attributes

3.2.1 IDM Model

The classical Intelligent Driver Model takes into account only the one-dimensional motion of vehicles. The model incorporates an acceleration tendency in free-flow conditions and a deceleration tendency considering the possibility of collision with

© The Author(s), under exclusive license to Springer Nature Singapore Pte Ltd. 2024
H. Qi, *Stochastic Two-Dimensional Microscopic Traffic Model*, Lecture Notes in Intelligent Transportation and Infrastructure,
https://doi.org/10.1007/978-981-97-3597-6_3

the leading vehicle, as expressed by the following equation:

$$a_f = a\left[1 - \left(\frac{v}{v_d}\right)^4 - \left(\frac{s^*}{s}\right)^2\right] \tag{3.1}$$

$$s^* = s_0 + max\left(0, vT + \frac{v\Delta v}{2\sqrt{ab}}\right) \tag{3.2}$$

In above equations, parameter a_f represents the follow-up acceleration of the vehicle; a denotes the maximum desired acceleration; b indicates the desired minimum acceleration; v signifies the velocity of the vehicle; v_d represents the desired velocity of the vehicle; s denotes the gap between the leading and following vehicles; T signifies the headway time; and s_0 represents the minimum safe distance.

3.2.2 Model for Steering of Front Wheel

In this section, the bicycle dynamics model is employed. In the aforementioned bicycle model, the front wheel steering angle δ is an unknown variable. Due to the difficulty in obtaining real-time precision values for δ, it cannot be directly applied in practical driving scenarios. To address this issue, a Front Wheel Steering Model is proposed to describe the front wheel steering behavior in the context of two-dimensional car-following process. Additionally, the two-dimensional car-following process is discussed, and its detailed procedure is depicted in Fig. 3.1.

In Fig. 3.1, Vehicle A serves as the leading vehicle, while Vehicle B represents a virtual vehicle that corresponds to the desired position of Vehicle C during the car-following process. Vehicle B is laterally aligned with Vehicle A, and their distance is equal to the minimum safe distance, indicating that $\delta_k = \delta_j = 0$ and $\psi_k = \psi_j = 0$. Initially, the values of δ and the heading angle ψ for Vehicle C are unknown and randomly determined. In order for Vehicle C to reach the position of Vehicle B, it

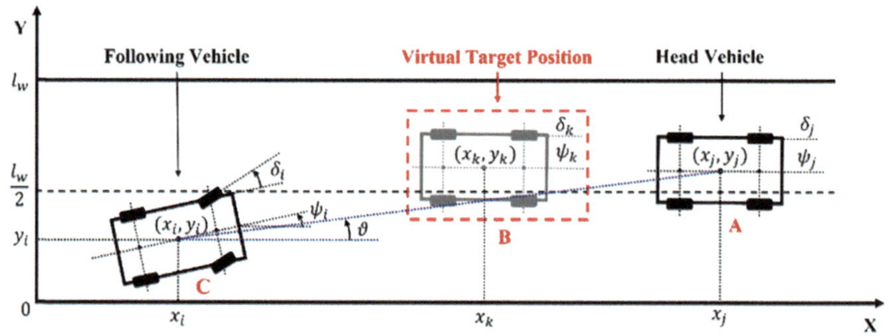

Fig. 3.1 The car following process for two dimensional vehicle

needs to adjust its position by steering the front wheels. Therefore, the front wheels of Vehicle C undergo angular acceleration $\ddot{\delta}$ and angular velocity $\dot{\delta}$. In Fig. 3.1, if the front wheels steer in the counterclockwise direction ($\delta > 0$), Vehicle C will move downwards. On the other hand, if $\delta < 0$, Vehicle C will move upwards. The stability of a car during acceleration or high-speed driving is maintained by keeping the wheels as parallel as possible to the longitudinal axis. This is reflected in the difference between the desired steering angle δ_d and the actual steering angle δ, denoted as $\Delta\delta = \delta_d - \delta = -\delta$.

Assuming that the vehicle has an expected angular steering acceleration δ^* and an expected angular velocity $\dot{\delta}_d$, the front wheel turning rate of the vehicle is denoted as δ. By comparing δ with the desired angular velocity $\dot{\delta}_d$, this chapter introduces a Forward Wheel Steering Model (FWSM) with a similar form to IDM, as shown in Eq. (3.3).

$$\ddot{\delta} = \ddot{\delta}^* \left[1 - \left(\frac{\dot{\delta}}{\dot{\delta}_d} \right)^4 - e^{-\Delta\delta} \right] \tag{3.3}$$

The FWSM model consists of an acceleration term and a deceleration term, which depend on $\dot{\delta}/\dot{\delta}_d$ and $\Delta\delta$, respectively. For instance, if the front wheels of the vehicle steer counterclockwise, resulting in $\Delta\delta < 0$ and $-e^{-\Delta\delta} < -1$, the value of δ will be less than zero, causing the front wheels to steer clockwise in order to decrease the absolute value of $\Delta\delta$. Conversely, if $\Delta\delta$ is greater than zero, the front wheels will steer counterclockwise. Through these adjustments, $\Delta\delta$ approaches zero.

Although Eq. (3.3) ensures $\Delta\delta = 0$, indicating that the vehicle's wheels are aligned parallel to the lane direction when the wheels reach a state of dynamic balance, the equilibrium state does not guarantee that the vehicle's lateral position aligns with that of the leading vehicle. Therefore, this study proposes a parameter ϑ to improve Eq. (3.3). In Fig. 3.1, ϑ can describe the lateral motion of vehicle C, defined as the vector $\overrightarrow{CA} = (x_j - x_k, y_j - y_k)$ and the vector $\overrightarrow{e_x} = (1,0)$. $\overrightarrow{CA}$ represents the vector between the centroids of vehicle A and vehicle C, which is given by $\overrightarrow{CA} = (x_j - x_k, y_j - y_k)$; whereas $\overrightarrow{e_x}$ denotes the vector corresponding to the longitudinal direction of the road, which typically takes the form of (1,0) in most scenarios. The improved version of the FWSM is shown as:

$$\ddot{\delta} = \ddot{\delta}^* \left[1 - \left(\frac{\dot{\delta}}{\dot{\delta}_d} \right)^4 - e^{\lambda_1 \delta} - \lambda_2 \cdot \tan(-\vartheta) \right] \tag{3.4}$$

In Eq. (3.4), λ_1 and λ_2 are sensitivity coefficients that describe the rate of lateral displacement change for vehicle C. They are calibrated using real driving trajectories. As λ_1 and λ_2 increase, the lateral acceleration and velocity of vehicle C during motion increase, resulting in a faster maintenance of lateral displacement with respect to the leading vehicle (vehicle A).

3.2.3 Two Dimensional Vehicle Dynamics Model

The conventional Intelligent Driver Model (IDM) only considers the one-dimensional motion (longitudinal) of vehicles. The previously discussed Follower Stop With Safety Margin (FWSM) model describes the lateral motion of vehicles. Therefore, based on the classic IDM and FWSM, a two-dimensional kinematic intelligent driver model (2D-KIDM) has been developed to accurately describe the longitudinal and lateral motion of vehicles. The following Eq. (3.5) demonstrates the relationship between them:

$$a_f = a\left[1 - \left(\frac{v}{v_d}\right)^4 - \left(\frac{s^*}{s}\right)^2 - \left(\frac{\dot{\delta}}{\dot{\delta}*}\right)^2\right] \tag{3.5}$$

Next, Eq. (3.2) is substituted into Eq. (3.1), and then the variable a_f is replaced with $\dot{v}$, resulting in the following equation:

$$\dot{v} = a\left[1 - \left(\frac{v}{v_d}\right)^4 - \left(\frac{s_0 + vT + \frac{v\Delta v}{2\sqrt{ab}}}{s}\right)^2 - \left(\frac{\dot{\delta}}{\dot{\delta}*}\right)^2\right] \tag{3.6}$$

Next, the variable of time, t, is considered. Assuming the initial direction of the vehicle (ψ_0) is 0, and the simulation time step is dt. The steering angle of the front wheels, δ, is a function of time, denoted as δ(t). Its initial value, δ_0, is a random number within the range of [−70°, 70°]. Additionally, ϑ(t) is also a function of time and depends on the relative position of the vehicle $[x(t), y(t)]$ and the leading vehicle $[x_f(t), y_f(t)]$. It is computed using the following formula:

$$\vartheta(t) = \tan^{-1}\left(\frac{y_f(t) - y(t)}{x_f(t) - x(t)}\right) \tag{3.7}$$

Then based con Eq. (3.4), we calculate $\dot{\delta(t)}$ as follows:

$$\dot{\delta(t)} = \int_0^t \ddot{\delta}*\left[1 - \left(\frac{\dot{\delta(t)}}{\dot{\delta}_d}\right)^4 - e^{\lambda_1\delta(t)} + \lambda_2\tan(\vartheta(t))\right]\mathrm{d}t \tag{3.8}$$

The steering angle $\delta(t)$ can be calculated via Eqs. (3.4) and (3.8):

$$\delta(t) = \sum_{i=1}^{t/\mathrm{dt}}\left\{\dot{\delta(t)}\mathrm{dt} + \frac{\ddot{\delta}*\left[1 - \left(\frac{\dot{\delta(t)}}{\dot{\delta}_d}\right)^4 - e^{\lambda_1\delta(t)} + \lambda_2\tan(\vartheta(t))\right]\mathrm{dt}^2}{2}\right\} \tag{3.9}$$

The slip angle of the vehicle, β(t) is a function based on Eqs. (3.8). The angular velocity of the vehicle direction is determined by β(t), δ(t), as follows:

$$\dot{\psi} = \frac{v\cos\left(\tan^{-1}\left(\frac{l_r\tan(\delta(t))}{l_r+l_f}\right)\right)}{l_r+l_f}\cdot\tan(\delta(t)) \tag{3.10}$$

The heading angle $\psi(t)$ can also be calculated based on the above Eq. (3.10):

$$\psi(t) = \sum_{i=1}^{t/\mathrm{dt}}\left[\frac{v\cdot\cos\left(\tan^{-1}\left(\frac{l_r\tan(\delta(t))}{l_r+l_f}\right)\right)}{l_r+l_f}\tan(\delta(t))\cdot\mathrm{dt}\right] \tag{3.11}$$

Except that, if the initial speed v_0 is given, then the speed v(t) can be computed based on Eq. (3.6) and (3.8):

$$v(t) = \sum_{i=1}^{t/\mathrm{dt}} a\left[1-\left(\frac{v(t)}{v_d}\right)^4-\left(\frac{s_0+v(t)T+\frac{v(t)\Delta v}{2\sqrt{ab}}}{s}\right)^2-\left(\frac{\dot{\delta}(t)}{\dot{\delta}*}\right)^2\right]\mathrm{dt}+v_0 \tag{3.12}$$

If the CoG (center of gravity) at the initial moment (x_0, y_0) is given, then the real time location of CoG $\left[x(t), y(t)\right]$ can be computed based Eq. (3.6) and (3.12):

$$x(t) = \sum_{i=1}^{t/\mathrm{dt}}\left[\left(v(t)\mathrm{dt}+\frac{\dot{v(t)}\mathrm{dt}^2}{2}\right)\cos(\psi(t)+\beta(t))\right]+x_0$$

$$y(t) = \sum_{i=1}^{t/\mathrm{dt}}\left[\left(v(t)\mathrm{dt}+\frac{\dot{v(t)}\mathrm{dt}^2}{2}\right)\sin(\psi(t)+\beta(t))\right]+y_0 \tag{3.13}$$

Through the above analysis and derivation, it is concluded that at moment t, the following variables are equal: (t)、$\beta(t)$ and $\psi(t)$. Besides, the vehicle state can be indicated by $\boldsymbol{S}(t) = \left[v(t), x(t), y(t), \psi(t), \delta(t)\right]^{\mathrm{T}}$, which can be expressed as follows:

$$\boldsymbol{S}(t) = \left[v(t), x(t), y(t), \psi(t), \delta(t)\right]^{\mathrm{T}}$$

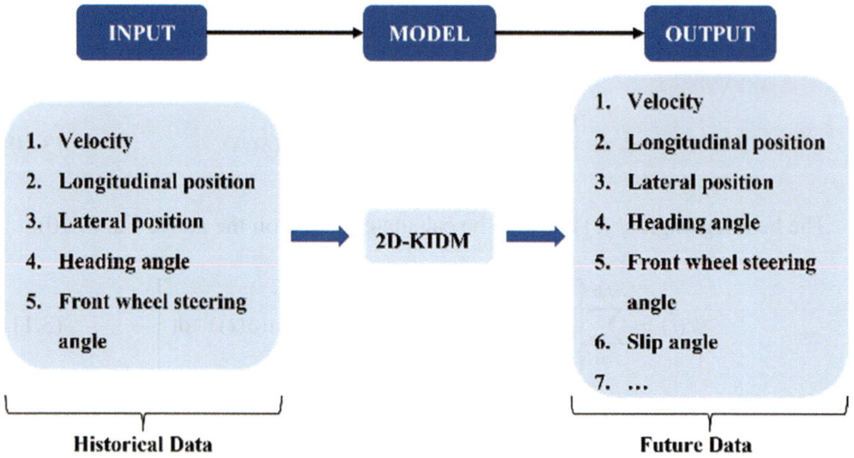

Fig. 3.2 Procedure of the prediction

$$= \begin{bmatrix} \sum_{i=1}^{t/\mathrm{dt}} a\left[1-\left(\frac{v(t)}{v_d}\right)^4-\left(\frac{s_0+v(t)T+\frac{v(t)\Delta v}{2\sqrt{ab}}}{s}\right)^2-\left(\frac{\delta(t)}{\delta*}\right)^2\right]\mathrm{dt}+v_0 \\ \sum_{i=1}^{t/\mathrm{dt}}\left[\left(v(t)\mathrm{dt}+\frac{\dot{v}(t)\mathrm{dt}^2}{2}\right)\cdot\cos(\psi(t)+\beta(t))\right]+x_0 \\ \sum_{i=1}^{t/\mathrm{dt}}\left[\left(v(t)dt+\frac{\dot{v}(t)dt^2}{2}\right)\cdot\sin(\psi(t)+\beta(t))\right]+y_0 \\ \sum_{i=1}^{t/\mathrm{dt}}\left[\frac{v\cdot\cos\left(\tan^{-1}\left(\frac{l_r\tan(\delta(t))}{l_r+l_f}\right)\right)}{l_r+l_f}\tan(\delta(t))\cdot\mathrm{dt}\right]+\psi_0 \\ \sum_{i=1}^{t/\mathrm{dt}}\left\{\dot{\delta(t)}+\frac{\ddot{\delta}*\left[1-\left(\frac{\dot{\delta(t)}}{\dot{\delta}_d}\right)^4-e^{\lambda_1\delta(t)}+\lambda_2\tan(\vartheta(t))\right]\mathrm{dt}}{2}\right\}\mathrm{dt}+\delta_0 \end{bmatrix} \tag{3.14}$$

The prediction procedure of the future movement of the vehicle is sketched as follows given in Fig. 3.2

3.2.4 *Simulation Validation*

The highD dataset is used to validate the model.

3.2.4.1 Simulation for Free-Flow State

In the free-flowing state, the vehicle is designed with $l_r = 2\,\mathrm{m}$, $l_f = 2\,\mathrm{m}$ and is traveling on a single-lane road. The maximum acceleration is $\mathrm{a} = 2.95\ \mathrm{m/s^2}$, the comfortable acceleration is $\mathrm{b} = 2.91\ \mathrm{m/s^2}$, and the desired velocity is $\mathrm{v_d} = 12.1\ \mathrm{m/s}$.

The time headway and minimum safety distance of the vehicle are set as T = 1.3 s and $s_0 = 2.58$ m, respectively. These parameters are referenced from the work of Sharath and Velaga [1], who calibrated the aforementioned parameters using 1050 different NGSIM trajectories. Additionally, $\ddot{\delta}^* = 5.81°/s^2$ and $\dot{\delta}^* = 5°/s$. If there are no other vehicles ahead in the free-flowing state, the gap s is considered infinite. Furthermore, the initial velocity, direction, and front-wheel steering angle of the vehicle are set as $v_0 = 0$, $\psi_0 = 0$, $\delta_0 = 1.5^{\circ}$. The initial state of the vehicle at time t = 0 is represented as $\boldsymbol{S}(t = 0) = \left[0, 0, 2.0\,\mathrm{m}, 0, 1.5^{\circ}\right]^{\mathrm{T}}$. Moreover, the simulation time is set as 10 s, and the simulation time step is dt = 0.04 s. The simulation results are shown in Fig. 3.3, while the lateral position results are depicted in Fig. 3.4.

In Fig. 3.3, the vehicle consistently travels along the centerline of the lane. Specifically, the trajectory of the vehicle exhibits noticeable fluctuations in the first 20 m,

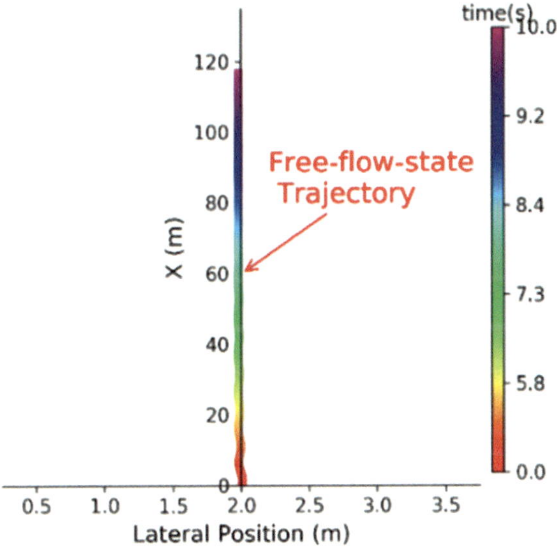

Fig. 3.3 Simulation under free flow

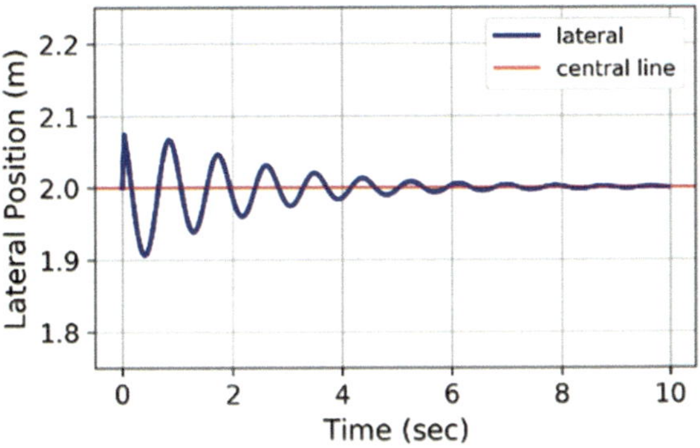

Fig. 3.4 The lateral simulation result

followed by a tendency to stabilize over time, which is also evident in Fig. 3.4. Additionally, in Fig. 3.4, the lateral position of the vehicle demonstrates the process of transitioning from fluctuations to stability, ultimately converging towards the center lane. This is because the vehicle has an initial steering angle, which is continuously modified by the 2D-KIDM during the driving process. The simulation results indicate the effectiveness of the proposed 2D-KIDM in the free-flowing state, as it accurately describes the longitudinal and lateral motion of the vehicle.

3.2.4.2 Simulation of a Platoon

In the platoon scenario, the driving behavior of following vehicles, including car-following and lane-changing, is influenced by the leading vehicle. Firstly, the car-following behavior is tested and evaluated. In this study, two vehicles, namely the leading vehicle and the following vehicle, are randomly selected from the highD database. The parameter values for this study, such as $l_r = 2\,\text{m}$, $l_f = 2\,\text{m}$, $a = 2.95\,\text{m/s}^2$, $b = 2.91\,\text{m/s}^2$, $\text{T} = 1.3$ s, and $s_0 = 2.58\,\text{m}$, are referenced from previous research [1]. According to the highD dataset, the desired velocity v_d is set to 25 m/s. Additionally, the values for $\ddot{\delta}^*$, $\dot{\delta}_d$, λ_1 and λ_2 are the same as those obtained from simulations under free-flow conditions. If the vehicle is the leading vehicle, the value of s is set to infinity. Otherwise, s is determined by the distance between the center of mass of the vehicle and the center of mass of the leading vehicle, minus the length of the vehicle. Furthermore, the initial state of the vehicle is the same as the state observed in the highD platoon, represented by $\boldsymbol{S}(t = 0) = [23.09\,\text{m/s}, 3.29\,\text{m}, 24.73\,\text{m}, 0, 2°]^\top$. The simulation time depends on the vehicles operational time. The results are given in Figs. 3.5 and 3.6.

In Fig. 3.5, the black line represents the lane boundary. The simulated vehicles, including the following vehicle and the leading vehicle, are all located in the same lane. The simulated vehicle closely follows the leading vehicle and exhibits the same trend as the following vehicle. The results demonstrate the effectiveness of the proposed 2D-KIDM model in the car-following process. Furthermore, Fig. 3.6a depicts the front wheel steering angle of the simulated vehicle, with the black curve consistently close to the baseline throughout the driving process. Additionally, Fig. 3.6b illustrates the heading angle of the simulated vehicle. The simulation results indicate that the proposed 2D-KIDM and FWSM models can accurately describe and predict the car-following behavior of vehicles.

Furthermore, the proposed 2D-KIDM model was utilized to evaluate the lane-changing behavior of vehicles. Similarly, two vehicle trajectories from the highD dataset were selected as the leading vehicle and the following vehicle. Then, we predicted the trajectory of the following vehicle. Apart from the initial states of the vehicles, the model parameters remained the same as in the aforementioned car-following experiments. The initial state of the simulated vehicle was set to the initial state of the following vehicle, i.e., $\boldsymbol{S}(t = 0) = [39.03\,\text{m/s}, 2.56\,\text{m}, 22.4\,\text{m}, 0, -2°]^\top$. The simulation results are shown in Figs. 3.7 and 3.8.

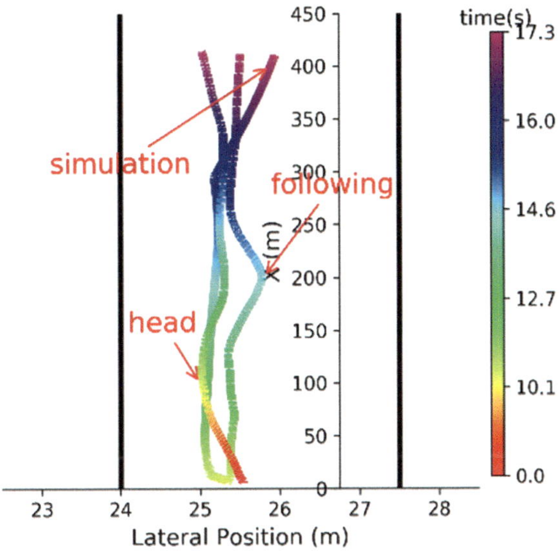

Fig. 3.5 Simulation of two-dimensional trajectories of a platoon for lane keeping

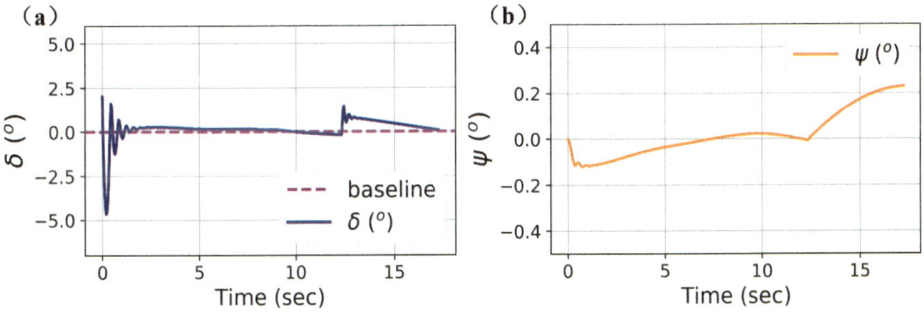

Fig. 3.6 Simulation **a** steering angle; **b** heading angle

3.2.4.3 Error Analysis

In order to evaluate the accuracy of the prediction results, the defined DTW, MAE, MSE, RMSE, and MAPE error metrics were used to calculate the metrics for the aforementioned experiments as shown in Table 3.1. Furthermore, to assess the reliability of these metric values, 110 different vehicle trajectories were randomly selected from the highD dataset, and their trajectories were predicted using the enhanced two-dimensional intelligent driver model proposed by Sharath and Velaga [1]. The error indices between the ground truth trajectories and the predicted trajectories were also calculated for these 110 sets, and their average values were taken as the baseline, the baseline for each index is provided in Table 3.1. Comparing these metrics in Table 3.1, we can observe minimal differences between the predicted trajectories

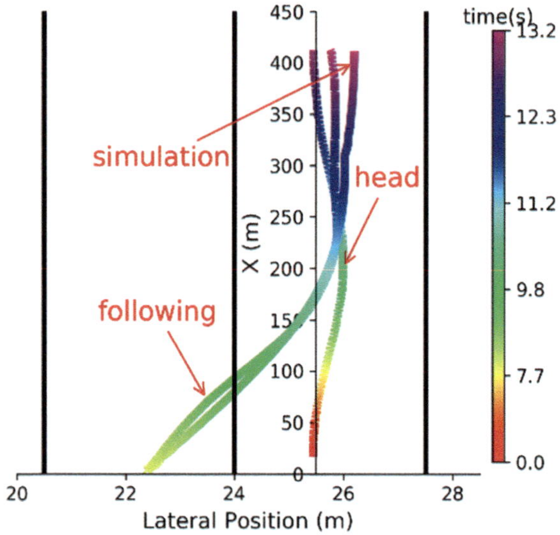

Fig. 3.7 Simulation of two dimensional trajectories of a platoon for lane changing

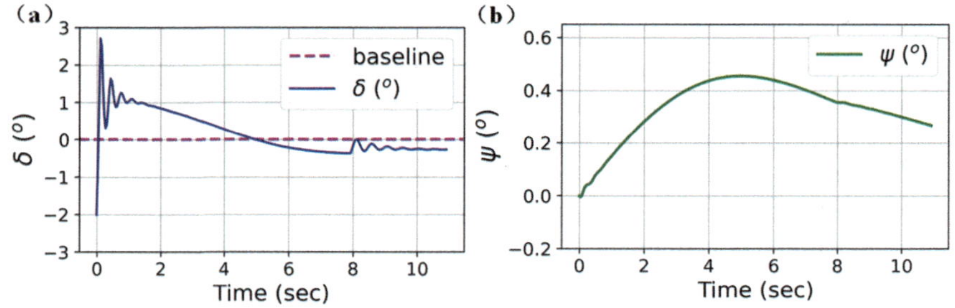

Fig. 3.8 Simulation result; **a** steering angle; **b** heading angle

and the real trajectories, indicating that the 2D-KIDM can accurately forecast the vehicle trajectories.

Table 3.1 Error of the platoon simulation

Error	Platoon simulation		
	CF	LC	Benchmark
DTW	9.31	17.29	72.04
MAE	1.03	3.21	11.51
MSE	1.30	21.18	215.26
RMSE	1.14	4.60	13.43
MAPE	1.75%	1.91%	6.33%

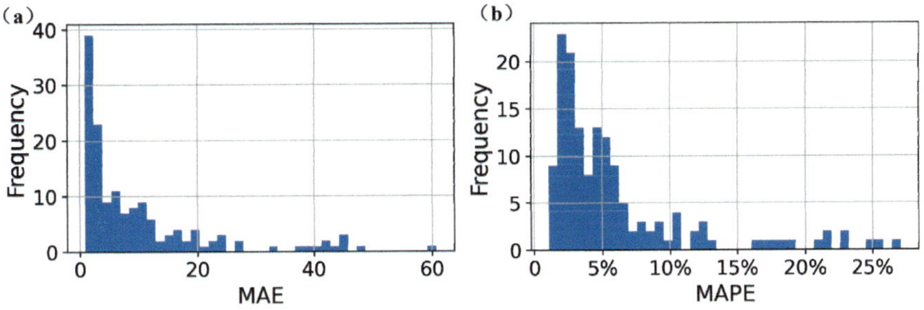

Fig. 3.9 Histogram of MAE and MAPE of 147 trajectories from highD

To further validate the effectiveness of the proposed model, an additional 147 different trajectories were randomly selected from the highD dataset to evaluate the 2D-KIDM by comparing the predicted trajectories with the real-world trajectories. Their MAE and MAPE were calculated, and the frequency histogram plots are also shown in Fig. 3.9.

In Fig. 3.9 above, it can be observed that 90.48% of the trajectories have an MAE less than 25, and 91.16% of the trajectories have an MAPE less than 15%. These results indicate that the proposed 2D-KIDM is capable of accurately describing and predicting vehicle behavior.

3.3 Conclusion

In this chapter, the two-dimensional driving behavior of vehicles is analyzed from a microscopic level based on vehicle dynamics. The following are the main findings regarding the study of vehicle following and lane-changing behaviors using vehicle dynamics models: Firstly, an analysis of the front wheel dynamics of automobiles is conducted, and a front wheel steering model is proposed. Incorporating this model with the traditional Intelligent Driver Model (IDM), a two-dimensional dynamic driving model is proposed, which can accurately describe and predict the longitudinal and lateral motion of vehicles and forecast their future states. Various simulations are performed to verify the performance of the proposed model, and several common error metrics are computed. The results demonstrate that the 2D-KIDM can accurately predict the two-dimensional motion of vehicles.

Reference

1. Sharath MN, Velaga NR (2020) Enhanced intelligent driver model for two-dimensional motion planning in mixed traffic. Transp Res Part C: Emerg Technol 120:102780. https://doi.org/10.1016/j.trc.2020.102780

Chapter 4
Two-Dimensional Intelligent Driving Model for Complex Road Topology

4.1 Introduction

Vehicle movement is the core of the microscopic traffic flow model. The movement of the vehicles is influenced by many factors, including the surrounding vehicles, the road infrastructure, etc. The road infrastructure includes many types, such as the highway, on and off ramp, etc. Current literatures classify the vehicle movement into two basic types: car following and lane changing. The former one describes the behavior when a vehicle follows another vehicle while the latter one considers the movement between neighboring lanes.

The above two methods work well for homogenous road. However, majority of traffic problems occur at infrastructure where the road topology is inhomogeneous. Such scenarios typically include the on and off ramp, intersections, etc. Under such circumstance, the vehicle's trajectory seldom follows a pre-specified line (as we done in the CF model, where the vehicle is assumed to follow the lane middle line). Therefore, a recipe is necessary to describe the vehicle movement when the inhomogeneous environment is presented. In this chapter, we propose two novel concepts, virtual boundary and virtual potential field, to describe the movement of the vehicle in the heterogeneous settings.

4.2 Basic Concept of the Virtual Boundary

The virtual boundary field studied in this chapter is an expansion and complement to the current research on two-dimensional microscopic traffic behavior of vehicles. Currently, many scholars focus on the following behavior of a trailing vehicle with respect to a leading vehicle or the influence of surrounding vehicles on the target vehicle. At highway entrances and exits, vehicles consider their own direction of travel. In addition to the influence of other surrounding vehicles, vehicles also

© The Author(s), under exclusive license to Springer Nature Singapore Pte Ltd. 2024
H. Qi, *Stochastic Two-Dimensional Microscopic Traffic Model*, Lecture Notes in Intelligent Transportation and Infrastructure,
https://doi.org/10.1007/978-981-97-3597-6_4

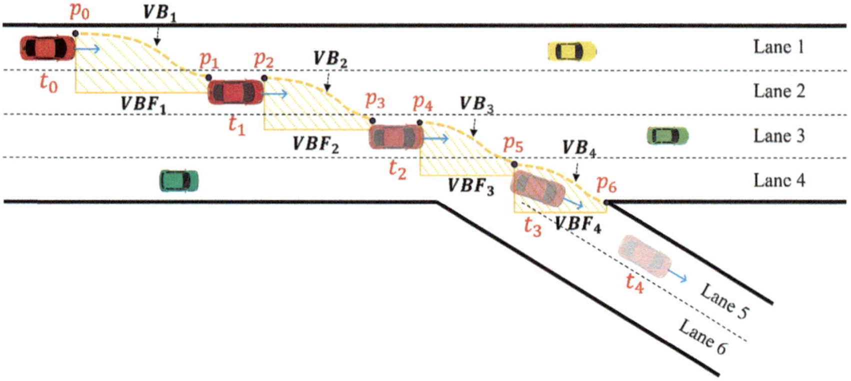

Fig. 4.1 Vehicle movement under the influence of virtual boundary potential

consider the topological factors of the road. For example, in order to smoothly exit the main road, vehicles leaving the ramp will change lanes one by one to the outermost lane before reaching the exit ramp. In order to characterize this behavior, this article introduces the concept of a "virtual boundary field". If a vehicle is located within the boundary field, its driving behavior will be influenced by the virtual boundary field, which is generated by the virtual boundary field line, defined over the physical boundary of the road. The virtual boundary field line defines the driving range of the vehicle and is the boundary range that the vehicle's boundary points can reach (theoretically not allowed to exceed). As shown in Fig. 4.1, under the influence of the defined virtual boundary field line (the thick yellow dashed line in the figure), the red vehicle gradually moves from Lane 1 to Lane 5 and eventually exits the highway via the ramp. The following sections discuss the virtual boundary line model, the effect of the boundary field on vehicles, and the construction of a two-dimensional intelligent driving model.

The parameters used in this study and their explanations are as follows: p_i and p_{i+1} represent the starting and ending points of the virtual boundary field line, and are defined as $p_i = [x_i, y_i]$, $p_{i+1} = [x_{i+1}, y_{i+1}]$; VB_j represents the jth virtual boundary field line, which is generated using Bezier curves; p_n represents any point on the virtual boundary field line, with $p_n = [x_n, y_n]$, $p_n \in VB_j$; VBF_j represents the jth virtual boundary field; V_w represents the width of the vehicle; V_L represents the length of the vehicle; $p_{v,t}$ represents the center coordinates of the vehicle at time t, with $p_n = [x_n, y_n]$, $p_n \in VB_j$; represents the boundary coordinates of the vehicle at time t, with $p_{v,t} = [x_{v,t}, y_{v,t}]$; $h_{v,t}$ represents the direction of travel of the vehicle at time t, with $h_{v,t} = \overrightarrow{p_{v,t-1}p_{v,t}}$; d^t_{margin} represents the distance between the boundary coordinates of the vehicle at time t and the virtual boundary field line; d^*_{margin} represents the ideal distance between the boundary coordinates $p'_{v,t}$ of the vehicle and the virtual boundary field line VB_j; v_t^{Long} and v_t^{Lat} represent the longitudinal and lateral velocities of the vehicle at time t, respectively; $a_{\text{Long}}^{\text{Eff}}$ and $a_{\text{Lat}}^{\text{Eff}}$ represent the effective accelerations of the vehicle in the longitudinal and lateral directions, respectively.

4.3 Derivation of Virtual Boundary Line

According to the previous discussion, the driving behavior of a vehicle is only affected when the boundary point $p'_{v,t}$ of the vehicle is within the boundary field VBF_j. Due to the advantages of easy optimization and continuous curvature variation, Bezier curves have been successfully applied in trajectory planning and road design by selecting appropriate parameters to maximize passenger comfort. Here, Bezier curves are used to characterize the virtual boundary line. As shown in Fig. 4.2, the boundary field refers to the blue shaded area in the figure, which consists of four parts: the starting control point (p_i) and ending control point (p_{i+1}) of the virtual boundary field line, the boundary field line (VB_j), and the ideal distance (d^*_{margin}). The boundary field line (VB_j) is generated using a third-order Bezier curve [4], determined by four points, with two points determining its position and the other two points determining the shape and direction of the curve. By selecting the appropriate positions, the generated curve can vividly depict the vehicle's motion trajectory. In this chapter, the driving trajectory of an autonomous vehicle from leaving the highway to the off-ramp process can be divided into two parts: (1) the lane-changing process on the highway, as shown in Fig. 4.1, where the vehicle moves from Lane 1 to Lane 2 and then to Lane 3, and the slope of the vehicle trajectory at the starting and ending points is "0"; (2) the transition from the highway to the off-ramp, as shown in Fig. 4.1, where the vehicle moves from Lane 4 to Lane 5, and the slope of the vehicle trajectory at the starting and ending points is not "0" and depends on the geometry of the off-ramp. Therefore, through the use of p_i and p_{i+1}, the coordinates of the two additional control points $p_{i,1}^{\text{control}}$ and $p_{i,2}^{\text{control}}$ related to the shape and direction can be calculated.

Scenario (1): the lane changing process at the highway main road:

$$\begin{cases} p_{i,1}^{\text{control}} = \left[0.5(x_i + x_{i+1}), y_i\right] \\ p_{i,2}^{\text{control}} = \left[0.5(x_i + x_{i+1}), y_{i+1}\right] \end{cases} \tag{4.1}$$

Scenario (2): driving from the highway main stream to the exit ramp:

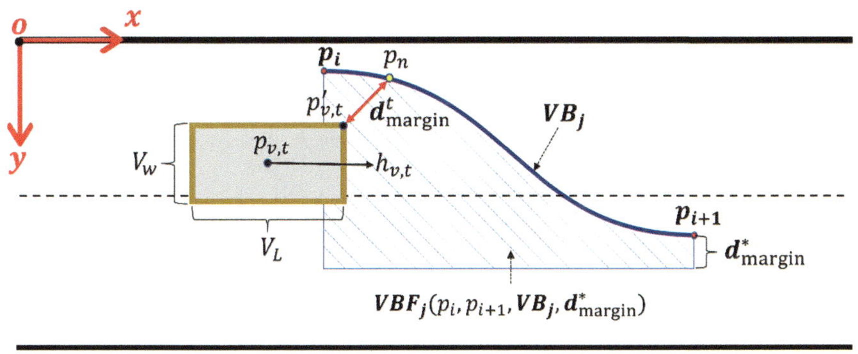

Fig. 4.2 Impact of potential on vehicle movement

$$\begin{cases} p_{i,1}^{\text{control}} = \left[0.5(x_i + x_{i+1}), y_i\right] \\ p_{i,2}^{\text{control}} = \left[\alpha(x_i + x_{i+1}), \beta(y_i + y_{i+1})\right] \end{cases}, \alpha, \beta \in (0, 1) \tag{4.2}$$

The virtual boundary line can be calculated based on the four points:

$$VB_j = (1-\lambda)^3 p_i + 3\lambda(1-\lambda)^2 p_{i,1}^{\text{control}} + 3\lambda^2(1-\lambda) p_{i,2}^{\text{control}} + \lambda^3 p_{i+1}, \quad \lambda \in [0,1] \tag{4.3}$$

At different time instances, the effect of the boundary field on the vehicle is realized through the actual shortest distance d_{margin}^t between the vehicle's boundary coordinate $p'_{v,t}$ and the boundary coordinate field line VB_j, as well as the vehicle's acceleration. The magnitude of d_{margin}^t is determined as follows:

$$d_{margin}^t = \left\|\overrightarrow{p'_{v,t} p_n}\right\| = \begin{bmatrix} \left(x_n - x_{v,t} + 0.5V_w\right)^2 + \\ \left(y_n - y_{v,t} - 0.5V_L\right)^2 \end{bmatrix}^{1/2}, (x_n, y_n) \in VB_j \tag{4.4}$$

The two terms $(x_{v,t}, y_{v,t})$ in (4.4) can be expressed as follows:

$$\begin{cases} x_{v,t} = x_{v,0} + \sum_t v_t^{\text{Long}} \cdot t + 0.5 \cdot a_{\text{Long}}^{\text{Eff}} \cdot t^2 \\ y_{v,t} = y_{v,0} + \sum_t v_t^{\text{Lat}} \cdot t + 0.5 \cdot a_{\text{Lat}}^{\text{Eff}} \cdot t^2 \end{cases} \tag{4.5}$$

where $\left[x_{v,0}, y_{v,0}\right]$ represents the initial velocities of the vehicle in the lateral and longitudinal directions respectively. Assuming that under the influence of the boundary field, the ideal distance between the vehicle and the boundary field line is d_{margin}^*. If d_{margin}^t is greater than d_{margin}^*, , the vehicle needs to accelerate in the lateral direction (ox direction in Fig. 4.2), but decelerate in the longitudinal direction (oy direction in longitudinal). Conversely, if d_{margin}^t is smaller than d_{margin}^*, the vehicle needs to decelerate in the lateral direction and accelerate in the longitudinal direction. If d_{margin}^t is equal to d_{margin}^*, the vehicle neither needs to accelerate nor decelerate. Therefore, the lateral and longitudinal accelerations can be calculated using the following formulas:

$$a_{\text{Long}}^{\text{Eff}} = a_{\text{Long}} \left[1 - \left(\frac{d_{margin}^*}{d_{margin}^t}\right)^2\right] \tag{4.6}$$

$$a_{\text{Lat}}^{\text{Eff}} = a_{\text{Lat}} \left[1 - \left(\frac{d_{margin}^t}{d_{margin}^*}\right)^2\right] \tag{4.7}$$

4.4 Two Dimensional IDM Model with Virtual Boundary Line

In practical driving environments, the driving behavior of a vehicle is also influenced by the surrounding vehicles. In Sharath and Velaga [3], the authors comprehensively consider the impact of surrounding vehicles on autonomous vehicles and propose an enhanced two-dimensional Intelligent Driver Model that effectively characterizes the vehicle's motion in both lateral and longitudinal directions. This study simultaneously considers the virtual boundary field and the surrounding vehicles, and proposes a two-dimensional IDM model based on the virtual boundary field. The lateral dimensional movement of the vehicle can be expressed as follows:

$$a_{\text{Long}}^{\text{Eff}} = \begin{cases} a_{\text{Long}} \left\{ 1 - \left(\frac{v_O^{\text{Long}}}{v_d^{\text{Long}}} \right)^4 - \left[\left(\frac{s_{\text{Long}}^*}{s_{\text{Front}}^{\text{Eff}}} \right)^2 - \left(\frac{s_{\text{Long}}^*}{s_{Rear}^{\text{Eff}}} \right)^2 \right] + \left[1 - \left(\frac{d_{margin}^*}{d_{margin}^t} \right)^2 \right] \right\}, p'_{v,t} \in VBF_j; \\ a_{\text{Long}} \left\{ 1 - \left(\frac{v_O^{\text{Long}}}{v_d^{\text{Long}}} \right)^4 - \left[\left(\frac{s_{\text{Long}}^*}{s_{\text{Front}}^{\text{Eff}}} \right)^2 - \left(\frac{s_{\text{Long}}^*}{s_{Rear}^{\text{Eff}}} \right)^2 \right] \right\}, p'_{v,t} \notin VBF_j. \end{cases} \tag{4.8}$$

where:

$$s_{\text{Long}}^* = s_O^{\text{Long}} + \max\left(0,\ v_O^{\text{Long}} T^{\text{Long}} + \frac{v_O^{\text{Long}} \Delta v^{\text{Long}}}{2\sqrt{a_{\text{Long}} b_{\text{Long}}}}\right) \tag{4.9}$$

In formulas 4.8 and 4.9, s_{Long}^* represents the ideal dynamic gap of the vehicle in the lateral direction; $s_{\text{Front}}^{\text{Eff}}$ represents the effective gap between the vehicle and other vehicles in the forward direction; s_{Rear}^{Eff} represents the effective gap between the vehicle and other vehicles in the rear direction; v_d^{Long} represents the ideal velocity of the vehicle in the lateral direction; v_O^{Long} represents the actual velocity of the vehicle in the lateral direction; Δv^{Long} represents the average velocity difference between the vehicle and other vehicles in the lateral direction; T^{Long} represents the time gap of the vehicle in the lateral direction; $a_{\text{Long}'}$ and b_{Long} represent the maximum acceleration and ideal acceleration of the vehicle in the lateral direction, respectively.

The longitudinal movement of the vehicle can be expressed as follows:

$$a_{\text{Lat}}^{\text{Eff}} = \begin{cases} a_{\text{Lat}} \left\{ \left(\frac{s_{\text{Lat}}^*}{s_{\text{Right}}^{\text{Eff}}} \right)^2 - \left(\frac{s_{\text{Lat}}^*}{s_{\text{Left}}^{\text{Eff}}} \right)^2 + \left[1 - \left(\frac{d_{margin}^t}{d_{margin}^*} \right)^2 \right] \right\}, p'_{v,t} \in VBF_j; \\ a_{\text{Lat}} \left[\left(\frac{s_{\text{Lat}}^*}{s_{\text{Right}}^{\text{Eff}}} \right)^2 - \left(\frac{s_{\text{Lat}}^*}{s_{\text{Left}}^{\text{Eff}}} \right)^2 \right], p'_{v,t} \notin VBF_j. \end{cases} \tag{4.10}$$

where:

$$s_{\text{Lat}}^* = s_O^{\text{Lat}} + \max\left(0,\ v_O^{\text{Lat}} T^{\text{Lat}} + \frac{v_O^{\text{Lat}} \Delta v^{\text{Lat}}}{2\sqrt{a_{\text{Lat}} b_{\text{Lat}}}}\right) \tag{4.11}$$

In formulas 4.10 and 4.11, s_{Lat}^{*} represents the ideal dynamic gap of the vehicle in the longitudinal direction; $s_{\text{Right}}^{\text{Eff}}$ represents the effective gap between the vehicle and other vehicles in the right direction; $s_{\text{Left}}^{\text{Eff}}$ represents the effective gap between the vehicle and other vehicles in the left direction; v_{d}^{Lat} represents the ideal velocity of the vehicle in the longitudinal direction; v_{O}^{Lat} represents the actual velocity of the vehicle in the longitudinal direction; Δv^{Lat} represents the average velocity difference between the vehicle and other vehicles in the longitudinal direction; T^{Lat} represents the time gap of the vehicle in the longitudinal direction; a_{Lat} and b_{Lat} represent the maximum acceleration and ideal acceleration of the vehicle in the longitudinal direction, respectively.

The values of s_{Rear}^{Eff}, $s_{\text{Right}}^{\text{Eff}}$, $s_{\text{Left}}^{\text{Eff}}$, Δv^{Lat}, Δv^{Long}, $s_{\text{Front}}^{\text{Eff}}$ in formulas 4.8–4.11 are jointly determined by the current states of the autonomous vehicle and the surrounding vehicles. The specific values can be calculated through the pseudo code algorithm flowchart shown in Fig. 4.3.

Fig. 4.3 Pseudo code

```
Algorithm GetPara(vehicle, Vobstacles)
INPUT: Status of the target vehicle, and obstacle vehicles
OUTPUT: Δv^Lat, Δv^Long, s^Eff_Front, s^Eff_Rear, s^Eff_Right, s^Eff_Left
If len(Vobstacles) is "0" do
    Δv^Lat, Δv^Long, s^Eff_Front, s^Eff_Rear, s^Eff_Right, s^Eff_Left all equal to Infinity
Else:
    Initialize six sets: dV_Lat, dV_Long, S_Front, S_Rear, S_Right, S_Left ← 0
    Initialize two unit vectors: e_x, e_y
    for V ∈ Vobstacles do
        Vector ← <vehicle.direction, V.direction>
        Distance ← || Vector ||
        dV ← V.speed - vehicle.speed
        dV_Lat.Insert(dV *Vector *e_y)
        dV_Long.Insert(dV *Vector *e_x)
        if Vector same as vehicle.direction do
            S_Front.Insert(Vector* vehicle.direction)
            S_Rear.Insert(Infinity)
        else Reverse.
        if Vector same as -vehicle.direction do
            S_Left.Insert(Vector* -vehicle.direction)
            S_Right.Insert(Infinity)
        else Reverse.
    Δv^Lat ← mean(dV_Lat), Δv^Long ← mean(dV_Long), s^Eff_Left ← min(S_Left),
    s^Eff_Front ← min(S_Front), s^Eff_Rear ← min(S_Rear), s^Eff_Right ← min(S_Right)
```

4.5 The Optimization of the Virtual Boundary Based on DTW

In this section, the DTW algorithm is used to compare the differences between the theoretical trajectory and the actual trajectory. The key parameters of the model are calibrated by selecting the coordinates of the starting point (0, −0.1) closest to the inside of the straight lane on the road, and the coordinates of the outermost point (7.6, 145) at the intersection of the straight lane and the ramp in the direction of vehicle travel, as the starting and ending control points of the virtual boundary field line. The coordinates of the end control point can also be easily determined at other ramps. The initial moment coordinates (1.05, 6.05) and the speed of 20.1 m/s of a randomly selected vehicle trajectory from the dataset are taken as the initial coordinates and initial velocity of the vehicle. The vehicle dimensions, referring to the dimensions of a private car, are 4.0 m for the length and 1.8 m for the width. The other key parameters, as referenced in Delpiano et al. [2], are as follows: $\mathrm{a}_{\mathrm{Long}} = 2.95\frac{\mathrm{m}}{\mathrm{s}^2}$, $a_{\mathrm{Lat}} = 0.63\frac{\mathrm{m}}{\mathrm{s}^2}$, $b_{\mathrm{Long}} = 2.21\frac{\mathrm{m}}{\mathrm{s}^2}$, $v_d^{\mathrm{Long}} = 12.1\frac{\mathrm{m}}{\mathrm{s}}$ $b_{\mathrm{Lat}} = 2.66\frac{\mathrm{m}}{\mathrm{s}^2}$, $T^{\mathrm{Long}} = 1.3\,\mathrm{s}$, $T^{\mathrm{Lat}} = 1.18\,\mathrm{s}$. In the study of Delpiano et al. [2], the above parameters were calibrated using 1050 different NGSIM data, and their means and standard deviations were provided. Therefore, in this study, it is only necessary to calibrate the boundary coordinates of the vehicle and the ideal distance d^*_{margin} from the virtual boundary field line.

Considering the physical significance of d^*_{margin}, its value should be determined based on the specific driving scenario of the vehicle. For example, in the driving scenario from Lane 1 to Lane 3, the range of values for d^*_{margin} is [0.01 m, 1.00 m]. In this study, a simulation step size of 0.01 m is set, while keeping other parameters unchanged. This value is then substituted into the model for simulation and solution, resulting in the theoretical trajectory. Subsequently, the DTW algorithm is applied to compute the differences between the theoretical and actual trajectories. Through calculations, the optimal value for d^*_{margin} is determined to be 0.31 m.

To determine the coordinates of the optimal virtual boundary lane, we employs the Particle Swarm Optimization (PSO) algorithm [1] to determine the optimal coordinate points. PSO is a stochastic search algorithm based on collective collaboration, developed by simulating the foraging behavior of bird flocks. The objective function utilizes the proposed model in this section to predict the off-ramp trajectories in 75% of the Mirror-Traffic database. Subsequently, the DTW algorithm is employed to calculate the differences between the predicted and actual trajectories and obtain their mean. The particle swarm size is set to 50, with a total of 500 iterations. The learning factors and inertia weights are set to 0.5, 0.3, and 1.5, respectively.

4.6 Model Analysis

In order to validate the effectiveness of the model, this section utilizes the off-ramp data from the Mirror-Traffic open dataset as presented. The traffic scenario of the highway off-ramp is shown in Fig. 4.4. The dataset consists of a total of 556 vehicle trajectories, with 246 straight-through vehicle trajectories and 290 off-ramp vehicle trajectories.

4.6.1 The Location of the Virtual Boundary Line and the Convergence of the PSO

In the driving environment depicted in Fig. 4.4, for the driving scenario of "from Lane 1 to Lane 3," it is necessary to determine the optimal position of the virtual boundary lane. According to the theoretical analysis mentioned earlier, the position of the virtual boundary lane is uniquely determined by three parameters: p_i, p_{i+1}, and d^*_{margin}. In this particular scenario, d^*_{margin} is calibrated to be 0.31 m, and the coordinates of p_{i+1} are the intersection point of Lane 2 and Lane 3, (145, 7.6). The only uncertain value is the starting control point p_0 of the virtual boundary lane. By determining its coordinates using the PSO algorithm mentioned earlier, the position of the boundary lane is uniquely determined. The optimization iteration curve is shown in Fig. 4.5, and the optimal coordinates of the starting control point p_0 are found to be (21.58, 0). Subsequently, the virtual boundary lane is visualized in Fig. 4.6.

4.6.2 The Prediction of Trajectory of Maintaining at Main Stream and Exit the Highway

According to the previous section, the coordinates of the optimal boundary lane's start and end control points, as well as the boundary coordinates of the vehicle $p_{v,t'}$ and the desired distance d^*_{margin} to the virtual boundary lane VB_j, have been obtained. Next, the trajectory of the off-ramp vehicles is predicted. The results are shown in

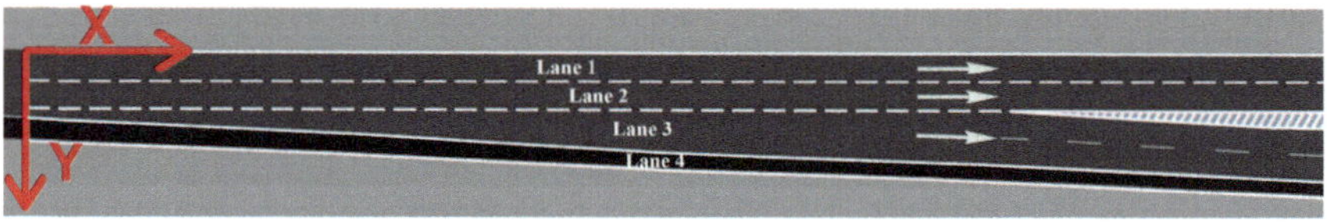

Fig. 4.4 Data source: mirror dataset

Fig. 4.5 PSO convergence

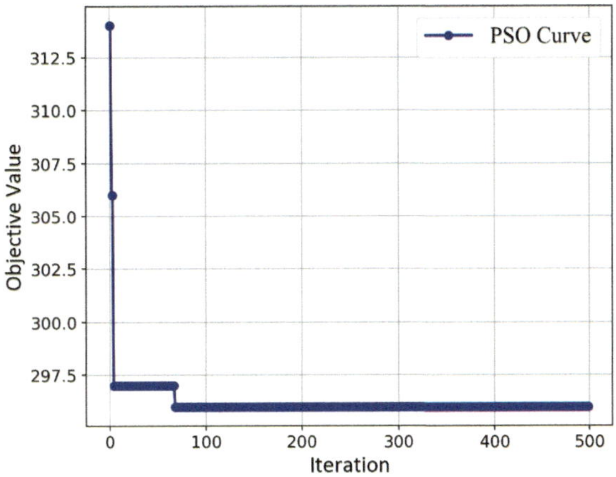

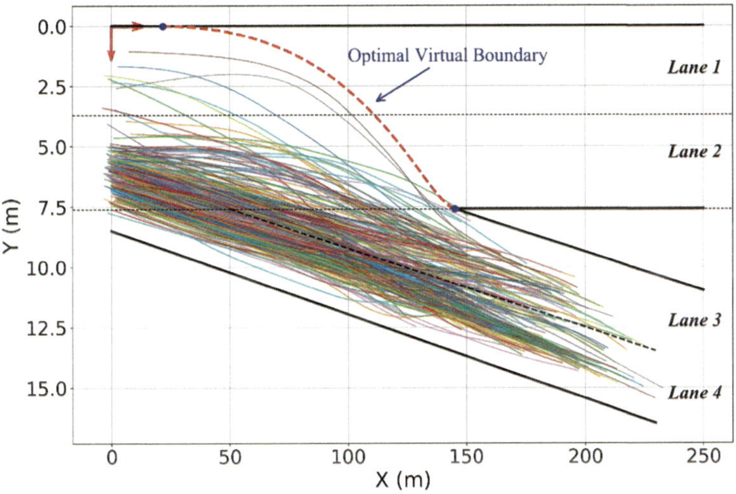

Fig. 4.6 Optima boundary and trajectory of 75% vehicles

Fig. 4.7. Additionally, a comparison is made between the predicted trajectories and the remaining 25% of the off-ramp vehicle trajectories.

In order to further validate the accuracy of the model, the present study utilizes the proposed model to predict the trajectories of four different types of off-ramp vehicles in the Mirror dataset. Additionally, surrounding vehicles of the target vehicle are selected as obstacle vehicles. All parameters are inputted into the model for simulation calculations, and the vehicle trajectories are visualized as shown in Fig. 4.8. Subsequently, two additional sets of experiments are conducted, and the results are illustrated in Figs. 4.9 and 4.10 respectively.

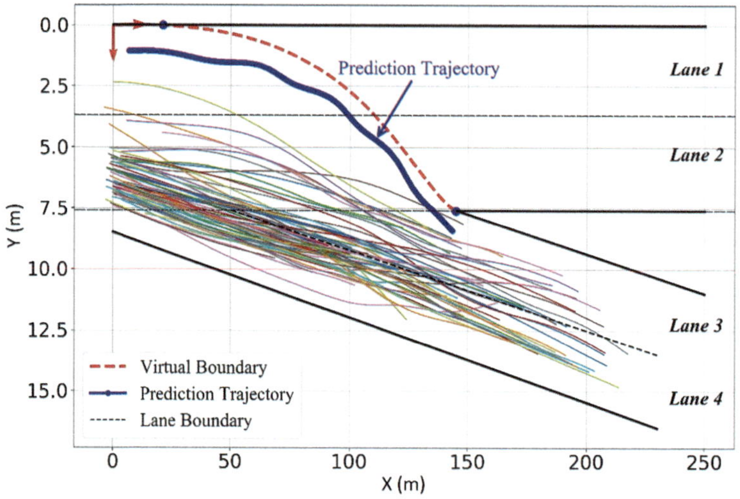

Fig. 4.7 Trajectory prediction under the optimal boundary

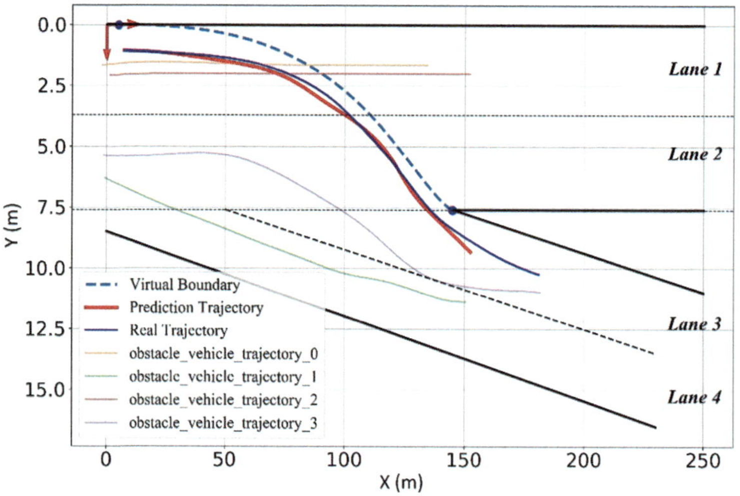

Fig. 4.8 Case study 1

The experimental results in Figs. 4.8, 4.9 and 4.10 respectively validate three different scenarios wherein the vehicles transition from Lane 1 to Lane 2, from Lane 2 to Lane 3, and from Lane 2 to Lane 4 when exiting the highway. In the aforementioned figures, the dashed lines in dark green represent the virtual boundary lanes, the blue curves depict the actual vehicle trajectories, and the red curves represent the vehicle trajectories predicted by the model mentioned earlier. By comparing the blue and red curves, it is evident that they exhibit a high degree of similarity and trend.

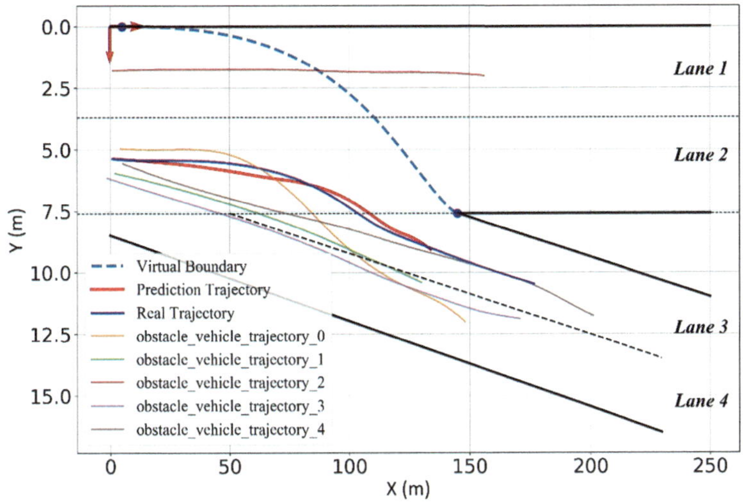

Fig. 4.9 Case study 2

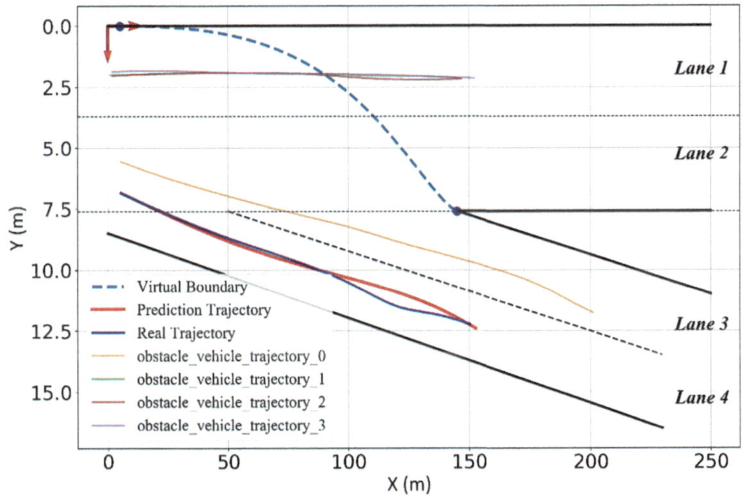

Fig. 4.10 Case study 3

In addition to its applicability to vehicles exiting off-ramps, the model proposed in this study is also suitable for straight-going vehicles. To verify this, simulations were carried out, and the results are depicted in Fig. 4.11. In Fig. 4.11, the yellow curve represents an off-ramp trajectory predicted by the model, the red dashed line represents a straight trajectory predicted based on the yellow curve and other surrounding obstacle vehicles, and the blue curve represents the actual vehicle trajectory. Comparing the trajectories, it can be observed that the red dashed line aligns

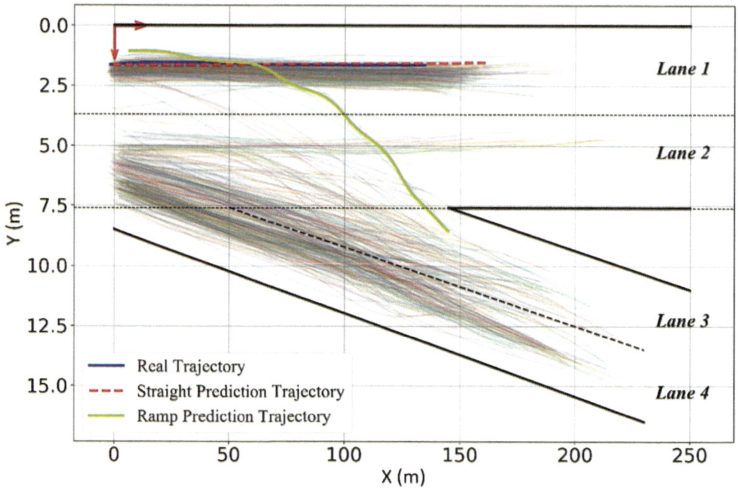

Fig. 4.11 Case study 4

closely with the blue curve, indicating that the proposed model is also applicable to straight-going vehicles.

The Mirror-Traffic dataset is specific to road environments in China, taking into account factors such as traffic conditions and driver characteristics that may differ from those in other countries. Therefore, in order to validate the model proposed in this study, the NGSIM dataset of US highways was also selected. After preprocessing the data, the model was used to predict the trajectory of one of the trajectories, as shown in Fig. 4.12, where the red curve aligns closely with the blue curve. Additionally, several other datasets were selected for validation, and the ultimate prediction results demonstrate.

4.6.3 Error Analysis

In the previous sections, the proposed two-dimensional IDM model based on the virtual boundary field was validated through five simulation experiments, which demonstrated its effectiveness and accuracy. However, there still exists a certain level of error between the predicted trajectory and the actual trajectory, necessitating an error analysis. The errors in this study can be classified into two categories: (1) Y_{Error}, which refers to the difference in the Y direction between the actual trajectory and the predicted trajectory as the vehicle trajectory changes along the X axis, i.e., $Y_{\text{Error}} = Y_{\text{real}} - Y_{\text{predict}}$; (2) $Y_{\text{Absolute Error}}$, which refers to the absolute difference in the Y direction between the actual trajectory and the predicted trajectory as the vehicle trajectory changes along the X axis, i.e., $Y_{\text{Absolute Error}} = abs(Y_{\text{Error}})$. Next, the errors

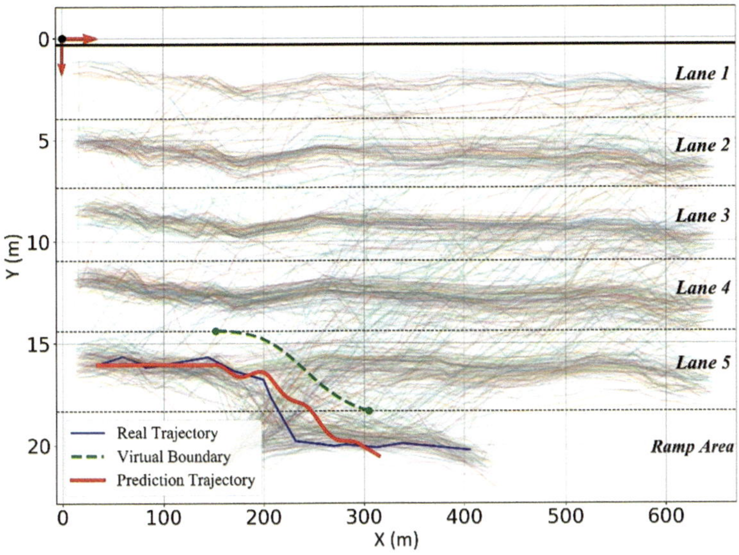

Fig. 4.12 Case study 5

for the four simulation experiments of the Mirror dataset are calculated, as shown in Fig. 4.13.

In Fig. 4.13, for the four experiments, as X varies, the values of Y_{Error} and $Y_{\text{Absolute Error}}$ remain within the range of [0, 0.8 m], with the smallest errors observed

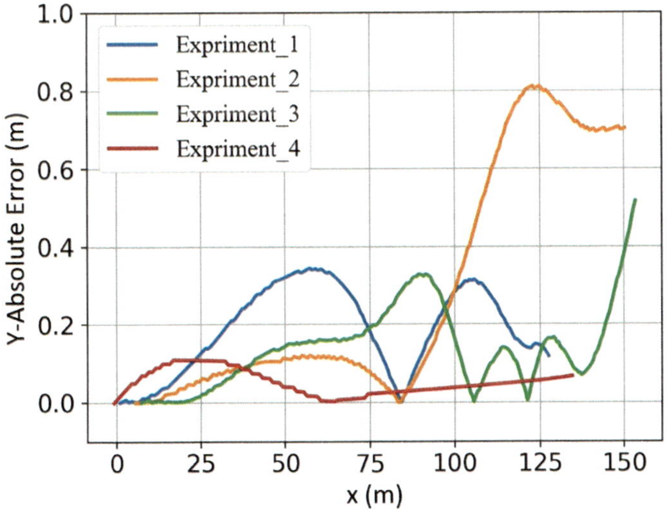

Fig. 4.13 Longitudinal (Y) prediction error with X in Four cases

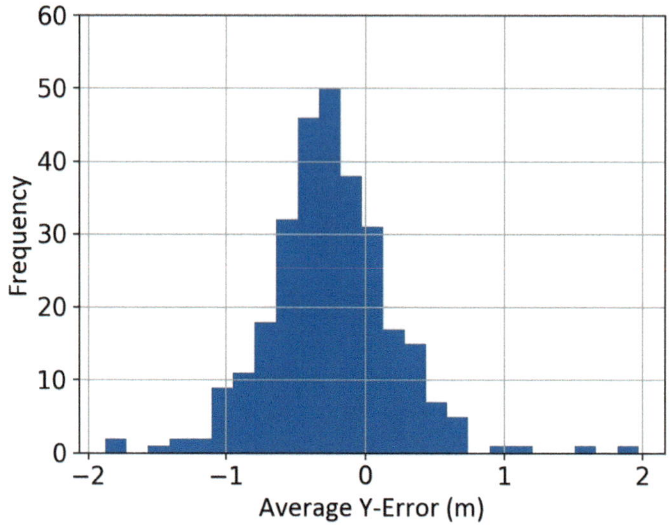

Fig. 4.14 Histogram of Y_{Erro}

in the trajectory predictions of straight scenarios. Subsequently, the prediction of all exit-ramp trajectories in the Mirror-Traffic dataset (a total of 290 trajectories) was conducted, and the average Y_{Error} and $Y_{\text{Absolute Error}}$ for each trajectory were calculated. The frequency histograms of these average errors are depicted in Figs. 4.14 and 4.15, respectively. In Fig. 4.14, approximately 95.5% of the data fall within the range of [−1.0 m, 1.0 m], while in Fig. 4.15, around 94.8% of the data exhibit errors within the range of [0, 1.0 m]. These results further demonstrate the high accuracy of the proposed two-dimensional IDM model based on the virtual boundary field in predicting the trajectories of vehicles exiting the ramp.

The following accuracy evaluation metrics are employed: Mean Absolute Error (MAE), Mean Squared Error (MSE), Root Mean Squared Error (RMSE), and Mean Absolute Percentage Error (MAPE), in order to further validate the accuracy of the model predictions. The model is used to predict 30 randomly selected Mirror-Traffic data trajectories and 45 NGSIM data trajectories for different time scales (short-term, medium-term, and long-term). The evaluation metrics are presented in Table 4.1. In Table 4.1, the MAE, MSE, and RMSE values for the 30 Mirror data trajectories are all less than 1 m, which aligns with the results depicted in the histogram distribution in Fig. 4.14. This strongly demonstrates the high accuracy of the proposed model. Additionally, the computation results for the 45 NGSIM trajectories are within an acceptable range; however, their values are significantly larger than those of the Mirror data. This discrepancy is primarily due to the fact that the NGSIM data was collected 15 years ago and was limited by the technological advancements at that time, resulting in higher levels of noise in the data and larger errors in the predicted trajectories.

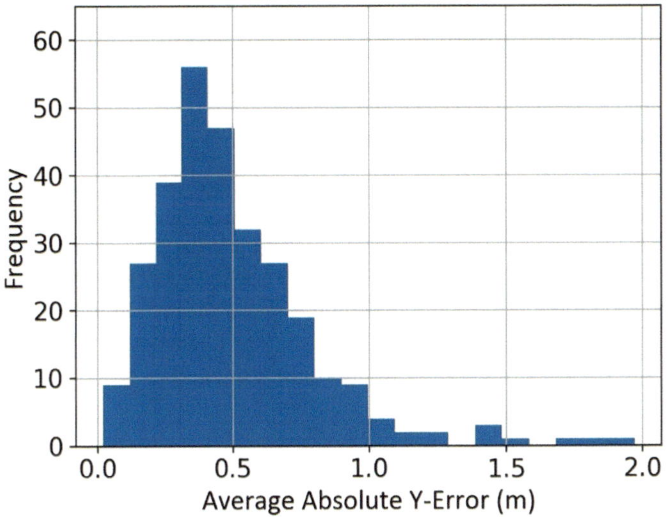

Fig. 4.15 Histogram of $Y_{\text{Absolute Error}}$

Table 4.1 Comparison of the prediction results between Mirror-Traffic and NGSIM

Datasets	Number of trajectories	Time scale (s)	MAE	MSE	RMSE	MAPE (%)
Mirror-Traffic	30	3	0.17	0.068	0.25	3.1
		6	0.42	0.34	0.58	6.4
		9	0.62	0.81	0.90	8.2
NGSIM	45	20	0.60	3.51	1.87	3.8
		45	1.13	17.34	4.16	9.13
		70	1.84	52.89	7.27	28.24

Considering the temporal characteristics of the two datasets, the longest vehicle trajectory timescale in both Mirror-Traffic data and NGSIM data is selected as a reference (Mirror-Traffic: 13.16 s; NGSIM: 99.2 s). Their respective 20%, 50%, and 70% durations are chosen as the short-term, medium-term, and long-term timescales. Consequently, the three determined timescales are 3, 6, and 9 s for Mirror-Traffic data, and 20, 45, and 70 s for NGSIM data. From the table, it can be observed that for both datasets, as the timescale increases, the errors in the predicted trajectories also increase, but they all remain within a certain range.

References

1. Bratton D, Kennedy J (2007) Defining a standard for particle swarm optimization. In: 2007 IEEE swarm intelligence symposium. IEEE, pp 120–127
2. Delpiano R, Herrera JC, Laval J, Coeymans JE (2020) A two-dimensional car-following model for two-dimensional traffic flow problems. Transp Res Part C: Emerg Technol 114:504–516
3. Sharath MN, Velaga NR (2020) Enhanced intelligent driver model for two-dimensional motion planning in mixed traffic. Transp Res Part C: Emerg Technol 120:102780
4. Sun L, Li K, Wang H, et al (2016) Phonetic posteriorgrams for many-to-one voice conversion without parallel data training. In: 2016 IEEE international conference on multimedia and expo (ICME). IEEE, pp 1–6

Chapter 5
A Stochastic Two-Dimensional IDM with Vehicular Dynamics

5.1 Incapability of Current Microscopic Traffic Flow Model

The research on microscopic traffic flow models, specifically car-following (CF) models, has been an enduring area of study. These models are crucial for simulating traffic flow and driver behavior, as well as autonomous vehicle simulations. Vehicular dynamics, which are inherently stochastic, arise from the interaction between driver behavior, vehicle characteristics, road pavements, etc. However, current microscopic models lack consideration for two-dimensional noise and vehicle kinetic attributes. More specifically, current models suffer from the following drawbacks:

- Neglecting driver-vehicle unit dynamics, which overlooks the impact of factors like steering and acceleration on vehicle behavior.
- Absence of stochastic factors such as driver characteristics, road conditions, surrounding traffic flow, and environmental factors, which significantly influence real-world traffic dynamics.

As a result, the current models fail to capture realistic traffic flow dynamics and are inadequate for analyzing and reproducing complex real-world traffic scenarios. This study builds upon the intelligent driver model by integrating vehicle dynamics and two-dimensional noise, resulting in a stochastic two-dimensional IDM model based on stochastic differential equations (SDEs). The model incorporates two control inputs derived from the traditional IDM: steer rate and longitudinal acceleration. The model's stochastic stability condition is evaluated using Lyapunov theory. Numerical analysis is conducted to assess the model's performance in two scenarios: accelerating a stationary vehicle and a platoon of vehicles following a leader with a stop-and-go speed pattern, simulating congestion formation and dispersal. The findings demonstrate the model's capability to replicate stochastic two-dimensional vehicle movement. This proposed model shows potential for application in simulation platforms and behavior analyses of human drivers.

© The Author(s), under exclusive license to Springer Nature Singapore Pte Ltd. 2024
H. Qi, *Stochastic Two-Dimensional Microscopic Traffic Model*, Lecture Notes in Intelligent Transportation and Infrastructure,
https://doi.org/10.1007/978-981-97-3597-6_5

5.2 Objectives and Contributions

Microscopic vehicular modeling is a topic of significant interest within traffic flow theory, and it attracts attention from various disciplines such as physics [9], control theory [12], and traffic engineering [4], etc. This modeling approach establishes a connection between the microscopic behavior of individual vehicles and the macroscopic traffic flow phenomenon, including stop-and-go traffic [11]. In the era of connected and autonomous vehicles, microscopic modeling assumes an even greater importance [6]. In mixed traffic flow scenarios, CAVs have the capability to acquire precise information regarding the surrounding vehicles through high-precision detectors. Therefore, having an accurate model for each driver-vehicle unit would greatly enhance the analysis, inference, and interactive predictions related to human-driven vehicles, ultimately ensuring safer and more robust decision-making and control for CAVs. Consequently, the analysis of human-driven vehicles has become a highly relevant subject in the context of CAV technology.

Car following (CF) is a prominent microscopic behavior that examines how drivers follow other vehicles in the same lane. Theoretical frameworks employed in studying CF encompass differential equations, cellular models or CA [2], and lattice-gas models, among others. Within differential type CF models, a stimulus-reaction framework is established to map the system's state (including the ego vehicle state and the state of surrounding vehicles) to the driver's reaction, with the stimulus consisting of the states of the surrounding vehicles (position, speed, acceleration, etc.). The ego vehicle adjusts its acceleration based on the status of the surrounding vehicles. Noteworthy CF models include the Gipps model, GM model, and intelligent driver model [10], with IDM being particularly well-known and featured in major traffic flow simulation software. Although these achievements provide valuable insights into the evolution mechanism of traffic flow, they overlook certain dynamics within the driver-vehicle unit. These dynamics are encapsulated in the generative process of vehicle dynamics as depicted in Fig. 5.1, where a driver perceives the traffic flow environment (e.g., distance to leading vehicle), makes decisions regarding optimal acceleration/steering, and then executes those decisions to control the vehicle. Typically, steering and acceleration are employed by the driver to exert control, and if the vehicle deviates from the middle lane to the left, the driver tends to steer right, and vice versa. With control inputs, the vehicle moves and generates dynamic behavior, culminating in real-time trajectory output. In microscopic traffic flow analysis, the trajectory is conventionally defined as the midpoint of the vehicle's front bumper. Numerous stochastic factors influence this process, including: (1) driver characteristics such as age, driving proficiency, emotional state, etc.; (2) road randomness and the ensuing road-vehicle interaction, where the flatness of the road impacts vehicle dynamics; (3) the inherent stochastic nature of the surrounding traffic flow dynamics; (4) environmental factors like rain, snow, etc.; and (5) additional stochastic factors such as vehicle body aging.

To comprehensively analyze the ultimate movement of a vehicle, apart from considering driver behavior, it is essential to incorporate vehicle dynamics and

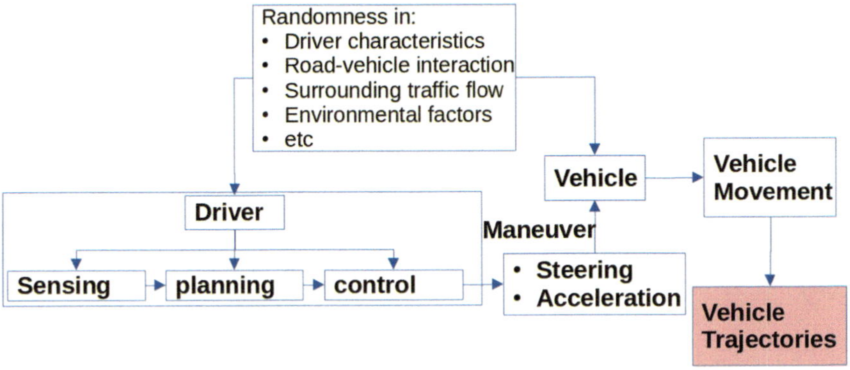

Fig. 5.1 Generation process of vehicular trajectories

stochastic factors. Neglecting the vehicle dynamics model may result in generated trajectories that are impracticable, while disregarding stochastic considerations may lead to unrealistic traffic flow dynamics. Figure 5.2 showcases the traffic flow simulation of vehicle dynamics and real-world data in the Carla framework. In Fig. 5.2b, the vehicle trajectory exhibits minimal fluctuations within the lane, which has a width of 3.7 m. The lane's middle line is defined at $y = 1.85$ m. Conversely, in Fig. 5.2d, the real-world highD dataset reveals that vehicles manifest stochastic movement even when maintaining their lanes. The deviation can reach approximately 3 m, and its distribution is illustrated in Fig. 5.2a. For an individual vehicle, the trajectory deviation is continuous. The driver continuously applies control inputs to the vehicle to maintain lateral and longitudinal equilibrium. Notably, present car-following models primarily describe unidirectional movement (i.e., longitudinal direction) by employing acceleration as the input. However, extensions towards the two-dimensional scenario often lack considerations for vehicle dynamics and stochasticity. Consequently, existing models cannot fully analyze or replicate real-world phenomena.

In this chapter, we propose a novel approach to address this issue by integrating the intelligent driver model framework with vehicle dynamics and stochastic factors to construct a two-dimensional stochastic intelligent driver car-following model with vehicle dynamics. This model is developed based on the principles of stochastic differential equation theory. The control inputs employed by the driver comprise steering and acceleration, which are derived from the concepts of an intelligent driver model and an optimal velocity model. Furthermore, we incorporate randomness in both lateral and longitudinal aspects to enhance the realism of the model.

This study offers the following contributions:

- Development of a two-dimensional driver-vehicle unit model that explicitly incorporates the noise factor.
- Incorporation of steering and acceleration as control inputs, derived from conventional IDM and optimal velocity models.

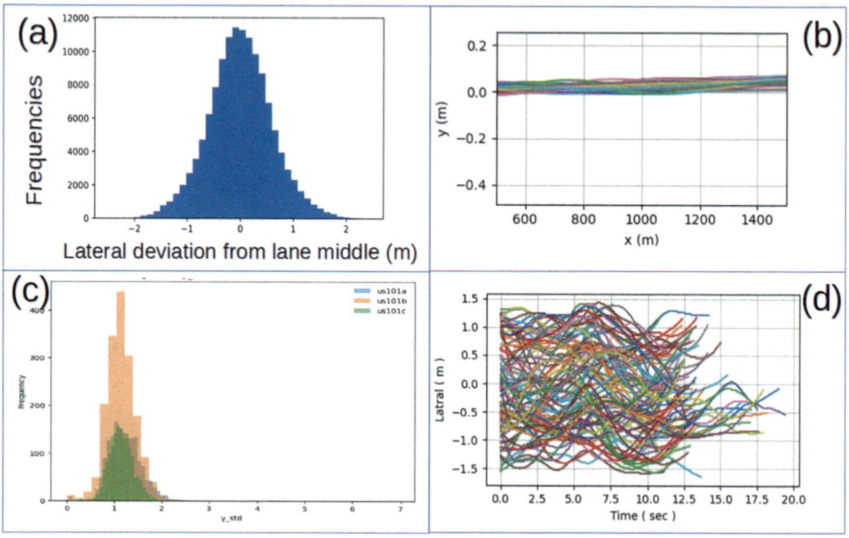

Fig. 5.2 Deviations of all lane keeping trajectories in the highD dataset. **a** lateral deviation distribution; **b** lateral evolution of the simulation in SUMO; **c** Lateral evolution in highD dataset

– Analysis of stochastic stability using the Lyapunov method.

5.3 Model Development

5.3.1 Notations

Symbols	
$l_f(l_r)$, $l_F(l_R)$	The distance of front (rear) wheel and front (rear) bump to the CG
$[x_i, y_i, \psi_i]$	Vehicle state vector for vehicle i. x and y are the coordinate of the CG, and ψ is the heading angle;
ψ_i^E, δ_i^*	The desired (or expected) heading angle of vehicle i; δ_i^* is the desired steering rate of vehicle i;
$[a_i, \gamma_i]^T$	The control input of vehicle i, speed and the steering rate of front wheel;
$Q = [q_1, q_2, q_3, q_4, q_5, q_6, q_7]$ $= [x, y, \psi, v, \delta, z_{lat}, z_{long}]$	Q is the extended state vector of one vehicle. z_{lat}, z_{long} is the two-dimensional noise
$dQ = F(Q)dt + L(Q)dW$	F and L are drift and diffusion terms. $F = [f_1, f_2, \ldots f_7]^T$ and $L = [l_1, l_2 \ldots l_7]^T$

(continued)

(continued)

Symbols	
θ, μ and σ	Parameters of the Ornstein–Uhlenbeck process
Γ_{acc} and Γ_{steer}	The function that maps the control input to a reasonable interval
K and Z_{bnd}	The meta noise and the bounded noise
σ_{lon} and σ_{lat}	The noise level of the two dimensions;
Z_{lat} and Z_{long}	The two-dimensional tire noise
η_{long} and η_{lat}	The sensitivity of the two-dimensional noise of the system state
s_i and s_i^*, s_0	s_i is the gap between vehicles i and its leading vehicle i–1. s_i^* is the desired gap. s_0 is the minimum gap
a_{MAX}, b, T, v_d	a_{MAX} is maximal acceleration, b is comfortable deceleration of the follower; T is the safe time headway; v_d is the desired speed
τ	The parameter in $IDM_{steering}$
IDM_{acc} and IDM_{steer}	The shortcut of the functions that calculate the acceleration and steering rate using an idea borrowed from the IDM model
$\mathcal{O}(\delta_i^*)$	The optimal steering angle. δ_i^* is the desired steering angle
$\Delta\xi_1$ and $\Delta\xi_2$	The angle difference between current position and desired position
V(Q) and LV(Q)	The Lyapunov function and the drift term of the V
Δt and ΔW	Δt is the time step for the simulation and ΔW is a bivariate Gaussian variable

5.3.2 Framework

Figure 5.3 presents the foundational setup of the model. Figure 5.3a portrays a single track model, which serves as a visual representation of the vehicle kinetics. Figure 5.3b depicts a scenario with a leading vehicle and a following vehicle within the same lane. Traditional models assume that these vehicles are positioned precisely on the middle line, aligned with a heading angle of zero, indicating the longitudinal direction. In contrast, our model framework permits unrestricted movement of vehicles on the x–y plane while adhering to physical constraints, such as the maximum steering angle. The control inputs for the single track model consist of the steering rate and acceleration, both of which are derived from the concepts integrated from the Intelligent Driver Model (IDM) and the optimal velocity model. Furthermore, we address the two-dimensional randomness by incorporating Brownian motion. These considerations culminate in the formulation of a stochastic two-dimensional IDM model with vehicular dynamics.

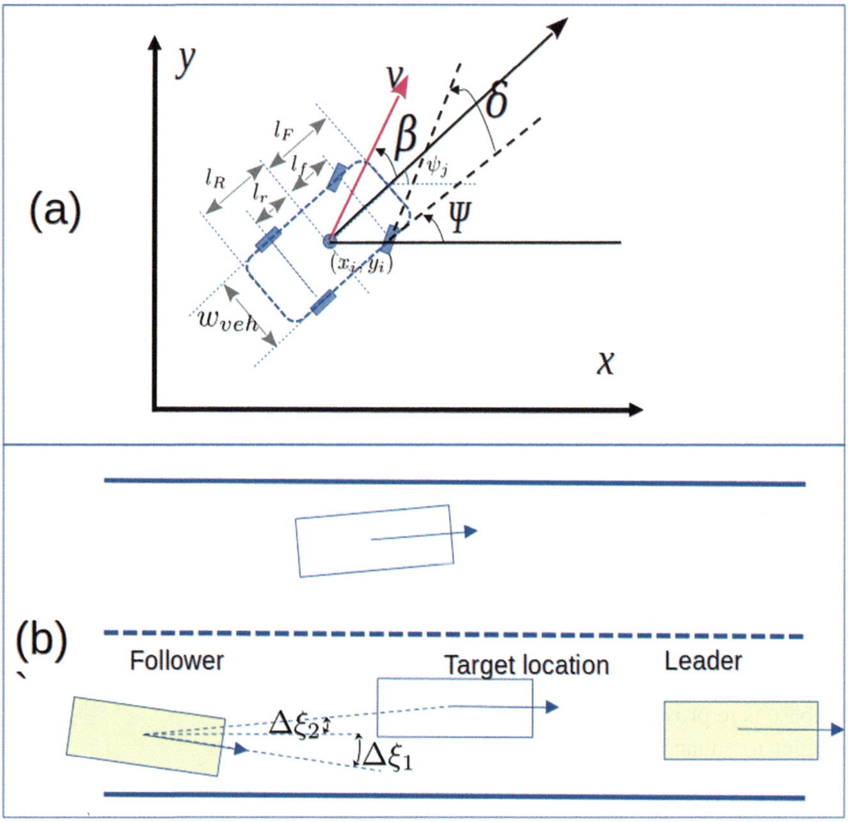

Fig. 5.3 Basic setting. **a** the definition of the state of a typical vehicle; **b** modeling scenario

5.3.3 Vehicle Model

5.3.3.1 Vehicle Dynamics Model

The lateral slip angles at front (α_f) and rear tire (α_r) are calculated as follows based on the illustration in Fig. 5.3:

$$\alpha_f = \arctan\left(\frac{v\sin(\beta) + \dot{\Psi} l_f}{v\cos(\beta)}\right) - \delta;\ \alpha_r = \arctan\left(\frac{v\sin(\beta) - \dot{\Psi} l_r}{v\cos(\beta)}\right) \tag{5.1}$$

The longitudinal slip for both wheels (front and rear) is given by:

$$s_f = 1 - \frac{R_w \omega_f}{u_{w,f}};\ s_r = 1 - \frac{R_w \omega_r}{u_{w,r}} \tag{5.2}$$

R_w is denotes the effective tire radius; $u_{w,f}$ and $u_{w,r}$ are the individual tire velocities. They are be calculated as following by assuming the rear axis is not steerable:

$$\begin{cases} u_{w,f} = vcos(\beta)cos(\delta) + \left(v\sin(\beta) + \dot{\Psi}l_f\right)sin(\delta) & (a) \\ u_{w,r} = vcos(\beta) & (b) \end{cases} \tag{5.3}$$

The longitudinal tire forces $F_{l,f}$, $F_{l,r}$ and lateral tire forces $F_{s,f}$ and $F_{s,r}$ are computed using the Pacejka magic formula for combined slip with the corresponding slip variable α as input:

$$Force(\alpha) = D\cdot\sin(C\cdot\arctan(B(\alpha+S_H) - E(B\cdot(\alpha+S_H)\cdot arctan(B(\alpha+S_H))))) + S_V \tag{5.4}$$

S_v and S_H are horizontal shift and vertical shift. Using the variables and parameters introduced above, the differential equations for the vehicles dynamics of the single-track drift model are:

$$\begin{cases} \dot{\delta} = v_\delta & (a) \\ \dot{\beta} = -\dot{\Psi} + \frac{1}{mv}\begin{pmatrix} F_{s,f}cos(\delta-\beta) + F_{s,r}cos(\beta) \\ -F_{l,r}sin(\beta) + F_{l,f}sin(\delta-\beta) \end{pmatrix} & (b) \\ \ddot{\Psi} = \frac{1}{I_z}\left(F_{s,f}\cos(\delta)l_f - F_{s,r}l_r + F_{l,f}\sin(\delta)l_f\right) & (c) \\ \dot{v} = \frac{1}{m}\left(-F_{s,f}\sin(\delta-\beta) + F_{s,r}\sin(\beta) + F_{l,r}\cos(\beta) + F_{l,f}\cos(\delta-\beta)\right) & (d) \\ \dot{s_x} = v\cdot\cos(\beta+\Psi) & (e) \\ \dot{s_y} = v\cdot\sin(\beta+\Psi) & (f) \end{cases} \tag{5.5}$$

Additionally, the wheel dynamics are modeled as

$$\begin{cases} \dot{\omega_f} = \frac{1}{I_{y,w}}\left(-R_wF_{l,f} + T_{s,b}T_B + T_{s,e}T_E\right) & (a) \\ \dot{\omega_r} = \frac{1}{I_{y,w}}\left(-R_wF_{l,r} + \left(1-T_{s,b}\right)T_B + \left(1-T_{s,e}\right)T_E\right) & (b) \end{cases} \tag{5.6}$$

$I_{y,w}$ is the wheel inertia and $T_{s,b}$, $T_{s,e}$ are the split parameters between front and rear axle for the brake and engine torque, respectively. As in the multi-body model, the brake torque T_B and engine torque T_E are determined from the acceleration input a_{long}:

$$T_B = \begin{cases} 0, & for\ \ a_{long} > 0 \\ mR_wa_{long}, & otherwise \end{cases};\ T_E = \begin{cases} mR_wa_{long}, & for\ \ a_{long} > 0 \\ 0, & otherwise \end{cases} \tag{5.7}$$

The system state vector is defined as:

$$[x_1, x_2, x_3, x_4, x_5, x_6, x_7, x_8, x_9] = \left[s_x, s_y, \delta, v, \Psi, \dot{\Psi}, \beta, \omega_f, \omega_r\right] \tag{5.8}$$

s_x and s_y are the position of the vehicle's CG; ψ is the heading angle; β is the slip angle at the CG and v is the speed at the CG; δ is the steer angle at the front tire; ω_f and ω_r are the angular speed of the front and rear tire. Using the re-defined state variables, the model then can be re-expressed as follows:

$$\begin{cases} \dot{x_1} = x_4 \cos(x_5 + x_7) & (a) \\ \dot{x_2} = x_4 \sin(x_5 + x_7) & (b) \\ \dot{x_3} = f_{steer}(x_3 + u_1) & (c) \\ \dot{x_4} = \frac{1}{m}\left(-F_{s,f} \sin(x_3 - x_7) + F_{s,r} \sin(x_7) + F_{l,r} \cos(x_7) + F_{l,f} \cos(x_3 - x_7)\right) & (d) \\ \dot{x_5} = x_6 & (e) \\ \dot{x_6} = \frac{1}{I_z}\left(F_{s,f} \cos(x_3) l_f - F_{s,r} l_r + F_{l,f} \sin(x_3) l_f\right) & (f) \\ \dot{x_7} = -x_6 + \frac{1}{m x_4}\left(F_{s,f} \cos(x_3 - x_7) + F_{s,r} \cos(x_7) - F_{l,r} \sin(x_7) + F_{l,f} \sin(x_3 - x_7)\right) & (g) \\ \dot{x_8} = \frac{1}{I_{y,w}}\left(-R_w F_{l,f} + T_{s,b} T_B + T_{s,e} T_E\right) & (h) \\ \dot{x_9} = \frac{1}{I_{y,w}}\left(-R_w F_{l,r} + (1 - T_{s,b}) T_B + (1 - T_{s,e}) T_E\right) & (i) \end{cases} \quad (5.9)$$

The control input of the vehicle include the steering rate (i.e. $v_\delta \stackrel{\text{def}}{=} \dot{\delta}$) and longitudinal acceleration a_{long}. We use the control input $u_1 \stackrel{\text{def}}{=} v_\delta$ and $u_2 \stackrel{\text{def}}{=} a_{long}$ to denote the two controls. The control input and state variable are bounded. The steer rate, steer angle and speed is bounded from both sides:

$$v_\delta \in \left[\underline{v_\delta}, \overline{v_\delta}\right], \quad \delta \in \left[\underline{\delta}, \overline{\delta}\right], \quad v \in \left[\underline{v}, \overline{v}\right] \quad (5.10)$$

The function f_{steer} is defined as:

$$v_\delta = f_{steer}\left(\delta, v_{\delta,d}\right) = \begin{cases} 0, & \text{for } \left(\delta \le \underline{v_\delta} \wedge v_{\delta,d} \le 0\right) \vee \left(\delta \ge \overline{v_\delta} \wedge v_{\delta,d} \ge 0\right) (C1), \\ \underline{v_\delta}, & \text{for } \neg C1 \wedge v_{\delta,d} \le \underline{v_\delta}, \\ \overline{v_\delta}, & \text{for } \neg C1 \wedge v_{\delta,d} \ge \overline{v_\delta}, \\ v_{\delta,d}, & \text{otherwise} \end{cases} \quad (5.11)$$

5.3.3.2 A Simpler Vehicular Kinetic Model

The dynamics model developed in the last section is complex. This section develops a kinetic model with a simpler form. The developed models are applicable to a specific vehicle that adheres to the characteristic of being front-wheel-driven. Figure 5.3a illustrates the physical configuration of this vehicle, with its coordinate denoted by the location of the center of gravity (CG), represented as $\left(x_i, y_i\right)$. The distances from the CG to the front and rear bump are designated as l_F and l_R, while the distances from the CG to the front and rear axles are denoted as l_f and l_r. The heading angle of the vehicle is represented as ψ_i. For the sake of simplicity, in instances where there is no ambiguity, the vehicle index j or the time dimension may be omitted, resulting in expressions such as x(t), y(t), ψ(t), or x_i, y_i and $ψ_i$. To describe the dynamic characteristics of the vehicle, we utilize the vehicle dynamic model as presented in [8]. Consequently, the state variables of the vehicle can be represented by the vector [x, y, ψ]. Building upon the concepts borrowed from the IDM model, the original control inputs of the vehicle encompass the front wheel steering angle δ and the speed v. The specific formulation of these control inputs will be detailed later. The expression of the Rajamani model is provided in Eq. 5.12.

$$\begin{cases} \dot{x} = vcos(\psi + \beta) \ (a) \\ \dot{y} = vsin(\psi + \beta) \ (b) \\ \dot{\psi} = \frac{vcos(\beta)\tan(\delta^f)}{l_f + l_r} \ (c) \end{cases} \tag{5.12}$$

where:

$$\beta = \tan^{-1}\left(\frac{l_r \tan \delta^f}{l_f + l_r}\right) \tag{5.13}$$

The aforementioned model incorporates the speed and steering angle as control inputs. These control inputs can be differentiated to obtain the acceleration, denoted as a_{long}, and the steering rate, denoted as $v_δ$. Subsequently, these two variables can be utilized as control inputs in the model. The formulation of the model, as described in Eq. 5.12, is presented as follows:

$$\begin{cases} \dot{x} = v \cdot cos\left(\psi + \tan^{-1}\left(\frac{l_r \tan\delta}{l_f + l_r}\right)\right) \ (a) \\ \dot{y} = v \cdot sin\left(\psi + \tan^{-1}\left(\frac{l_r \tan\delta}{l_f + l_r}\right)\right) \ (b) \\ \dot{\psi} = \frac{v \cdot cos\left(\tan^{-1}\left(\frac{l_r \tan\delta}{l_f + l_r}\right)\right)\tan(\delta)}{l_f + l_r} \ (c) \\ \dot{v} = a_{long} \ (d) \\ \dot{\delta} = v_\delta \ (e) \end{cases} \tag{5.14}$$

Within the vehicle control domain, the inputs for control encompass the steering rate, represented by $v_δ$ (equivalently $v_\delta \overset{\text{def}}{=} \dot{\delta}$), and the longitudinal acceleration, denoted as a_{long}. These control inputs, along with the state variables, are subjected to bounds. Specifically, the steering angle, speed, acceleration, and steering rate are all bounded, with constraints imposed from both lower and upper limits.

$$\delta \in [\underline{\delta}, \overline{\delta}], v \in [\underline{v}, \overline{v}], a_{long} \in [\underline{a}, \overline{a}], v_\delta \in [\underline{v_\delta}, \overline{v_\delta}] \tag{5.15}$$

The ultimate determination of the acceleration input, denoted as a, shall conform to the aforementioned constraints on maneuverability. The formulation of this acceleration input can be expressed as follows:

$$a = \Gamma_{acc}\left(v, a_{long}\right) = \begin{cases} 0, & \text{for } \left(v \leq \underline{v} \wedge a_{long} \leq 0\right) \vee \left(v \geq \overline{v} \wedge a_{long} \geq 0\right)(C2), \\ \underline{a}, & \text{for } C2 \wedge a_{long} \leq \underline{a}, \\ \overline{a}(v), & \text{for } C2 \wedge a_{long} \geq \overline{a}(v), \\ a_{long}, & \text{otherwise} \end{cases} \tag{5.16}$$

Analogously, the formulation for the steering rate, represented by γ, can be determined by taking into account the associated maneuverability constraints.

$$\gamma = \Gamma_{steer}(\delta, v_\delta) = \begin{cases} 0, & \text{for } \left(\delta \leq \underline{\delta} \wedge v_\delta \leq 0\right) \vee \left(\delta \geq \overline{\delta} \wedge v_\delta \geq 0\right)(C1), \\ \underline{v_\delta}, & \text{for } \neg C1 \wedge v_\delta \leq \underline{v_\delta}, \\ \overline{v_\delta}, & \text{for } \neg C1 \wedge v_\delta \geq \overline{v_\delta}, \\ v_\delta, & \text{otherwise} \end{cases} \tag{5.17}$$

5.3.4 Control Input Formulation: Acceleration

The driver manipulates the vehicle's motion by adjusting both the steering and acceleration parameters, with the aim of ensuring the vehicle remains centered within its designated lane while maintaining an appropriate distance from the leading vehicle. Given the inherent stochastic nature of the system, the vehicle is persistently perturbed from its equilibrium state, necessitating a continuous implementation of control inputs. It is assumed that the underlying theoretical frameworks governing the two control inputs, namely the steering rate and the acceleration, share similar characteristics. In the case of acceleration, the conventional IDM model has been augmented to account for the two-dimensional scenario, while for steering, inspiration has been drawn from the optimal velocity model in order to formulate the steering rate. The subsequent sections explore these two behaviors in greater detail. To devise the acceleration control input, we have incorporated the conventional form of the Intelligent Driver Model, albeit with specific modifications to accommodate the two-dimensional setting. The standard expression for IDM is as follows [3]:

$$\begin{cases} a_i = a_{MAX}\left[1 - \left(\frac{v}{v_d}\right)^4 - \left(\frac{s^*}{s_i}\right)^2\right] & (a) \\ s^* = s_0 + vT + \frac{v\Delta v}{2\sqrt{ab}} & (b) \end{cases} \tag{5.18}$$

Assuming a vehicle length of $l_F + l_R$ and without any loss of generality, we consider the scenario where all vehicles share the same length. However, it should be noted that this assumption can be readily modified as needed. In this context, we define the gap, denoted as s_i, and the relative speed within the two-dimensional scenario using the following expressions:

$$\begin{cases} s = \sqrt{(x_i - x_{i-1})^2 + (y_i - y_{i-1})^2} - (l_F + l_R) \ (a) \\ \Delta v = \mathrm{v_i} - v_{i-1} \qquad (b) \end{cases} \tag{5.19}$$

Equation 5.19 represents an extension of the gap concept to the context of a two-dimensional scenario. By substituting the original Intelligent Driver Model formulation with updated expressions for the gap and relative speed, we arrive at a novel formulation for acceleration within the two-dimensional scenario:

$$\mathrm{a} = a_{MAX}\left[1 - \left(\frac{v}{v_d}\right)^4 - \left(\frac{s_0 + v_i T + \frac{v_i(\mathrm{v_i} - v_{i-1})}{2\sqrt{a_{MAX} b}}}{\sqrt{(x_i - x_{i-1})^2 + (y_i - y_{i-1})^2} - (l_F + l_R)}\right)^2\right]$$
$$\stackrel{\text{def}}{=} \mathrm{IDM}_{\mathrm{acc}}(x_i, y_i, v_i) \tag{5.20}$$

$\mathrm{IDM}_{\mathrm{acc}}(\cdot, \cdot, \cdot)$ is shorthand for the function that calculates the acceleration input.

5.3.5 Control Input Formulation: Steering

The primary objective of steering control is to ensure that a vehicle remains centered within its designated lane and does not deviate significantly from the lane's midpoint. Specifically, when a vehicle veers to the left of the centerline, corrective steering action towards the right is required, and vice versa. Within the framework of stimulus-reaction analysis, we consider two fundamental stimuli that prompt the steering maneuver. This is depicted in Fig. 5.3b. Moreover, we posit the existence of an ideal or target location for the vehicle, denoted as the target position in Fig. 5.3b. This target position represents the equilibrium state when the speed of the leading vehicle (referred to as the leader in Fig. 5.3b) remains unchanged. The speed of the leader vehicle is denoted by $\mathrm{v_{i-1}}$. The equilibrium headway is expressed as:

$$\mathrm{s^{equi}} = (s_0 + v_{i-1} T)\left(1 - \frac{v_{i-1}}{v_d}\right)^{-\frac{1}{2}} \tag{5.21}$$

The target position can be calculated as $\mathrm{x_{i-1}} + \mathrm{s^{equi}}$

$$\mathrm{x_i^d} = \mathrm{x_{i-1}} + (s_0 + v_{i-1} T)\left(1 - \frac{v_{i-1}}{v_d}\right)^{-\frac{1}{2}} \tag{5.22}$$

The target position for vehicle i is characterized by the heading angle ψ_i^E, with the superscript "E" denoting expectation. In our specific scenario where the lane's middle line aligns with the x-axis, ψ_i^E coincides with zero. The steering maneuver is influenced by two distinct stimuli, designated as $\Delta\xi_1$ and $\Delta\xi_2$ in Fig. 5.3b. Firstly, under ideal circumstances, the driver's steering operation would cause ψ_i to converge towards ψ_i^E, and the disparity between the two angles constitutes the first stimulus. Secondly, there exists a discrepancy in the angle between the anticipated heading angle and the line connecting the center of gravity (CG) of the current and target positions, which should ideally align with the lane's middle line. The difference between these two angles serves as the second stimulus. When either of these stimuli deviates from the ideal conditions, the driver is motivated to steer the vehicle either to the left or to the right. The definitions for $\Delta\xi_1$ and $\Delta\xi_2$ are as follows:

$$\begin{cases} \Delta\xi_1 = \psi_i^E - \psi_i & (a) \\ \Delta\xi_2 = \psi_i^E - \mathrm{acos}\left(\frac{x_i - x_i^d}{\sqrt{(x_i - x_i^d)^2 + (y_i - y_i^d)^2}}\right) & (b) \end{cases} \tag{5.23}$$

Equation 5.23-a represents the disparity in heading angles, while Eq. 5.23-b denotes the distinction between the lane's middle line and the line connecting the center of gravity (CG) of the two positions. In cases where $\Delta\xi_1$ equals zero and $\Delta\xi_2$ equals zero, the driver lacks any incentive for initiating a steering maneuver. As a result, we establish the following definition for the ideal steering angle:

$$\delta_i^* = 1 - e^{\frac{\Delta\xi_1 + \Delta\xi_2}{\tau}} \tag{5.24}$$

τ is a parameter. Inspired by the optimal velocity model [1], we adopt the concept and define the steering rate γ_i in the subsequent manner:

$$\gamma_i = \mathcal{O}(\delta_i^*) - \delta_i \stackrel{\text{def}}{=} \mathrm{IDM}_{\text{steer}}(x_i, y_i, \psi_i, \delta_i) \tag{5.25}$$

The formulation for the function $\mathcal{O}(\delta_i^*)$ follows the construction $\mathcal{O}(\delta_i^*) = \tanh(\delta_i^*)$, similar to [1], albeit with a slight modification. In the original formulation, $\tanh(\delta_i^* - 2) - \tanh(2)$, the result would yield an asymmetrical steering maneuver (where steering to the left or right should ideally be symmetrical about the lane's middle line). Consequently, we arrive at the following formulation for the steering rate:

$$\gamma_i = \mathrm{IDM}_{\text{steer}}(x_i, y_i, \psi_i, \delta_i) = \tanh\left(1 - e^{\frac{2\psi_i^E - \psi_i - \mathrm{acos}\left(\frac{x_i - x_i^d}{\sqrt{(x_i - x_i^d)^2 + (y_i - y_i^d)^2}}\right)}{\tau}}\right) \tag{5.26}$$

5.4 Stochastic Noise

5.4.1 Stochastic Noise Type 1: OU Process

Within our framework, we assume the presence of noise in both dimensions, arising from various factors such as road-vehicle interaction, air drag, and similar sources. To account for this stochasticity, we employ the Ornstein–Uhlenbeck (OU) process as the designated noise source. The OU process, denoted as Z, can be expressed as follows:

$$dZ = \theta(\mu - Z)dt + \sigma dW \tag{5.27}$$

The parameters associated with the aforementioned Ornstein–Uhlenbeck (OU) process possess distinct physical interpretations. Specifically, θ represents the rate at which the process reverts, with its value being strictly positive. Meanwhile, μ signifies the long-term equilibrium value of the process, while σ is indicative of its volatility. In our case, we aim for the process to ultimately converge to a state of equilibrium at zero, signifying that the driver would guide the vehicle towards the center of the lane. Consequently, we assign μ the value of zero. As a result, the OU process can be reformulated as $\mathrm{dZ} = -\theta\mathrm{Zdt} + \sigma\mathrm{dW}$, where Z will experience fluctuations around zero.

Given that the noise encountered during the driving process remains within certain bounds, it is imperative to undertake a specific transformation from the aforementioned Ornstein–Uhlenbeck (OU) noise. This transformed noise will subsequently be incorporated into the system model. To accomplish this, our approach involves utilizing the hyperbolic tangent function to convert the OU process. Afterwards, the resulting noise is added to the control input. We express this two-dimensional noise (longitudinal and lateral) formulation in terms of the OU process as follows:

$$\begin{cases} dZ_{long} = -\theta_{long} Z_{long} dt + \sigma_{long} dW & (a) \\ dZ_{lat} = -\theta_{lat} Z_{lat} dt + \sigma_{lat} dW & (b) \end{cases} \tag{5.28}$$

With regards to the vehicular dynamics model presented in Eq. 5.14, we proceed to incorporate the noise into it. To denote the system's sensitivity to the noise, we introduce the symbols η_{lat} and η_{long}. Given that the acceleration primarily governs longitudinal maneuvers while the steering predominantly controls lateral adjustments, we incorporate the noise into the model in the following manner:

$$\begin{cases} \mathrm{dv_i} = \Gamma_{acc}\big(IDM_{acc}(x_i, y_i, v_i) + \eta_{long} \tanh\big(Z_{long}\big)\big)dt & (a) \\ d\delta_i = \Gamma_{steer}(\mathrm{IDM_{steer}}(x_i, y_i, \psi_i, \delta_i) + \eta_{lat} \tanh(Z_{lat}))dt & (b) \end{cases} \tag{5.29}$$

5.4.2 Synthesis of the Model

Put the vehicles dynamics model, the two-dimensional noise and the IDM logic for steer and acceleration, we finally have the formulation of the stochastic two dimensional IDM model with vehicle dynamics as follows:

$$\begin{cases} dx_1 = x_4 \cos(x_5 + x_7)dt & (a) \\ dx_2 = x_4 \sin(x_5 + x_7)dt & (b) \\ dx_3 = f_{steer}(x_3 + u_1)dt & (c) \\ dx_4 = \frac{1}{m}\begin{pmatrix} -(\mathbb{M}_{s,r}(x_4, x_6, x_7))\sin(x_3 - x_7) + (\mathbb{M}_{s,r}(x_4, x_6, x_7) + \eta_{lat}x_{10})\sin(x_7) \\ +(\mathbb{M}_{l,f}(x_3, x_4, x_6, x_7, x_8) + \eta_{long}x_{10})\cos(x_7) + (\mathbb{M}_{l,r}(x_7, x_9) + \eta_{long}x_{10})\cos(x_3 - x_7) \end{pmatrix}dt & (d) \\ dx_5 = x_6 dt & (e) \\ dx_6 = \frac{1}{I_z}\begin{pmatrix} (\mathbb{M}_{s,r}(x_4, x_6, x_7))\cos(x_3)l_f - (\mathbb{M}_{s,r}(x_4, x_6, x_7) + \eta_{lat}x_{10})l_r \\ +(\mathbb{M}_{l,f}(x_3, x_4, x_6, x_7, x_8) + \eta_{long}x_{10})\sin(x_3)l_f \end{pmatrix}dt & (f) \\ dx_7 = -x_6 + \frac{1}{mx_4}\begin{pmatrix} (\mathbb{M}_{s,r}(x_4, x_6, x_7))\cos(x_3 - x_7) + (\mathbb{M}_{s,r}(x_4, x_6, x_7) + \eta_{lat}x_{10})\cos(x_7) \\ (-\mathbb{M}_{l,f}(x_3, x_4, x_6, x_7, x_8) + \eta_{long}x_{10})\sin(x_7) + (\mathbb{M}_{l,r}(x_7, x_9) + \eta_{long}x_{10})\sin(x_3 - x_7) \end{pmatrix}dt & (g) \\ dx_8 = \frac{1}{I_{y,w}}(-R_w(\mathbb{M}_{l,f}(x_3, x_4, x_6, x_7, x_8) + \eta_{long}x_{10}) + T_{s,b}T_B + T_{s,e}T_E)dt & (h) \\ dx_9 = \frac{1}{I_{y,w}}(-R_w(\mathbb{M}_{l,r}(x_7, x_9) + \eta_{long}x_{10}) + (1 - T_{s,b})T_B + (1 - T_{s,e})T_E)dt & (i) \\ d\mathrm{x}_{10} = -\sigma_{lat}^2(1 - \mathrm{x}_{10}^2)\mathrm{x}_{10}dt + \sigma_{lat}(1 - \mathrm{x}_{10}^2)dW & (j) \\ \mathrm{dx}_{11} = -\sigma_{long}^2(1 - \mathrm{x}_{11}^2)\mathrm{x}_{11}dt + \sigma_{long}(1 - \mathrm{x}_{11}^2)dW & (k) \end{cases} \tag{5.30}$$

By integrating the vehicle kinetic model, the two-dimensional noise, and the Intelligent Driver Model (IDM) logic pertaining to steering and acceleration, we arrive at a comprehensive formulation for a stochastic two-dimensional IDM model applicable to vehicle. This formulation can be expressed as follows:

$$\begin{cases} \mathrm{dx_i} = \left[v_i \cdot cos\left(\psi_i + \mathrm{a}\tan\left(\frac{l_r \tan\delta_i}{l_f + l_r}\right)\right)\right]dt & (a) \\ \mathrm{dy_i} = \left[v_i \cdot \sin\left(\psi_i + \mathrm{a}\tan\left(\frac{l_r \tan\delta_i}{l_f + l_r}\right)\right)\right]dt & (b) \\ \mathrm{d}\psi_\mathrm{i} = \left[\frac{v_i \cdot cos\left(\mathrm{a}\tan\left(\frac{l_r \tan\delta_i}{l_f + l_r}\right)\right)\tan(\delta_i)}{l_f + l_r}\right]dt & (c) \\ \mathrm{dv_i} = \Gamma_{acc}\left(IDM_{acc}(x_i, y_i, v_i) + \eta_{long}\tanh\left(Z_{long}\right)\right)dt & (d) \\ d\delta_i = \Gamma_{steer}(\mathrm{IDM_{steer}}(x_i, y_i, \psi_i, \delta_i) + \eta_{lat}\tanh(Z_{lat}))dt & (e) \\ dZ_{long} = -\theta_{long}Z_{long}dt + \sigma_{long}dW & (f) \\ \mathrm{d}Z_{lat} = -\theta_{lat}Z_{lat}dt + \sigma_{lat}dW & (g) \end{cases} \tag{5.31}$$

Within the domain of undetermined parameters, with the exception of the parameters within the Intelligent Driver Model (IDM) that can be acquired from previous research, we encompass variables such as θ_{long}, θ_{lat}, σ_{long}, σ_{lat}, η_{long} and η_{lat}. To render the aforementioned intricate equation more succinctly, we reconfigure it into a matrix representation, as shown below:

$$\mathrm{dQ} = \mathrm{F}(Q)dt + L(Q)dW = \begin{bmatrix} f_1 \\ \cdots \\ \mathrm{f}_7 \end{bmatrix}dt + \begin{bmatrix} l_1 \\ \cdots \\ \mathrm{l}_7 \end{bmatrix}d\mathbf{W} \tag{5.32}$$

The functions denoted as $f_1 \sim f_7$ and $l_1 \sim l_7$ can be readily inferred from Eq. 5.31. In the Numerical Study Section, in order to simulate the dynamics of the system, a discretization scheme becomes necessary. In this particular case, we employ the Euler discretization scheme for Stochastic Differential Equations in the following manner:

$$Q(k+1) = Q(k) + F(Q(k))\Delta t + L(Q(k))\Delta W \tag{5.33}$$

The time step duration, denoted as Δt, represents the duration between consecutive steps. Meanwhile, ΔW signifies a two-dimensional Gaussian variable with a mean value of zero. The two dimensions correspond to the longitudinal and lateral directions, respectively. The covariance matrix can be expressed as $\mathrm{diag}(\Delta t, \Delta t)$. In the presence of an initial state $Q(0)$, the system's evolution can be simulated accordingly.

5.5 Stochastic Lyapunov Stability Analysis

5.5.1 Lyapunov Function

The SDE presented in Eq. 5.31 characterizes the movement of the vehicle in a two-dimensional space, with its dynamics being influenced by Brownian motion. Subsequently, we delve into the analysis of the SDE's stability. Unlike deterministic Ordinary Differential Equations, the stability of an SDE is apprehended through a probabilistic perspective. To begin with, we examine the equilibrium point by assuming the presence of a leader vehicle, referred to as i–1, which maintains a constant speed represented as v_{i-1}. Under equilibrium conditions, the follower vehicle, denoted as vehicle i, also moves at a speed of v_{i-1}, while exhibiting zero acceleration and steering rates. Additionally, the heading angles ψ_i and y_i are required to be zero as well. Accordingly, the equilibrium points can be expressed as follows under the presumption of a leading vehicle traveling at a constant speed:

$$Q^{EQU} = \left[v_{i-1}t, 0, 0, 0, 0, 0, 0\right]^T \tag{5.34}$$

Let $M = Q - Q^{EQU} \overset{def}{=} [m_1, m_2, \ldots m_7]$ represent the error vector in our analysis. Given that Q is a stochastic process, described by Eq. 5.31, it follows that M is also a stochastic process. It is evident that each component of M possesses second-order differentiability. As a result, employing the Ito formula, we can deduce the formulation of M represented in Eq. 5.35. For brevity, we omit the common derivation process. Notably, Eq. 5.35b–e can directly inherit from Eq. 5.31 since the function components in M correspond to the variables themselves. However, the primary distinctions between Eqs. 5.31 and 5.35 become apparent in Eq. 5.35a.

$$\begin{cases} dm_1 = \left[-v_{i-1} + (m_4 - v_{i-1}) \cdot cos\left(m_3 + \mathrm{a}\tan\left(\frac{l_r \tan m_5}{l_f + l_r}\right)\right)\right]dt & (a) \\ dm_2 = \left[(m_4 - v_{i-1}) \cdot \sin\left(m_3 + \mathrm{a}\tan\left(\frac{l_r \tan m_5}{l_f + l_r}\right)\right)\right]dt & (b) \\ \mathrm{dm}_3 = \left[\frac{(m_4 - v_{i-1}) \cdot cos\left(\mathrm{atan}\left(\frac{l_r \tan m_5}{l_f + l_r}\right)\right) \tan(m_5)}{l_f + l_r}\right]dt & (c) \\ \mathrm{dm}_4 = \Gamma_{acc}\left(IDM_{acc}(m_1, m_2, m_4) + \eta_{long} \tanh(m_6)\right)dt & (d) \\ dm_5 = \Gamma_{steer}(\mathrm{IDM}_{\mathrm{steer}}(m_1, m_2, m_3, m_5) + \eta_{lat} \tanh(m_7))dt & (e) \\ dm_6 = -\theta_{long} m_6 dt + \sigma_{long} dW & (f) \\ dm_7 = -\theta_{lat} m_7 dt + \sigma_{lat} dW & (g) \end{cases} \tag{5.35}$$

Through the matrix representation of Eq. 5.35 as $\mathrm{dM} = \mathcal{F}(M)dt + \mathcal{L}(M)dW$, where the functions F and L are appropriately evident from the context, we establish the basis for defining stochastic stability as follows:

Definition 1 (*Stochastic stability*) The stochastic stability of the system is established when there exists a $\delta = \delta(\mathrm{r},\xi) > 0$, for all $\mathrm{r} > 0$ and $\xi > 0$, such that for any initial condition $\mathrm{M}(0) \in \mathrm{S}_\delta$ (where $\mathrm{S}_\delta = \{|M| < \delta\}$), the subsequent condition is satisfied:

$$\mathrm{p}\{\omega : |M(t)| < \delta \ \forall t \geq 0\} > 1 - \xi \tag{5.36}$$

To analyze the stability, we construct the Lyapunov function $\mathrm{V}(M)$, as $V(\mathrm{M}) = \frac{\sum_i (m_i)^2}{2}$. The scale factor $\frac{1}{2}$ is just for simplifying derivation. The first order differential V_M and second order differential V_{MM} of the function $V(\mathrm{M})$ are:

$$V_M = (m_i)^{\mathrm{T}};\ V_{\mathrm{MM}} = \mathrm{E} \tag{5.37}$$

Considering E as an identity matrix, we acknowledge that M represents a stochastic process, as defined by Eq. 5.35. In turn, the Lyapunov function V also assumes the nature of a stochastic process, its derivation relying on the utilization of the Ito theorem in the following manner:

$$dV = LV(M)dt + V_{\mathrm{M}}(M) \cdot \mathcal{L}(M)dW \tag{5.38}$$

where:

$$LV(\mathrm{M}) = V_t + V_{\mathrm{M}}\mathcal{F} + \frac{1}{2}\mathrm{trace}\left(\mathcal{L}^{\mathrm{T}} V_{MM} \mathcal{L}\right) = \frac{1}{2}\sum_i \mathcal{L}_{\mathrm{i}}^2 + \sum_i m_i \mathcal{F}_i \tag{5.39}$$

By employing the Lyapunov function and LV(M), we are able to provide a precise specification for the stability condition. Let K represent the set of all continuous non-decreasing functions $\mu : \mathrm{R}_+ \to \mathrm{R}_+$, satisfying $\mu(0) = 0$ and $\mu(\mathrm{r}) > 0$ if $\mathrm{r} > 0$. Furthermore, let $\mathcal{K}_\infty$ denote the set of functions $\mu \in \mathcal{K}$ for which the limit as r approaches infinity of $\mu(\mathrm{r})$ equals infinity. For any given $\mathrm{h} > 0$, we define $\mathrm{S_h}$ as the set of M values in R^7 that satisfy $|\mathrm{M}| < \mathrm{h}$. According to Mao's work [5], if there exists a

Lyapunov function V and a function $\mu \in \mathcal{K}$ such that $V(0, t) = 0$, $\mu(|M|) \leq V(M, t)$, and $LV(M, t) \leq 0$ for all $(M, t) \in S_h \times R_+$, then the solution of the stochastic differential equation (SDE) is considered stochastically stable. In order to meet the stochastic stability requirements, we can set the function $\mu(r) = \frac{r}{3}$. It is important to ensure that LV(M) remains non-positive to preserve stability. Consequently, we impose the condition that LV(M) is non-positive, leading to:

$$LV(M) \leq 0 \Rightarrow \frac{1}{2}\sum_i \mathcal{L}_i^2 + \sum_i m_i \mathcal{F}_i \leq 0,\ \forall M \tag{5.40}$$

The parameters set (which includes θ_{long}, θ_{lat}, σ_{lon}, σ_{lat}, η_{lon}, η_{lat}) is adjusted so that the stability condition of the system is maintained. Because $\sin(\cdot)$, $\cos(\cdot)$, and $\tanh(\cdot)$ are smaller than 1, then we have the following relationship:

$$\begin{aligned}\sum_i m_i \mathcal{F}_i &\leq m_1 m_4 + m_2(m_4 - v_{i-1}) + m_3(m_4 - v_{i-1}) \\ &\quad + m_4\left(a_{MAX} + \eta_{long}\right) + m_5(1 + \eta_{lat}) + m_6\left(-\theta_{long} m_6\right) \\ &\quad + m_7(-\theta_{lat} m_7) \\ &= -\theta_{long} m_6^2 - \theta_{lat} m_7^2 + m_1(-v_{i-1} + m_4) + m_2 m_4 + m_3 m_4 \\ &\quad + m_4\left(a_{MAX} + \eta_{long}\right) + m_5(1 + \eta_{lat}) - m_2 v_{i-1} - m_3 v_{i-1}\end{aligned} \tag{5.41}$$

Besides,

$$\frac{1}{2}\sum_i \mathcal{L}_i^2 = \frac{1}{2}\left[\sigma_{long}^2 + \sigma_{lat}^2\right] \tag{5.42}$$

We conclude the following inequality:

$$\begin{aligned}&\frac{1}{2}\sum_i \mathcal{L}_i^2 + \sum_i m_i \mathcal{F}_i \\ &\quad \leq -\theta_{long} m_6^2 - \theta_{lat} m_7^2 + m_1 m_4 + m_2 m_4 + m_3 m_4 \\ &\quad + m_4\left(a_{MAX} + \eta_{long}\right) + m_5(1 + \eta_{lat}) - m_2 v_{i-1} \\ &\quad - m_3 v_{i-1} + \frac{1}{2}\left[\sigma_{long}^2 + \sigma_{lat}^2\right]\end{aligned} \tag{5.43}$$

The summation of the initial term present on the right-hand side (RHS) of the equation, denoted as $-\theta_{long} m_6^2 - \theta_{lat} m_7^2$, consistently yields a negative result. It is evident that the existence of suitably large positive constants for θ_{long} and θ_{lat} (given their positive default values and the noise's fluctuation around 0) ensures the negativity of the RHS. Consequently, this guarantees the negativity of LV(M) and ultimately establishes a state of stability.

5.5.2 Stability Results

In a subsequent section, we employ the state in simulation to examine the stability conditions. The state, denoted as $\left[y, \psi, v, \delta, Z_{long}, Z_{lat}\right] = [-1.01, 1, 16.7, 1, 0.97, -0.96]$, is subject to modifications of ψ and δ in order to investigate the behavior of the function LV(M), while maintaining the remaining parameters unchanged. The condition $y = -1.01$ signifies that the vehicle is situated to the right of the lane's centerline. A contour plot showcasing LV(M) is depicted in Fig. 5.4. In Fig. 5.4a–d, the parameters θ_{long} and θ_{lat} progressively increase from 0.5 to 3. Whenever the function LV(M) assumes negative values, the system remains stable. The figure enables the identification of the isogram exhibiting LV(M) = 0. Specifically, in Fig. 5.4a, when ψ is approximately -0.3 (equivalent to approximately 17°) and δ is approximately 0.25 (also approximately 17°), the system becomes unstable as LV(M) trends towards positivity. Consistent with Eq. 5.43, the increment of θ_{long} and θ_{lat} serves to stabilize the system by reducing the value of LV(M). This phenomenon is evident when observing Fig. 5.4b–d, where the isogram representing LV(M) = 0 progressively shifts towards the lower left quadrant. Consequently, the stable region enlarges alongside the two parameters θ_{long} and θ_{lat}.

5.6 Numerical Results

In this investigation, the following parameters are utilized: l_f and l_r assume values of 1.1 m and 1.3 m, respectively; a_{MAX} and b are assigned values of $3.0\,m/s^2$ and $3.21\,m/s^2$, respectively. T in the IDM model is configured to 1.5 s, while v_d is set at either 60 km/h or 16.6 m/s.

5.6.1 Single Vehicle Acceleration from Standstill

Initially, the vehicle is positioned at $x = -3\,km$, and undergoes acceleration from a standstill to reach the free-flow speed of 90 km/h (about 25m/sec). Figure 5.5 illustrates the states obtained from a simulation employing the Euler discretization scheme outlined in Eq. 5.33. The time step is set to 0.5 s, and the simulation encompasses a time horizon of 60 s, equivalent to 1 min. Figure 5.5a represents the two-dimensional trajectory, revealing slight deviations from the central line ($y = 0$) with intermittent fluctuations. Figure 5.5b displays the heading angle, while Fig. 5.5c depicts the speed profile. Notably, the speed profile exhibits a relatively smooth trend as the vehicle endeavors to accelerate towards reaching the imposed speed limit. Conversely, Fig. 5.5d illustrates the steering angle, which demonstrates consistent variations and fluctuations in the proximity of zero. This observation suggests that the simulated vehicle operator strives to maintain a central position within the lane.

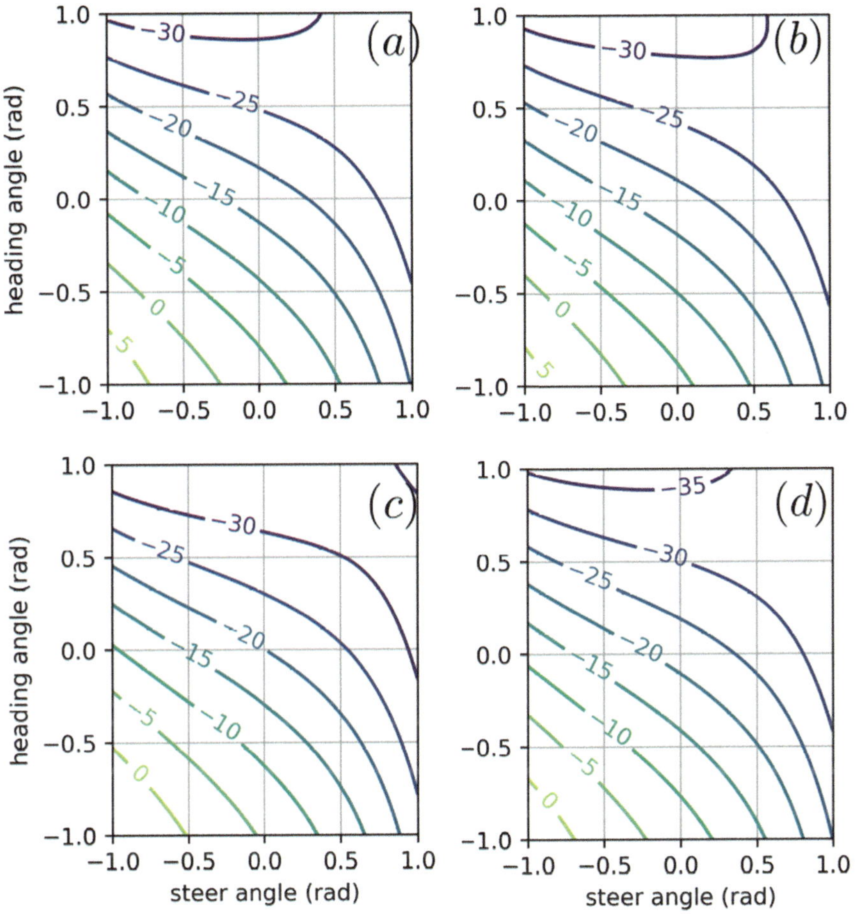

Fig. 5.4 States of evolution of single sample. **a** θ_long = 0.5, θ_lat = 0.5; **b** θ_long = 1, θ_lat = 1; **c** θ_long = 2, θ_lat = 2; **d** θ_long = 3, θ_lat = 3. Reprint from ref [7], copyright 2023, with permissions from Chinese Physics B

However, due to inherent random influences, the heading angle sporadically fluctuates, thereby aligning with the driver's intention to utilize the steering angle for realigning the vehicle towards the center of the lane.

To ensure adequate visualization and differentiation, we conduct 300 simulations of the system paths using an identical initial state for a duration of 1 min. Figure 5.6 portrays 400 selected simulations to avoid excessive overlap and indistinguishability of the curves. The configuration of this figure aligns with that of Fig. 5.5. Notably, Fig. 5.6c demonstrates relatively minimal fluctuations solely in respect to the acceleration. Conversely, the trajectories portrayed in the remaining sub-figures exhibit various degrees of fluctuations. At any given moment, the system state adheres to a specific probability distribution. Figure 5.7 provides insight into the distributions of

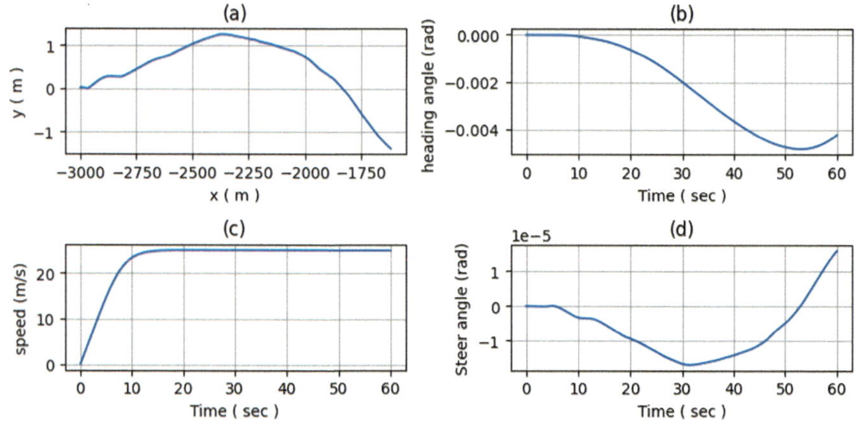

Fig. 5.5 States of evolution of a single sample. **a** spatial trajectory; **b** the heading angle; **c** the longitudinal speed; **d** the steer angle. Reprint from ref [7], copyright 2023, with permissions from Chinese Physics B

the lateral location (denoted as y). The initial state corresponds to y = 0, signifying that the vehicle is initially positioned precisely at the centerline of the lane. Approximately 50 s later (specifically 49.04 s as depicted in the plot), the lateral deviation amounts to approximately 0.6 m (±0.3 m). Notably, the distribution of the lateral location consistently evolves towards a more pronounced variance.

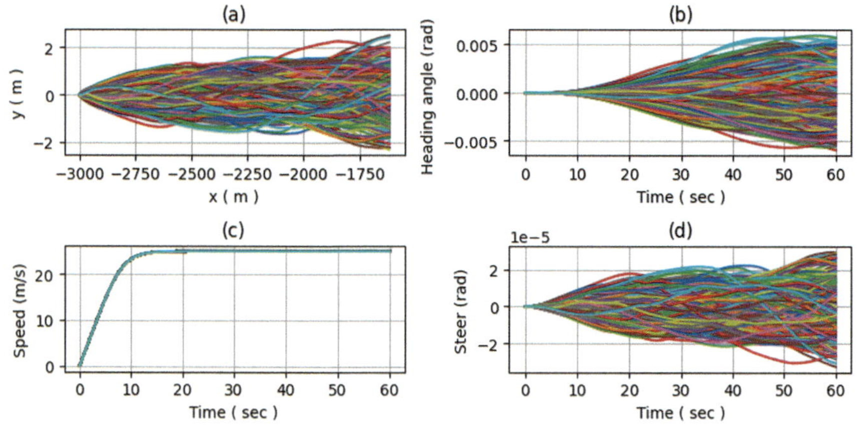

Fig. 5.6 States of evolution of 400 paths with the same initial state. **a** spatial trajectory; **b** the heading angle; **c** the longitudinal speed; **d** the steer angle. Reprint from ref [7], copyright 2023, with permissions from Chinese Physics B

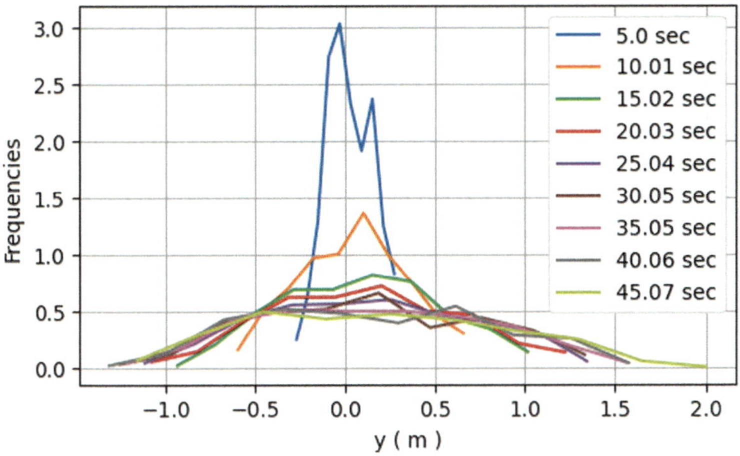

Fig. 5.7 Lateral distribution of the a model of 300 paths. Reprint from ref [7], copyright 2023, with permissions from Chinese Physics B

5.6.2 A Platoon with a Leading Vehicle

We employ a leading vehicle that emulates the stop-and-go traffic flow encountered during congested conditions. The leading vehicle, along with four additional vehicles forming a platoon. Figure 5.8 presents the speed profile of the leading vehicle. To generate the acceleration profile, we utilize a sinusoidal function, subsequently computing the speed profile and longitudinal location. The leading vehicle initiates from a stationary position, accelerates to reach the free-flow speed, and subsequently comes to a complete halt. This cyclic process repeats multiple times.

We conduct simulations involving a group of five vehicles, where the initial space headways between each consecutive pair of vehicles are set at 30 m. At the initial time instance $t = 0$, the leading vehicle is positioned at $x = 0$. Each vehicle commences from a state of rest and undergoes acceleration. The temporal evolution of the state

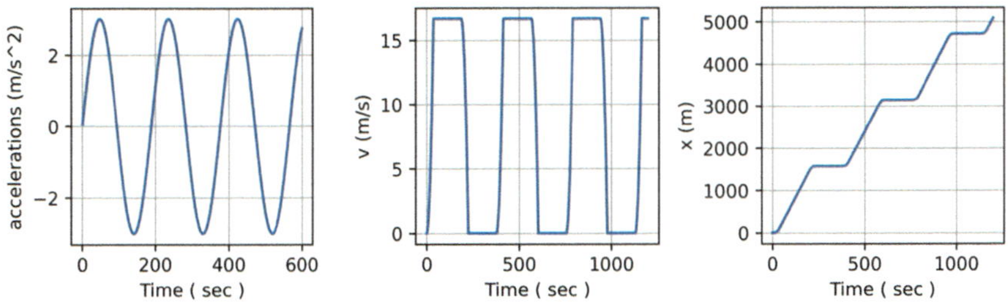

Fig. 5.8 Acceleration, speed, and location of the leading vehicle. Left: acceleration curve; middle: speed curve; right: longitudinal distance. Reprint from ref [7], copyright 2023, with permissions from Chinese Physics B

variables for the second vehicle (which follows directly behind the leading vehicle) and the fifth vehicle are depicted in Fig. 5.9 and Fig. 5.10 respectively. These figures consistently adhere to the same visual layout as Fig. 5.6. To avoid excessive overlap in the representation of the simulation paths, we present the paths of 30 simulations utilizing the Euler scheme, despite conducting 300 simulations. Figure 5.9a showcases the evolution of x–y trajectories. Figure 5.9b provides insights into the evolution of the heading angles. Notably, the behavior depicted in Fig. 5.9b–d noticeably deviates from that observed in the absence of a leading vehicle, as exemplified in Fig. 5.6. In Fig. 5.9c, the speed profile closely resembles that of the leading vehicle, as depicted in Fig. 5.8. When the speed of a vehicle attains zero, it remains stationary, and consequently, the steering angle remains unchanged. This is effectively presented in Fig. 5.9d. For instance, within the time interval spanning approximately 300 s to 400 s, when the vehicle speed approaches zero, the steering angle remains nearly constant without any fluctuations (each line in Fig. 5.9d corresponds to an almost unchanging steering angle during this interval; specifically from 300 to 400 s). This consistency arises as the calculation is based on a pair of vehicles that are both stationary. Consequently, due to the zero speed, the heading angle illustrated in Fig. 5.9b also remains unchanged. A similar pattern can be observed for the fifth vehicle, as demonstrated in Fig. 5.10.

Figure 5.11 presents a comprehensive visualization of the trajectories of the five vehicles under consideration. Of particular interest is the propagation of the speed profiles along the platoon, as depicted in Fig. 5.11c. Notably, the speed profile of each follower vehicle consistently lags behind that of the leading vehicle. Moreover, the lateral positions of the five vehicles exhibit continual variation and do not align with the middle line of the corresponding lane, aligning with empirical observations.

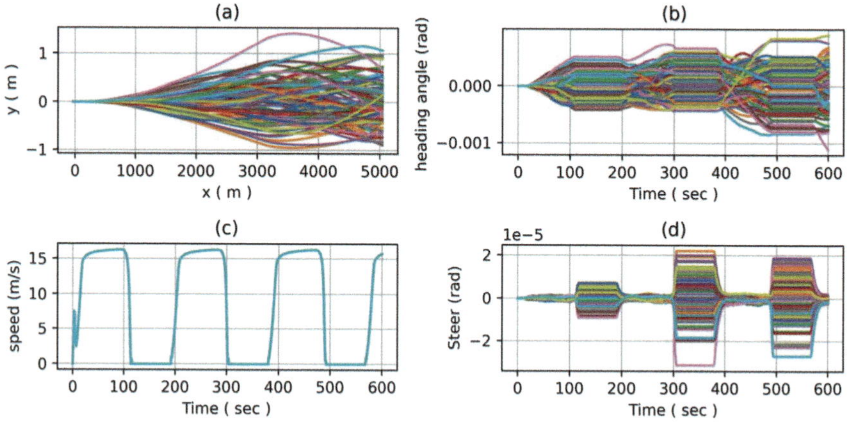

Fig. 5.9 The state of the first vehicle in the platoon. **a** spatial trajectory; **b** the heading angle; **c** the longitudinal speed; **d** the steer angle. Reprint from ref [7], copyright 2023, with permissions from Chinese Physics B

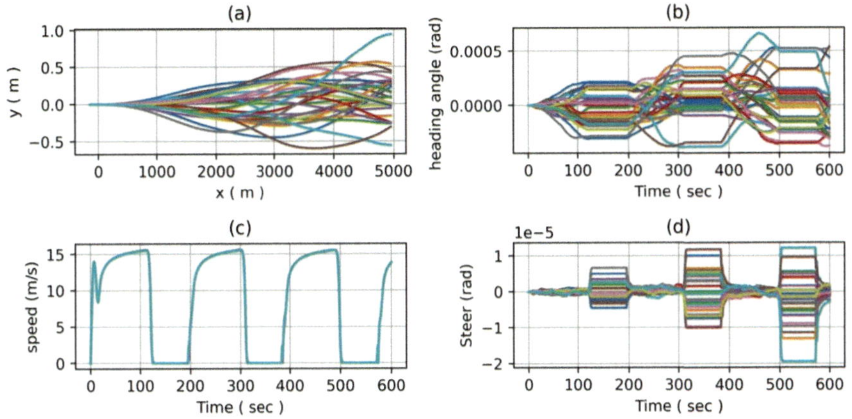

Fig. 5.10 The state of the fifth vehicle in the platoon. **a** spatial trajectory; **b** the heading angle; **c** the longitudinal speed; **d** the steer angle. Reprint from ref [7], copyright 2023, with permissions from Chinese Physics B

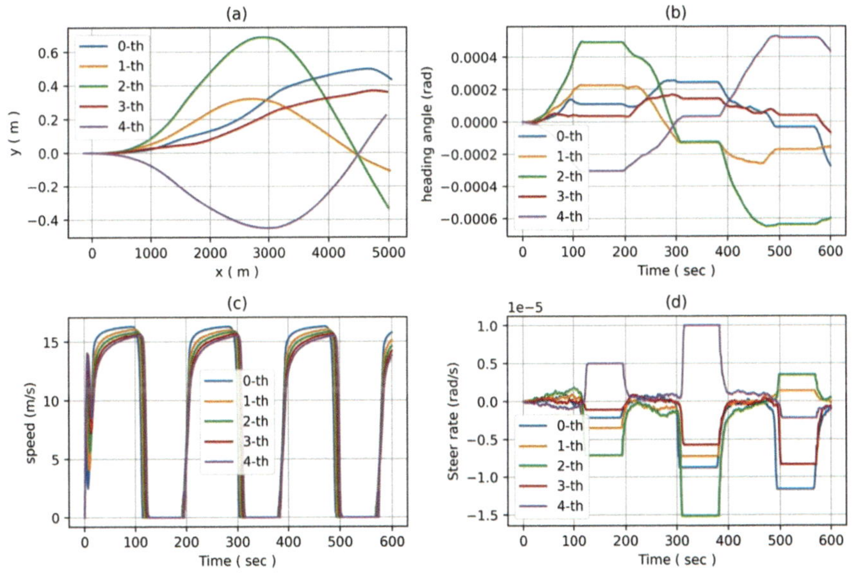

Fig. 5.11 Platoon dynamics evolution. **a** spatial trajectory; **b** the heading angle; **c** the longitudinal speed; **d** the steer angle. Reprint from ref [7], copyright 2023, with permissions from Chinese Physics B

5.7 Notes

Existing microscopic level models commonly focus solely on driver responses to leading vehicles while neglecting the consideration of vehicle dynamics and the influence of two-dimensional stochasticity. Additionally, several well-known traffic flow and autonomous vehicle simulators, which rely on driver behavior models, struggle to accurately emulate real-world characteristics. To address these limitations, we leverage the advantages of the widely recognized intelligent driver model (IDM) and propose the development of a stochastic two-dimensional IDM model that accounts for vehicular dynamics. The control inputs of this model encompass longitudinal acceleration and steering rate, both derived from concepts presented in the IDM and optimal velocity models. Furthermore, both longitudinal and lateral noises are incorporated into the model. Through numerical simulations, we demonstrate that this model successfully reproduces stochastic driver-vehicle unit behavior.

There are several potential extensions and applications for this model. Firstly, in the era of connected and autonomous vehicles (CAVs), information can be acquired from both a lead vehicle and roadside infrastructure, enabling the driver of the vehicle to receive insights regarding downstream traffic flow conditions. Consequently, a more intricate formulation for stimulus–response relationships can be derived. Secondly, the proposed model holds promise for applications in stochastic vehicular trajectory inference or prediction. Existing prediction methods often rely on deep networks, necessitating pre-training prior to deployment. In contrast, our model allows for parameter adjustments, facilitating more detailed predictions of a vehicle's dynamics. Lastly, while our model incorporates noise generated from Brownian motion, it is worth exploring the consideration of more complex noise structures, such as fractal Brownian motion.

References

1. Bando M, Hasebe K, Nakayama A et al (1995) Dynamical model of traffic congestion and numerical simulation. Phys Rev E 51:1035–1042. https://doi.org/10.1103/PhysRevE.51.1035
2. Jiang Y, Wang S, Yao Z et al (2021) A cellular automata model for mixed traffic flow considering the driving behavior of connected automated vehicle platoons. Phys Stat Mech Appl 582:126262. https://doi.org/10.1016/j.physa.2021.126262
3. Kesting A, Treiber M, Helbing D (2010) Enhanced intelligent driver model to access the impact of driving strategies on traffic capacity. Philos Trans R Soc Math Phys Eng Sci 368:4585–4605. https://doi.org/10.1098/rsta.2010.0084
4. Li T, Ngoduy D, Hui F, Zhao X (2020) A car-following model to assess the impact of V2V messages on traffic dynamics. Transp B Transp Dyn 8:150–165. https://doi.org/10.1080/21680566.2020.1728591
5. Mao X (1991) Stability of stochastic differential equations with respect to semimartingales (Longman)
6. Navas F, Milanés V (2019) Mixing V2V- and non-V2V-equipped vehicles in car following. Transp Res Part C Emerg Technol 108:167–181. https://doi.org/10.1016/j.trc.2019.08.021
7. Qi H-S, Ying Y-Y (2023) A stochastic two-dimensional intelligent driver car-following model with vehicular dynamics. Chin Phys B 32:044501

8. Rajamani R (2012) Vehicle dynamics and control. Springer, US, Boston, MA
9. Sun Y, Ge H, Cheng R (2019) A car-following model considering the effect of electronic throttle opening angle over the curved road. Phys Stat Mech Its Appl 534:122377. https://doi.org/10.1016/j.physa.2019.122377
10. Treiber M, Kesting A (2017) The intelligent driver model with stochasticity -new insights into traffic flow oscillations. Transp Res Procedia 23:174–187. https://doi.org/10.1016/j.trpro.2017.05.011
11. Wang Z, Ge H, Cheng R (2020) An extended macro model accounting for the driver's timid and aggressive attributions and bounded rationality. Phys Stat Mech Its Appl 540:122988. https://doi.org/10.1016/j.physa.2019.122988
12. Xu L, Li Y, Feng H (2018) Non-lane-discipline-based car-following model considering the effects of full lateral gaps. in: 2018 chinese automation congress (CAC). IEEE, Xi'an, China, pp 1146–1150

Chapter 6
Lateral Dimension Modeling and Application

6.1 Why Lateral Dimension Matters?

A vehicle's motion can be analytically separated into its longitudinal and lateral components. While the longitudinal component has received considerable attention in the literature, the stochastic nature of the lateral component during both lane keeping and lane changing processes has been comparatively overlooked, thus limiting the capabilities of existing traffic models. In Fig. 6.1, the vehicle trajectories from the highD dataset is presented. The vehicles traverse the road with length of 400 m. Both the lane keeping and lane changing trajectories are displayed. It is observed that the lateral deviation exists along the whole process and cannot be ignored.

In order to address this gap, a stochastic differential model is formulated to describe the lateral movement in both lane keeping and lane changing scenarios. Initially, a lateral noise model is established based on the mathematical framework of the Wiener process. Subsequently, a stochastic model for lateral vehicle movement is developed, with its parameters possessing clear physical interpretations. The stability conditions of the model are thoroughly analyzed, and the solution of the model is derived through the utilization of the Fokker–Planck equation. The obtained results demonstrate the model's efficacy in several aspects: (1) accurately representing the marginal distribution of lateral displacement, as evidenced by good fits with real-world mean and standard deviation; (2) achieving a zero false negative and false positive rate when applied to discriminate between HDVs/CAVs (human driven vehicles and connected and autonomous vehicles); (3) successfully reproducing the observed distribution of lane changing duration (LCD); (4) quantifying a lateral deviation sensitivity for lane changing cases that is approximately seven times higher than that for lane keeping scenarios.

© The Author(s), under exclusive license to Springer Nature Singapore Pte Ltd. 2024
H. Qi, *Stochastic Two-Dimensional Microscopic Traffic Model*, Lecture Notes in Intelligent Transportation and Infrastructure,
https://doi.org/10.1007/978-981-97-3597-6_6

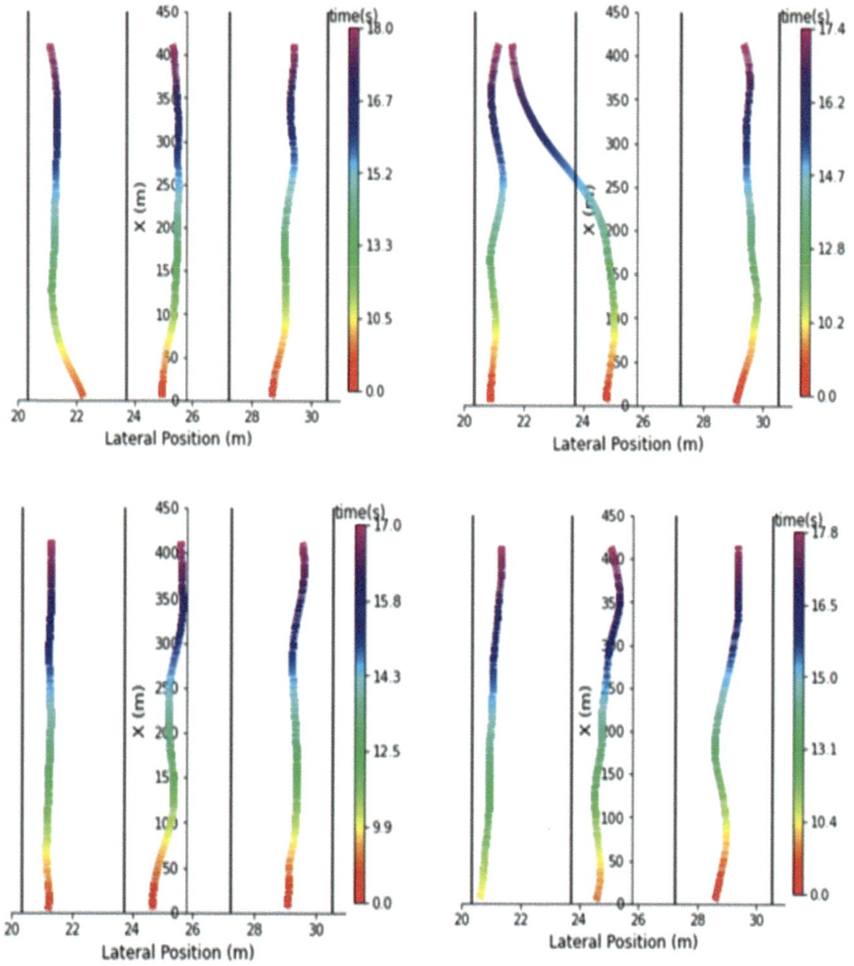

Fig. 6.1 Lateral movements of the vehicle in highD dataset

6.2 Background and Motivation

Microscopic traffic behavior has long been the focus of traffic flow theory. Traditionally, the microscopic behavior is classified into car following and lane changing. The former one is modeled by car following model and the latter one is modeled by lane changing model, which consists of many sub-models such as lane changing decision, lane changing execution, etc. The car following and lane changing process can be decomposed by the longitudinal and lateral movement. The longitudinal part receives considerable attention, while the lateral movement, no matter it occurs during lane keeping or lane changing period, receive relatively little attention. Here, the definition of lateral movement is illustrated in Fig. 6.2a. There is an ideal middle line of the lane, and the lateral shift from this middle line is defined as the lateral displacement. It is denoted by Y. The reference lateral point (which is the middle line of the current lane) can be adjusted as needed. For instance, vehicle would converge to the middle line at target lane, and the reference point can be set as the target lateral coordinate. The independent variable is either time or the longitudinal distance, or Y(t) and Y(X), where X is the longitudinal distance, depending on the context. In CF theory, X is modeled in the stimulus-reaction framework, while the lateral movement Y is ignored in most CF models. Due to this reason, many traffic flow or CAV simulation software cannot produce realistic lateral movement, as they rely on traditional car following model theories, where the vehicle is assumed to follow the center line nearly exactly when driving. The comparison of SUMO, Carla and real-world trajectory is given in Fig. 6.4. It is shown that in SUMO or Carla, vehicles nearly locate at the middle with little fluctuation, while in reality, it is very common that the vehicles would fluctuate around the middle-line of the lane. Figure 6.2b and c present two demonstrations of the lateral movement. The former one is non-congested condition while the latter one is the jam condition. In both cases, the lateral coordinate deviate from the ideal middle line is very common. The stochastic lateral movement is also very common in some famous open-source traffic flow dataset. In Fig. 6.3, multiple trajectories are extracted from SEU traffic data. Each line represents a two-dimension trajectory of a vehicle. The horizontal-axis is the lateral movement while the vertical axis is the longitudinal movement. The trajectories on two lanes are shown. For illustration purpose, we translate the beginning of the longitudinal distance of each trajectory to zero. The trajectories show that the fluctuation of Y reaches at least 2 m during the driving, and changes continuously. The above real world example indicate that lateral movement cannot be totally neglected.

The stochastically nature of the lateral movement is evident in Fig. 6.5, which showcases two distinct trajectories originating from the same initial lateral position. These trajectories have been extracted from the highD dataset, specifically from the sub-dataset where locationId equals 1. In order to facilitate comparison, we have translated the lateral coordinate of all trajectories to the interval $[-\frac{l_w}{2}, \frac{l_w}{2}]$, with l_w denoting a value of 3.5. The scaling of the lateral trajectory is determined by the ratio between its lane width and l_w (3.5). Figure 6.5a corresponds to the scenario in which the initial lateral location Y(0) falls within the interval [−0.1, 0.1], while Fig. 6.5b represents the case when Y(0) ∈ [0.8, 0.9]. The evolution of these trajectories exhibits significant diversity, thereby suggesting that approaches that overlook the temporal dimension (such as distribution fitting or regression) may prove inadequate. As depicted in Fig. 6.5, the deviation reaches almost 2 m after 5 s. This behavior can be attributed to the combined influence of the vehicle, driver, road, and traffic components comprising the vehicle-driver unit. Throughout this process, various sources of randomness coexist, encompassing factors like driver experiences and road surface conditions. Therefore, to gain a comprehensive understanding of the stochastic nature of the lateral movement, an approach grounded in stochastic research is essential. Figure 6.5c and d present the trajectories from other initial locations. All the four plots present that, even if the vehicle starts from the same

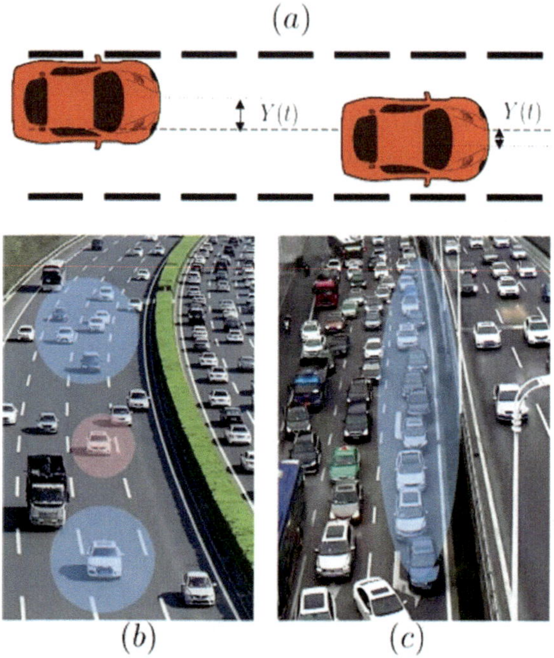

Fig. 6.2 The definition of lateral movement

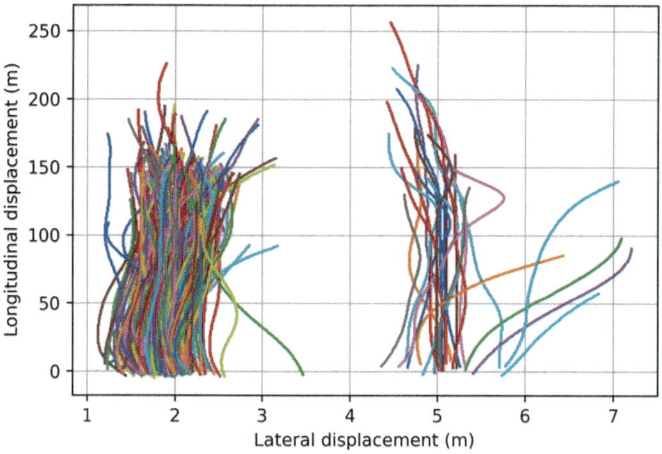

Fig. 6.3 Real world lateral movement data for two lanes. Reprint from Ref. [6], copyright 2023, with permissions from Taylor & Francis

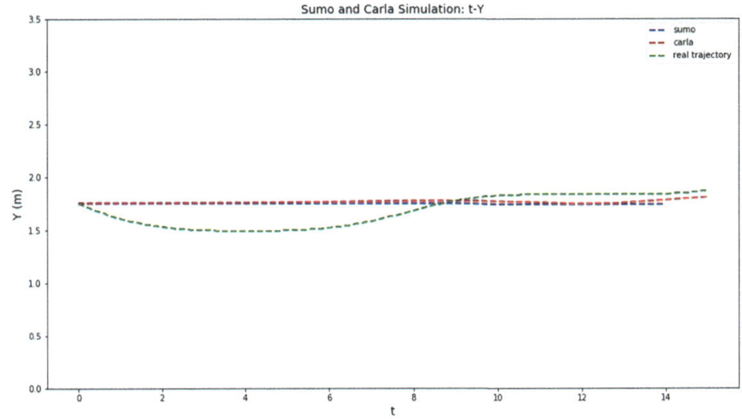

Fig. 6.4 The comparison among lateral movement

lateral location, the paths present considerable diversity. This is due to the fact that the vehicle suffers from continuous lateral forces by many originations.

Figure 6.6 presents the trajectories of both lane keeping and lane changing for the highD dataset of Location 1. In left subplot, the two dimensional trajectories are presented. These trajectories cover the whole road space, rather than the assumed lane middle line. In right subplot, the lane changing trajectories are presented. Note that the lane changing trajectories are normalized. They both start from the right lane to the left lane.

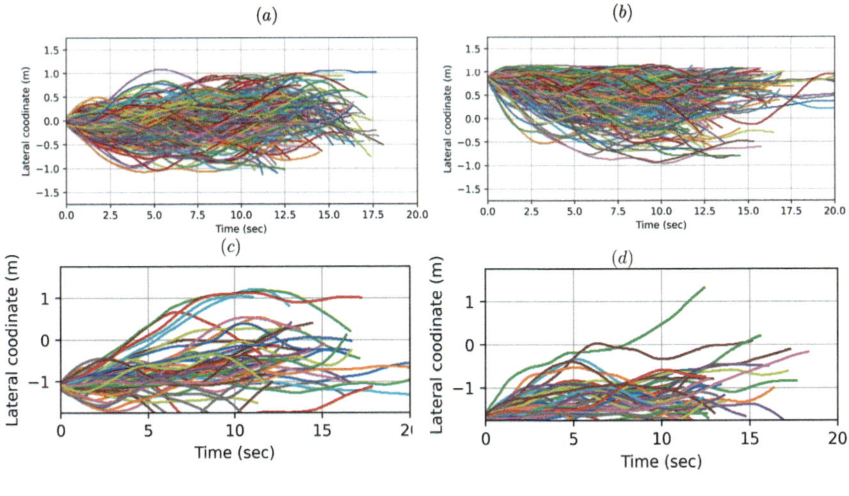

Fig. 6.5 Lateral trajectories from the same initial lateral location in highD dataset; **a** initial location is within [−0.1, 0.1]; **b** initial location is within [0.8, 0.9]

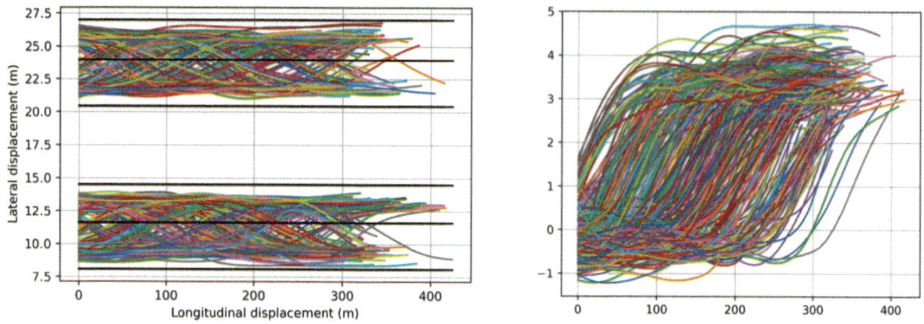

Fig. 6.6 The two dimensional trajectories of the highD; left: all trajectories; right: lane changing trajectories

The above features of the real world lateral stochasticity have long been out of focus of the model development. Numerous scholars acknowledge the importance of incorporating lateral component modeling into traffic analysis and applications. The need for this modeling is evidenced by several analyses and applications involving lateral movement. First, Delpiano [3, 4] identified the presence of lateral impact between neighboring lanes, which leads to lateral friction and reduces capacity. Second, as Zhou et al. [8] noted, interactions between CAVs and human drivers are dynamic, and the lateral behavior of CAVs differs from that of HDVs. Knowledge of HDV's stochastic lateral movement could aid in CAV's behavioral inference. Third, major simulation packages such as traffic flow simulation or CAV simulation tools have weakly modeled the stochastic lateral component. As Delpiano [3, 4] (2021) demonstrated, this leads to results that may not be robust. Fourth, CAVs typically follow a specified trajectory, namely the middle line of the lane, with minimal deviation. This causes bias in pavement stress and resulting fatigue [1, 2]. Therefore, the stochastic lateral modeling can assess the stress of mixed CAV traffic flow. Additionally, due to the occurrence of many accidents during lateral movement stages, a theoretical lateral model could benefit traffic safety analysis.

In response to the aforementioned research requirements, various remedial approaches have been suggested for traditional microscopic traffic models to enhance their ability to describe lateral dynamics. These approaches encompass the weak-lane-discipline based CF model, the two-dimensional CF model, and data-driven statistical methods, among others. These models may either be deterministic or address the lane keeping and changing processes separately. A more comprehensive treatment of the stochastic lateral movement would facilitate the above-discussed applications in a manner that is both realistic and coherent.

To achieve this objective, we examine the lateral movement from a stochastic perspective. The core of our investigation is a stochastic differential equation (SDE) formulation of the noise, which is then integrated into a lateral movement model. The primary advantages of this model include: (1) its capacity to simultaneously capture lateral movement during both lane keeping and changing; (2) its parameters, such as the noise intensity, the driver's responsiveness to lateral deviation, and the presence of noise disturbances, have explicit physical interpretations and can effectively depict vehicle-specific behavior.

6.3 Literature Review

6.3.1 Lateral Movement in Lane Keeping Stage

As previously stated, microscopic modeling of traffic behavior can be dichotomized into the CF theory and LC theory. Lateral movement is typically disregarded in most CF models, whereas it is not independently addressed in the LC formulation. To explore lateral movement, it is possible to extend either the conventional CF or LC model, or alternatively, formulate a statistical model that is based on observational data. This stochastic model can involve techniques such as distribution fitting, or linear/nonlinear regression (as elucidated by [3]).

When considering the lateral component in theoretical car-following models, the non-lane-based approach may be the first to take into account the vehicle's lateral displacement. This type of model includes the impact of the lateral gap between the pair of the lead and following vehicles. However, the stochastic evolution of the lateral movement is often neglected in these non-lane-based models, as the focus lies in capturing vehicle-following behaviors. Other theoretical models aim to incorporate the lateral movement into account with the development of two-dimensional car-following models. Delpiano et al. [4] developed a two-dimensional car-following model that considers the lateral movement while accounting for the acceleration force, repulsive force, and lane force that influence the vehicle's movement. However, all aforementioned models are deterministic, and may therefore not be capable of uncovering or reproducing the randomness in the lateral dimension.

Literature covering real-world instances of lateral movement can be found in several works, which consider the presence of factors such as mixed traffic flow involving motorcycles or weak lane discipline. However, it should be highlighted that the majority of real-world observations of lateral movement occur at fixed locations (fixed section or fixed segment) and do not include a temporal factor.

6.3.2 Lateral Movement in Lane Changing Stage

The process of executing lane changes involves lateral movement and can be categorized into discretionary and mandatory types. Broadly, the overall lane change procedure can be divided into distinct stages: lane change intention, lane change decision, and lane change execution (LCE). During LCE, both the longitudinal dimension and the corresponding lateral dimension (including parameters such as lateral speed, acceleration, and position) follow a certain stochastic evolution. Researchers have employed various data-driven approaches to analyze the stochastic characteristics of lane changes. These methodologies encompass clustering, regression, time series analysis, survival analysis, and deep networks Some literature investigated the distribution of lateral shift duration during lane changes. They discovered that this duration ranged from 0.5 to 15 s and followed a Weibull distribution, which was influenced by factors such as vehicle types and the space gap with leading vehicles.

The lane changing event is characterized by a duration known as lane changing duration (LCD). Numerous factors contribute to the LCD, and scholarly literature presents various definitions and observations of this phenomenon. For instance, the beginning of the event is defined as the moment when the vehicle initiates lateral movement and the end as when the vehicle is centered in the target lane. Besides, the LCD can be defined as the difference between two moments based on half the vehicle's width. Also the onset and end of the event when the lateral speed reaches zero are used to defined LCD. Consequently, reported LCD values vary.

6.4 Model Development

6.4.1 Notation List

For easy reference, we put the notations and abbreviations together as follows.

Y: the lateral displacement
X: longitudinal coordinate
Doerf: derivate of error function
$\Gamma(Y)$: deterministic function when the noise is not considered
α: parameter in $\Gamma(Y)$
Y_{base}: target lateral position
Z: the lateral noise
Z' and Z'': first and second derivatives of Z;
K: raw noise used to construct the lateral noise Z
W: Standard Brownian noise, used to construct the K
θ_Y, θ_Z, and σ: the parameters in the SDE model;
l_w: the lane width;
Δt: the time interval in the discretization of SDE-LM
$LV(Y, Z|\theta_Y, \theta_Z, \sigma)$: lyapunov function of the developed SDE
$\mathcal{K}(\theta_Y, \theta_Z, \sigma)$: defined as $\max_{Y,Z} LV(Y, Z|\theta_Y, \theta_Z, \sigma)$.
LM: lateral movement

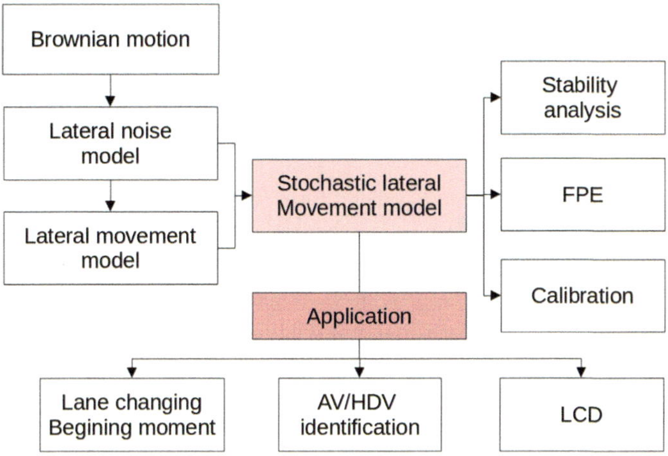

Fig. 6.7 Framework

6.4.2 Framework

Our approach establishes separate models for lateral noise and lateral movement to account for their distinct characteristics. Notably, lateral noise arises from diverse sources, including the stochastic nature of collected lateral data due to data collection methods, the effects of pavement and materials on the road, vehicle dynamics, driving behaviors, and characteristics of the traffic flow environment. Conversely, lateral movement represents a continuous process that is prone to the influence of lateral noise. The model framework is given in the following Fig. 6.7.

6.4.3 Noise Modeling

The first question is that which type of the noise is suitable for the lateral modeling. In order to answer this question, we firstly examine the noise in the real world dataset as in Fig. 6.8. Figure 6.8a is the lateral deviation (measured from the middle point of the front bump to the lane middle line). The distribution is represented as a histogram. It is concluded that the noise is very like the normal distribution. In Fig. 6.8b, the QQ plot is presented. It is shown that the distribution is well fitted by the normal distribution. Therefore, we use the Brownian motion as the start of the modeling.

In order to capture the noise component, various modeling approaches have been proposed in the literature. These include classical Brownian motion, geometrical Brownian motion, fractal Brownian motion, among others. When examining the distribution of lateral deviation, which displays similarities to a normal distribution as illustrated in Fig. 6.8, we opt to employ a common Wiener process to represent the stochasticity. To accommodate this, we introduce a meta noise denoted as K.

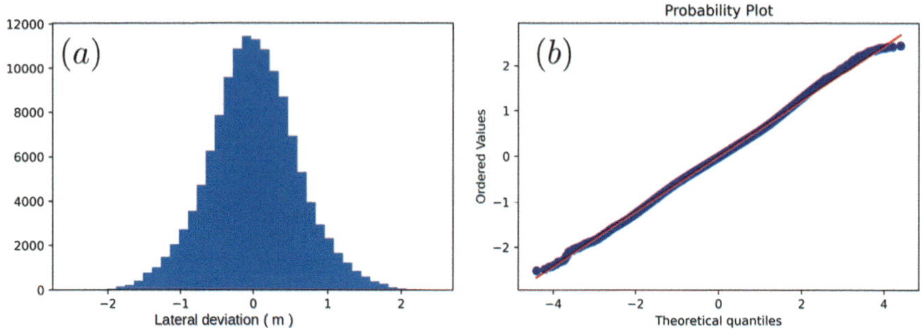

Fig. 6.8 **a** Distribution of the lateral deviation from the lane middle line and **b** QQ plot. The data source is the lane keeping trajectories in NGSIM US-101 dataset

$$\mathrm{dK} = \sigma \cdot \mathrm{dW} \tag{6.1}$$

Given that the lateral noise Z is confined within the interval of $[-\frac{l_w}{2}, \frac{l_w}{2}]$, where the midpoint of the current lane serves as the lateral origin, a transformation is required to convert the general stochastic process into a representation for the lateral noise. In this regard, we opt to leverage the scaled hyperbolic tangent function as the chosen transformation function. To elaborate, let W denote a standard Wiener process, and let σ denote a parameter that specifies the noise intensity level. The conversion becomes:

$$\mathrm{Z} = \frac{l_w}{2}\tanh(\mathrm{K}) = \frac{l_w}{2}\frac{e^K - e^{-K}}{e^K + e^{-K}} \tag{6.2}$$

The hyperbolic tangent function, by default, has a range of [−1,1]. To ensure compatibility with the lateral noise Z, which operates within the domain of $[-\frac{l_w}{2}, \frac{l_w}{2}]$, it is necessary to scale the function by multiplying it by $\frac{l_w}{2}$. Additionally, we can obtain the meta noise K by evaluating the expression $\mathrm{K} = \frac{1}{2}\ln\frac{l_w+2Z}{l_w-2Z}$. Notably, the specified domain of Z is $[-\frac{l_w}{2}, \frac{l_w}{2}]$. Finally, it is relevant to consider the first and second derivatives of Z.

$$\begin{cases} Z' = \frac{l_w\left(1-\tanh^2(K)\right)}{2} \\ Z'' = -l_w\left(1-\tanh^2(K)\right)\tanh(K) \end{cases} \tag{6.3}$$

By applying the Ito rule, an expression for the noise Z can be derived as follows:

$$\begin{aligned} \mathrm{dZ} &= -\frac{l_w\sigma^2\left(1-\tanh^2(K)\right)\tanh(K)}{2}dt + \frac{l_w\sigma\left(1-\tanh^2(K)\right)}{2}dW \\ &= -\frac{l_w\sigma^2\left(1-\frac{4Z^2}{l_w^2}\right)\frac{2Z}{l_w}}{2}dt + \frac{l_w\sigma\left(1-\frac{4Z^2}{l_w^2}\right)}{2}dW \end{aligned}$$

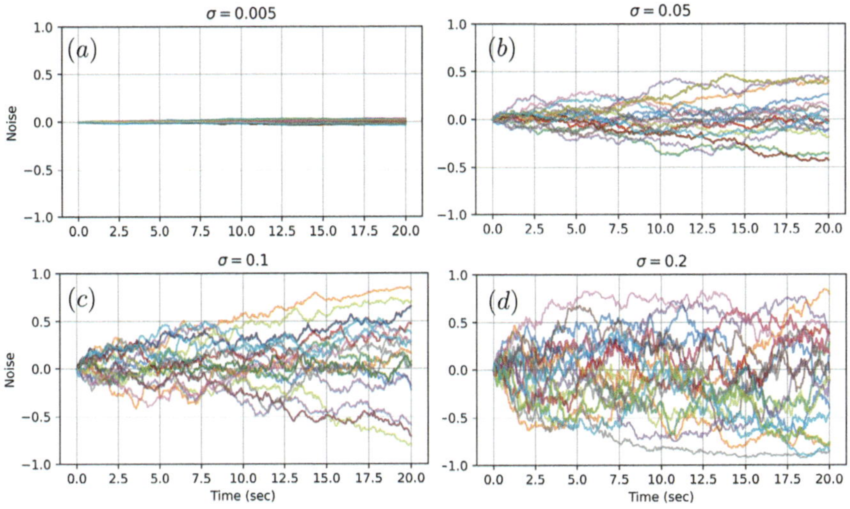

Fig. 6.9 Simulation paths of the bounded noise

$$= -\frac{2Z \cdot \sigma^2 \left(l_w^2 - 4Z^2\right)}{2l_w^2} dt + \frac{l_w \sigma \left(l_w^2 - 4Z^2\right)}{2l_w^2} dW \tag{6.4}$$

The second derivative makes use of the established relationship between $\tanh(K) = \frac{2Z}{l_w}$. It is worth noting that while the lane width l_w remains fixed, the noise Z is dependent on the parameter σ. We simulate the path of Z and the results are given in Fig. 6.9. When $\sigma = 0.005$, the variance is much lower than that of $\sigma = 0.01$. Both simulation is within [−1,1].

6.4.4 Lateral Movement Modeling

Following the characterization of the noise component, our attention turns towards modeling the lateral movement. The process of lane changing exhibits an S-shaped curve, wherein the rate of change initially accelerates and then decelerates upon nearing the middle line of the target lane. Conversely, during the lane keeping process, any deviation from the lane center prompts a corrective steering action. Notably, the magnitude of the deviation influences the speed of the steering response. This behavior can be accurately described using well-known mathematical functions such as the logistic function, sigmoid function, or the error function. Given its simplicity in terms of differentiation and successful application in analyzing microscopic traffic behavior [5], we adopt the error function as the starting point for our model development. The standard error function and its derivative, which is a simple exponential expression, can be represented as:

$$\begin{cases} erf(x) = \frac{1}{\sqrt{\pi}} \int\limits_{-x}^{x} e^{-t^2} dt \\ \frac{d(\text{erf}(t))}{dt} = e^{-t^2} \end{cases} \tag{6.5}$$

To simplify the analysis, we adopt the Doerf function, defined as $\mathrm{e}^{-\mathrm{t}^2}$, as a representation of our model. The standard Doerf function is illustrated in the left subplot of Fig. 6.10. During lane keeping, the driver maintains their position within the lane. However, if there is a deviation from the middle lane, the driver steers back to correct their position. This corrective steering action follows an S-shaped curve, characterized by the Doerf function's derivative. In order to replicate this behavior, we shift the Doerf function both to the left and to the right, and express the deterministic lateral speed as a function of the lateral deviation, as shown in Eq. 6.6.

$$\begin{aligned} \mathrm{dY} &= \left(\mathrm{e}^{-\left(\frac{Y+\frac{l_W}{2}}{\frac{l_W}{\alpha}}\right)^2} - \mathrm{e}^{-\left(\frac{Y-\frac{l_W}{2}}{\frac{l_W}{\alpha}}\right)^2}\right) dt \\ &= \left(\mathrm{e}^{-\left(\alpha\frac{Y}{l_W}+\frac{\alpha}{2}\right)^2} - \mathrm{e}^{-\left(\alpha\frac{Y}{l_W}-\frac{\alpha}{2}\right)^2}\right) dt \overset{\text{def}}{=} \Gamma(Y)dt \end{aligned} \tag{6.6}$$

Figure 6.10 serves to illustrate both the standard error function and the function on the right-hand side of Eq. 6.6. The adjustment factor α in Eq. 6.6 serves to modify the function. To describe lane changing, we express Y as a deviation from the base location Y_{base}, which can be interpreted as the targeted lateral location during the process.

$$\mathrm{dY} = \left(\mathrm{e}^{-\left(\alpha\frac{Y-\mathrm{Y}_{\text{base}}}{l_W}+\frac{\alpha}{2}\right)^2} - \mathrm{e}^{-\left(\alpha\frac{Y-\mathrm{Y}_{\text{base}}}{l_W}-\frac{\alpha}{2}\right)^2}\right) dt \tag{6.7}$$

The representation of the lateral dynamics can be formulated by combining the noise variable Z with the lateral displacement variable Y, expressed as a set of

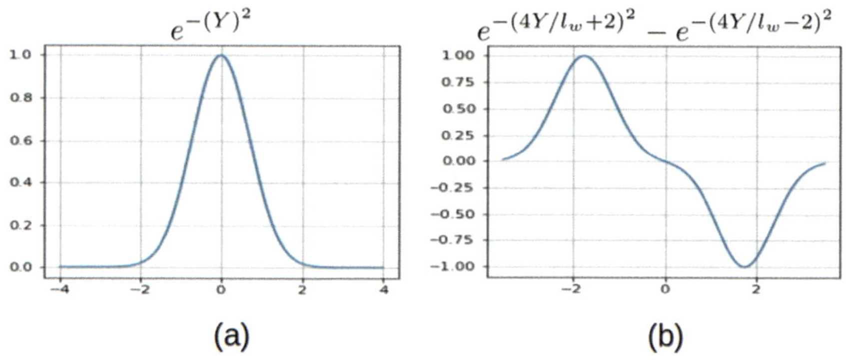

Fig. 6.10 Original standard error function and transformed derivate of error function

stochastic differential equations[1]:

$$\begin{cases} dY = \left(\theta_Y\left(e^{-\left(\alpha\frac{Y-Y_{base}}{l_W}+\frac{\alpha}{2}\right)^2} - e^{-\left(\alpha\frac{Y-Y_{base}}{l_W}-\frac{\alpha}{2}\right)^2}\right) - \theta_Z \cdot Z\right)dt \quad (a) \\ dZ = -\frac{2Z\cdot\sigma^2\left(l_w^2-4Z^2\right)}{2l_w^2}dt + \frac{l_w\sigma\left(l_w^2-4Z^2\right)}{2l_w^2}dW \quad (b) \end{cases} \tag{6.8}$$

The lateral dynamics are influenced by three parameters: σ, θ_Y and θ_Z. The parameter σ denotes the intensity of the noise, which is contingent upon factors such as the road surface condition, vehicle depreciation level, and driver attributes including their driving skills. The second parameter, θ_Y, signifies the responsiveness to lateral deviation, while the third parameter, θ_Z. , indicates the sensitivity to noise Z. These two parameters collectively characterize the influence of both derivation and noise on the lateral speed. To conduct numerical simulations, we employ Euler discretization on Eq. 6.8, resulting in the subsequent equation.

$$\begin{cases} Y_{i+1} = Y_i + \Delta t\theta_Y\left(e^{-\left(\frac{Y_i-Y_{base}}{l_W}+\frac{\alpha}{2}\right)^2} - e^{-\left(\frac{Y_i-Y_{base}}{l_W}-\frac{\alpha}{2}\right)^2}\right) - \Delta t\theta_Z \cdot Z_i \quad (a) \\ Z_{i+1} = Z_i - \frac{2Z_i\cdot\sigma^2\left(l_w^2-4Z_i^2\right)}{2l_w^2}\Delta t + \frac{l_w\sigma\left(l_w^2-4Z_i^2\right)}{2l_w^2}\Delta W \quad (b) \end{cases} \tag{6.9}$$

In Eq. 6.9, the lateral location and noise at step i is represented by Y_i and Z_i respectively, wherein ΔW is a Gaussian variable characterized by zero mean and Δt variance. Simulation results depicting lateral evolution during lane changing when Y_{base} is set to 3.5 m have been presented in Fig. 6.11, wherein the initial lateral position is 0. The blue curves represent the path of Z while the black curves represent the paths of Y. The first column of Fig. 6.11 corresponds to a scenario where σ is set to 0.03, while the second column corresponds to a greater noise level where σ is set to 0.08. The blue lines (path of the noise Z) exhibit a wider divergence in the second column. The value of θ_Y increases from 1 to 1.5 to 2, from the first row to the last row in the figure. The results indicate that higher sensitivity to Y facilitates quicker convergence of the vehicle to the middle line of the target lane. Stochastic lane changing trajectory has been illustrated, while Fig. 6.12 shows the stochastic trajectory for the lane-keeping case based on 50 paths for noise and displacement. The trajectories exhibit scattering features akin to those observed in the real-world case presented in Fig. 6.5. Several examples for lane keeping and lane changing trajectories with different parameters are also given in Figs. 6.13 and 6.14.

[1] Note that in the equation, the derivatives are taken with respect to time t. In reality, when the vehicle speed is zero, the lateral noise and displacement both don't change. Thus alternatively, we can interpret the equation as the derivatives with respect to longitudinal coordinate, x. In this way, the equations are reasonable.

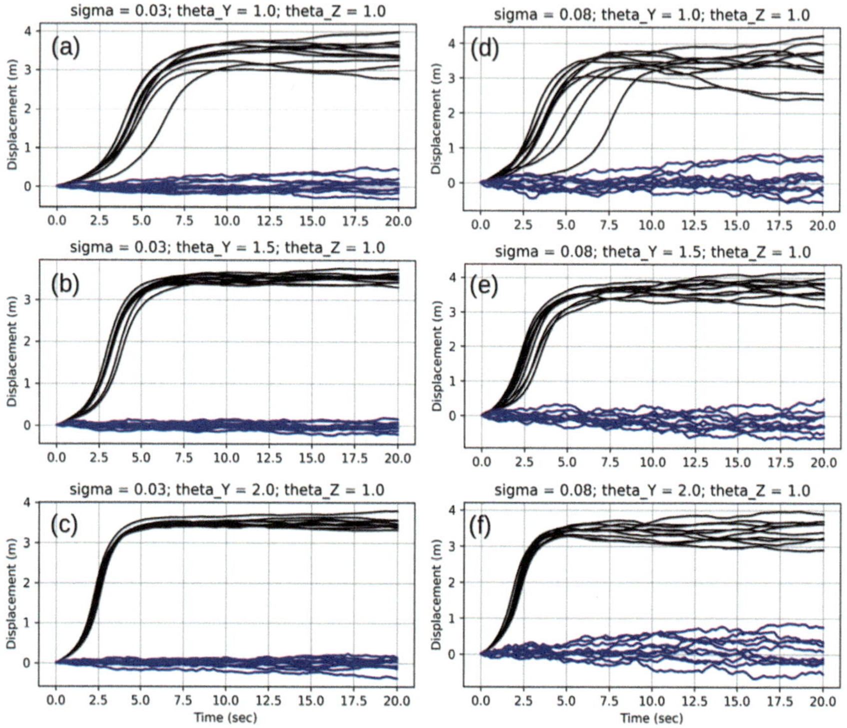

Fig. 6.11 Simulation of lateral noise and displacement during the lane changing scenario. Initial Y and Z are all 0

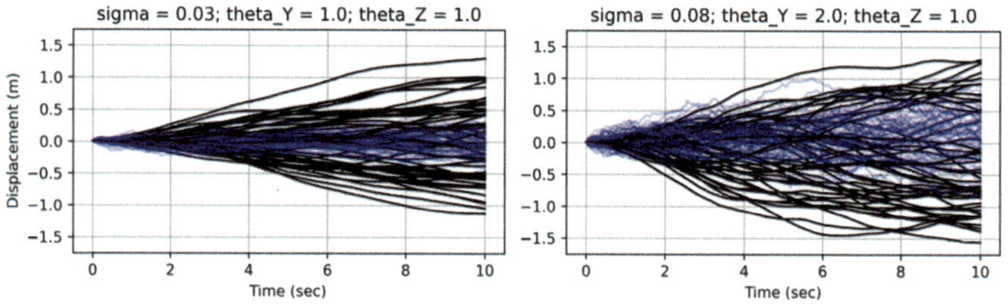

Fig. 6.12 Simulation of lateral noise and displacement during the lane keeping scenario

6.4.5 *Fokker–Planck Equation (FP Equation) of the Lateral Movement*

The solution to a SDEs is a stochastic process. Given a fixed time, a marginal distribution is defined. Hence the solution can be interpreted as an evolved distribution. This distribution is described by the Fokker–planck equation. In order to formulate

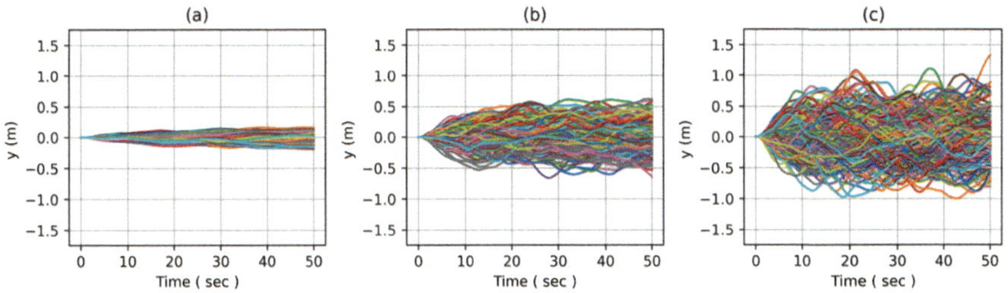

Fig. 6.13 Simulated trajectory for lane keeping with different parameters

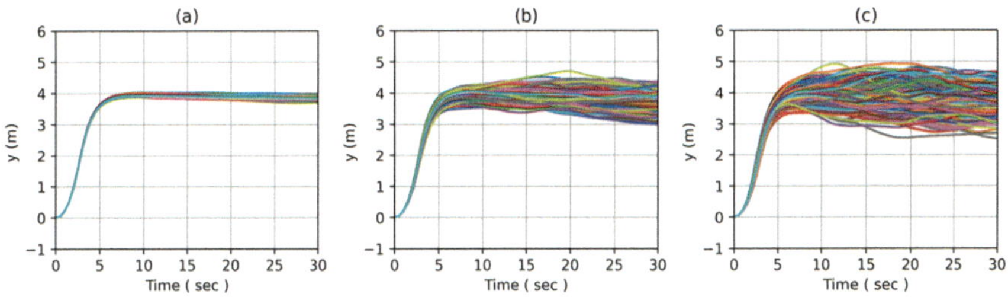

Fig. 6.14 Simulated trajectory for lane changing with different parameters

the distribution of the lateral coordinate Y, i.e., P(Y, t), we first formulate the distribution of the noise P(Z, t), and then obtain the distribution of P(Y, Z, t). Finally, the distribution of P(Y, t) can be calculated numerically by $\int P(Y, Z, t)dZ$.

6.4.5.1 Theoretical Distribution P(Z, t)

Note the that the expression of Z is independent of Y, and we rewrite the dZ as:

$$\mathrm{dZ} = f(Z, t)dt + L(Z, t)dW \tag{6.10}$$

where $f(Z, t) = -\frac{2Z\cdot\sigma^2(l_w^2-4Z^2)}{2l_w^2}$ and $L(Z, t) = \frac{l_w\sigma(l_w^2-4Z^2)}{2l_w^2}$. Then according to the Fokker–Planck-Kolmogorov theorem, the distribution of Z, which is p(Z, t), can solve the following second order PDE:

$$\frac{\partial P(Z, t)}{\partial t} = -\frac{\partial(f(Z, t)\cdot P(Z, t))}{\partial Z} + \frac{1}{2}\left[\frac{\partial^2\left((L(Z, t))^2 P(Z, t)\right)}{\partial Z^2}\right] \tag{6.11}$$

The first term in the RHS of Eq. 6.11 is:

$$\frac{\partial(f(Z, t)\cdot P(Z, t))}{\partial Z} = P(Z, t)\frac{\partial(f(Z, t))}{\partial Z} + f(Z, t)\frac{\partial P(Z, t)}{\partial Z} \tag{6.12}$$

$\frac{\partial(f(Z,t))}{\partial Z}$ in the above formula can be calculated as:

$$\frac{\partial(f(Z,t))}{\partial Z} = \frac{8Z^2\sigma^2}{l_w^2} - \frac{\sigma^2\left(-4Z^2 + l_w^2\right)}{l_w^2} \tag{6.13}$$

Plugin Eq. 6.13 into Eq. 6.12:

$$\begin{aligned}\frac{\partial(f(Z,t)\cdot P(Z,t))}{\partial Z} &= -P(Z,t)\left(\frac{8Z^2\sigma^2}{l_w^2} - \frac{\sigma^2\left(-4Z^2 + l_w^2\right)}{l_w^2}\right)\\ &\quad + \frac{2Z\cdot\sigma^2\left(\mathrm{l}_\mathrm{w}^2 - 4Z^2\right)}{2\mathrm{l}_\mathrm{w}^2}\frac{\partial P(Z,t)}{\partial Z}\end{aligned} \tag{6.14}$$

On the other hand, in order to formulate $\frac{\partial^2\left((L(Z,t))^2 P(Z,t)\right)}{\partial Z^2}$, we first calculate the first partial derivate:

$$\frac{\partial\left((L(Z,t))^2 P(Z,t)\right)}{\partial Z} = P(Z,t)\frac{\partial\left((L(Z,t))^2\right)}{\partial Z} + (L(Z,t))^2\frac{\partial P(Z,t)}{\partial Z} \tag{6.15}$$

Then for the second partial derivate:

$$\begin{aligned}\frac{\partial^2\left((L(Z,t))^2 P(Z,t)\right)}{\partial Z^2} &= \left[P(Z,t)\frac{\partial^2\left((L(Z,t))^2\right)}{\partial Z^2} + \frac{\partial\left((L(Z,t))^2\right)}{\partial Z}\frac{\partial P(Z,t)}{\partial Z}\right]\\ &\quad + \left[\frac{\partial\left((L(Z,t))^2\right)}{\partial Z}\frac{\partial P(Z,t)}{\partial Z} + (L(Z,t))^2\frac{\partial^2 P(Z,t)}{\partial Z^2}\right]\\ &= P(Z,t)\frac{\partial^2\left((L(Z,t))^2\right)}{\partial Z^2} + 2\frac{\partial\left((L(Z,t))^2\right)}{\partial Z}\frac{\partial P(Z,t)}{\partial Z}\\ &\quad + (L(Z,t))^2\frac{\partial^2 P(Z,t)}{\partial Z^2}\end{aligned} \tag{6.16}$$

For the $\frac{\partial\left((L(Z,t))^2\right)}{\partial Z}$ in RHS of Eq. 6.16, we have:

$$\frac{\partial\left((L(Z,t))^2\right)}{\partial Z} = -\frac{4Z\sigma^2\left(-4Z^2 + l_w^2\right)}{l_w^2} \tag{6.17}$$

For the $\frac{\partial^2\left((L(Z,t))^2\right)}{\partial Z^2}$ in RHS of Eq. 6.16, we have:

$$\frac{\partial^2\left((L(Z,t))^2\right)}{\partial Z^2} = \frac{32\cdot Z^2\sigma^2}{l_w^2} - \frac{4\sigma^2\left(-4Z^2 + l_w^2\right)}{l_w^2} \tag{6.18}$$

Plug Eqs. 6.17 and 6.18 into Eq. 6.16 we have:

$$
\begin{aligned}
\frac{\partial^2\left((L(Z,t))^2 P(Z,t)\right)}{\partial Z^2} &= P\left(\frac{32 \cdot Z^2\sigma^2}{l_w^2} - \frac{4\sigma^2\left(-4Z^2 + l_w^2\right)}{l_w^2}\right) \\
&\quad + 2\left(-\frac{4Z\sigma^2\left(-4Z^2 + l_w^2\right)}{l_w^2}\right)\frac{\partial P}{\partial Z} \\
&\quad + \left(\frac{\sigma^2 l_w^2\left(\mathrm{l}_\mathrm{w}^2 - 4Z^2\right)^2}{4 l_w^4}\right)\frac{\partial^2 P}{\partial Z^2}
\end{aligned} \tag{6.19}
$$

Plug Eqs. 6.16, 6.14 into Eq. 6.11, we finally get the distribution of Z as follows (the details are omitted):

$$
\begin{aligned}
\frac{\partial P(Z,t)}{\partial t} &= -P(Z,t)\left(\frac{8Z^2\sigma^2}{l_w^2} - \frac{\sigma^2\left(-4Z^2 + l_w^2\right)}{l_w^2}\right) \\
&\quad + \frac{2Z \cdot \sigma^2\left(\mathrm{l}_\mathrm{w}^2 - 4Z^2\right)}{2\mathrm{l}_\mathrm{w}^2}\frac{\partial P(Z,t)}{\partial Z} \\
&\quad + \frac{1}{2}\left[P(Z,t)\left(\frac{32 \cdot Z^2\sigma^2}{l_w^2} - \frac{4\sigma^2\left(-4Z^2 + l_w^2\right)}{l_w^2}\right)\right. \\
&\quad + 2\left(-\frac{4Z\sigma^2\left(-4Z^2 + l_w^2\right)}{l_w^2}\right)\frac{\partial P(Z,t)}{\partial Z} \\
&\quad \left. + \left(\frac{l_w\sigma\left(\mathrm{l}_\mathrm{w}^2 - 4Z^2\right)}{2\mathrm{l}_\mathrm{w}^2}\right)^2\frac{\partial^2 P(Z,t)}{\partial Z^2}\right]
\end{aligned} \tag{6.20}
$$

The boundary condition for the above PDE can be expressed as follows, if we assume that the initial noise is zero:

$$
\begin{cases} P(0,0) = 1(a) \\ P(Z,0) = 0, \forall Y \neq 0(b) \end{cases} \tag{6.21}
$$

6.4.5.2 Theoretical Distribution of $\mathbf{P(Y, Z, t)}$

For the theoretical distribution of lateral displacement Y, as there is no standalone equation of Y, it must be derived with the noise Z. Equation 6.8 can be re-written as the matrix form in Eq. 6.34, where $dW = [dW_1, dW_2]^T$ and we have:

$$\begin{cases} \mathbb{F} = \begin{bmatrix} F_1 \\ F_2 \end{bmatrix} = \begin{bmatrix} \left(\theta_Y\left(e^{-\left(\frac{\alpha Y}{lw}+\frac{\alpha}{2}\right)^2} - e^{-\left(\frac{\alpha Y}{lw}-\frac{\alpha}{2}\right)^2}\right) - \theta_Z \cdot Z\right) \\ -\frac{2Z\cdot\sigma^2(l_w^2-4Z^2)}{2l_w^2} \end{bmatrix} \\ \mathbb{L} = \begin{bmatrix} \mathcal{L}_{11} & \mathcal{L}_{12} \\ \mathcal{L}_{21} & \mathcal{L}_{22} \end{bmatrix} = \begin{bmatrix} 0 & 0 \\ 0 & \frac{l_w\sigma(l_w^2-4Z^2)}{2l_w^2} \end{bmatrix} \end{cases} \tag{6.22}$$

The covariance of W is $Q = \begin{bmatrix} 1 & 0 \\ 0 & 1 \end{bmatrix}$ and thus

$$\mathbb{L}\mathbb{Q}\mathbb{L}^T = \begin{bmatrix} 0 & 0 \\ 0 & \frac{\sigma^2 l_w^2 (l_w^2-4Z^2)^2}{4l_w^4} \end{bmatrix} \stackrel{\text{def}}{=} \begin{bmatrix} 0 & 0 \\ 0 & L^2 \end{bmatrix} \tag{6.23}$$

According to the Fokker–Planck-Kolmogorov theorem, the distribution of (Y, Z), which is p(Y, Z, t), can solve the following second order PDE:

$$\frac{\partial P}{\partial t} = -\left[\frac{\partial(F_1(Y,Z,t)\cdot P)}{\partial Y} + \frac{\partial(F_2(Y,Z,t)\cdot P)}{\partial Z}\right] + \frac{1}{2}\left[\frac{\partial^2(L^2\cdot P)}{\partial Z^2}\right] \tag{6.24}$$

The second term at RHS of Eq. 6.24 is the same as the second term in RHS of Eq. 6.11, while the term $\frac{\partial(F_2(Z,t)\cdot P)}{\partial Z}$ (second term in the 1st bracket at RHS of Eq. 6.24) is the same as the first term at RHS of Eq. 6.11. Thus we need just to formulate $\frac{\partial(F_1(Z,t)\cdot P)}{\partial Y}$.

$$\begin{aligned} \frac{\partial(F_1(Y,Z,t)\cdot P)}{\partial Y} &= P\frac{\partial(F_1(Y,Z,t))}{\partial Y} + F_1\cdot\frac{\partial P}{\partial Y} \\ &= P\cdot\left(\frac{98Y-49l_w}{l_w^2}e^{-\left(\frac{\alpha Y}{lw}+\frac{\alpha}{2}\right)^2} - \frac{98Y+49l_w}{l_w^2}e^{-\left(\frac{\alpha Y}{lw}-\frac{\alpha}{2}\right)^2}\right) \\ &\quad + \left(\theta_Y\left(e^{-\left(\frac{\alpha Y}{lw}+\frac{\alpha}{2}\right)^2} - e^{-\left(\frac{\alpha Y}{lw}-\frac{\alpha}{2}\right)^2}\right) - \theta_Z\cdot Z\right)\cdot\frac{\partial P}{\partial Y} \end{aligned} \tag{6.25}$$

Substitute the above results into Eq. 6.24, and we have the distribution of P(Y,Z,t):

$$\begin{aligned} \frac{\partial P}{\partial t} = -\Bigg[& P\cdot\left(\frac{98Y-49l_w}{l_w^2}e^{-\left(\frac{\alpha Y}{lw}+\frac{\alpha}{2}\right)^2} - \frac{98Y+49l_w}{l_w^2}e^{-\left(\frac{\alpha Y}{lw}-\frac{\alpha}{2}\right)^2}\right) \\ & + \left(\theta_Y\left(e^{-\left(\frac{\alpha Y}{lw}+\frac{\alpha}{2}\right)^2} - e^{-\left(\frac{\alpha Y}{lw}-\frac{\alpha}{2}\right)^2}\right) - \theta_Z\cdot Z\right)\cdot\frac{\partial P}{\partial Y} \\ & + P\left(\frac{8Z^2\sigma^2}{l_w^2} - \frac{\sigma^2(-4Z^2+l_w^2)}{l_w^2}\right) - \frac{2Z\cdot\sigma^2(l_w^2-4Z^2)}{2l_w^2}\frac{\partial P}{\partial Z}\Bigg] \end{aligned}$$

$$+\frac{1}{2}\left[P\left(\frac{32\cdot Z^2\sigma^2}{l_w^2}-\frac{4\sigma^2(-4Z^2+l_w^2)}{l_w^2}\right)\right.$$
$$\left.+2\left(-\frac{4Z\sigma^2(-4Z^2+l_w^2)}{l_w^2}\right)\frac{\partial P}{\partial Z}+\left(\frac{\sigma^2 l_w^2(\mathrm{l}_\mathrm{w}^2-4Z^2)^2}{4l_w^4}\right)\frac{\partial^2 P}{\partial Z^2}\right] \tag{6.26}$$

The boundary condition of the above PDE is:

$$\begin{cases} \mathrm{P}(0,0,0)=1 & \text{(a)} \\ \mathrm{P}(Y,Z,0)=0, \forall \mathrm{Y}, \mathrm{Z}\neq 0 & \text{(b)} \end{cases} \tag{6.27}$$

6.4.6 LC Duration Model and the CLM Moment Distribution

We present two distributions: the lane changing duration (LC duration) distribution and the cross-lane-mark duration (CLM duration) distribution. A lane changing behavior is considered finished when the lateral displacement of the vehicle reaches lane width (without loss of generality we assume that the lane width is a constant). The cross-lane-mark moment is the moment when the trajectory crosses the lane mark and the duration which is counted from the onset moment of lane changing is named CLM duration. They are defined as follows:

$$\begin{cases} LC\ duration\ distribution: P(Y=l_w,t)=\int P(Y=l_w,Z,t)dZ & (a) \\ CLM\ duration\ distribution: P\left(Y=\frac{l_w}{2},t\right)=\int P\left(Y=\frac{l_w}{2},Z,t\right)dZ & (b) \end{cases} \tag{6.28}$$

Thus in order to formulate the two distributions, the FP equation needs to be solved and integrated. Unfortunately, as the FP equation solution cannot be expressed into analytical form, we need to resort to numerical methods or Monte Carlo method.

6.4.7 Stability Condition

Prior to applying a differential model, such as a car following model (Wang et al. 2020) or a CACC model (Zheng et al. 2015), the stability of the model needs to be examined. Similar to ordinary differential equations (ODEs), the stability of a system described by stochastic differential equations (SDEs) has multiple definitions with corresponding physical implications. In this study, we adopt the stability definition for stochastic systems proposed by Mao (1991). Assuming that the equilibrium point of the system is denoted as (Y_e, Z_e), the system is considered stochastically stable if there exists a positive value of δ, denoted as $\delta = \delta(\mathrm{r}, \xi)$ for any $\mathrm{r} > 0$ and $\xi > 0$, such that for any initial condition [Y(0); Z(0)] belonging to the set S_δ (where S_δ is defined as $S_\delta = \left\{ \left| \begin{matrix} Y - Y_e \\ Z - Z_e \end{matrix} \right| < \delta \right\}$, the following condition holds:

$$p\left\{ \omega : \left| \begin{matrix} Y(t) \\ Z(t) \end{matrix} \right| < \delta \, \forall t \geq 0 \right\} > 1 - \xi \tag{6.29}$$

In order to ensure the stability of the system, it is essential to carefully adjust the parameters within the appropriate domain. The behavior of the equation outlined in Eq. 6.8 is contingent upon three key parameters: θ_Y, θ_Z and σ. To thoroughly investigate the condition required for stochastic stability, the equilibrium points of the system are obtained by setting the drift and diffusion components in Eq. 6.8 to zero. By doing so, the system is analyzed under the assumption of no net flow and absence of random fluctuations, providing valuable insights into its steady-state behavior.

$$\begin{cases} \theta_Y \left(e^{-\left(\frac{\alpha Y}{l_w} + \frac{\alpha}{2}\right)^2} - e^{-\left(\frac{\alpha Y}{l_w} - \frac{\alpha}{2}\right)^2} \right) - \theta_Z \cdot Z = 0 \text{ (a)} \\ \frac{2Z \cdot \sigma^2 (l_w^2 - 4Z^2)}{2l_w^2} = 0; \ \frac{l_w \sigma (l_w^2 - 4Z^2)}{2l_w^2} = 0 \quad \text{(b)} \end{cases} \tag{6.30}$$

The equilibrium point for Z, denoted as $Z_e = \pm\frac{l_w}{2}$, can be readily obtained. On the other hand, the equilibrium points for Y can be numerically determined given the values of the parameters θ_Z and θ_Y. To visualize the behavior of the function $e^{-\left(\frac{\alpha Y}{l_w} + \frac{\alpha}{2}\right)^2} - e^{-\left(\frac{\alpha Y}{l_w} - \frac{\alpha}{2}\right)^2}$, which is relevant to the equilibrium points, Fig. 6.15 presents a contour plot with varying α and Y (with the understanding that Y lies within the range of $\left[-\frac{l_w}{2}, \frac{l_w}{2}\right]$. Denoting the equilibrium point for Y as Y_e, it does not have an explicit representation as it relies on the specific values of θ_Z and θ_Y. The values of Y_e and Z_e can be obtained by solving the equation outlined below.

$$\begin{cases} e^{-\left(\frac{\alpha Y_e}{l_w} + \frac{\alpha}{2}\right)^2} - e^{-\left(\frac{\alpha Y_e}{l_w} - \frac{\alpha}{2}\right)^2} = \frac{\theta_Z}{\theta_Y}\left(\pm\frac{l_w}{2}\right) \text{ (a)} \\ Z_e = \pm\frac{l_w}{2} \quad \text{(b)} \end{cases} \tag{6.31}$$

To guarantee the stability, we resort to the Lyapunov function method. We construct the Lyapunov function as follows:

$$V(Y, Z) = (Y - Y_e)^2 + (Z - Z_e)^2 \tag{6.32}$$

The first and second derivate of $V(Y, Z)$ reads:

$$V_{YZ} = [2(Y - Y_e), 2(Z - Z_e)]; \ V_{YZ,YZ} = \begin{bmatrix} 2 & 0 \\ 0 & 2 \end{bmatrix} \tag{6.33}$$

According to Ito rule, we can derive the expression of $V(Y, Z)$ as a stochastic differential equation. Equation 6.8 can be reorganized into the following type:

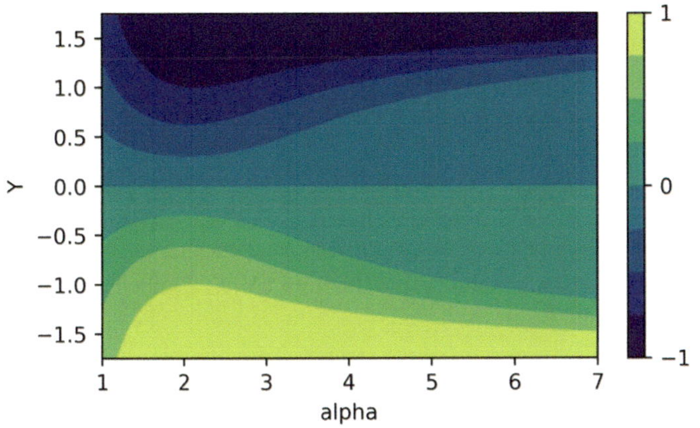

Fig. 6.15 Contour plot of the function $\Gamma(Y)$. Reprint from ref [6], copyright 2023, with permissions from Taylor & Francis

$$\mathrm{d}\begin{bmatrix} Y \\ Z \end{bmatrix} = \mathbb{F}(\mathrm{Z, Y, t})dt + \mathbb{L}(Z, Y, t)d\mathbf{W} = \begin{bmatrix} F_1 \\ F_2 \end{bmatrix} dt + \mathbb{L}(Z, Y, t)d\mathbf{W} \tag{6.34}$$

Then according to Ito rule, we write the equation for V(Y, Z) as:

$$\mathrm{dV}(Y, Z) = \mathrm{LV}(Y, Z) \cdot dt + [2Y_E, 2Z_E] \cdot \mathbb{L} \cdot dW \tag{6.35}$$

where $\mathrm{LV}(Y, Z)$ is obtained via:

$$\begin{aligned} \mathrm{LV}(Y, Z) &= [2(Y - Y_e), 2(Z - Z_e)] \begin{bmatrix} F_1 \\ F_2 \end{bmatrix} + \frac{1}{2} trace\left(\mathbb{L}^{\mathrm{T}} \mathrm{V}_{\mathrm{YZ,YZ}} \mathbb{L}\right) \\ &= [2(Y - Y_e), 2(Z - Z_e)] \begin{bmatrix} F_1 \\ F_2 \end{bmatrix} + \frac{1}{2} trace\left(\mathbb{L}^{\mathrm{T}} \mathrm{V}_{\mathrm{YZ,YZ}} \mathbb{L}\right) \end{aligned} \tag{6.36}$$

The second term in Eq. 6.36 is derived as follows:

$$\begin{aligned} \frac{1}{2} trace\left(\mathbb{L}^{\mathrm{T}} \mathrm{V}_{\mathrm{YZ,YZ}} \mathbb{L}\right) &= \frac{1}{2} trace\left(\begin{bmatrix} 0 & 0 \\ 0 & \frac{l_w \sigma \left(\mathrm{l}_{\mathrm{w}}^2 - 4\mathrm{Z}^2\right)}{2\mathrm{l}_{\mathrm{w}}^2} \end{bmatrix} \begin{bmatrix} 2 & 0 \\ 0 & 2 \end{bmatrix} \begin{bmatrix} 0 & 0 \\ 0 & \frac{l_w \sigma \left(\mathrm{l}_{\mathrm{w}}^2 - 4\mathrm{Z}^2\right)}{2\mathrm{l}_{\mathrm{w}}^2} \end{bmatrix} \right) \\ &= \frac{1}{2} trace\left(\begin{bmatrix} 0 & 0 \\ 0 & \frac{l_w \sigma \left(\mathrm{l}_{\mathrm{w}}^2 - 4\mathrm{Z}^2\right)}{\mathrm{l}_{\mathrm{w}}^2} \end{bmatrix} \begin{bmatrix} 0 & 0 \\ 0 & \frac{l_w \sigma \left(\mathrm{l}_{\mathrm{w}}^2 - 4\mathrm{Z}^2\right)}{2\mathrm{l}_{\mathrm{w}}^2} \end{bmatrix} \right) \\ &= \frac{1}{2} \frac{l_w^2 \sigma^2 \left(\mathrm{l}_{\mathrm{w}}^2 - 4\mathrm{Z}^2\right)^2}{2 \cdot \mathrm{l}_{\mathrm{w}}^4} = \frac{\sigma^2 \left(\mathrm{l}_{\mathrm{w}}^2 - 4\mathrm{Z}^2\right)^2}{4 \cdot \mathrm{l}_{\mathrm{w}}^2} \end{aligned} \tag{6.37}$$

And finally, the function of $\mathrm{LV}(Y, Z)$ is:

$$\begin{aligned}
\mathrm{LV}(Y,Z) &= [2(Y-Y_e), 2(Z-Z_e)]\begin{bmatrix} F_1 \\ F_2 \end{bmatrix} + \frac{\sigma^2\left(\mathrm{l}_\mathrm{w}^2-4\mathrm{Z}^2\right)^2}{4\cdot \mathrm{l}_\mathrm{w}^2} \\
&= 2(Y-Y_e)\left(\theta_\mathrm{Y}\left(\mathrm{e}^{-\left(\frac{\alpha Y}{lw}+\frac{\alpha}{2}\right)^2}-\mathrm{e}^{-\left(\frac{\alpha Y}{lw}-\frac{\alpha}{2}\right)^2}\right)-\theta_Z\cdot Z\right) \\
&\quad -(2\mathrm{Z}-l_w)\frac{\mathrm{Z}\cdot\sigma^2\left(\mathrm{l}_\mathrm{w}^2-4\mathrm{Z}^2\right)}{\mathrm{l}_\mathrm{w}^2}+\frac{\sigma^2\left(\mathrm{l}_\mathrm{w}^2-4\mathrm{Z}^2\right)^2}{4\cdot \mathrm{l}_\mathrm{w}^2}
\end{aligned} \tag{6.38}$$

To guarantee stochastic stability, the function $\mathrm{LV}(Y,Z)$ needs to be non positive for all possible Y and Z, i.e.:

$$\mathrm{LV}(Y,Z)\le 0 \Rightarrow \theta_\mathrm{Y} \le \frac{8\cdot(Y-Y_e)\mathrm{l}_\mathrm{w}^2\cdot\theta_\mathrm{Z}\cdot \mathrm{Z}+4(2\mathrm{Z}-l_w)\mathrm{Z}\cdot\sigma^2\left(\mathrm{l}_\mathrm{w}^2-4\mathrm{Z}^2\right)+\sigma^2\left(\mathrm{l}_\mathrm{w}^2-4\mathrm{Z}^2\right)^2}{\left(\mathrm{e}^{-\left(\frac{\alpha Y}{lw}+\frac{\alpha}{2}\right)^2}-\mathrm{e}^{-\left(\frac{\alpha Y}{lw}-\frac{\alpha}{2}\right)^2}\right)} \tag{6.39}$$

We define the maximum of $\mathrm{LV}(Y,Z|\theta_Y,\theta_Z,\sigma)$ as $\mathcal{K}(\theta_Y,\theta_Z,\sigma)$:

$$\mathcal{K}(\theta_Y,\theta_Z,\sigma)=\max_{Y,Z}\mathrm{LV}(Y,Z|\theta_Y,\theta_Z,\sigma)\le 0 \tag{6.40}$$

Thus to guarantee the stochastic stability, the condition of Eq. 6.40 must be maintained. Equation 6.40 then is the hard constraint that the parameters settings should satisfy.

6.5 Model Calibration, and Validation

6.5.1 Calibration Method

Within Eq. 6.8, the model incorporates several parameters that necessitate calibration through the utilization of real-world data. These parameters include σ, θ_Y and θ_Z, each of which possesses distinct physical implications. Specifically, σ denotes the level of noise, θ_Y reflects the sensitivity to lateral deviation (some drivers might not prioritize precise alignment with the middle lane and thus prefer a smaller value for θ_Y), and θ_Z denotes the sensitivity to lateral noise (a smaller value for θ_Z corresponds to reduced driver sensitivity to uneven road surfaces). The samples for estimation are obtained from the series of lateral displacements represented as $\mathrm{Y}_0, \mathrm{Y}_1, \ldots \mathrm{Y}_\mathrm{n}$ (for simplicity, we discuss the parameter estimation procedure assuming the availability of a single trajectory with a length of n; however, extension to multiple trajectories is a natural progression. Additionally, we assume a constant time interval, Δt, for simplicity, while acknowledging that accommodating variable time intervals is also a natural generalization). Subsequently, the likelihood of the data can be formulated as follows:

$$L(\theta_Y, \theta_Z, \sigma) = p_{\theta_Y,\theta_Z,\sigma}(y_1)\prod_{2}^{N} p_{\theta_Y,\theta_Z,\sigma}(y_i|y_{i-1}) \overset{def}{=} p(y_1)\prod_{2}^{N} p(y_i|y_{i-1}) \tag{6.41}$$

We adopt the Euler discretization technique outlined in Eq. 6.9.When provided with Z_i the subsequent value Z_{i+1} conforms to a Gaussian distribution, characterized by its mean and variance as expressed below:

$$\begin{cases} \mu_{Z_{i+1}} = Z_i - \frac{2Z_i \cdot \sigma^2 \left(l_w^2 - 4Z_i^2\right)}{2l_w^2}\Delta t \\ \sigma^2_{Z_{i+1}} = \left(\frac{l_w \sigma \left(l_w^2 - 4Z_i^2\right)}{2l_w^2}\right)^2 \Delta t \end{cases} \tag{6.42}$$

Consequently, the probability of transitioning to Z_{i+1}, denoted as $p(Z_{i+1}|Z_i)$, can be formulated as follows:

$$p(Z_{i+1}|Z_i) = \frac{1}{\sqrt{2\pi \cdot \left(\frac{l_w \sigma \left(l_w^2 - 4Z_i^2\right)}{2l_w^2}\right)^2 \Delta t}} e^{\left\{-\frac{1}{2} \cdot \frac{Z_{i+1} - \left(Z_i - \frac{2Z_i \cdot \sigma^2 \left(l_w^2 - 4Z_i^2\right)}{2l_w^2}\Delta t\right)}{\left(\frac{l_w \sigma \left(l_w^2 - 4Z_i^2\right)}{2l_w^2}\right)^2 \Delta t}\right\}} \tag{6.43}$$

Conversely, the value of Y_{i+1} relies on both Y_i and Z_i in accordance with Eq. 6.9, wherein Z_i follows a Gaussian distribution given Z_{i-1}. Consequently, Y_{i+1} also conforms to a Gaussian distribution when conditioned on Z_{i-1} and Y_i, characterized by its respective mean and variance:

$$\begin{cases} \mu_{Y_{i+1}} = Y_i + \Delta t \theta_Y \left(e^{-\left(\frac{\alpha Y_i}{l_w} + \frac{\alpha}{2}\right)^2} - e^{-\left(\frac{\alpha Y_i}{l_w} - \frac{\alpha}{2}\right)^2}\right) - \Delta t \theta_Z \cdot \left(Z_{i-1} - \frac{2Z_{i-1} \cdot \mathbb{V}^2 \sigma^2 \left(l_w^2 - 4Z_{i-1}^2\right)}{2l_w^2}\Delta t\right) \\ \sigma^2_{Y_{i+1}} = \left(\frac{l_w \sigma \left(l_w^2 - 4Z_{i-1}^2\right)}{2l_w^2}\right)^2 \Delta t \cdot (\theta_Z \cdot \Delta t)^2 \end{cases} \tag{6.44}$$

We represent the conditional probability $p(Y_{i+1}|Y_i, Z_{i-1})$ as $\mathcal{A}(Y_{i+1}|Y_i, Z_{i-1})$. Subsequently, the computation of the likelihood function can be expressed as follows (Fig. 6.16):

$$\begin{aligned} L(\theta_Y, \theta_Z, \sigma) &= p(Y_1)\prod_{1}^{n} p(Y_{i+1}|Y_i) \\ &= p(y_1) \int\int..\int \mathcal{A}(Y_2|Y_1, Z_0) \cdot \mathcal{A}(Y_3|Y_2, Z_1)..\mathcal{A}(Y_N|Y_{N-1}, Z_{N-2}) dZ_0 dZ_1 \ldots dZ_{N-2} \end{aligned} \tag{6.45}$$

Considering an observation Y of length n, there exist n-1 sources of noise (herein represented by Z_0~Z_{n-2}. The product $\mathcal{A}(Y_2|Y_1, Z_0) \cdot \mathcal{A}(Y_3|Y_2, Z_1)..\mathcal{A}(Y_N|Y_{N-1}, Z_{N-2})$ can be denoted as:

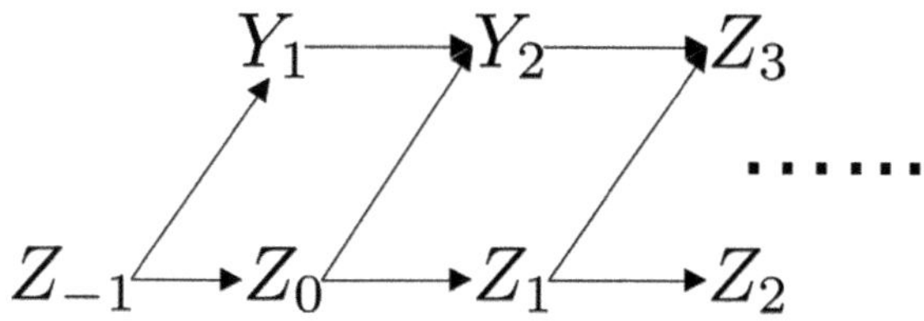

Fig. 6.16 Dependency among variables

$$\mathcal{A}(Y_2|Y_1, Z_0) \cdot \mathcal{A}(Y_3|Y_2, Z_1)..\mathcal{A}(Y_N|Y_{N-1}, Z_{N-2})$$
$$\overset{\text{def}}{=} \mathbb{A}(Z_0, Z_1, ..Z_{N-2}, \ldots Y_2, Y_3, \ldots Y_N) \tag{6.46}$$

Each noise value (denoted as Z_i) exists within the interval of $[-\frac{l_w}{2}, \frac{l_w}{2}]$. The resolution of Eq. 6.45 can be accomplished through the implementation

$$\Theta_Y^*, \theta_Z^*, \sigma = \underset{\theta_Y, \theta_Z, \sigma}{\text{argmin}}\, L(\theta_Y, \theta_Z, \sigma)$$
$$\text{s.t.}\mathcal{K}(\theta_Y, \theta_Z, \sigma) = \max_{Y,Z} LV(Y, Z|\theta_Y, \theta_Z, \sigma) \leq 0 \tag{6.47}$$

The aforementioned optimization problem can be efficiently tackled through the implementation of intelligent optimization methodologies. Among them, PSO (particle swarm optimization) is selected to calibrate the parameters. The stability condition, expressed in Eq. 6.47, can be represented as a penalty term in the objective.

6.5.2 *Validation*

6.5.2.1 Data Source

The study utilizes trajectory data collected from the highly-regarded NGSIM and highD data sets in conjunction with data generated from vehicle simulation tool Carla. To accomplish this, we first extract the lane keeping and lane-changing data and reposition the lateral coordinates of the trajectories relative to a reference lane middle point. Additionally, we implement lateral scaling, applying a scaling factor based on the benchmark lane width of 3.5 m. A total of 12,747 lane keeping data from highD are utilized in our research. However, the NGSIM dataset exhibits considerable noise contamination and relatively lower temporal resolution (at 0.1 s intervals). The transformed lateral trajectories scaled according to this methodology are provided in Fig. 6.18.

6.5.2.2 Validation Results

The present study conducts a comparison of the marginal distribution of lateral location at a particular moment t, denoted as P(Y, t), from the Monte Carlo simulations of the proposed model calibrated by the observed data. Specifically, the effectiveness of the proposed model is demonstrated by comparing the mean and standard deviation of Y. In order to obtain the marginal distribution Monte of the model, Monte Carlo simulations are performed based on an initial lateral coordinate Y(0). An exemplar marginal distribution of both the noise variable Z and the lateral location Y, with initial values of 0, is presented in Fig. 6.19 through histograms. To maintain scaling, the histograms are not normalized, which implies that the integration of the histogram curve is not equal to 1. The simulation results reveal that the lateral coordinate spreads out over time, which aligns with the observation depicted in Fig. 6.5. Notably, at approximately 15 s, the marginal distributions of lateral location almost completely overlap. Figure 6.19c and d are for the case of lane changing. In Fig. 6.19c, the noise of the lane changing process is presented while in Fig. 6.19d, the marginal distribution of the lane changing is presented.

In Fig. 6.20, the simulated data and the real-world data are compared in terms of the initial lateral location $Y(0) \in [-0.2, 0.2]$. This comparison is depicted in Fig. 6.20a~c, which represent the simulation, highD dataset, and NGSIM data, respectively. For the simulation, a total of 2000 paths are generated, with the initial lateral location uniformly distributed in the range of $[-0.2, 0.2]$. The histograms in Fig. 6.20a~c are normalized, ensuring that the integral of each curve equals 1. The observations from Fig. 6.20a~c are as follows: (1) The simulated distributions display regularity and smoothness; (2) The NGSIM data, with a lower resolution of 0.1 s, exhibits more noise, resulting in a wider distribution compared to the simulation and highD data, which exhibit close resemblance; (3) Across all three cases, the marginal distributions almost completely overlap after 30 s, indicating a relatively stable state. To further analyze the results, we modify the initial lateral location to $Y(0) \in [-0.7, -0.5]$ and repeat the simulation. The new comparison is presented in Fig. 6.20d~f, where similar outcomes are observed. It is important to note that due to a limited number of samples for the lateral location interval in the real-world data, the temporal horizon is constrained to 40 s. As a result, the vehicles tend to regress towards the middle line when there is deviation.

6.6 Model Applications

The introduced model effectively characterizes the lateral dynamics of HDVs, making it suitable for implementation in the inference processes of Connected and Automated Vehicles (CAVs). Specifically, the inference procedure of a CAV encompasses several steps. Initially, the CAV detects the presence of nearby vehicles, followed by inferring whether they are operated by human drivers or by automated systems. Given the potential variability in vehicle trajectories, the CAV further assesses whether the detected vehicle executes a lane change maneuver. Finally, the proposed model is employed to analyze the duration of lane changes within this context.

6.6.1 *Identification of the CAV*

Given the proposed model's ability to capture lateral vehicle movements, divergences between CAVs and HDVs in this regard are presumed. Given an observation set $y_{observe} = y_1, \ldots y_N$, two distinct classification outcomes are possible, delineated as $C \overset{def}{=} \{C_{hdv}, C_{CAV}\}$. Leveraging the Bayes classifier, we attempt to differentiate between the two classes.

$$p\left(C_{\#}|y_{observe}\right) = \frac{p(C_{\#})p(y_{observe}|C_{\#})}{p\left(y_{observe}\right)}, C_{\#} \in \{C_{hdv}, C_{CAV}\} \tag{6.48}$$

The classification of the class C^* can be approximated using the expression$C^* = \underset{C_{\#}}{\operatorname{argmax}} p(C_{\#})p\left(y_{observe}\middle|C_{\#}\right)$, where the denominator does not depend on$C_{\#}$. The argmax function consists of two terms: $p(C_{\#})$ represents the prior probability distribution of the vehicle type #, and $p\left(y_{observe}\middle|C_{\#}\right)$ denotes the conditional probability of the observation. Since there is no prior information available, the first term is considered equal for all vehicle classes (HDV or CAV). Therefore, the greater value between the two computed values is selected.

Carla simulations were conducted to obtain lane keeping trajectories, with a portion of the data presented in Fig. 6.21. It should be noted that the lateral domain of the data is encompassed therein. The vehicles' movement is governed by a PID algorithm. A comparison with real-world data (i.e., Fig. 6.17) reveals that the fluctuations in the simulations are insignificant. To evaluate the identification performance, the following procedure was adopted: (1) for each trajectory (whether sourced from Carla simulations or highD data), the trajectories were trimmed to a duration of 1 s, and the logarithm of the probability of the trimmed observation was calculated using CAV parameters and HDV parameters; (2) the vehicle type (HDV or CAV) was determined; (3) the false positive and false negative errors of the results were computed. A false positive refers to situations where the true type is HDV, but an estimate of CAV is provided, while a false negative occurs when the true type is CAV, but it is incorrectly inferred as HDV.

We randomly choose 200 Carla data and 200 highD dataset to validate the inference method. All trajectories are trimmed to 1 s. The false negative and false positive errors are both zero. The ratios of $\frac{\log(observation|\text{cav para})}{\log(\text{observation }|\text{hdv para})}$ are plotted for Carla data and highD data in Fig. 6.22. In Fig. 6.22-left subplot, the ratio for Carla dataset is given. As log of a probability must be negative, the ratio which is smaller than 1 indicate a correct estimation. The maximal one is 0.452. For the highD dataset, a ratio

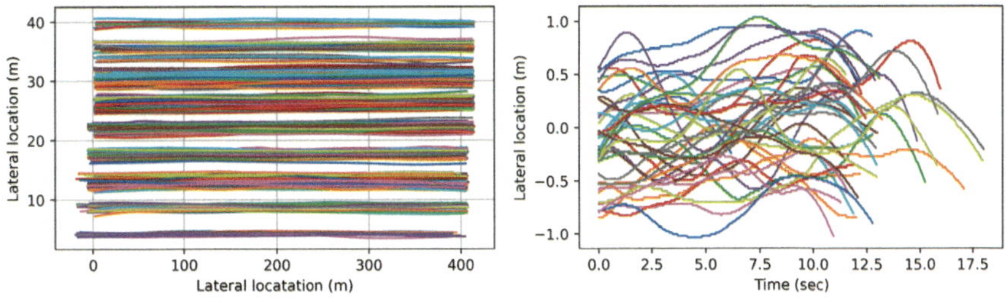

Fig. 6.17 Raw data and translated data; Left: Raw trajectory; Right: translated/scaled data (part). Reprint from Ref. [6], copyright 2023, with permissions from Taylor & Francis

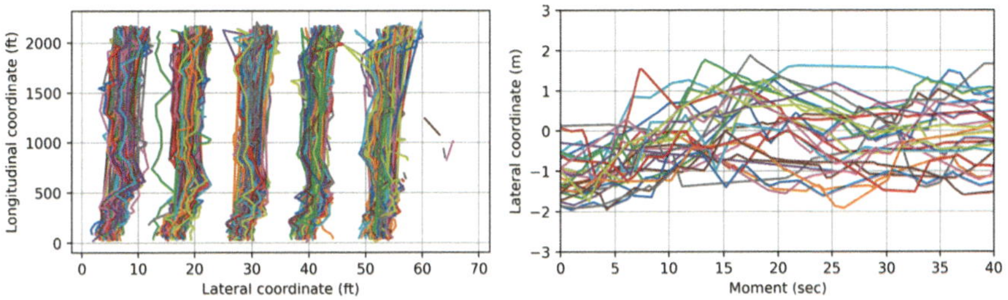

Fig. 6.18 Raw data and translated NGSIM data; Left: Raw trajectory; Right: translated/scaled data (part). Reprint from Ref. [6], copyright 2023, with permissions from Taylor & Francis

which is greater than 1 indicates a correct estimation. The minimal one is 1.618. The estimation on both datasets performs well.

6.6.2 Lane Changing Initiation Moment (LCIE) Identification

When analyzing vehicle behavior based on trajectory data, it is challenging to accurately determine the moment when a vehicle initiates a lane change due to the continuous and stochastic nature of the trajectories. Existing methods for identifying lane change moments rely on either lateral speed measurements or predefined lateral thresholds. However, these methods do not account for the inherent noise present in both the trajectory data and driving behavior, making them less robust. To address this issue, a probabilistic approach is proposed to truncate the trajectory data and accurately identify the lane change initiation event (LCIE). This approach considers the noise in the trajectory data and models the lateral variation before and after a lane change. It distinguishes between three lane change behaviors: lane keeping (no lane change), lane change to the left, and lane change to the right. To capture these behaviors, an additional parameter ρ is introduced in the formula for lateral displacement (dY):

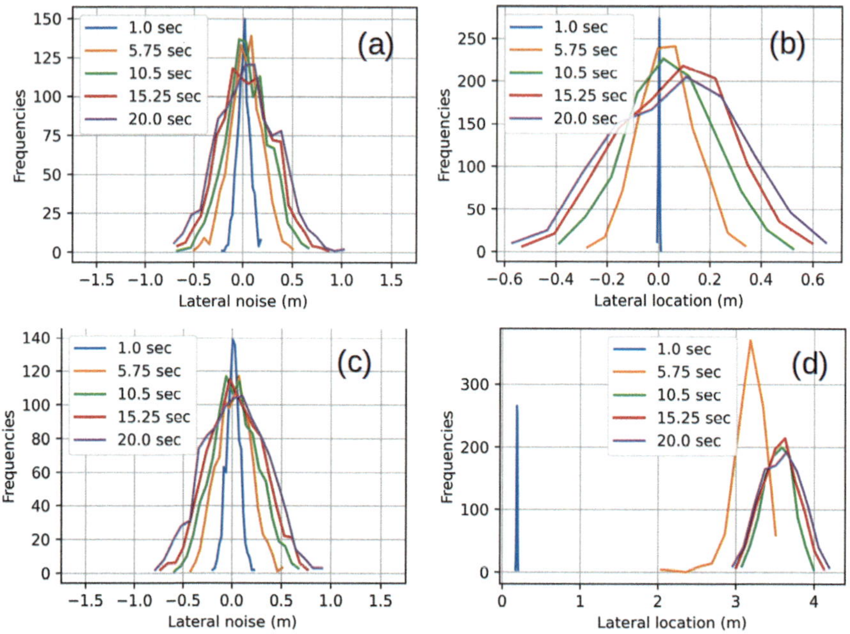

Fig. 6.19 Simulation of lane keeping scenario; **a** noise of lane keeping; **b**: marginal distribution of lane keeping; **c** noise for lc; **d** marginal distribution for lane changing. Reprint from Ref. [6], copyright 2023, with permissions from Taylor & Francis

$$\begin{cases} \mathrm{dY} = \left(\theta_{\mathrm{Y}}\left(\mathrm{e}^{-\left(\alpha\frac{Y-\rho * l_W}{l_W}+\frac{\alpha}{2}\right)^2} - \mathrm{e}^{-\left(\alpha\frac{Y-\rho * l_W}{l_W}-\frac{\alpha}{2}\right)^2}\right) - \theta_Z \cdot Z\right)dt \quad \text{(a)} \\ \mathrm{dZ} = -\frac{2\mathrm{Z}\cdot\mathbb{V}^2\sigma^2\left(\mathrm{l}_{\mathrm{w}}^2-4\mathrm{Z}^2\right)}{2\mathrm{l}_{\mathrm{w}}^2}dt + \frac{l_w\mathbb{V}\sigma\left(\mathrm{l}_{\mathrm{w}}^2-4\mathrm{Z}^2\right)}{2\mathrm{l}_{\mathrm{w}}^2}dW \quad \text{(b)} \end{cases} \tag{6.49}$$

In the context of lane control, the parameter ρ can assume three distinct values: 1, -1, or 0. A value of 0 corresponds to the scenario of lane-keeping, whereas the remaining two values indicate instances of lane-changing, contingent upon the direction of lane transition. Upon the initiation of the lane-changing event, denoted as the onset of the Lane Change Event (LCE) stage, the parameter ρ undergoes an abrupt alteration. Thus, the identification of the Lane Change Initiation Event (LCIE) can be effectively rephrased as the detection of the inflection point in the parameter ρ.

6.6.2.1 Identification Method

We use Euler discretization again and get:

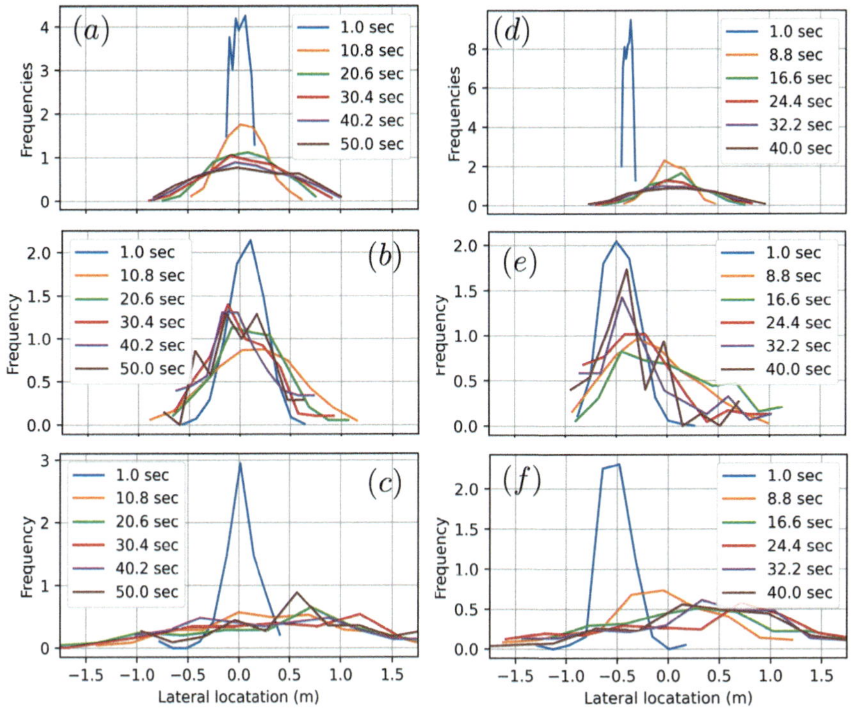

Fig. 6.20 Marginal distribution; **a** and **d** simulation; **b** and **e** highD; **c** and **f** NGSIM; **a** ~ **c**: $Y(0) \in [-0.2, 0.2]$; **d** ~ **f** $Y(0) \in [-0.7, 0 - 0.5]$; $\sigma = 0.05$, $\theta_Y = 0.8$, $\theta_Z = 0.4$. Reprint from Ref. [6], copyright 2023, with permissions from Taylor & Francis

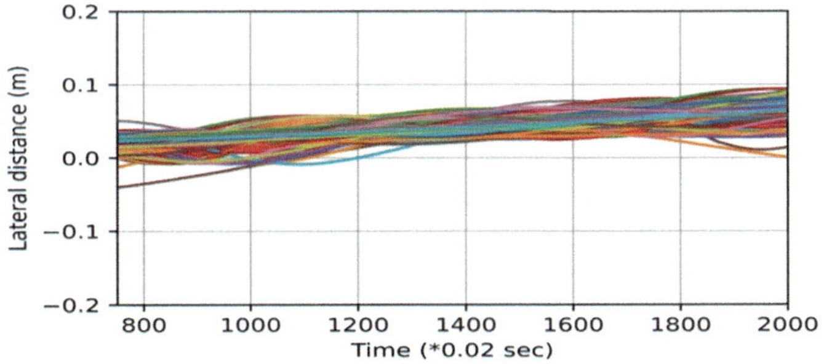

Fig. 6.21 The lateral trajectories of the Carla simulation[2]

[2] Carla Version 0.9.11.

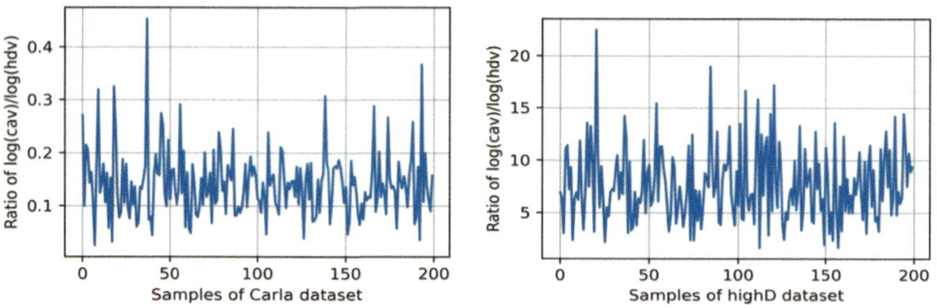

Fig. 6.22 The ratio of the log of two dataset; Left: Carla dataset; Right: highD dataset

$$\begin{cases} \mathrm{Y_{i+1}} = \mathrm{Y_i} + \Delta\,\mathrm{t}\,\theta_\mathrm{Y}\left(\mathrm{e}^{-\left(\alpha\frac{Y_i-\rho*l_{width}}{l_w}+\frac{\alpha}{2}\right)^2} - \mathrm{e}^{-\left(\alpha\frac{Y_i-\rho*l_{width}}{l_w}-\frac{\alpha}{2}\right)^2}\right) - \Delta t\theta_Z\cdot Z_i \quad (a) \\ \mathrm{Z_{i+1}} = Z_i - \frac{2Z_i\cdot\sigma^2\left(\mathrm{l_w^2}-4Z_i^2\right)}{2\mathrm{l_w^2}}\Delta t + \frac{l_w\sigma\left(\mathrm{l_w^2}-4Z_i^2\right)}{2\mathrm{l_w^2}}\Delta W \quad (b) \end{cases} \tag{6.50}$$

The probability distribution of $\mathrm{Z_{i+1}}$ remains Gaussian even in the absence of the parameter ρ, and is characterized by its mean and variance as prescribed in Eq. 6.42. Conversely, the probability distribution of $\mathrm{Y_{i+1}}$ is conditional upon the previous variables $\mathrm{Y_i}$ and $\mathrm{Z_i}$ that are further dependent on $\mathrm{Z_{i-1}}$. Despite the constant variance, the mean of $\mathrm{Y_{i+1}}$ is contingent upon the value of the parameter ρ as stipulated in Eq. 6.50.

$$\begin{cases} \mu_{\mathrm{Y_{i+1}}} = \mathrm{Y_i} + \Delta\,\mathrm{t}\,\theta_\mathrm{Y}\left(\mathrm{e}^{-\left(\alpha\frac{Y_i-\rho*l_{width}}{l_w}+\frac{\alpha}{2}\right)^2} - \mathrm{e}^{-\left(\alpha\frac{Y_i-\rho*l_{width}}{l_w}-\frac{\alpha}{2}\right)^2}\right) - \Delta t\theta_Z\cdot\left(Z_i - \frac{2Z_i\cdot\sigma^2\left(\mathrm{l_w^2}-4Z_i^2\right)}{2\mathrm{l_w^2}}\Delta t\right) \\ \sigma^2_{\mathrm{Y_{i+1}}} = \left(\frac{l_w\sigma\left(\mathrm{l_w^2}-4Z_i^2\right)}{2\mathrm{l_w^2}}\right)^2\Delta t\cdot(\theta_\mathrm{Z}\cdot\Delta t)^2 \end{cases} \tag{6.51}$$

Considering the occurrence of a lane change event at a specific time instant k, the parameter ρ is represented as ρ^- prior to k and ρ^+ subsequent to k. To simplify the analysis, we set $\rho^- = 0$, $\rho^+ \in \{-1, 0, 1\}$, while ρ^+ can take values from the set $\{-1, 0, 1\}$. Here, the value '0' indicates that the vehicle maintains its current lane, while the values '$\pm$ 1' imply a transition to the left or right lane, based on the definition of the Y axis direction. The probability of the observed series of $y_1, \ldots y_N$, given the occurrence of a lane change event at moment k, can be mathematically expressed as:

$$\mathfrak{L}_\mathrm{k} = p_{\rho^-}(\mathrm{y}_1)\prod_{2}^{k} p_{\rho^-}(y_i|y_{i-1})\prod_{k+1}^{N} p_{\rho^+}(y_i|y_{i-1}) \tag{6.52}$$

The optimal initiation moment for a lane change event, denoted as k* or equivalently as the optimal k, can be determined by employing the following method:

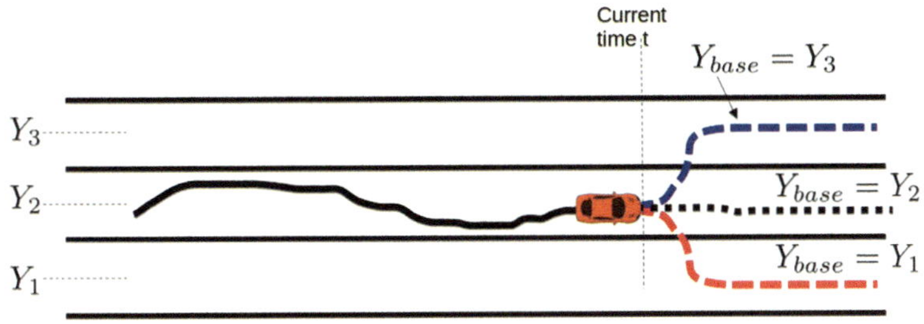

Fig. 6.23 Illustration of the lane changing identification. Reprint from Ref. [6], copyright 2023, with permissions from Taylor & Francis

$$\mathrm{k}^* = \underset{k\in 2\sim N}{\operatorname{argmax}} \mathfrak{L}_\mathrm{k} \tag{6.53}$$

The rationale behind commencing the analysis from moment 2 is that the first term of the expression, denoted as $p_{\rho^-}(\mathrm{y}_1)$, remains constant across different values of k and can be disregarded during the optimization process. This term does not affect comparison among different values of k, yielding the following equation. Every distribution on the right-hand side of the aforementioned equation follows a normal distribution. Regrettably, in order to solve the aforementioned problem, it is necessary to enumerate over all potential values of k (Fig. 6.23).

$$\mathrm{k}^* = \underset{k\in 2\sim N}{\operatorname{argmax}} \prod_{2}^{k} p_{\rho^-}(y_i|y_{i-1}) \prod_{k+1}^{N} p_{\rho^+}(y_i|y_{i-1}) \tag{6.54}$$

The model in Eq. 6.54 is implemented given the whole trajectory. It can be easily extended to the calculation of the probabilities of lane keeping, lane changing to left and lane changing to right during the case of field application when new data arrives in a sequential way. The lane changing state estimation method in field application can be implemented by applying following steps sequentially: (1) obtaining new data (the lateral displacement) and (2) calculate the three possibilities (lane keeping, lane changing to left and lane changing to right) given the new lateral displacement data; (3) select the greatest among them as the inference result. At moment t (suppose that we count the time from 0 to t), the probability of lane changing or keeping (ρ^+, which can be 0, 1 or −1) is calculated as

$$\mathfrak{L}(\rho^+) = p_{\rho^-}(\mathrm{y}_0)\left(\prod_{1}^{t-1} p_{\rho^-}(y_i|y_{i-1})\right) p_{\rho^+}(y_t|y_{t-1}) \tag{6.55}$$

Then we can pick the ρ^+ which lead to the greatest $\mathfrak{L}(\rho^+)$ as the inference result.

6.6.2.2 Example

An illustration of the proposed identification technique is demonstrated via an example. The excerpt is depicted in the uppermost subplot of Fig. 6.24, where the temporal axis measures 0.04 s, resulting in a time horizon of approximately 14 s. The initial lateral location is situated at roughly 1.2 m, and the target lane corresponds to the left lane, which boasts a width of 3.5 m. Figure 6.24 displays the probability distributions of $\rho = 1$ and $\rho = -1$ for a plethora of switching intervals (ρ is assumed to be equal to 0 before the lane switch and 1 or -1 after the lane switch) with respect to the input observation. Notably, the aforementioned trajectory indicates that the vehicle switched to the left lane; thus, the logarithmic probability of $\rho = 1$ ought to surpass that of $\rho = -1$, which is confirmed by both the intermediate subplot and the lower subplot of Fig. 6.24.

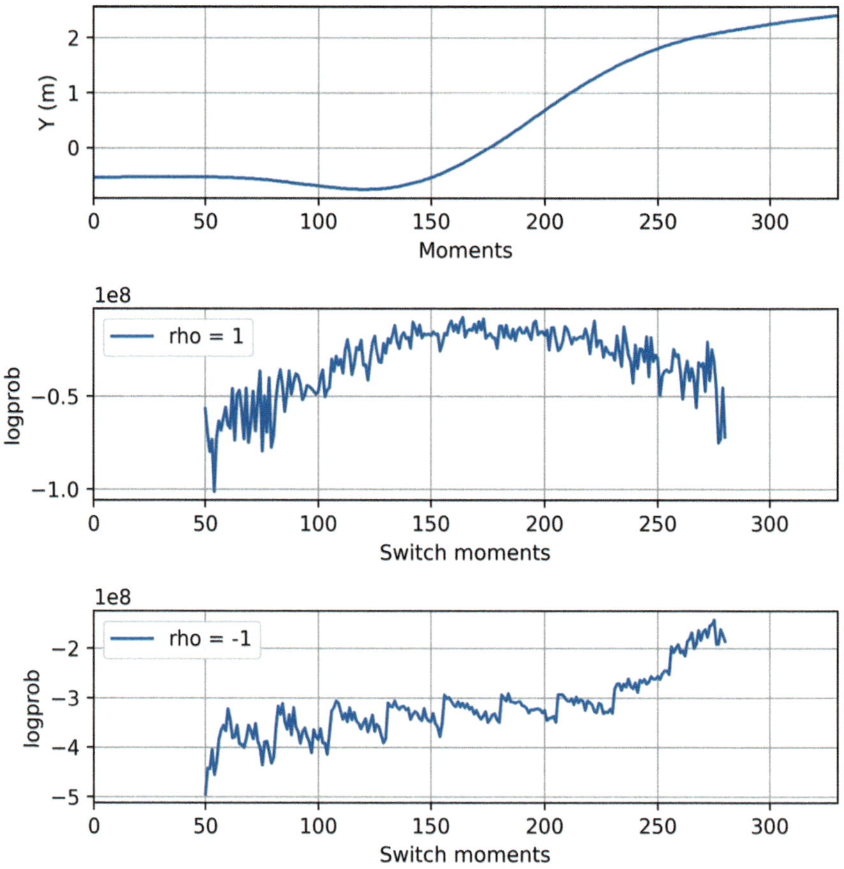

Fig. 6.24 Illustration of the lane changing identification. The Monte Carlo samples points number is 1e5

For the case of $\rho = 1$, it is shown that the log probabilities firstly increase and then decrease. The maximal one (moment 164 in this case) indicates the best match of the assumed switch moment.

The above inference assumes that the lane boundaries are known. There are certain cases when the lane marking is missing, or the lane discipline is not well restricted. As the proposed model only needs the relative lateral difference between current and target lateral condition, theoretically the model should still work when the lane boundary is not clearly defined or marked. To prove this, we translate the observation by a randomly generated lateral distance. The initial Y_{base} of the translated observation is set to the average over certain time horizon (for instance 5 s). Then we apply our model again, by assuming that the new target is $Y_{base} + l_{w'}$, where $l_{w'}$ is not exactly known by should be within certain interval. We set several $l_{w'}$ (2.7m, 3m, 3.3m respectively). The switching moment identification is given in Fig. 6.25.For different $l_{w'}$, the trend of the log probability is the same, and local maxima is very like. The results show that the proposed model is tolerant to the lane definitions or lane width error.

Existing approaches often rely on lateral speed measurements to determine the onset of the lane-change stage. Commonly, this moment is identified when the lateral speed of the vehicle reaches zero. Nevertheless, both the recorded data and the behavior of the vehicle are subject to stochasticity, while different sources contribute to varying levels of sensor noise. Consequently, the lateral speed values result in fluctuating measurements, leading to an unreliable analysis outcome.

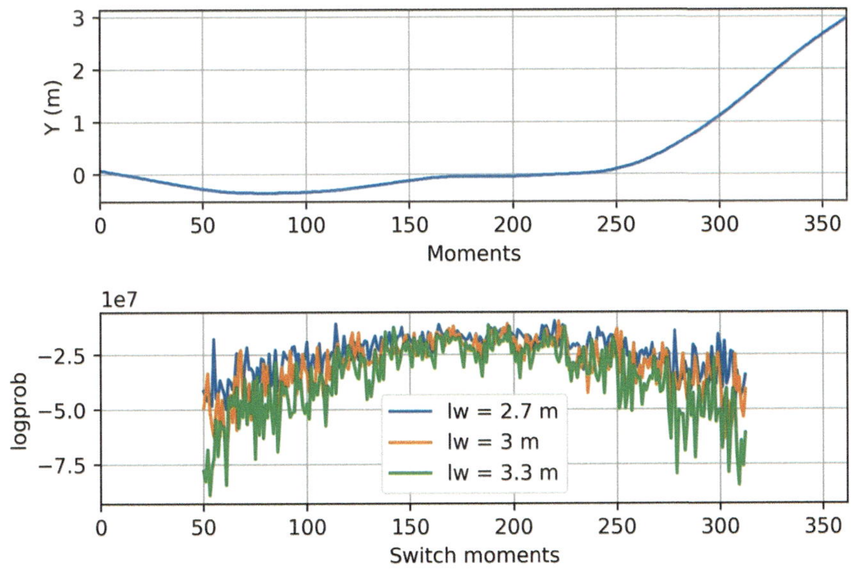

Fig. 6.25 lane changing initiation moment for unknown lane mark. Reprint from Ref. [6], copyright 2023, with permissions from Taylor & Francis

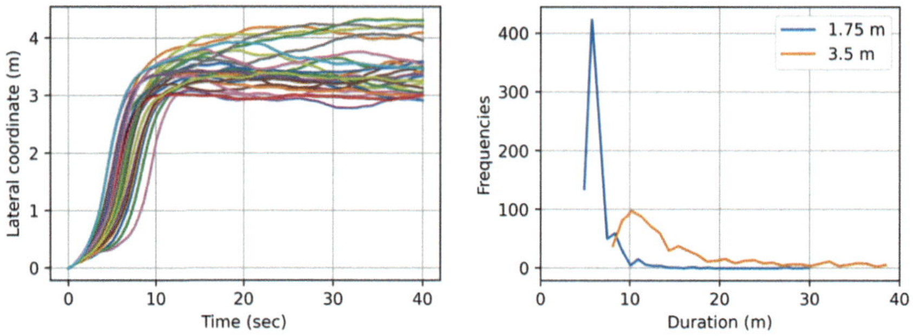

Fig. 6.26 Trajectories for lane changing; Left: lane changing trajectories; Right: Marginal distribution. $\sigma = 0.05, \theta_Y = 0.8, \theta_Z = 0.4$

6.6.3 The Analysis of Lane Changing Duration Distribution (LCD)

The focus in this section is on the calculation of the time taken for a vehicle to change lanes, known as Lane Change Duration (LCD), and the comparison of the obtained distribution with existing literature observations. The simulation is conducted with a fixed initial lateral coordinate within a given range. Figure 6.26 illustrates the trajectories in the left subplot and presents the distributions at the lane mark (1.75 m) and the LCD (3.5 m) in the right subplot. While the distribution pattern aligns with previous studies, the average LCD in the simulation exceeds the empirical observation of 5~6 s. Instead, it reaches an average value of over 10 s. This discrepancy suggests that the calibration results obtained for the lane keeping case cannot be directly applied to lane changing scenarios.

The relationship between vehicle lateral deviation from Y_{base} and Lane Change Duration (LCD) is evident in this analysis. Noise, represented by σ, is assumed to be the same for both lane changing and lane keeping scenarios, thus focusing solely on θ_Y. Simulations are conducted to determine the LCD for different values of θ_Y, and the resulting findings are depicted in Fig. 6.27. Notably, when θ_Y exceeds 6, the LCD averages approximately 4.5 s, which is comparatively shorter than the observed real-world range of 5–6 s. Accordingly, a suitable value for θ_Y in the context of lane changing scenarios is suggested to be around 5.

We further study the LCD variation when the initial lateral location is changed. Three typical locations are chosen and simulated in Fig. 6.28. The three subplots are for the case of $Y(0) = 0.1$m, $Y(0) = 0.5$m and $Y(0) = 1$m, respectively. It is shown that the initial location have certain influence on the LCD. It is expected that a location nearer to the target lane corresponds to a shorter LCD. The simulated histogram of LCD for different initial lateral location is given in Fig. 6.29. In the figure, the lateral location '-1' means that the vehicle change lane when he locates at the lateral coordinate '-1', which means that the lateral distance to the middle line at target lane is 4.5 m ($3.5 - (-1) = 4.5$). From Fig. 6.29, the most probable LCD

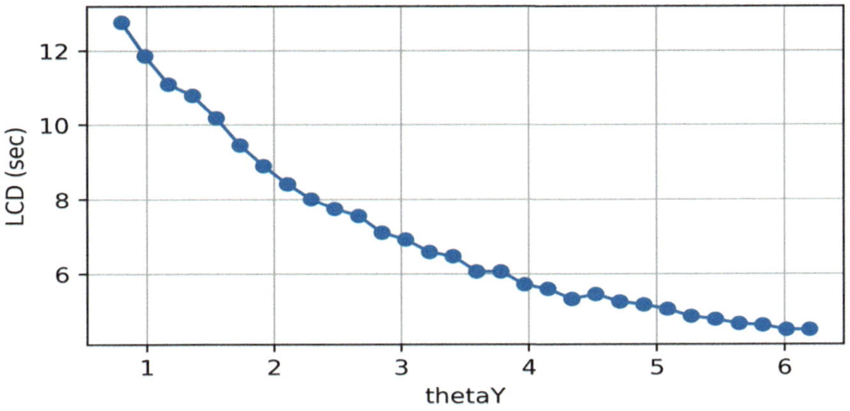

Fig. 6.27 LCD variation with θ_Y

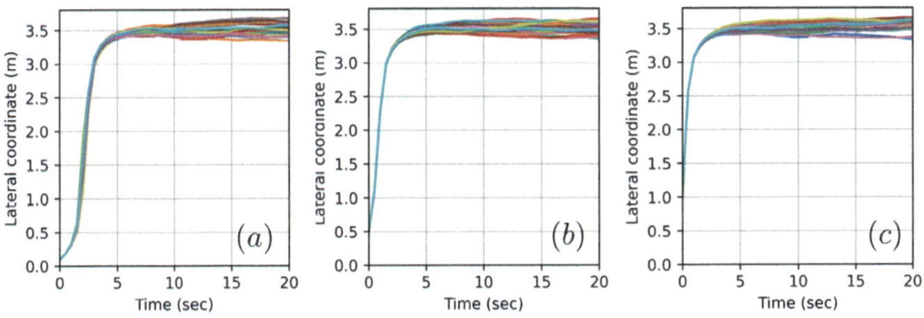

Fig. 6.28 100 lateral trajectories for different initial lateral location; **a**: Y(0) = 0.1; **b**: Y(0) = 0.5; **c**: Y(0) = 1.0

for $Y(0) = -1$ is about 20 s. The most probable LCD for $Y(0) = 0.2$m is about 5 s. While for $Y(0) = 1m$, although there are some samples with smaller LCD (3~4 s), the most probable LCD is about 6 s. Therefore it is concluded that the initial location also contribute to the LCD. The final estimated from lognormal distribution fitting is given in Fig. 6.30. When $Y(0) = 0$, a LCD of 5 s is expected. When vehicle is very near the lane mark at desired direction (initial location = 1 m in Fig. 6.30), a short LCD (about 3 s) is observed. While at opposite direction (initial location = −1 m in Fig. 6.30), a longer LCD (about 14 s) is observed.

6.7 Notes

The chapter is a detailed version of the research published in [7]. To advance the understanding and application of behavior inference for HDVs in the context of CAVs, an investigation into temporal lateral behavior is crucial. Addressing this

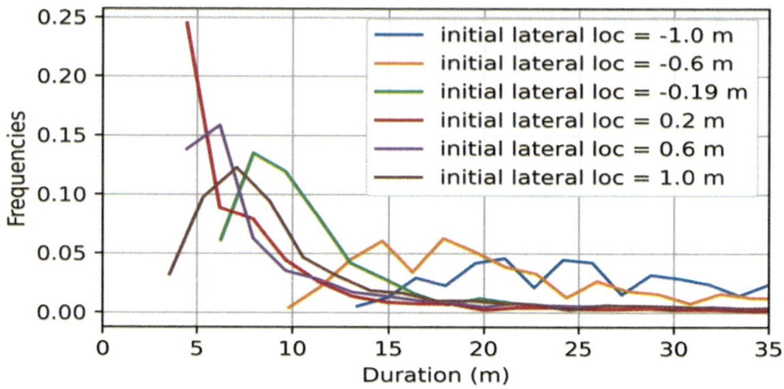

Fig. 6.29 LCD histogram for different initial lateral location. The vehicles are assumed to turn toward positive lateral axis

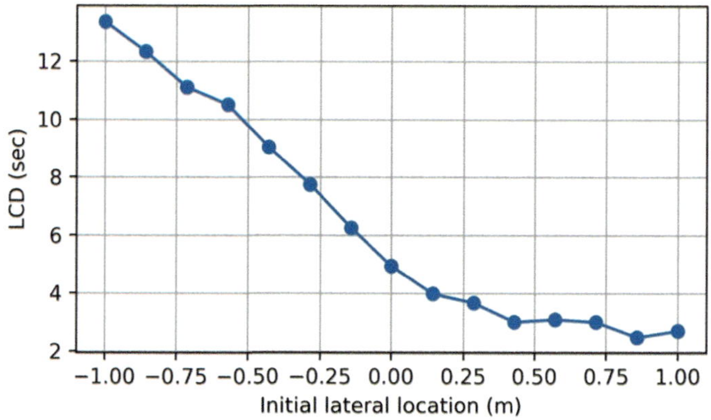

Fig. 6.30 LCD for different initial lateral location

research gap, the lateral movement is modeled utilizing a stochastic differential equation (SDE) framework. A dedicated lateral noise model is developed, and the lateral movement model is formulated by incorporating the lateral noise. Stability conditions are analyzed, and the Fokker–Planck equation, which characterizes the evolution of the marginal distribution, is derived. Real-world data can be utilized to calibrate the model parameters. The model is then applied to multiple applications, yielding insights into producing the marginal distribution of lateral location and facilitating tasks such as LCD modeling, HDV/CAV identification, and LCIE beginning moment identification.

Future research endeavors encompass the following areas:

- The current model framework does not directly capture the dynamics of surrounding vehicles. Recent studies (e.g., [4]) have demonstrated the existence of the side effect of these surrounding vehicles, which can be considered by enhancing the parameters in a functional form (e.g., σ, θ_Y and θ_Z as functions of the states of vehicles in adjacent lanes).

- In scenarios involving complex road layouts (e.g., on–off ramps, work zones), where lane widths change, it is necessary to incorporate the lateral stochastic mechanism.
- The current modeling framework primarily focuses on dY, which can be interpreted as the lateral speed. To further understand the stochastic lateral movement, an additional viewpoint incorporating lateral acceleration or even lateral jerk may be essential. For instance, the lateral dynamics can be encapsulated using $dY = v_y$ and $dv_y = f_{v_y}(Y, v_y, t)dt + L_{v_y}(Y, v_y, t)dW$.
- Exploring the functional form and attributes such as stability conditions will be pursued. The parameter σ in the model represents the total system noise, which can be decomposed as follows: data collection noise σ_{data}, driver behavior noise $\sigma_{behavior}$ (accounting for different driver levels and behavior preferences), road-vehicle interaction noise $\sigma_{road+veh}$ (which considers the response of tire dynamics and pavement influence), and traffic flow environment noise $\sigma_{traffic}$ (characterizing the states of surrounding vehicles and road-side environment). Therefore, theoretically, σ can be expressed as $\sigma = f(\sigma_{data}, \sigma_{behavior}, \sigma_{road+veh}, \sigma_{traffic})$. Accordingly, the comprehensive formulation may be represented as $dY = f(Y) + \sum_{\# \in \{data, behavior, road+vehicle, traffic\}} L(Y|\sigma_{\#})dW_{\#}$. Pursuing these research directions will be the focus of our future work.

References

1. Chen F, Song M, Ma X, Xingyi Z (2019) Assess the impacts of different autonomous trucks' lateral control modes on asphalt pavement performance. Transp Res Part C Emerg Technol 103. https://doi.org/10.1016/j.trc.2019.04.001
2. Chen F, Song M, Ma X (2020) A lateral control scheme of autonomous vehicles considering pavement sustainability. J Clean Prod 256:120669. https://doi.org/10.1016/j.jclepro.2020.120669
3. Delpiano R, Herrera MJC, Coeymans AJE (2015) Characteristics of lateral vehicle interaction. Transp Transp Sci 11:636–647. https://doi.org/10.1080/23249935.2015.1059377
4. Delpiano R, Herrera JC, Laval J, Coeymans JE (2020) A two-dimensional car-following model for two-dimensional traffic flow problems. Transp Res Part C Emerg Technol 114:504–516. https://doi.org/10.1016/j.trc.2020.02.025
5. Papageorgiou M, Mountakis K-S, Karafyllis I et al (2021) Lane-free artificial-fluid concept for vehicular traffic. Proc IEEE 109:114–121. https://doi.org/10.1109/JPROC.2020.3042681
6. Qi H (2023) Modelling the lateral dimension of vehicles movement: a stochastic differential approach with applications. Transportmetrica A: Transp Sci 1–25. https://doi.org/10.1080/23249935.2023.2239375
7. Qi H, Hu X (2023) Behavioral investigation of stochastic lateral wandering patterns in mixed traffic flow. Transp Res Part C: Emerg Technol 155:104310
8. Zhou F, Hu S, Chrysler S et al (2019) Optimization of lateral wandering of automated vehicles to reduce hydroplaning potential and to improve pavement life. Transp Res Rec J Transp Res Board 2673:81–89. https://doi.org/10.1177/0361198119853560

Chapter 7
Two Dimensional Jerk Modeling: Jump-Diffusion Approach

7.1 The Definition of Jerk, Its Importance and Our Approach

The differential of acceleration, known as jerk, is present in both longitudinal and lateral dimensions due to the ability of vehicles to move in these directions. Jerk is often used to indicate driving styles and is related to accidents or crashes. For instance, jerk has been used to identify extreme events or aggressive drivers, along with acceleration, where a critical jerk threshold of approximately 9.8 ms^{-3} is set. [15] suggest that negative jerk increases the likelihood of collision. Several studies have examined the relationship between dimensional behavior and crash performance [14]. Safety critical measurements utilize various indexes or measures, including two-dimensional acceleration, jerk, and time to collision (TTC) or "time to crash". The analysis of oscillation formation and propagation, triggered by driving behaviors such as timid or aggressive driving, lane changing, or bottlenecks, can be conducted through temporal domain and frequency domain approaches. Aggressive, risky, reckless, irregular, and dangerous driving behaviors are strongly associated with crash events [3, 4]. Modeling aggressive driving involves adjusting parameters such as headway and desired speed in simulation packages [6]. Numerous studies have proposed enhancements to conventional car-following models to replicate aggressive driving behavior, including the incorporation of anticipative terms [9] and the integration of data-driven car-following models [10]. Volatility in driving, which encompasses speed variations, can be quantified using various indexes such as standard deviation and variation quintiles [16, 18]. It is worth noting that acceleration alone does not accurately represent aggressiveness, with jerk being considered a better indicator of volatility [14]. Some studies have defined volatility using threshold values, such as the mean plus one standard deviation [17]. Data-driven approaches, such as XGBoost, have been utilized to analyze different types of data sources. The behavior of drivers can be influenced by the presence of autonomous vehicles (AVs) [16]. Unsafe driving behaviors can be categorized and differentiated, influenced by

© The Author(s), under exclusive license to Springer Nature Singapore Pte Ltd. 2024
H. Qi, *Stochastic Two-Dimensional Microscopic Traffic Model*, Lecture Notes in Intelligent Transportation and Infrastructure,
https://doi.org/10.1007/978-981-97-3597-6_7

multiple factors. The jerk during acceleration and deceleration can vary significantly under different scenarios, including crashes, near-crashes, and baseline situations [14]. Heterogeneity among drivers has also been observed [7]. Additionally, thrill-seeking behavior can contribute to unsafe driving. The intrusion of human-driven vehicles (HDVs into the lane of AVs is responsible for approximately 13.11% of AV-related accidents. Lateral and longitudinal acceleration, along with speed, have been used to identify unsafe driving behavior. Simulated approaches have been employed to study aggressive driving behavior [3].

In traffic flow analysis, microscopic behavior plays a central role, and substantial efforts have been devoted to its development in both the longitudinal and lateral dimensions. Empirical observations demonstrate that jerk (the differential of acceleration) has a significant impact on traffic safety, and a speed-dependent jerk profile exists for both longitudinal and lateral movement. Replicating the speed-dependent jerk profile is crucial for analyzing vehicle safety using microscopic models. However, this research reveals that current stochastic microscopic models fail to accurately describe speed-dependent jerks and are inadequate for describing driving behavior with significant jerk profiles. Therefore, this study first derives the jerk distribution for a general stochastic car-following model and subsequently demonstrates that several car-following models, combined with a lateral movement model, fail to generate a realistic jerk distribution. To address this limitation, a compound Poisson formulation is proposed as a remedy. The proposed model consists of a diffusion component and a jump component. The diffusion component describes the stochasticity of normal driving, while the jump component represents driving involving high jerk. Numerical studies demonstrate that the proposed model can successfully replicate the speed-dependent jerk phenomenon. Furthermore, the propagation of this behavior in traffic flow is investigated.

7.2 Can Current Models Describ Realisitc Sped-Dependent Jerk?

In this section, we firstly use the empirical data to investigate the jerk behaviors. Then a theoretical model of the jerk distribution is derived, by applying the Euler–Maruyama scheme and then Markov theory to several mainstream stochastic CF models. The results show that current stochastic models cannot replicate the real work jerk distributions, which motivate us to pursue an appropriate model formulation to describe the longitudinal and lateral jerk exising in real world dataset.

7.2.1 Empirical Observation

We utilize the existing study and the empirical dataset to demonstrate the inherent state-dependent characteristic of jerk. Throughout the process of car following, drivers adapt their behavior (jerk, acceleration, and speed) in order to achieve specific objectives. Each driver exhibits distinct attributes, with some displaying a preference for a moderate acceleration profile, while others opt for sudden vehicle control, leading to a more erratic jerk profile. Moreover, drivers may encounter unique circumstances, such as driving under the influence of distractions, resulting in risky or hazardous driving behavior. Numerous studies have attested to the heterogeneity of driving behavior. In Wang et al. [15], a comprehensive dataset is employed to investigate driving volatility, with the jerk distribution graphically represented in Fig. 5 in their manuscript. Evidently, diversity is prevalent in driving behavior.

The driving behavior exhibits a significant and diverse range with the x-axis representing longitudinal speed and the y-axis denoting jerk. Notably, various driving styles are discernible from the findings. Furthermore, the jerk exhibits a dependence on speed. Specifically, the peak jerk is observed at approximately 10 mph, while at both low and high speeds, the jerk boundary is inconsequential. Essentially, the jerk profiles capture the extremes of physical dynamics during vehicle operation. Subsequently, our investigation leverages a real-world dataset to examine the jerk profiles in both the longitudinal and lateral dimensions. Additionally, we explore the temporal intervals between successive instances where the jerk surpasses or falls below certain thresholds. The dataset is derived from the dajiang, which is a Chinese drone company's release that bears a notable resemblance to the renowned NGSIM dataset. Firstly, we extract the lane-keeping dataset, which pertains to cases where the vehicle does not change lanes. Subsequently, we compute the longitudinal and lateral jerk values. The obtained longitudinal jerk is presented in Fig. 7.1 while the lateral jerk is shown in Fig. 7.2 The x-axis in both figures represents the longitudinal speed, expressed in m/s.

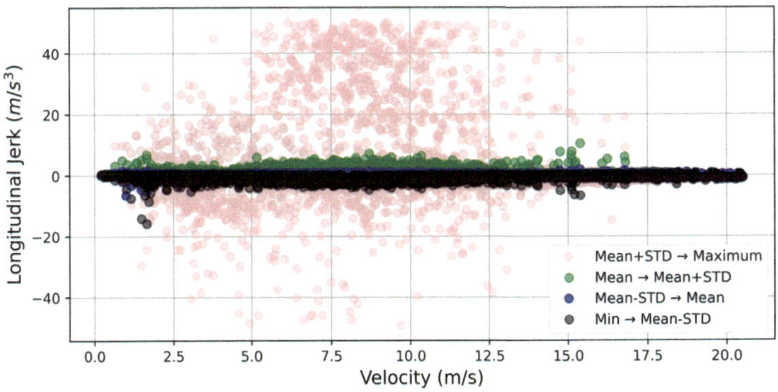

Fig. 7.1 Speed-dependent longitudinal jerk profiles

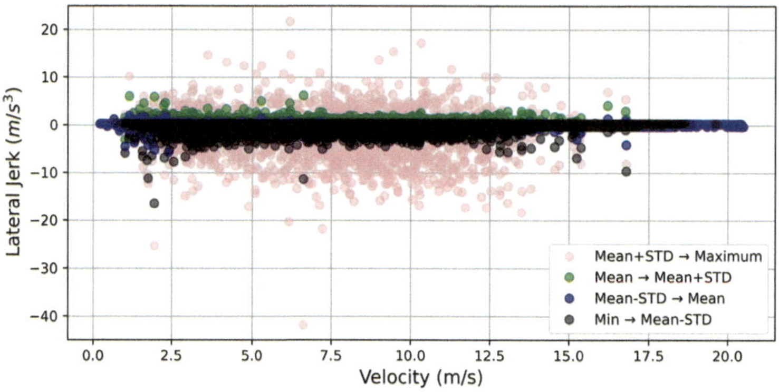

Fig. 7.2 Speed-dependent lateral jerk profiles

Regarding the longitudinal jerk displayed in Fig. 7.1 both the maximum and minimum jerk values are approximately ± 50 ms^{-3}. Similarly, a speed-dependent jerk pattern is observed, with the highest jerk occurring at a longitudinal speed of 7.5 m/s. The considerable magnitude of jerk can be attributed to noise present in the dataset. Turning to the lateral jerk illustrated in Fig. 7.3, the maximal and minimal jerk values are roughly ± 20 ms^{-3}. Analogously, a speed-dependent jerk pattern is discernible. Notably, the longitudinal speed corresponding to the maximal lateral jerk is remarkably similar to that associated with the peak longitudinal jerk.

We also investigate the jerk for longitudinal and lateral dimension based on NGSIM dataset. However, due to the noise exsiting in the dataset, the jerk is very greater than the previous dataset. The results are given in Fig. 7.3 Jerks in both dimensional also display speed-dependent style. Regarding the longitudinal, the maximal jerk occurs at the speed of 40 m/s. We examine the temporal intervals between consecutive instances in which the jerk surpasses predefined threshold values. The distributions of these intervals are illustrated in Fig. 7.4. In particular, Fig. 7.4a

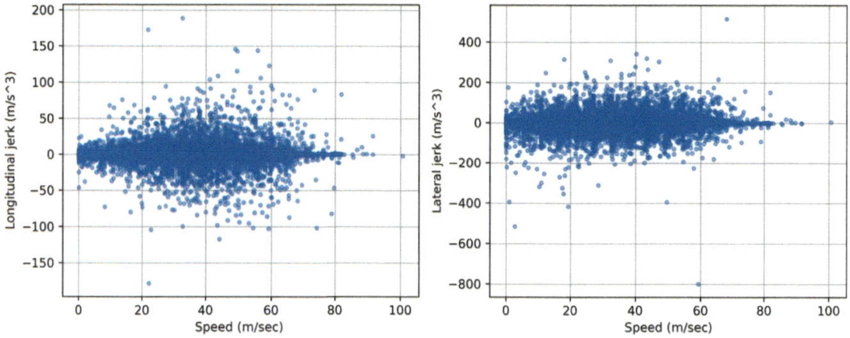

Fig. 7.3 Speed-dependent lateral jerk profiles for NGSIM dataset

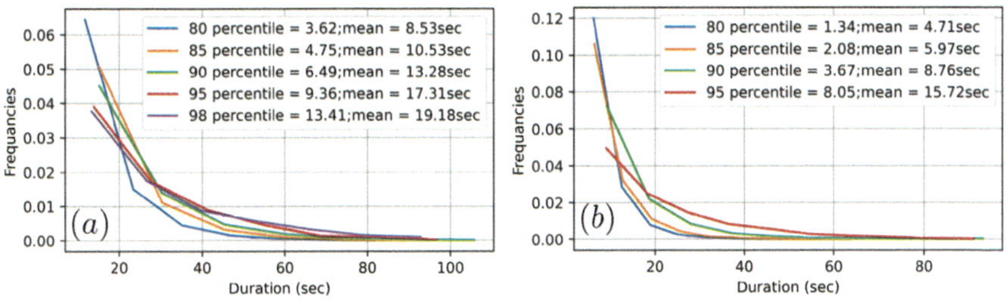

Fig. 7.4 Distributions of duration between neighboring moments when jerk exceeds certain percentile

represents the distribution of durations between successive occurrences of longitudinal jerk exceeding distinct percentiles. Conversely, Fig. 7.4b pertains to the lateral counterpart. The observed distributions of these intervals suggest that they may conform to an exponential law. This characteristic distribution is one of the factors that justify the utilization of the compound Poisson process (CPP) as a model for jerk. In the subsequent section, we demonstrate the inadequacy of existing stochastic car-following models in replicating the speed-dependent jerk patterns discussed earlier. Our approach involves initially deriving the theoretical distribution of jerk and then simulate them via numerical methods.

7.2.2 Theotiretical Deduction of the Longitudinal Jerk Distribution: Markov Chain Approach

The conventional car-following models are developed on the assumption of deterministic normal driving behaviors, which proves to be inadequate for accurately simulating irregular driving behaviors [10]. Hence, efforts have been made to enhance these models by incorporating stochastic elements to account for phenomena like stop-and-go or oscillatory patterns, [5, 13]. However, driving behavior exhibits inherent volatility characteristics, as evidenced by the presence of speed-dependent jerk profiles [15] Consequently, when utilizing a car-following model to simulate scenarios in which the vehicle is pushed towards its physical limits, it becomes crucial to accurately approximate the jerk profiles, given that longitudinal and lateral jerk serve as frequent indicators of anomalous driving behavior. In our study, we undertake an examination of the jerk distributions within a stochastic car-following model, which can be generally characterized by the following set of stochastic differential equations [5, 8, 13]):

$$\begin{cases} \mathrm{dx} = \mathrm{v} \ (\mathrm{a}) \\ \mathrm{dv} = \mathrm{f}(\delta x, \delta v, \mathrm{v}|\mathrm{v_c})\mathrm{dt} + \mathrm{LdW} \quad (\mathrm{b}) \end{cases} \tag{7.1}$$

The desired speed, denoted as v_c, is a known parameter. To acquire the distribution of jerk from the model, the following steps are employed: (1) The Euler–Maruyama scheme is initially implemented on the model, resulting in the construction of the Markov process governing the system. (2) A transition matrix for jerk is subsequently formulated, taking into consideration that jerk can be represented as the second difference of speed. (3) Once the transition matrix for jerk is derived, the distribution of jerk can be computed. Applying the aforementioned steps, we first apply the Euler–Maruyama scheme to Eq. 7.1 in the model.

$$\begin{bmatrix} x(k+1) \\ v(k+1) \end{bmatrix} = \begin{bmatrix} x(k) \\ v(k) \end{bmatrix} + \begin{bmatrix} v(k) \\ f(\delta xk, \delta v(k), v(k)) \end{bmatrix} \delta t + \begin{bmatrix} 0 \\ L \end{bmatrix} \delta w \tag{7.2}$$

δt is the time step, and δW is a normal distributed variable with variance δt. The above discretization scheme results in the following one-step Markov process, i.e. $\mathrm{p}\left(\begin{bmatrix} x(k+1) \\ v(k+1) \end{bmatrix} \middle| \begin{bmatrix} x(k) \\ v(k) \end{bmatrix}\right)$:

$$\mathrm{p}\left(\begin{bmatrix} \mathrm{x(k+1)} \\ \mathrm{v(k+1)} \end{bmatrix} \middle| \begin{bmatrix} \mathrm{x(k)} \\ \mathrm{v(k)} \end{bmatrix}\right) \sim \mathcal{N}\left(\mu_{\left|\begin{bmatrix} \mathrm{x(k)} \\ \mathrm{v(k)} \end{bmatrix}\right.}, \Sigma_{\left|\begin{bmatrix} \mathrm{x(k)} \\ \mathrm{v(k)} \end{bmatrix}\right.}\right) \tag{7.3}$$

The mean and variance are:

$$\begin{cases} \mu_{\left|\begin{bmatrix} x(k) \\ v(k) \end{bmatrix}\right.} = \begin{bmatrix} v(k) \\ f(\delta xk, \delta v(k), v(k)) \end{bmatrix} & (a) \\ \Sigma_{\left|\begin{bmatrix} x(k) \\ v(k) \end{bmatrix}\right.} = \begin{bmatrix} 0 & 0 \\ 0 & L^T L \end{bmatrix} \delta t & (b) \end{cases} \tag{7.4}$$

The transition probability from speed $\mathrm{v}(k)$ at step k to $\mathrm{v}(k+1)$ at step $\mathrm{k}+1$ can be expressed as follows:

$$p(v(k+1)|v(k)) = \sum_{x(k+1),x(k)} \mathrm{p}\left(\begin{bmatrix} x(k+1) \\ v(k+1) \end{bmatrix} \middle| \begin{bmatrix} x(k) \\ v(k) \end{bmatrix}\right) \tag{7.5}$$

The jerk at step k $\omega(k)$ is the differential of acceleration, and it can be calculated from the speed using the following formula:

$$\omega(k) = \frac{a(k) - a(k-1)}{\delta t} = \frac{\frac{v(k)-v(k-1)}{\delta t} - \frac{v(k-1)-v(k-2)}{\delta t}}{\delta t} \\ = \frac{v(k) + v(k-2) - 2v(k-1)}{\delta t^2} \tag{7.6}$$

Hence given $[v(k), v(k-1), v(k-2)]$, the probability that jerk at step $k + 1$ is $\omega(k+1)$ can be calculated as:

$$\mathrm{p}(\omega(\mathrm{k}+1)|\mathrm{v}(\mathrm{k}), \mathrm{v}(\mathrm{k}-1), \mathrm{v}(\mathrm{k}-2)) \\ = \begin{cases} \mathrm{p}(\mathrm{v}(\mathrm{k}+1)|\mathrm{v}(\mathrm{k})), \frac{\mathrm{v}(\mathrm{k}+1)+\mathrm{v}(\mathrm{k}-1)-2\mathrm{v}(\mathrm{k})}{\delta t^2} = \mathrm{w}(\mathrm{k}+1) \\ 0, \mathit{else} \end{cases} \tag{7.7}$$

In the above formulation, we use the Markov property of the speed, i.e.:

$$\mathrm{p}([\mathrm{v}(\mathrm{k}+1), \mathrm{v}(\mathrm{k}), \mathrm{v}(\mathrm{k}-1)]|\mathrm{v}(\mathrm{k}), \mathrm{v}(\mathrm{k}-1), \mathrm{v}(\mathrm{k}-2)) \\ = \mathrm{p}(\mathrm{v}(\mathrm{k}+1)|\mathrm{v}(\mathrm{k})) \tag{7.8}$$

Therefore, the transition probability from $\omega(k)$ to $\omega(k+1)$ is calculated as follows:

$$p(\omega(k+1)|\omega(k)) = \sum_{v(k+1),v(k)} p(v(k+1)|v(k)), \\ \forall \frac{v(k+1) + v(k-1) - 2v(k)}{\delta t^2} = \omega(k+1), \frac{v(k) + v(k-2) - 2v(k-1)}{\delta t^2} \\ = \omega(k) \tag{7.9}$$

The transition matrix for every assignable value of jerk can be denoted using $\mathbb{P}_{\omega \times \omega}$. By obtaining the product of the stable distribution of jerk π_ω with $\mathbb{P}_{\omega \times \omega}$, we are able to ascertain said distribution. It should be noted that the state space must first be discretized prior to engaging in these aforementioned procedures. This established procedure is based on a predefined desired speed value, v_c. By manipulating distinct values of v_c, we can construct graphical representations of the jerk distribution ω.

7.2.3 Results for Several Typical Car Following Models

The following stochasticity-enhanced car following models are considered in this study:

$$\begin{cases} M1: dv = f^{IDM}(\delta x, \delta v, v|v_c)dt + \sigma_{M1}dW & (a) \\ M2: dv = \beta\big(V^{opt}(s) - v\big)dt + \sigma_{M2}\sqrt{v}dW & (b) \\ M3: dv = \alpha_{opt}\big[V^{opt}(s) - v + \lambda s + (2p-1)\alpha V^{opt'}(s)\big]dt + \sigma_{M3}\sqrt{v}dW & (c) \\ M4: dv = f^{IDM}(\delta x, \delta v, v|v_c)dt + \sigma_{M4}\sqrt{v}dW & (d) \end{cases} \tag{7.10}$$

Within the aforementioned set of three models (M1~M4), the Wiener process is denoted as W. Model M1 incorporates a noise component into the IDM model $f^{IDM}(\cdot)$, following the approach of Treiber and Kesting (2018). Meanwhile, model M2, as introduced by [8], expands upon this by incorporating the same stochastic term. M3, which is the car following model devised by [9], accounts for both anticipation time and delay to differentiate between timid and aggressive behavior. This distinction is represented by the parameter p in the model. Finally, M4 is the model put forth in [2].

The present study implements the Euler–Maruyama scheme to analyze the aforementioned four models and establish a Markovian process. In the instance of a typical stochastic differential equation (SDE) of the form $dx = fdt + LdW$, application of the Euler–Maruyama scheme gives rise to $x(k+1) = x(k) + f(k)\delta t + L(k)\Delta W$, where ΔW denotes a zero-mean Gaussian variable with variance equaling δt, a specified time step. The values assigned to the parameters (e.g., σ_{M1}, σ_{M2}, α) in the models are extracted from relevant prior studies. Thus, the Euler–Maruyama scheme for each of the four equations is expressed as follows, in order:

$$\begin{cases} M1: v(k+1) = v(k) + f^{IDM}(\delta x(k), \delta v(k), v(k)|v_c)\delta t + \sigma_{M1}\delta W & (a) \\ M2: v(k+1) = v(k) + \beta\big(V^{opt}(s) - v(k)\big)\delta t + \sigma_{M2}\sqrt{v}\delta W & (b) \\ M3: v(k+1) = v(k) + \alpha_{opt}\Big[V^{opt}(s) - v(k) + \lambda s + (2p-1)\alpha V^{opt'}(s)\Big]dt + \sigma_{M3}\sqrt{v(k)}\delta W & (c) \\ M4: v(k+1) = v(k) + f^{IDM}(\delta x, \delta v, v|v_c)\delta t + \sigma_{M4}\sqrt{v(k)}\delta W & (d) \end{cases} \tag{7.11}$$

The set of four distinctive discretization techniques, alongside the relationship $x(k+1) = x(k) + v(k)\Delta t$, collectively constitute four distinct discrete systems characterized by the structure specified in Eq. 7.2

Each candidate longitudinal speed (v) was assigned to a leading vehicle with the same velocity, and the subsequent vehicle followed one of four driving models (M1~M4) for an extended period spanning 10 min. Figure 7.5 illustrates the experimental outcomes. Specifically, model M1 resulted in stochastic jerks that ranged between $\left[-6\,\text{ms}^{-3}, 6\,\text{ms}^{-3}\right]$. Model M2 produced significantly greater jerks that exceeded 10 ms^{-3}. Although model M3 had a similar profile to M1, it differed in terms of intervals. Despite the inclusion of an aggressive driving model (M3), all tested models failed to generate a realistic jerk distribution, implying that traditional stochastic car following models may be inadequate when describing driving behaviors that exhibit risky maneuvers around speed-jerk boundaries. Additionally, the jerk produced by model M4 was identified as independent of longitudinal speed and inconsistent with empirical evidence.

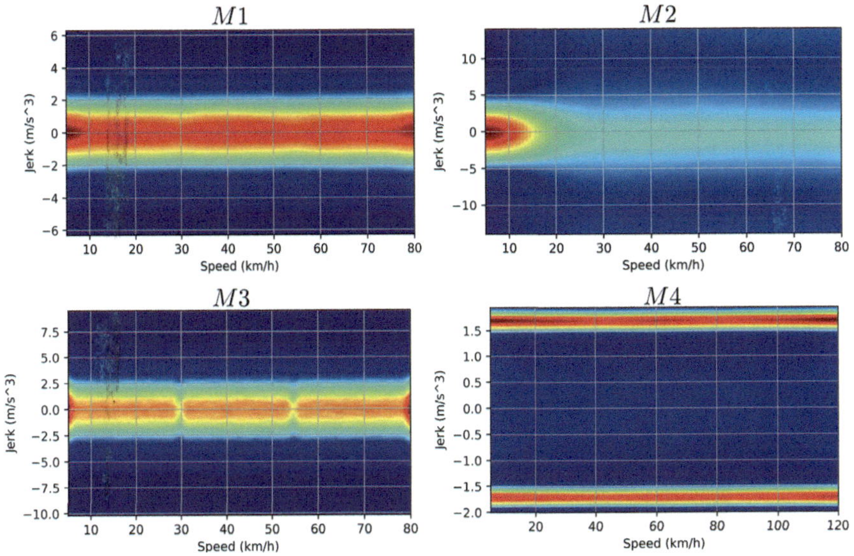

Fig. 7.5 Speed-jerk relationship for M1~M4 respectively

7.2.4 Results for the Lateral Jerks

With regard to the lateral dynamics, extant research indicates that the development of stochastic lateral models has not achieved significant progress. A recent contribution made by Qi [11] includes the creation of a novel stochastic model for lateral movement. Specifically, this model involves the consideration of the lateral coordinate variable, y, and a designated target lateral coordinate, y_{base}, which is typically defined as the lateral midpoint of a given lane. Additionally, the variable z_i^{lat} represents lateral noise. The lateral dynamics can thus be expressed mathematically according to the following equation:

$$\begin{cases} dy_i = \left[\theta_y\left(e^{-\left(\alpha\frac{y-y_{base}}{l_w}+\frac{\alpha}{2}\right)^2} - e^{-\left(\alpha\frac{y-y}{l_w}-\frac{\alpha}{2}\right)^2}\right) - \theta_Z \cdot Z\right]dt \quad (a) \\ dz_i^{lat} = -\frac{2z_i^{lat}\sigma^2\left(l_w^2-4z_i^{lat^2}\right)}{2l_w^2}dt + \frac{l_w\sigma\left(l_w^2-4z_i^{lat^2}\right)}{2l_w^2}dW \quad (b) \end{cases} \tag{7.12}$$

The adjustable parameters in the present study include θ_y, θ_Z, σ, and α. By employing the calibrated values obtained from the manuscript, it becomes possible to simulate the lateral trajectories denoted as $y_i(k)$, where k represents the discrete time steps (k = 0, 1, 2, and so forth). The lateral speed and acceleration can subsequently be calculated as $\frac{y_i(k+1)-y_i(k)}{\delta t}$ and $\frac{y_i(k+2)-2y_i(k+1)+y_i(k)}{(\delta t)^2}$, and finally the jerk can be expressed as follows:

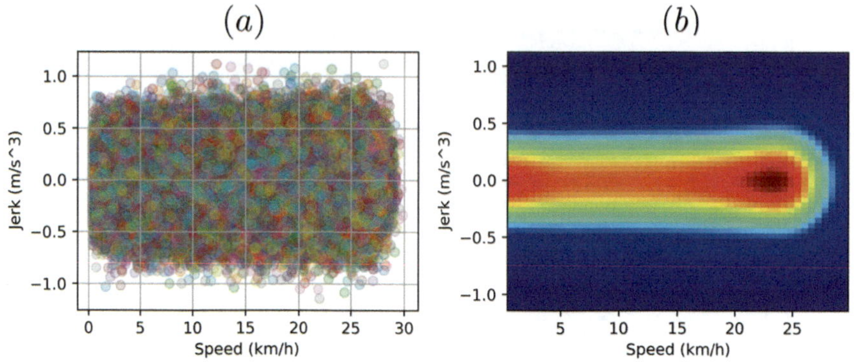

Fig. 7.6 Speed-jerk relationship for stochastic lateral model in Qi 11

$$jerk(k) = \frac{\mathrm{y_i}(k+3) - 3\mathrm{y_i}(k+2) + 3\mathrm{y_i}(k+1) - y_i(k)}{(\delta t)^3} \tag{7.13}$$

Figure 7.6 provides a graphical representation of the association between the lateral jerk and the longitudinal speed. Specifically, Fig. 7.6a is presented as a scatter plot, while Fig. 7.6b is demonstrated as a density heatmap. The outcomes show that the majority of the lateral jerk measurements fall within the confines of the interval ranging from $\left[-1\frac{\mathrm{m}}{\mathrm{s}^3}, 1\frac{\mathrm{m}}{\mathrm{s}^3}\right]$. Moreover, the distributional variance of the jerk profiles exhibits limited alteration concerning various longitudinal jerks. Through comparison with the theoretical results presented in Fig. 7.6 and the empirical observations provided in Fig. 7.2, it is apparent that the theoretical model fails to capture the speed-dependent jerk profiles. Therefore, the development of an enhanced stochastic model that possesses the capability of replicating the appropriate real world jerk is necessary.

7.3 Current Practice Regarding Microscopic Modeling and Jerk Description

Microscopic vehicular modeling is a multifaceted topic within the realm of traffic flow theory, attracting attention across various disciplines such as physics, control theory, and traffic engineering. This modeling approach seeks to establish a connection between microscopic individual behaviors and the macroscopic phenomenon of traffic flow, specifically addressing situations like stop-and-go traffic. As we move towards an era dominated by connected and autonomous vehicles, the significance of microscopic level modeling becomes even more pronounced.

Microscopic movement can be classified into two dimensions: longitudinal and lateral. Numerous stochastic models have been devised to address these two-dimensional movements. With regards to longitudinal movement, the car following model stands out as one of the most critical models for describing behavior when a driver follows another vehicle without any involvement of lane changing. Extensive research efforts have been dedicated to the development of car following models, with notable examples including the intelligent driver model and the Newell car following model. In contrast, the lateral dimension has received relatively less attention. However, due to the emergence of autonomous vehicles, the significance of lateral movement models has grown, leading to the proposal of various such models in recent years.

Empirical observations reveal the heterogeneity of drivers, with variations in driving behavior observed during normal driving, crashes, and near-crash events [14]. The concept of driving volatility [4] is employed to capture the degree of speed variations exhibited by a vehicle during operation [19]. Several volatility indicators have been established, encompassing metrics such as standard deviation, variation quintiles [16] of observations (e.g., speed, acceleration, jerk, and speed recovery time [1]), and threshold-based definitions (e.g., mean plus one standard deviation). Notably, acceleration alone cannot adequately signify aggressiveness [7], whereas jerk proves to be a better indicator of volatility [17]. Some studies employ a threshold-based definition, setting it at mean plus one standard deviation.

The aforementioned research highlights the need to fine-tune car following models to account for specific driving behaviors, including aggression, riskiness, recklessness, irregularity, and dangerous driving, all of which are associated with crash events [3, 4]. Previous studies have employed thresholds of $+0.83$ ms^{-3} and -1.57 ms^{-3} to identify particular driving behaviors. Tailgating and speeding are commonly used indicators for identifying aggressive driving in existing literature. Simulation packages often adjust specific parameters such as headway and desired speed to model aggressive drivers.

Efforts have been made to enhance conventional car following models in order to replicate the aforementioned driving behaviors. For example, the optimal velocity (OV) model was modified in [9] to incorporate an anticipative term, enabling the description of aggressive driving behavior. Additionally, literature [10] estimated the risk effects of distracted driving by integrating a dynamic, data-driven car following model within an algorithmic framework.

Despite these efforts, accurately characterizing specific stochastic car following behaviors, such as risky driving involving significant jerking, remains a challenge. This complexity arises due to the fact that the car following process constitutes a complex stochastic differential system with a wide range of states, spanning from free flow to standstill conditions. Moreover, the model's predicted car following behavior should align with real-world observations. In addition to accurately representing instantaneous jerks, the statistical jerk distribution across the entire speed spectrum should also match empirical observations. We demonstrate that existing stochastic car following models fail to replicate the empirical speed-dependent jerk distributions. Consequently, the development of a new car following model is necessary to accurately replicate risky driving behavior. This model should encompass both the normal stochasticity of car following and the extreme jerk conditions that are dependent on speed.

7.4 A Compound Poisson Approach to Car Following

7.4.1 Notations

For easy reference, the notations are listed:

λ: the density of the jumps of the longitudinal and lateral acceleration;

x, v, y, u, z: longitudinal coordinate, longitudinal speed, lateral coordinate, lateral speed, and noise.

z_i^{lon}: the longitudinal noise for i-th vehicle;

ω: the jerk.

π_ω: stable distribution of the jerk of the Markov chain.

$\mathbb{P}_{\omega\times\omega}$: transition matrix of the jerk.

$x_{LE(i)}$: the longitudinal coordinate of the leader of vehicle i;

C_t: compound Poisson process.

$F_i^{IDM}(\cdot)$, $F_i^{lat}(\cdot)$: the short hand of the function expressing the.

$S = \left[x_i, v_i, y_i, u_i, z_i^{lon}, z_i^{lat}\right]^T$: state variable of vehicle labeled i.

$\mathcal{N}(m; \mu, \sigma)$: normal distribution value at point m, when the mean and std are μ and σ respectively;

$g(v_i)$: function that scale the jump.

$\mathcal{N}(\mu, \sigma)$: normal distribution.

σ_{jump}: the parameter of the compound poisson process.

ξv_c: desired speed, or speed limit of the road.

δt, δx: discretization of the time and space.

$\sigma_{M1} \sim \sigma_{M4}$: parameters of the four selected car following models;

θ_y θ_Z, α: parameters in the lateral model.

F_i^{IDM}: function that represents the intelligent driver model;

K_i: the i-th random number.

ξ,: certain distribution law;

N_t.: a counting process;

θ_{lon} and θ_{lat}: parameters of the jump-diffusion formulation;

Y_{base}: target lateral location.

$\mathbb{F}$, $\mathbb{L}$ and $\mathbb{G}$: matrix in the jump-diffusion equation.

J_n: jump moments;

7.4.2 Longitudinal Component

The aforementioned findings demonstrate the inadequacy of current models in generating realistic jerk values that depend on speed. However, it is imperative to consider the driving behavior that operates at the boundary of jerk, as it plays a critical role in crash and accident analysis due to its potential indication of hazardous driving patterns. Consequently, it becomes necessary to develop a stochastic car following model that incorporates distributions of jerk that are dependent on speed. In this pursuit, we enhance the well-known Intelligent Driver Model by introducing diffusion and jump terms. The former term captures the inherent stochasticity, while the latter term encapsulates the boundary jerk behavior. As for the deterministic component, we opt for the straightforward adoption of the renowned Intelligent Driver Model. Specifically, for a given vehicle (i), its leading vehicle on the same lane is denoted as LE(i). By employing the Intelligent Driver Model and augmenting it with additional diffusion and jump terms, we effectively model both normal driver behavior and risky longitudinal driving behavior.

$$\begin{aligned} dv_i = & \left[\mathrm{F}_\mathrm{i}^\mathrm{IDM}\left(x_i, v_i, x_{LE(i)}, v_{LE(i)}\right)\right. \\ & \left.+z_i^{lon}\right]dt + \mathrm{g}(v_i)\mathrm{dC}_\mathrm{t} \end{aligned} \quad (7.14)$$

Within the preceding formulation, the deterministic element denoted as $\mathrm{F}_\mathrm{i}^\mathrm{IDM}$ serves to embody the essence of intelligent driving behavior. The stochastic noise component, denoted as $\mathrm{z}_\mathrm{i}^\mathrm{lon}$, contributes an element of randomness. Additionally, the function $\mathrm{g}(\mathrm{v}_\mathrm{i})$ introduces adjustable jump magnitudes. Furthermore, the compound Poisson process C_t, can be succinctly expressed using the subsequent equation:

$$\mathrm{C}_\mathrm{t} = C_0 + \sum_{i=1}^{N_t} K_i = C_0 + \int_0^t K_{N_s} dN_s \quad (7.15)$$

The distribution K_i adheres to a common law governed by ξ. In this study, we make the assumption that law follows a normal distribution characterized by a zero mean and a variance denoted as σ_{Jump}, represented by $\mathcal{N}\left(0, \sigma_{\mathrm{jump}}\right)$. Specifically, the value at point $\mathfrak{K}$ is denoted as $\mathcal{N}\left(\mathfrak{K}; 0, \sigma_{\mathrm{jump}}\right)$. Meanwhile, N_s corresponds to a Poisson process featuring a density parameter λ, only exhibiting jumps of magnitude $+ 1$. The value of λ signifies the average time interval at which these jumps occur. Its expression at time t can be defined as:

$$N_t = \sum_{s \geq 1} \mathbb{I}_{[T_s,\infty]}(t), t \geq 0 \tag{7.16}$$

In the above equation, $\mathbb{I}_{[T_s,\infty]}(t)$ is defined via the following expression:

$$\mathbb{I}_{[T_s,\infty]}(t) = \begin{cases} 1, \textit{if} t \geq T_s \\ 0, \textit{if} 0 \leq t < T_s \end{cases} \tag{7.17}$$

The model equationEq. (7.14) encompasses certain indeterminate elements, namely the diffusion term denoted as z_i^{lon} and the adjustment function $g(v_i)$. Numerous alternatives exist for the diffusion term z_i^{lon}, such as Brownian motion or geometric Brownian motion. In this study, we opt for the utilization of the Ornstein–Uhlenbeck process (OU process), also known as the Vasicek model in financial modeling. This choice is primarily motivated by its applicability in microscopic wave propagations. The expression of the OU process is as follows:

$$dz_i^{lon} = \theta_{lon}\left(\mu - z_i^{lon}\right)dt + \sigma_{lon}dW \tag{7.18}$$

The aforementioned OU process is governed by parameters that possess distinct physical interpretations. Specifically, θ represents the reversion speed, while μ signifies the long-run equilibrium value of the process. Furthermore, σ is elucidated as the volatility. In our study, we assign μ a value of zero, resulting in the simplified form of the OU process: $dz_i^{lon} = -\theta z_i^{lon} dt + \sigma dW$. Consequently, the resulting Z exhibits fluctuating behavior centered around zero.

Regarding the function $g(v_i)$, it exerts influence over the jump magnitude observed in the process of v_i. Empirical observations represented in Wang et al. [15] Fig. 5 denote that the jerk boundary is contingent upon the speed of the process. Therefore, we formulate g as a function of speed, denoted as v_i. In order to express the jerk boundary as depicted in Wang et al. [15] Fig. 5, we employ the lognormal distribution. Subsequently, the function g assumes the following formulation (Fig. 7.7):

$$g(v_i) = 150 * \left[\frac{1}{v_i \sigma_g \sqrt{2\pi}} e^{\left(-\frac{\ln(v_i)-\mu}{2\sigma_g^2}\right)}\right] \tag{7.19}$$

Within the aforementioned equation, the parameter σ_g refers to the standard deviation of the underlying normal distribution, while μ denotes the mean Fig. 7.7 showcases the function $g(v_i)$ under various values of σ_g and μ. It is worth noting that through empirical observations displayed in Wang et al. [15] Fig. 5, the maximum jerk is observed at an approximate speed of 10mph. To ensure the maximum jerk approximates the speed-jerk relationship boundary, we introduce the coefficient 150, which scales the function accordingly.

The car following model expressed in (7.14) encompasses three distinct terms: deterministic, diffusion, and jump terms. If we were to eliminate the latter two terms, we would obtain the deterministic intelligent driving model. Similarly, by discarding

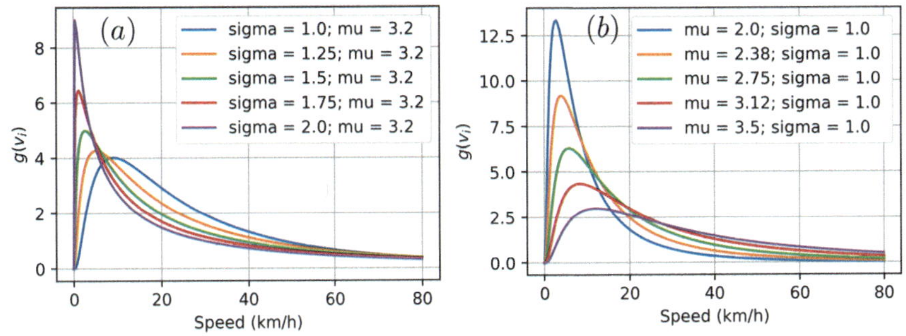

Fig. 7.7 The function $g(v_i)$

the jump term, we arrive at the conventional stochastic car following model. Furthermore, if the diffusion term is omitted, we are left with a deterministic car following model disrupted solely by random jumps occurring at arbitrary moments. Therefore, the formulation presented in Eq. 7.14 represents a comprehensive behavioral model capable of describing the entire spectrum of driving behavior, ranging from common driving scenarios with minimal stochasticity to more risky driving behaviors featuring abrupt changes. In summary, we elucidate the longitudinal driving model as follows:

$$\begin{cases} dx_i = v_i dt & (a) \\ dv_i = \left[\mathrm{F}_{\mathrm{i}}^{\mathrm{IDM}}\left(x_i, v_i, x_{LE(i)}, v_{LE(i)}\right) + z_i^{lon}\right]dt + \mathrm{g}(v_i)\mathrm{dC_t} & (b) \\ dz_i^{lon} = \theta_{lon}\left(\mu - z_i^{lon}\right)dt + \sigma_{lon}dW & (c) \end{cases} \tag{7.20}$$

7.4.3 *Lateral Component*

As for the latter component, the lateral model expressed in Eq. (7.12) emerges as a potential candidate model presently. However, it solely addresses the modeling of lateral speed, rather than lateral acceleration. To incorporate the modeling of lateral acceleration, we differentiate the lateral speed accordingly. Initially, we extract the deterministic component of the lateral speed model stated in Eq. (7.12) in the subsequent manner:

$$\mathrm{dy_i} = \mathrm{u_i} = \theta_y\left(\mathrm{e}^{-\left(\alpha\frac{Y - \mathrm{Y_{base}}}{lw} + \frac{\alpha}{2}\right)^2} - \mathrm{e}^{-\left(\alpha\frac{Y - \mathrm{Y_{base}}}{lw} - \frac{\alpha}{2}\right)^2}\right)dt \tag{7.21}$$

Subsequently, we proceed with the differentiation of the right-hand side of Eq. 7.21, yielding the subsequent expression for lateral acceleration, denoted as $\mathrm{u_i}$.

$$\begin{aligned}\frac{\mathrm{du_i}}{\mathrm{dt}} &= \frac{du_i}{dy_i}\frac{\mathrm{dy_i}}{\mathrm{dt}} \\ &= \theta_y\Bigg(-\frac{2\alpha}{l_w}\left(\alpha\frac{Y-\mathrm{Y_{base}}}{l_w}+\frac{\alpha}{2}\right)\mathrm{e}^{-\left(\alpha\frac{Y-\mathrm{Y_{base}}}{l_w}+\frac{\alpha}{2}\right)^2} \\ &+\frac{2\alpha}{l_w}\left(\alpha\frac{Y-\mathrm{Y_{base}}}{l_w}-\frac{\alpha}{2}\right)\mathrm{e}^{-\left(\alpha\frac{Y-\mathrm{Y_{base}}}{l_w}-\frac{\alpha}{2}\right)^2}\Bigg)u_i \\ &\stackrel{\mathrm{def}}{=} \mathrm{F_i}^{lat}(y_i,u_i)\end{aligned} \tag{7.22}$$

Analogous to the longitudinal component, we incorporate a lateral noise process, denoted as $\mathrm{z_i^{lat}}$, to augment the lateral acceleration. Consequently, the resulting expression for the lateral component is as follows. Additionally, we introduce a jump term, represented as $\mathrm{g_{lat}(v_i)dC_t}$, to the lateral acceleration. The interpretation of this jump term aligns with that of the jump term within the longitudinal component.

$$\begin{cases} dy_i = u_i dt & (a) \\ \mathrm{du_i} = \left[\mathrm{F_i}^{lat}(y_i,u_i)+z_i^{lat}\right]dt+\mathrm{g_{lat}}(v_i)dC_t & (b) \\ dz_i^{lat} = -\theta_{lat}z_i^{lat}dt+\sigma_{lat}dW & (c)\end{cases} \tag{7.23}$$

7.4.4 Synthesis

The microscopic traffic flow model for one-lane traffic flow integrates the longitudinal and lateral components, which can be represented as follows:

$$\begin{cases} dx_i = v_i dt & (a) \\ dv_i = \left[\mathrm{F_i^{IDM}}\left(x_i,v_i,x_{LE(i)},v_{LE(i)}\right)+z_i^{lon}\right]dt+g_{lon}(v_i)dC_t & (b) \\ dy_i = u_i dt & (c) \\ \mathrm{du_i} = \left[\mathrm{F_i}^{lat}(y_i,u_i)+z_i^{lat}\right]dt+\mathrm{g_{lat}}(v_i)dC_t & (d) \\ dz_i^{lon} = -\theta_{lon}z_i^{lon}dt+\sigma_{lon}dW & (e) \\ dz_i^{lat} = -\theta_{\mathrm{lat}}\, z_i^{lat}dt+\sigma_{lat}dW & (f)\end{cases} \tag{7.24}$$

$\mathrm{C_t}^{lon}$ and $\mathrm{C_t}^{lat}$ denote the longitudinal and lateral jumps respectively. To transform the above formulations into matrix form, the following is obtained:

$$\mathrm{d}\left(\begin{bmatrix} x_i \\ \mathrm{v_i} \\ y_i \\ u_i \\ z_i^{lon} \\ z_i^{lat}\end{bmatrix}\right) = \mathbb{F}\left(\begin{bmatrix} x_i \\ \mathrm{v_i} \\ y_i \\ u_i \\ z_i^{lon} \\ z_i^{lat}\end{bmatrix}\right)dt+\mathbb{L}\mathrm{dW}+\mathbb{G}\left(\begin{bmatrix} x_i \\ \mathrm{v_i} \\ y_i \\ u_i \\ z_i^{lon} \\ z_i^{lat}\end{bmatrix}\right)dC_t \tag{7.25}$$

The three vector in the formulation are defined as follows:

$$\begin{cases} \mathbb{F}\left(\begin{bmatrix} x_i \\ \mathrm{v_i} \\ \mathrm{y}_i \\ u_i \\ z_i^{lon} \\ z_i^{lat} \end{bmatrix}\right) = \begin{bmatrix} v_i \\ \mathrm{F_i^{IDM}}\left(x_i, v_i, x_{LE(i)}, v_{LE(i)}\right) + z_i^{\mathrm{lon}} \\ u_i \\ \mathrm{F_i^{lat}}\,(y_i, u_i) + z_i^{\mathrm{lat}} \\ -\theta_{lon} z_i^{lon} \\ -\theta_{lat} z_i^{lat} \end{bmatrix} \overset{\mathrm{def}}{=} \begin{bmatrix} \mathcal{F}_2 \\ \mathcal{F}_3 \\ \mathcal{F}_4 \\ \mathcal{F}_5 \\ \mathcal{F}_6 \end{bmatrix} & (a) \\ \mathbb{L} = [0, 0, 0, 0, \sigma_{\mathrm{lon}}, \sigma_{\mathrm{lat}}]^T \overset{\mathrm{def}}{=} [\mathcal{L}_1, \mathcal{L}_2, \mathcal{L}_3, \mathcal{L}_4, \mathcal{L}_5, \mathcal{L}_6]^T & (b) \\ \mathbb{G}\left(\begin{bmatrix} x_i \\ \mathrm{v_i} \\ \mathrm{y}_i \\ u_i \\ z_i^{lon} \\ z_i^{lat} \end{bmatrix}\right) = \left[0, \mathrm{g_{lon}}\,(v_i), 0, \mathrm{g_{lat}}\,(v_i), 0, 0\right]^T \overset{\mathrm{def}}{=} [\mathcal{G}_1, \mathcal{G}_2, \mathcal{G}_3, \mathcal{G}_4, \mathcal{G}_5, \mathcal{G}_6]^T & (c) \end{cases} \tag{7.26}$$

In this formulation, the behavior of the vehicle state is characterized by the deterministic term, diffusion term, and jump term. Similarly to the car following model in Eq. 7.20 the jump term can be omitted if necessary to describe general driving behavior.

7.5 Numerical Solution

7.5.1 Inifinitesmonial Generator

The infinitesimal generator is a key concept in the analysis of continuous-time stochastic processes, particularly for Markov processes. It characterizes the local behavior of the process at each state in time and is closely related to transition rates or probabilities. In our case, we simplify the state representation as $\mathrm{S(t)} = \left[\mathrm{x_i(t), v_i(t), y_i(t), u_i(t), z_i^{lon}(t), z_i^{lat}(t)}\right]^{\mathrm{T}}$, and p(S) represents the probability of the vehicle state. The state equation is given by either Eqs. 7.24 or 7.25. With this, the infinitesimal generator A for the jump-diffusion type equation can be expressed as follows:

$$\begin{aligned} \mathcal{A}\mathrm{p(S)} &= \sum_{\mathrm{j}} \frac{\partial \mathrm{p}}{\partial \mathrm{S_j}} \mathcal{F}_{\mathrm{j}} + \frac{1}{2} \sum_{\mathrm{j}} \sum_{\mathrm{k}} \frac{\partial^2 \mathrm{p}}{\partial \mathrm{S_j} \partial \mathrm{S_k}} \left(\mathbb{L}\mathbb{L}^{\mathrm{T}}\right)_{\mathrm{jk}} \\ &+ \lambda \int_{-\infty}^{+\infty} \left[p(S + \mathbb{G}(\mathrm{S}) \cdot \mathrm{m}) - p(S)\right] \mathcal{N}(\mathrm{m}; 0, 1)\mathrm{dm} \end{aligned} \tag{7.27}$$

The formulation above introduces $\mathcal{F}_j$ as defined in Eq. 7.26. Additionally, $\left(\mathbb{L}\mathbb{L}^T\right)_{jk}$ denotes the value of the element in the j-th row and k-th column of the $\mathbb{L}\mathbb{L}^T$ matrix. This value can be computed using the following calculation:

$$\mathbb{L}\mathbb{L}^T = \begin{bmatrix} 0_{4\times4} & 0_{4\times2} \\ 0_{2\times4} & \begin{matrix} \sigma_{lon}^2 & 0 \\ 0 & \sigma_{lat}^2 \end{matrix} \end{bmatrix} \tag{7.28}$$

The term in the last integrant in (7.27) is expressed as follows:

$$S + \mathbb{G}(S) \cdot m = \left[x_i, v_i + m \cdot g_{\text{lon}}(v_i), y_i, u_i + m \cdot g_{\text{lat}}(v_i), z_i^{lon}, z_i^{lat}\right]^T \tag{7.29}$$

Substitute Eq. 7.28 into Eq. 7.27, the operator can be reformulated as:

$$\begin{aligned} \mathcal{A}p(S) = & \sum_j \frac{\partial p}{\partial S_j}\mathcal{F}_j + \frac{1}{2}\left[\sigma_{lon}^2 \frac{\partial^2 p}{\partial z_i^{lon}\partial z_i^{lon}} + \sigma_{lat}^2 \frac{\partial^2 p}{\partial z_i^{lat}\partial z_i^{lat}}\right] \\ & + \lambda \int_{-\infty}^{+\infty} \left[p(S + \mathbb{G}(S) \cdot m) - p(S)\right]\mathcal{N}(m; 0, 1)dm \end{aligned} \tag{7.30}$$

7.5.2 Numerical Scheme

The jump-diffuion type stochastic differential equation is notorious difficult to solve. Only in special condition, the analysitcal solution exist. Thus to investiage the behavior of the model, the numerical treatment is necessary. On the other hand, unlike the diffusion model, the jump times (which is indicated by $\{E_n\}, n \geq 1$) is irregular, and may not coincide with the time grid (i.e.$\{nh\} \cup T$, n = 1,2,3..., where T is the time horizon and h is the time step) of common Euler scheme, thus a numerical scheme is developed to deal with this issue. We define a linear process $A_t^h = A_{\tau_j}^h + \Gamma\left(v_{\tau_j}^h\right)(t - \tau_j)$, $\tau_j \leq t < \tau_j$, where τ_j is either the j-th jump moment or the regular time grid (i.e. $\tau_j \in \{nh\} \cup T$). The prodecure is given in the following Table 7.1

7.6 Case Study

In this section, we investigate the behavior of the model using two specific scenarios. The first scenario involves studying the acceleration of a vehicle from a standstill state. In the second scenario, we examine the propagation of vehicle dynamics within the traffic flow.

Table 7.1 Pseudo code of the adapted Eueler scheme

Algorithm	Jump-diffusion discretization scheme Let $\{N_i\}_{i\geq 1}$ and $\{E_n\}_{n\geq 1}$ be sequences of i.i.d. standard normal vectors and Poisson arrival times; let $\{K_n^h\}_{n\geq 1}$ be an i.i.d. sequence of random variables. Initialize $h > n = i = l = 0, \tau_0 = 0, v^h(0) = v(0)$
	Set $\tau_{j+1} = \tau_j + \frac{E_{n+1} - A_{\tau_j}^h}{\Gamma(\cdot)} \cup (l+1)h \cup T$
	Set $v^*(\tau_{j+1}) = v(\tau_j) + (F^{IDM}(\cdot) + \Delta z_{lon})h$
	If $\tau_{j+1} \in (l+1)h \cup T$, then $v(\tau_{j+1}) \leftarrow v^*(\tau_{j+1})$, and $l \leftarrow l+1$
	Else $v(\tau_{j+1}) = v^*(\tau_{j+1}) + g(u_i)K_{\tau_{j+1}}^h$, $n \leftarrow n+1$
	If $\tau_{j+1} = T$, return $v(\tau_{j+1})$

For the first scenario, we analyze the case in which the vehicle accelerates from zero speed. The free flow speed is fixed at 120 km/h or 33.3 m/s. We compare three distinct sets of trajectories: one with the diffusion term only (as defined in Eq. 7.11), one with the jump term only, and one with both terms included. These three cases are depicted in Figs. 7.8, 7.9 and 7.10 respectively. Each figure consists of four subplots representing the longitudinal location, speed profiles, accelerations, and jerks, respectively. In all three cases, the number of simulated paths is 100. Figure 7.8 illustrates the scenario in which both the diffusion and jump terms are considered. The acceleration profile (lower left subplot) displays a combination of smooth variations and sudden jumps. The jerk distribution demonstrates a narrow span at both low and high speeds, but widens when the speed approaches 10 km/h (around 10 s), consistent with the empirical observation highlighted in Wang et al. [15] Fig. 5. Figure 7.9 excludes the diffusion term, with the stochastic source solely originating from the Poisson process. Consequently, the acceleration profile in the lower left subplot exhibits significant volatility due to the jump behavior. However, the increased occurrence of higher jerk values observed in Fig. 7.9 compared to Fig. 7.8 may not accurately reflect the behavior of regular drivers, making this form of the stochastic system less suitable for describing common driving behavior. In Fig. 7.10 only the diffusion term is taken into consideration. As the diffusion term remains constant, the amplitude of jerk becomes independent of the speed. As a result, the resulting jerk profile does not align with empirical observations and may not offer a realistic representation when extreme jerk values are involved (Fig. 7.11).

Subsequently, an analysis is conducted on the interaction between risky drivers and heterogeneous drivers within the traffic flow. A road of considerable length is established, and traffic flow is simulated using periodic boundary conditions. A subset of vehicles is randomly designated as risky vehicles, characterized by a jump term, while the remaining vehicles are categorized as regular vehicles, characterized by solely a diffusion term. The road density is fixed at 80 vehicles per kilometer.

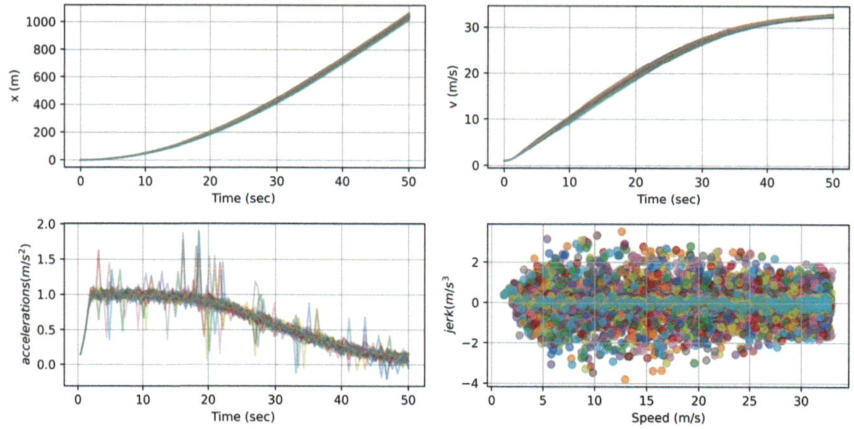

Fig. 7.8 With both diffusion and jump term

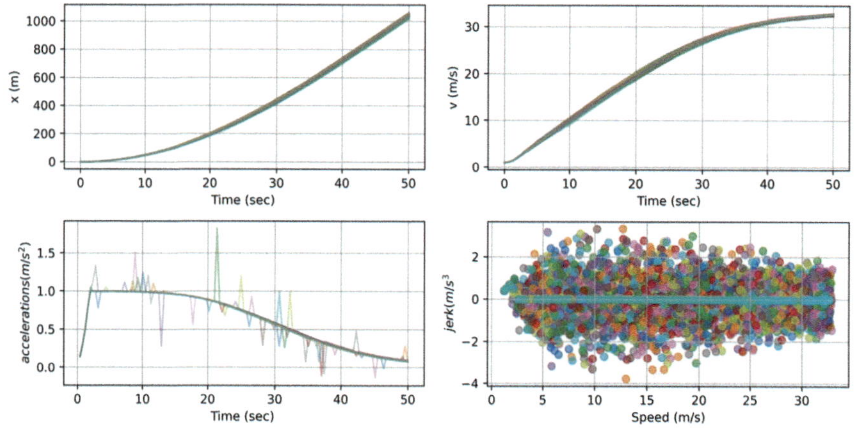

Fig. 7.9 No diffusion term but with jump term

The outcomes are depicted in Fig. 7.12. The behavior of the risky vehicles is contingent on the jump size distribution parameter, σ_{jump}. For $\sigma_{\text{jump}} = 0$, the behavior of the vehicle aligns with that of regular drivers. The highlighted trajectories in the left subplot of Fig. 7.12 are presented merely for comparative purposes. The right subplot of Fig. 7.12 effectively demonstrates the influence of jerk. The substantial jerk necessitates larger gaps to accommodate for its effects, resulting in pronounced heterogeneity within the traffic flow dynamics.

The presence of vehicles exhibiting jump behavior can affect headway, possibly leading to an alteration in capacity. In order to investigate this influence, simulations are conducted to generate a fundamental diagram and subsequently calculate the capacity, representing the maximum flow rate exhibited by the fundamental diagram.

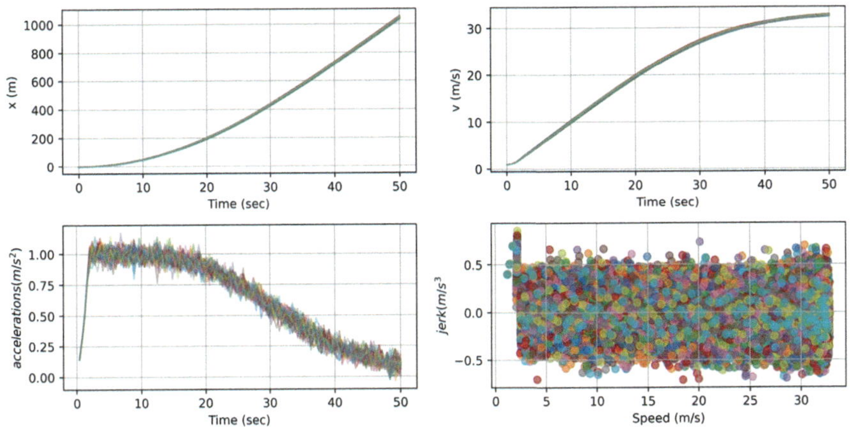

Fig. 7.10 With diffusion term but no jump term

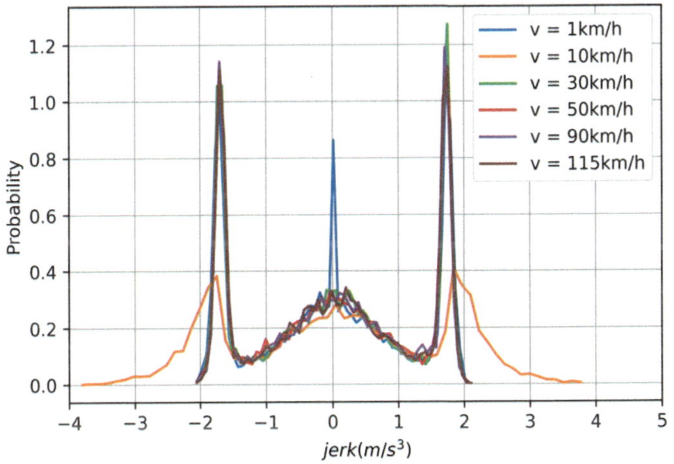

Fig. 7.11 Distribution of the jerk for jump-diffusion

Various attributes are set as follows to assess capacity: (1) different standard deviations of the jump parameter σ_{jump}; (2) the average density of jump occurrence, denoted as λ; (3) varying numbers of vehicles with the jump term within the traffic flow. Fig. 7.13 illustrates the results. The findings reveal the following tendencies: (1) an increase in the number of vehicles with the jump term within the traffic flow leads to a decrease in capacity; (2) an increase in σ_{jump} yields a positive impact on capacity; (3) the effect of jump density on capacity is found to be relatively insignificant.

To investigate the lateral dynamics, we establish a 500-m long road with periodic boundary conditions, ensuring that vehicles re-enter the road at the entrance (i.e., x = 0) upon leaving the road (i.e., the location where x = 500). Figure 7.14 presents

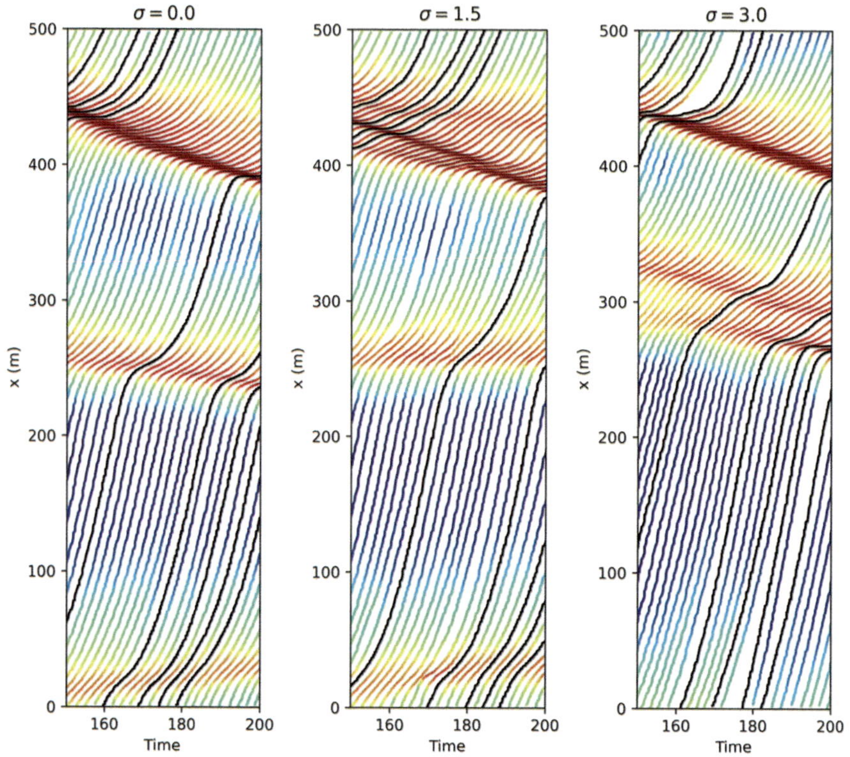

Fig. 7.12 Simulation results for single lane road with period boundary for different σ_{jump}

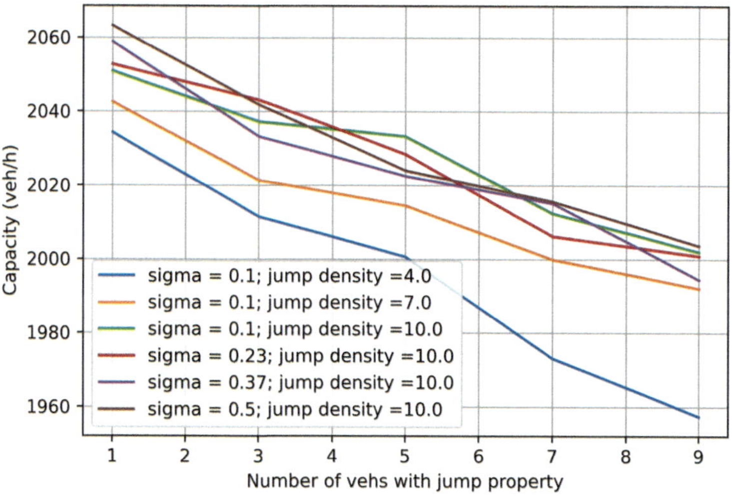

Fig. 7.13 Fundamental diagram when the vehicle with jump is mixed with regular vehicles

the lateral dynamics, including lateral location, lateral speed, lateral acceleration and finally longitudinal speed-lateral jerk relationship. The results show that the lateral dimension experiences continuous disturbance and the lateral jectories are very look likes the real world lateral trajectory. Figure 7.15 displays the vehicle trajectories. Figure 7.15a illustrates the two-dimensional trajectories of the model defined by Eq. 7.24, which exclusively incorporates the diffusion term (excluding jump terms). The bold black lines represent the lane boundaries, with the lane middle line positioned at $y = 1.75$ m. The color of the trajectories signifies the vehicle's speed. Starting from rest, the vehicle's longitudinal coordinate is presented in Fig. 7.15b. Figure 7.15c depicts the model defined by Eq. 7.24, incorporating both diffusion and jump terms. It is evident that the lateral deviation becomes substantial when the jump characteristic is present.

The distributions of lateral speed, acceleration, and jerk are illustrated in Fig. 7.16 for both scenarios: with and without jump terms. Figure 7.16a–c pertain to cases without jump terms, while Fig. 7.16d–f correspond to cases with jump terms. When

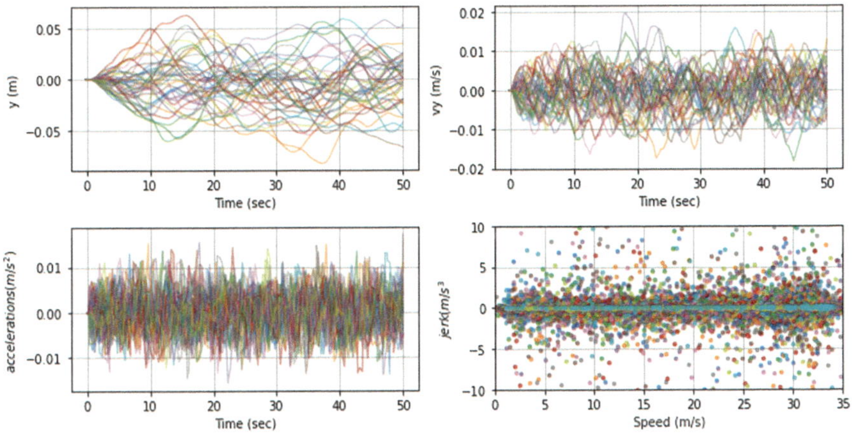

Fig. 7.14 Simulation path for the trajectories. **a** lateral coordinate; **b** lateral speed; **c** lateral acceleration; **d** speed jerk relationship

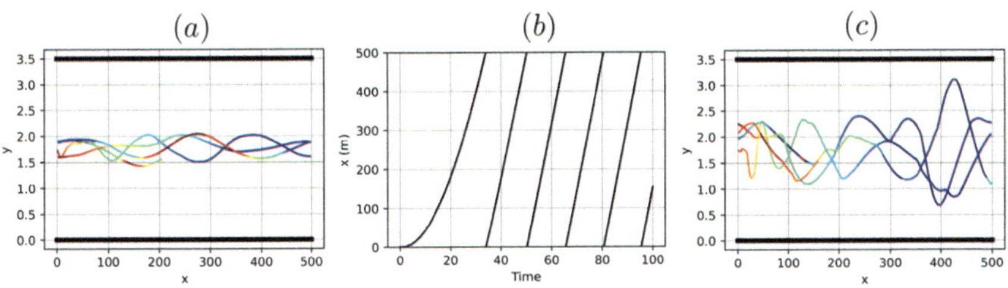

Fig. 7.15 Simulation path for the trajectories for without Poisson and with Poisson term. **a** two dimensional trajectory for the case of without Poisson term; **b** space–time trajectory; **c** two dimensional trajectory for the case of with Poisson term;

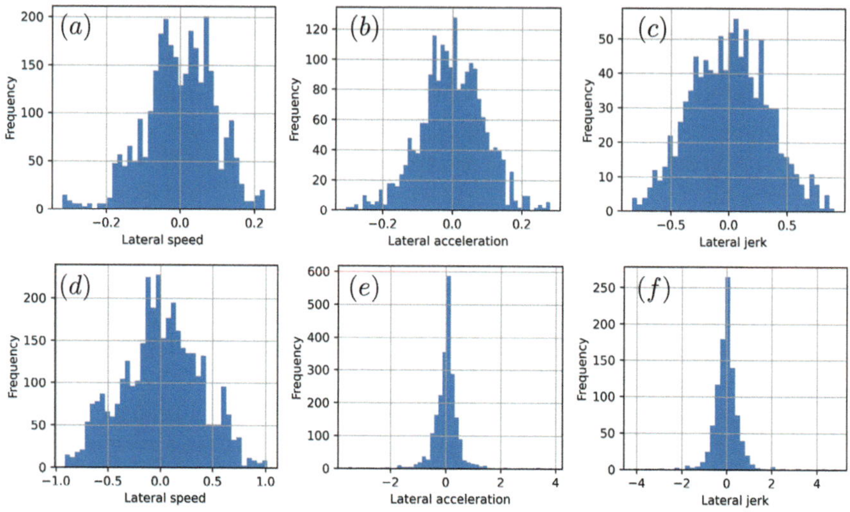

Fig. 7.16 The distribution of lateral speed, acceleration, jerk for the case of without Poisson term and with Poisson term

no jump terms are employed, the lateral speed, acceleration, and jerk demonstrate relatively moderate values. However, when jump terms are introduced, these three variables of lateral movement all exhibit noteworthy increases. The lateral speed almost doubles, while the acceleration and jerk rise by a factor of approximately ten.

7.7 Notes

This chapter is a detailed version of Qi [12]. The microscopic traffic model holds significant importance in the analysis of traffic flow, simulation, and testing of autonomous vehicles. Traditionally, car following models have been proposed to address longitudinal driving behaviors, while more recently, lateral behavior models have also emerged to capture the lateral movements of vehicles. In reality, both dimensions are subject to stochastic fluctuations, necessitating the introduction of stochastic enhancements to accurately capture the randomness in vehicle movements. These models have proven successful in characterizing the propagation of disturbances and the stochastic nature of driving behaviors. Most current microscopic models consider location and speed as state variables, with acceleration explicitly modeled. However, the jerk, which represents the differential of acceleration, may not be adequately described. Empirical observations have indicated the presence of a speed-dependent jerk profile in both the lateral and longitudinal dimensions. Certain vehicles exhibiting risky behaviors tend to prefer higher values of jerk. Existing stochastic car following models, including aggressive improved car following models, fail to capture the speed-dependent jerk profiles, leading to biased results when applied to driving behavior analysis. To address this gap, this research implements a jump-diffusion approach to incorporate speed-dependent jerk behavior. A car following model incorporating diffusion and jump terms is proposed, with the diffusion term representing the normal stochastic component and the jump term representing the risky component. Results demonstrate the model's capability to accurately describe speed-dependent jerk profiles.

Future extensions of this research may include the following:

(1) Developing a more comprehensive model that takes into account the correlation between the longitudinal and lateral dimensions. This could involve introducing special jump or diffusion terms to capture the correlation structure.
(2) Utilizing the jump-diffusion model for behavioral analysis, inference, and prediction of human-driven vehicles. For example, based on short time history trajectories, the probability of longitudinal and lateral jumps in accelerations or speeds can be calculated, aiding in the behavioral planning of autonomous vehicles.
(3) Exploring the explicit solution of the jump-diffusion formulation, considering nonlinear behaviors. This would enable the application of the proposed model to vehicle control in situations where the environment exhibits jump-behavior factors, such as volatile road pavements.

References

1. Adavikottu A, Velaga NR, Mishra S (2023) Modelling the effect of aggressive driver behavior on longitudinal performance measures during car-following. Transp Res Part F Traffic Psychol Behav 92:176–200. https://doi.org/10.1016/j.trf.2022.11.006
2. Bouadi M, Jia B, Jiang R et al (2022) Stochastic factors and string stability of traffic flow: analytical investigation and numerical study based on car-following models
3. Habtemichael FG, De Picado SL (2014) Crash risk evaluation of aggressive driving on motorways: microscopic traffic simulation approach. Transp Res Part F Traffic Psychol Behav 23:101–112. https://doi.org/10.1016/j.trf.2013.12.022
4. Khattak AJ, Wali B (2017) Analysis of volatility in driving regimes extracted from basic safety messages transmitted between connected vehicles. Transp Res Part C Emerg Technol 84:48–73. https://doi.org/10.1016/j.trc.2017.08.004; Kovaceva J, Isaksson-Hellman I, Murgovski N (2020) Identification of aggressive driving from naturalistic data in car-following situations. J Safety Res 73:225–234. https://doi.org/10.1016/j.jsr.2020.03.003
5. Laval JA, Toth CS, Zhou Y (2014) A parsimonious model for the formation of oscillations in car-following models. Transp Res Part B Methodol 70:228–238. https://doi.org/10.1016/j.trb.2014.09.004
6. Ma X, Chen S, Chen F (2017) Multivariate space-time modeling of crash frequencies by injury severity levels. Anal Methods Accid Res 15:29–40. https://doi.org/10.1016/j.amar.2017.06.001
7. Makridis MA, Anesiadou A, Mattas K et al (2023) Characterising driver heterogeneity within stochastic traffic simulation. Transp B Transp Dyn 11:725–743. https://doi.org/10.1080/21680566.2022.2125458
8. Ngoduy D, Lee S, Treiber M et al (2019) Langevin method for a continuous stochastic car-following model and its stability conditions. Transp Res Part C Emerg Technol 105:599–610. https://doi.org/10.1016/j.trc.2019.06.005

9. Peng G, He H, Lu W-Z (2016) A new car-following model with the consideration of incorporating timid and aggressive driving behaviors. Phys Stat Mech Its Appl 442:197–202. https://doi.org/10.1016/j.physa.2015.09.009
10. Przybyla J, Taylor J, Jupe J, Zhou X (2015) Estimating risk effects of driving distraction: A dynamic errorable car-following model. Transp Res Part C Emerg Technol 50:117–129. https://doi.org/10.1016/j.trc.2014.07.013
11. Qi H (2023a) Modelling the lateral dimension of vehicles movement: a stochastic differential approach with applications. Transp Transp Sci 1–25. https://doi.org/10.1080/23249935.2023.2239375
12. Qi H (2023b) Are current microscopic traffic models capable of generating jerk profile consistent with real world observations? Int J Transp Sci Technol
13. Tordeux A, Schadschneider A (2016) White and relaxed noises in optimal velocity models for pedestrian flow with stop-and-go waves. J Phys Math Theor 49:185101. https://doi.org/10.1088/1751-8113/49/18/185101
14. Wali B, Khattak AJ, Karnowski T (2019) Exploring microscopic driving volatility in naturalistic driving environment prior to involvement in safety critical events—Concept of event-based driving volatility. Accid Anal Prev 132:105277. https://doi.org/10.1016/j.aap.2019.105277
15. Wang X, Khattak AJ, Liu J et al (2015) What is the level of volatility in instantaneous driving decisions? Transp Res Part C Emerg Technol 58:413–427. https://doi.org/10.1016/j.trc.2014.12.014
16. Wen X, Cui Z, Jian S (2022) Characterizing car-following behaviors of human drivers when following automated vehicles using the real-world dataset. Accid Anal Prev 172:106689. https://doi.org/10.1016/j.aap.2022.106689
17. Yeo J, Lee J, Jang K (2021) The effects of rainfall on driving behaviors based on driving volatility. Int J Sustain Transp 15:435–443. https://doi.org/10.1080/15568318.2020.1756543
18. Yu R, Li S (2022) Exploring the associations between driving volatility and autonomous vehicle hazardous scenarios: Insights from field operational test data. Accid Anal Prev 166:106537. https://doi.org/10.1016/j.aap.2021.106537
19. Zhang J, Wang Y, Lu G (2019) Impact of heterogeneity of car-following behavior on rear-end crash risk. Accid Anal Prev 125:275–289. https://doi.org/10.1016/j.aap.2019.02.018

Chapter 8
Lateral Influence on Capacity Adjustment of Lanes Number for Mixed Autonomous Vehicles

8.1 Capacity Adjustment Factor (CAF) and Current Practice

8.1.1 CAF Introduction

Capacity is an essential metric in transportation engineering, playing a vital role in various traffic-related applications such as planning, design, and control (HCM 2022). Traditional capacity values are calculated by multiplying the base capacity using various adjustment factors, which are defined for a plethora of scenarios, including road conditions, pedestrian conflict, weather, and more [2]. However, conventional capacity methodologies are designed for human-driven vehicles, and as autonomous vehicles become more prevalent, the existing capacity adjustment factors are no longer suitable when HDV flow is mixed with AVs. Therefore, there is a need to investigate the behavioral differences between HDVs and AVs and further re-evaluate capacity characteristics as the market penetration rate (MPR) of AVs varies.

It is widely acknowledged that the introduction of AVs would increase capacity and improve traffic safety. This is due to the exchange of information through vehicle-to-everything (V2X) communication and accurate control algorithms, which allows AVs to drive with a relatively small headway compared to HDVs, increasing capacity. Moreover, the capacity should vary with market penetration rates (MPRs). Numerous studies have investigated capacity using either simulation [4] or theoretical deduction [6], with analytical capacity formulations and capacity distributions being deduced or capacity values determined considering specific MPRs or configurations.

Previous research has predominantly focused on the longitudinal behavior of autonomous vehicles and its impact on traffic flow, neglecting the lateral dimension and movement of both AVs and human-driven vehicles. AVs, equipped with high-precision sensors and low-latency controllers, tend to follow the desired path with

© The Author(s), under exclusive license to Springer Nature Singapore Pte Ltd. 2024
H. Qi, *Stochastic Two-Dimensional Microscopic Traffic Model*, Lecture Notes in Intelligent Transportation and Infrastructure,
https://doi.org/10.1007/978-981-97-3597-6_8

minimal lateral deviation compared to HDVs. This has two effects on mixed AVs flow:

- The reduced lateral deviation of AVs affects the lateral repulsive force exerted on neighboring HDVs, leading to decreased lateral friction phenomenon that affects capacity [7], thus improving capacity.
- AVs have the ability to accurately predict the behavior of HDVs [17], making them less affected by HDV's lateral movements on neighboring lanes. This reduces lateral interactions between lanes after AVs are introduced, resulting in capacity benefits.

Moreover, the lane utilization policy for multi-lane roads has become more diverse, with the introduction of dedicated lanes for AVs in addition to lanes shared by HDVs and AVs [19]. Adjustments in lane width and re-division of road segments for different lane usages alter lateral interactions, necessitating a re-evaluation of capacity. Therefore, when developing adjustment factors for different lane numbers and dedicated lanes, the lateral dynamics should be taken into account.

While there has been research on capacity and adjustment factors in scenarios involving autonomous vehicles, there is a lack of systematic exploration of the stochastic lateral dimension from a behavioral perspective. As a result, current methodologies may overestimate capacity or inaccurately calculate adjustment factors. To fill this gap, this study introduces a two-dimensional stochastic traffic flow model to investigate capacity on multilane roads with varying market penetration rates. The model is able to accurately replicate the lateral friction phenomenon.

8.1.2 Current Practice

Capacity is a central focus in traffic engineering, encompassing planning, operational analysis, and control. To evaluate capacity, the Highway Capacity Manual provides comprehensive guidelines and systematic procedures. However, the HCM predominantly addresses human-driven vehicle flow, and with the increasing market penetration rate of connected and automated vehicles, the HCM's outcomes may no longer be appropriate or accurate.

In capacity benefit analyses for CAVs, it is widely acknowledged that information exchange and efficient control algorithms contribute to reduced time headway between CAVs [6]. Common algorithms include adaptive cruise control (ACC) and cooperative adaptive cruise control (CACC) [9, 13], some of which maintain a constant time or space headway. In merge/diverge/work zone areas [3], AVs can collaborate to enhance efficiency, leading to increased capacity, which is inversely proportional to the headway.

The capacity assessment for mixed traffic flow including connected and automated vehicles (CAVs) involves the consideration of various car following scenarios, such as HDV following CAV, CAV following HDV, and CAV following CAV in ACC/CACC mode. To capture the spectrum of market penetration rate (MPR), Ghiasi et al. [8]

employed a Markov chain model to analyze heterogeneous and stochastic headway distributions. Chen et al. [6] offered analytical stochastic capacity and bounds, while Sala and Soriguera [20] developed capacity evaluation methods for platoons consisting of both regular vehicles and CAVs, incorporating cooperative and opportunistic platooning schemes. Numerous studies have explored adjustment factors to accommodate mixed autonomous vehicle and human-driven vehicle flows, examining freeway segments, signalized intersections, roundabouts, and other scenarios [1, 2, 10, 11, 18]. It should be noted that the capacity advantages of connected and autonomous vehicles extend beyond the longitudinal aspect of traffic flow. Research by Chen et al. [5] and Zhou et al. demonstrates the contribution of CAVs to capacity improvements in the lateral dimension.

Zhou discovered that CAVs exhibit significantly reduced lateral oscillation through precise controller design, such as the proportional-integral-derivative (PID) controller. The standard deviation of CAVs' lateral deviation is estimated to be one-tenth to one-third that of human-driven vehicles. This reduction in lateral deviation has two important effects:

- firstly, neighboring HDVs in adjacent lanes experience less influence from the CAV since the AVs accurately follow the lane centerline with minimal wandering. This decrease in "lateral friction," resulting from stochastic lateral deviation, is anticipated.
- Secondly, CAVs are equipped with high-precision algorithms that accurately predict the behavior of HDVs. Consequently, CAVs are less affected by the lateral fluctuations in HDVs' lane-keeping behavior in adjacent lanes.

When considering both of these effects, it becomes evident that the capacity benefits of CAVs are underestimated if only the longitudinal dimension is considered, as the introduction of AVs eliminates bidirectional interaction, leading to increased capacity.

The evaluation of the capacity of mixed autonomous vehicle and human-driven vehicle traffic, specifically concerning lateral interaction, remains a significant challenge according to the authors. Addressing this challenge requires the utilization of a stochastic two-dimensional traffic flow model. However, existing two-dimensional models, such as those developed by Delpiano et al. [7] and Sharath et al. [21], are deterministic in nature and do not consider the stochastic lateral wanderings of vehicles. To overcome this limitation, this study presents a two-dimensional stochastic model that examines the capacity of mixed AV and HDV traffic while considering lateral interaction. This approach enables a more accurate and comprehensive assessment of the capacity benefits of connected and autonomous vehicles (CAVs) in mixed traffic scenarios.

8.2 Symbols in the Chapter

x and y: general reference for the longitudinal and lateral coordinate.

$s_{HDV} = \left[x_i, u_i, y_i, v_i, z^{lon}_{HDV,i}, z^{lat}_{HDV,i}\right]^T$: the state variable for an HDV. u_i and v_i are the longitudinal and lateral speed; $z^{lon}_{HDV,i}$ and $z^{lat}_{HDV,i}$ are longitudinal and lateral noise.

$F^x(\cdot)$ and $F^y(\cdot)$: the longitudinal and lateral forces;

$NEIB(i)$: the set containing neighboring vehicles at neighboring lanes of vehicle i that influence the ego vehicle.

$s_{CAV} = \left[x_i, u_i, e_{i1}, \dot{e_{i1}}, e_{i2}, \dot{e_{i2}}, z_{CAV,i}\right]^T$: the state variable for an CAV.

α and β: the parameters used in the lane boundary potential and lane mark potential expressions.

γ: the scaling parameter corresponding to the lane middle line attraction force.

θ_{lon}, θ_{lat}, σ_{lon} and σ_{lat}: the parameters for the noise, which is a OU process;

l_w: the lane width

$F^{bnd}(\cdot), F^{mark}(\cdot), F^{middle}(\cdot)$, : the force of lane boundary, lane mark and lane middle line;

F^{IDM}: the longitudinal force from leader vehicle at the same lane;

a^x_{MAX} and a^y_{MAX}: the maximal longitudinal and lateral acceleration.

x^{ellip} and y^{ellip}: the longitudinal and lateral distance threshold that lateral repulsive force exist;

$F^{lon}_{i,j}(\cdot)$ and $F^{lat}_{i,j}(\cdot)$: the longitudinal and lateral force on vehicle i from the vehicle j at neighboring lanes;

A, B, Q, R: matrix used when deriving the AV's lateral control algorithm;

δ: the control input of AV's lateral control;

C_{base}: base capacity. Its unit is veh/h.

k_{desire}: the desired density. Its unit is veh/km

v_{avg}: the average speed over the road.

C_{adj}: adjusted capacity, which is the simulation results when certain road or traffic flow configurations are applied;

$\mathcal{N}(x; \mu_x, Var_x)$: the normal distribution value at point x, if the mean and variance of x are μ_x and Var_x respectively.

8.3 Two Dimensional Stochastic Microscopic Traffic Model

8.3.1 Model Framework

Two types of vehicles, namely, human-driven vehicles (HDV) and autonomous vehicles (AV), are considered in this study. The AVs can operate in two modes: Adaptive Cruise Control (ACC) mode and Cooperative Adaptive Cruise Control (CACC) mode. In ACC mode, the AVs adhere to predetermined following policies, while in

CACC mode, multiple AVs form a platoon and operate as a cohesive unit. Conversely, the behavior of HDVs is influenced by the leading vehicle, which can be either another HDV or an AV, as well as vehicles in adjacent lanes. The contrasting behaviors of HDV and AV are depicted in Fig. 8.1. Figure 8.1b describes the influence of one vehicle. Figure 8.1c illustrates the influence of HDVs, where the ego vehicle (Vehicle 2) is influenced by its leader in the ego lane and the leaders in the neighboring lanes (Vehicles 1 and 3). Similarly, Vehicle 4 is influenced by Vehicle 3. In Fig. 8.1d, the trajectories of AVs exhibit reduced variability, with most of them maintaining a centered path.

Despite their behavior differences, both types of vehicles can be characterized by the longitudinal and lateral components of their motion. For human-driven vehicles, the longitudinal movement is determined by the intelligent driver model, while the lateral movement is governed by a stochastic model, which will be discussed in detail later. On the other hand, for autonomous vehicles, the longitudinal movement is guided by the constant time headway policy, and the lateral movement is controlled by an ordinary feedback controller. The output of this controller, which represents the steering action, allows for the management of lateral deviations. The state variables for HDV are denoted as $\left[\mathrm{x_i}, \mathrm{u_i}, \mathrm{y_i}, \mathrm{v_i}, \mathrm{z}_{\mathrm{HDV,i}}^{\mathrm{lon}}, \mathrm{z}_{\mathrm{HDV,i}}^{\mathrm{lat}}\right]^{\mathrm{T}}$, encompassing the longitudinal location, longitudinal speed, lateral location, lateral speed, and two-dimensional noise source specific to HDV. The HDV movement model can be succinctly summarized as follows:

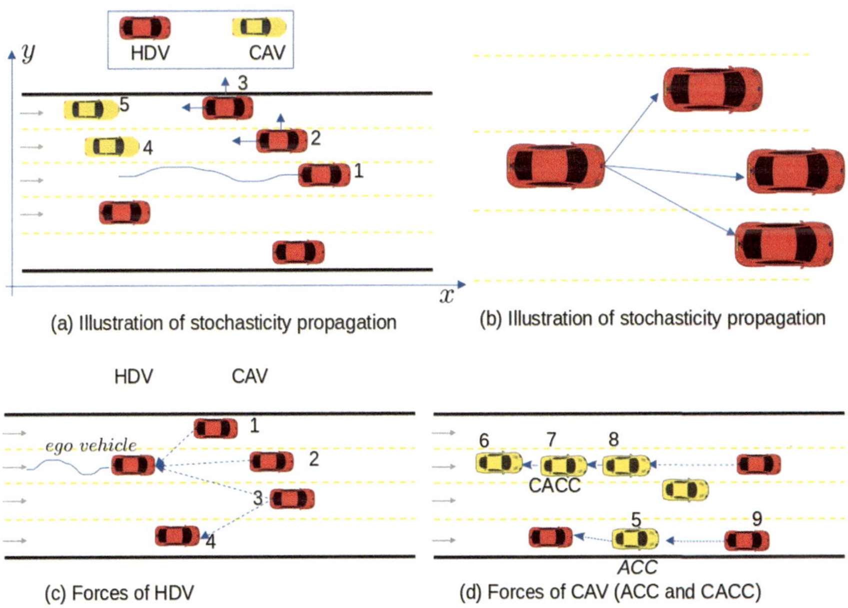

Fig. 8.1 Behavioral difference of HDV and AVs on multi-lanes road. Reprint from ref [16], copyright 2024, with permissions from American Society of Civil Engineers

$$
\begin{cases}
dx_i = u_i dt & (a) \\
dy_i = v_i dt & (b) \\
du_i = F^x\left(\left\{\Delta x_{ij}, \Delta y_{ij}\right\}, z_{HDV,i}^{lon}\right)dt, \forall j \in NEIB(i) & (c) \\
dv_i = F^y\left(\left\{\Delta x_{ij}, \Delta y_{ij}\right\}, z_{HDV,i}^{lat}\right)dt, \forall j \in NEIB(i) & (d) \\
dz_{HDV,i}^{lon} = F_{HDV}^{lon}\, dt + L_{HDV}^{lon}\, dW & (e) \\
dz_{HDV,i}^{lat} = F_{HDV}^{lat}\, dt + L_{HDV}^{lat}\, dW & (f)
\end{cases}
\tag{8.1}
$$

In the given formulation, the superscripts 'lon' and 'lat' denote the longitudinal and lateral dimensions, respectively. F^x, F^y, F^{HDV} and L^{HDV} represent functions that will be explained later. W refers to a standard Wiener process, while L_{HDV}^{lon} and L_{HDV}^{lat} represent the diffusion terms of the noise. NEIB(i) signifies the set of neighboring vehicles for vehicle i. It is worth noting that the noise is modeled separately in Eq. 8.1 for two reasons. First, the vehicle's trajectory should be sufficiently smooth to avoid jerky motions for the driver. Brownian motion, known as "everywhere continuous but nowhere differential," is integrated once in Eq. 8.1c and d to generate a smooth trajectory. Second, the noise carries clear physical significance, encompassing factors such as driving sensing noise, maneuver noise, and tyre-pavement couple noise. Therefore, it is natural to model the noise as a separate entity.

In the case of autonomous vehicles (AV), their motion is governed by longitudinal and lateral controllers. The state variable is represented as $\left[x_i, u_i, e_{i1}, \dot{e}_{i1}, e_{i2}, \dot{e}_{i2}, z_{CAV}\right]^T$, where x_i and u_i carry the same meaning as in the case of human-driven vehicles (HDV). The variables $e_{i1}, \dot{e}_{i1}, e_{i2}, \dot{e}_{i2}, z_{CAV}$ correspond to the lateral deviation, the derivative of lateral deviation, the angular deviation from the desired path, the derivative of angular deviation, and the noise source. A model for $\left[x_i, u_i, e_{i1}, \dot{e}_{i1}, e_{i2}, \dot{e}_{i2}, z_{CAV}\right]^T$ can be developed through the combination of an Adaptive Cruise Control controller and a lateral feedback controller.

8.3.2 HDV Two Dimensional Model

8.3.2.1 Lateral Component

Several factors contribute to a vehicle's lateral movement:

- the vehicle's inclination to remain close to the lane centerline. Any deviation from the centerline prompts the vehicle to steer back, aiming to mitigate the deviation;
- the vehicle's adherence to maintaining a certain distance from lane markings and road boundaries to ensure compliance with traffic regulations;
- the vehicle's tendency to avoid collisions with neighboring vehicles in adjacent lanes;
- the influence of stochastic noise on the vehicle's movements. This noise comprises various sub-factors, including driver sensing errors and noise generated by the interaction between the tires and the pavement.

Consequently, the lateral trajectory exhibits stochastic behavior. To describe the lateral dynamics, we adopt a concept from Newtonian mechanics. The lateral speed is denoted as v_i, and the lateral acceleration is represented as dv_i. The lateral acceleration, dv_i, is the sum of the lateral forces acting on the vehicle. The function $F^y(\cdot)$, which emulates the lateral force, consists of five components.

$$F_i^y(\cdot) = F_i^{bnd}(\Delta y) + F_i^{mark}(\Delta y) + F_i^{middle}(y) + z_{HDV,i}^{lat} + \sum_{j \in NEIB(i)} F_{i,j}^{lat}(\Delta x_{ij}, \Delta y_{ij}) \quad (8.2)$$

Five force components are distinct forces that contribute to the vehicle's lateral movement. These components include the boundary force, lane mark force, lane middle line force, noise, and the force exerted by neighboring lane vehicles. The variable Δy represents the distance between the vehicle and the boundary or lane mark. For the boundary force, it is imperative that the vehicle remains within the boundary. As the vehicle approaches the boundary, the force should approach infinity. Conversely, when the vehicle is positioned at the middle line of the side lane, the force from the boundary should be zero. To satisfy these requirements, the formulation $F^{bnd}(\Delta y) = \frac{1}{\left(\Delta y * \left(\frac{2}{l_w}\right)\right)^{\alpha}} - 1$ is utilized to represent this force. Here, Δy denotes the lateral distance between the vehicle and the road boundary, while l_w represents the lane width. Notably, when Δy equals $\frac{l_w}{2}$, $F^{bnd}(\Delta y)$ is zero, and as Δy approaches zero, $F^{bnd}(\Delta y)$ approaches infinity. The parameter α, an adjustable factor in the range of [0,10], regulates the rate of force variation. For instance, if the driver does not consider the boundary until the vehicle is about to touch it, a small value close to 0 can be employed. Regarding the lane mark force or the force from the lane boundary, the vehicle may occasionally traverse the lane mark, such as when avoiding a potential collision with nearby lateral vehicles. Similar to the boundary force, when the vehicle aligns precisely with the lane middle line, the force exerted by the lane marks should be zero. To express this force, we adopt the expression $F^{mark}(\Delta y) = e^{-(\beta \cdot \Delta y)^2} - e^{-\left(\beta \cdot \frac{l_w}{2}\right)^2}$.

The third force considered is the lane middle line force, which always has an attractive nature, with its direction from the vehicle towards the lane middle line. It is essential to formulate the mathematical expression for this force consistently, regardless of whether the vehicle is in a lane-keeping or lane-changing state. To achieve a reasonable form for the lane middle line force, the following factor is taken into account: the trajectory during lane-changing resembles an "S" shaped curve, with the lateral speed being close to zero at the initial and final stages. Various types of "S" shaped curves exist, such as the logistic function and the error function. Due to the simplicity of the error function and its application in microscopic traffic modeling, it is selected as the initial basis for constructing the lane middle line force. The error function, erf(y), and its derivative are defined as follows:

$$\begin{cases} erf(y) = \frac{1}{\sqrt{\pi}} \int\limits_{-y}^{y} e^{-t^2} dt \\ \frac{d(\mathrm{erf}(y))}{dt} = e^{-y^2} \end{cases} \tag{8.3}$$

Considering the tendency for the vehicle to steer right when the lane middle line is on the right and left when the middle line is on the left, we apply the derivative of the error function in both directions and obtain the nominal lateral speed as follows:

$$\mathrm{v_y} = \left(\mathrm{e}^{-\gamma^2\left(y+\frac{l_W}{2}\right)^2} - \mathrm{e}^{-\gamma^2\left(y-\frac{l_W}{2}\right)^2}\right)dt \tag{8.4}$$

The calculation of the lateral acceleration, which is the derivative of the lateral speed, can be expressed as follows:

$$\frac{d\mathrm{v_y}}{dt} = \frac{d\mathrm{v_y}}{dt} = -\gamma^2(2y + l_w)\mathrm{e}^{-\gamma^2\left(y+\frac{l_W}{2}\right)^2} + \gamma^2(2y - l_w)\mathrm{e}^{-\gamma^2\left(y-\frac{l_W}{2}\right)^2} \stackrel{\mathrm{def}}{=} \mathrm{F}^{\mathrm{middle}}(y) \tag{8.5}$$

y is the lateral distance from vehicle to the target lane middle line. The third force in consideration is characterized as the noise $z^{lat}_{HDV,i}$. Drawing upon the successful application of the Ornstein-Uhlenbeck process in microscopic modeling [22], we employ the OU process to construct this noise component. The OU process is represented as $\mathrm{d}Z = \theta(\mu - Z)dt + \sigma dW$, where θ denotes the speed of process regression, μ represents the equilibrium point, and σ indicates the noise level. In the absence of additional information, the equilibrium point of the noise is assumed to be zero, resulting in $\mu = 0$. Consequently, the formulation for the lateral noise is expressed as $\mathrm{d}z^{lat}_{HDV,i} = -\theta_{lat} z^{lat}_{HDV,i} dt + \sigma_{lat} dW$.

The force exerted by vehicles in neighboring lanes constitutes the final force considered. Let us consider the coordinates of the ego vehicle as $(\mathrm{x_i}(\mathrm{t}), \mathrm{y_i}(\mathrm{t}))$, and those of a neighboring vehicle as $(\mathrm{x_j}(\mathrm{t}), \mathrm{y_j}(\mathrm{t}))$. When the distance between them is sufficiently small, this force becomes significant. To represent this force zone, an ellipse shape is utilized, as demonstrated by Delpiano et al. [7]. The region is defined by $\frac{(\mathrm{x_j}-\mathrm{x_i})^2}{(\mathrm{x^{ellip}})^2} + \frac{(\mathrm{y_j}-\mathrm{y_i})^2}{(\mathrm{y^{ellip}})^2} \leq 1$. Here, $\frac{(\mathrm{x_j}-\mathrm{x_i})^2}{(\mathrm{x^{ellip}})^2} + \frac{(\mathrm{y_j}-\mathrm{y_i})^2}{(\mathrm{y^{ellip}})^2} \leq 1$ denotes the critical longitudinal distance required for the force to in effect. In order to determine a reasonable critical distance, an assumption is made that the ego vehicle will avoid a collision with the neighboring vehicle. If the ego vehicle's longitudinal speed is u_ i, then the distance covered when the ego vehicle comes to a stop with maximum longitudinal acceleration $\mathrm{a^x_{MAX}}$ is given by $\frac{\mathrm{u_i^2}}{2\mathrm{a^x_{MAX}}}$. Therefore, the critical longitudinal distance is defined using this distance, i.e., $\mathrm{x^{ellip}} = \frac{\mathrm{a \cdot t^2}}{2} = \frac{\mathrm{u_i^2}}{2\mathrm{a^x_{MAX}}}$. Additionally, $\mathrm{y^{ellip}}$ is set equal to the width of the lane, denoted as $\mathrm{l_w}$.

The presence of a neighboring vehicle within the area defined by the aforementioned ellipse function exerts an influence on the ego vehicle. The resultant force, denoted as a vector $\left(\mathrm{F}^{lon}_{\mathrm{i,j}}(\Delta x_{ij}, \Delta y_{ij}), \mathrm{F}^{lat}_{\mathrm{i,j}}(\Delta x_{ij}, \Delta y_{ij})\right)$, operates in the longitudinal and lateral dimensions. The total force, a scalar quantity, must satisfy the following conditions based on the longitudinal distance Δx_{ij} between the ego vehicle and its neighbor: (1) When the vehicles are almost parallel and the lateral distance Δy_{ij} equals $\frac{l_w}{2}$ (indicating that the neighbor encroaches into the ego lane), the force should approach infinity. (2) When the neighbor is located at the boundary of the ellipse, given by $y^{ellip}\sqrt{\left(1-\frac{(\Delta x_{ij})^2}{(x^{ellip})^2}\right)}$, the force should be zero. Therefore, the total force is expressed as $\left(\frac{1}{(\Delta y_{ij}-\frac{l_w}{2})}-\frac{1}{y^{ellip}\sqrt{\left(1-\frac{(\Delta x_{ij})^2}{(x^{ellip})^2}\right)}-\frac{l_w}{2}}\right)$. This total force is then partitioned into longitudinal and lateral components using a trigonometric relationship.

$$\begin{cases} F_{i,j}^{lon}(\Delta x_{ij}, \Delta y_{ij}) = -\frac{\Delta x_{ij}}{\sqrt{(\Delta x_{ij})^2+(\Delta y_{ij})^2}}\left(\frac{1}{(\Delta y_{ij}-\frac{l_w}{2})}-\frac{1}{y^{ellip}\sqrt{\left(1-\frac{(\Delta x_{ij})^2}{(x^{ellip})^2}\right)}-\frac{l_w}{2}}\right) \\ F_{i,j}^{lat}(\Delta x_{ij}, \Delta y_{ij}) = -\frac{\Delta y_{ij}}{\sqrt{(\Delta x_{ij})^2+(\Delta y_{ij})^2}}\left(\frac{1}{(\Delta y_{ij}-\frac{l_w}{2})}-\frac{1}{y^{ellip}\sqrt{\left(1-\frac{(\Delta x_{ij})^2}{(x^{ellip})^2}\right)}-\frac{l_w}{2}}\right) \end{cases} \tag{8.6}$$

8.3.2.2 Longitudinal Component

The longitudinal force can be divided into three distinct components: (1) the force exerted by the leading vehicle in the same lane, (2) the force resulting from neighboring vehicles in adjacent lanes, and (3) the noise component. The first component is modeled using the conventional intelligent driver model. The second component is represented by $F_{i,j}^{lon}(\Delta x_{ij}, \Delta y_{ij})$ in Eq. 8.6. As for the third component, it is assumed to exhibit a similar form as lateral noise. The formulation of the longitudinal noise is given by $dz_{HDV,i}^{lon} = -\theta_{lon} z_{HDV,i}^{lon}dt + \sigma_{lon}dW$. Hence, the longitudinal component can be expressed as follows:

$$F^x(\{\Delta x_{ij}, \Delta y_{ij}\}, z_{HDV,i}) = F_i^{IDM}(x_i, v_i, x_{LE(i)}, v_{LE(i)}) + \sum_{j\in NEIB(i)} F_{i,j}^{lon}(t) + z_{HDV,i}^{lon} \tag{8.7}$$

In the above equation, $F_i^{IDM}(\cdot)$ can use the conventional IDM model; $F_{i,j}^{lon}(t)$ has been formulated in Eq. 8.6. Thus all components in Eq. 8.7 are specified.

8.3.2.3 Synthesis of the Model

The two dimensional forces are synthesized as follows:

$$\begin{cases} dx_i = u_i dt & (a) \\ du_i = \left[\mathrm{F}_{\mathrm{i}}^{\mathrm{IDM}}\left(x_i, v_i, x_{LE(i)}, v_{LE(i)}\right) + \sum\limits_{j \in NEIB(i)} \mathrm{F}_{\mathrm{i,j}}^{lon}(t) + z_{HDV,i}^{lon} \right] dt & (b) \\ \mathrm{dy_i} = v_i dt & (c) \\ dv_i = \left[\mathrm{F}^{\mathrm{middle}}(y_i) + \mathrm{F}_{\mathrm{i}}^{mark}(y_i) + \mathrm{F}_{\mathrm{i}}^{bnd}(y_i) + \sum\limits_{j \in NEIB(i)} \mathrm{F}_{\mathrm{i,j}}^{lat}\left(x_i, y_i, x_j, y_j\right) + z_{HDV,i}^{lat} \right] dt & (d) \\ \mathrm{d}z_{HDV,i}^{lon} = -\theta_{lon} z_{HDV,i}^{lon} dt + \sigma_{lon}\, dW & (e) \\ \mathrm{d}z_{HDV,i}^{lat} = -\theta_{lat} z_{HDV,i}^{lat} dt + \sigma_{lat}\, dW & (f) \end{cases} \tag{8.8}$$

The model presented can be compactly represented in matrix form as $\mathrm{dS} = \mathrm{F}(S)dt + \mathrm{L}(S)dW$. Here, F is a vector of the same dimension as S, and L is a matrix with dimensions (6,2). The noise term, W, follows a bivariate Gaussian distribution, representing the longitudinal and lateral dimensions, respectively.

8.3.3 AV Stochastic Behavioral Model: ACC and CACC

Regardless of the operating mode, both require the same fundamental control inputs: longitudinal control and lateral control. Empirical observations of ACC and CACC [15] indicate that even with a well-designed and efficient control algorithm, there is still inherent two-dimensional stochasticity. Therefore, relying solely on a noiseless theoretical model would result in a biased estimation of capacity. The vehicle's motion can be classified into longitudinal and lateral dimensions, and the noise can be examined in both dimensions. Previous research by Zhou et al. delves into the investigation of lateral distributions of autonomous vehicles through AV field tests. The findings showcase that the noise level, represented by the standard deviation of the lateral location distribution, ranges from 10 to 30%. To analyze longitudinal noise, Makridis et al. [15] examines ACC/CACC scenarios using data from AstaZero. The data, with a time resolution of 0.1 s, includes two-dimensional vehicle locations. Calculation of the trajectory for the second to fifth vehicles in the platoon is possible based on the theoretical model and leader vehicle states. By comparing the theoretical results with observations, the noise of an ACC platoon can be determined. Figure 8.2 presents the noise for the second vehicle in the platoon, with a standard deviation of 8.48 cm. The noise exhibits approximate symmetry, primarily concentrated within ±20 cm, with a maximum noise level of approximately 70 cm. Similar observations apply to the third to fifth vehicles in the platoon, with standard deviations of 10.08 cm, 16.93 cm, and 11.73 cm, respectively. It is evident that the noise cannot be disregarded completely. The ACC/CACC model cannot be directly applied to analyze a mix of AV flows, where the presence of stochasticity is significant. To develop a stochastic

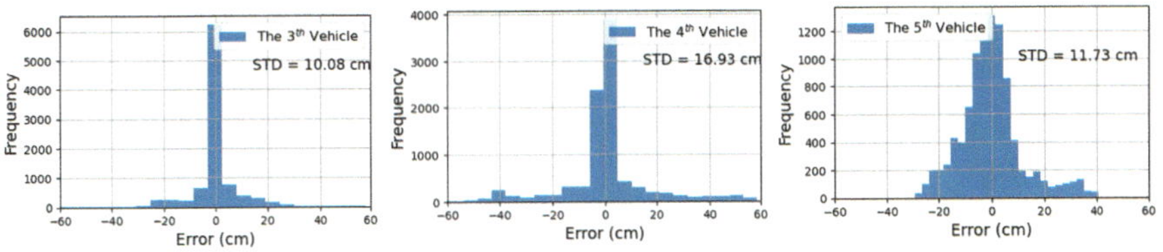

Fig. 8.2 The longitudinal noise of the CACC

AV model for capacity analysis, the noise source is integrated with the feedback controller, as elaborated in the subsequent sections.

8.3.3.1 Steering

The steering angle of the front tyre is denoted as δ. A common feedback controller is employed to represent both ACC and CACC. The state variable is defined as $\left[e_1, \dot{e_1}, e_2, \dot{e_2}\right]^T$. The track following feedback controller is characterized by the following equations:

$$\frac{d}{dt}\begin{bmatrix} e_1 \\ \dot{e_1} \\ e_2 \\ \dot{e_2} \end{bmatrix} = \begin{bmatrix} 0 & 1 & 0 & 0 \\ 0 & -\frac{2C_{af}+2C_{ar}}{mV_x} & \frac{2C_{af}+2C_{ar}}{m} & \frac{-2C_{af}l_f+2C_{ar}l_r}{mV_x} \\ 0 & 0 & 0 & 1 \\ 0 & -\frac{2C_{af}l_f-2C_{ar}l_r}{I_zV_x} & \frac{2C_{af}l_f-2C_{ar}l_r}{I_z} & -\frac{2C_{af}l_f^2+2C_{ar}l_r^2}{I_zV_x} \end{bmatrix}\begin{bmatrix} e_1 \\ \dot{e_1} \\ e_2 \\ \dot{e_2} \end{bmatrix} dt + \begin{bmatrix} 0 \\ \frac{2C_{af}}{m} \\ 0 \\ \frac{2C_{af}l_f}{I_z} \end{bmatrix}\delta$$

$$\stackrel{\text{def}}{=} \frac{dS}{dt} = A \cdot S + B \cdot \delta \tag{8.9}$$

Let's consider a control gain matrix G, which allows us to rewrite the system as $\frac{dS}{dt} = (A - BG)\delta$. Assuming our control objective is $J = \int \left(S^TQS + R \cdot \delta^2\right) dt$, we can obtain the control gain G using $G = \left(B^TXB + R\right)^{-1}B^TXA$. Here, X is obtained by solving the Riccati equation $X = Q + A^T\left(I + XBR^{-1}B^T\right)^{-1}XA$. Considering the stochastic nature of the system, we introduce a noise term to the system, utilizing the same noise source as in HDV but with a different parameter σ_{AV}. As a result, we extend the system state to $\left[e_1, \dot{e_1}, e_2, \dot{e_2}, z_{AV}\right]^T$. The system equation can be expressed as follows:

$$\frac{d}{dt}\begin{bmatrix} e_1 \\ \dot{e_1} \\ e_2 \\ \dot{e_2} \\ z_{av} \end{bmatrix} = \begin{bmatrix} 0 & 1 & 0 & 0 & 0 \\ 0 & -\frac{2C_{af}+2C_{ar}}{mV_x} & \frac{2C_{af}+2C_{ar}}{m} & \frac{-2C_{af}l_f+2C_{ar}l_r}{mV_x} & 0 \\ 0 & 0 & 0 & 1 & 0 \\ 0 & -\frac{2C_{af}l_f-2C_{ar}l_r}{I_zV_x} & \frac{2C_{af}l_f-2C_{ar}l_r}{I_z} & -\frac{2C_{af}l_f^2+2C_{ar}l_r^2}{I_zV_x} & 0 \\ 0 & 0 & 0 & 0 & -\theta \end{bmatrix}\begin{bmatrix} e_1 \\ \dot{e_1} \\ e_2 \\ \dot{e_2} \\ z_{AV} \end{bmatrix} dt$$

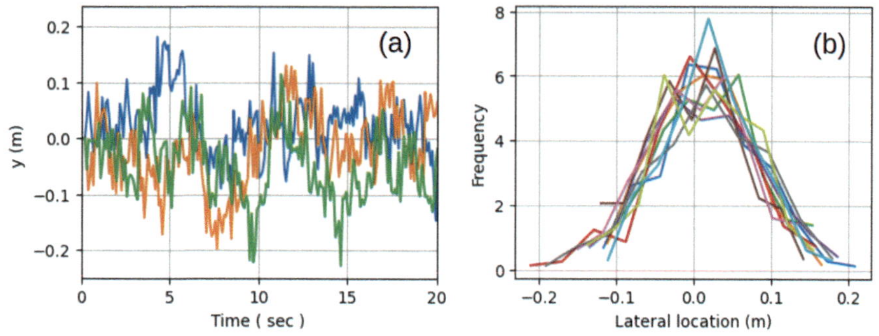

Fig. 8.3 Behaviors of the AV's lateral wandering; **a** tracks; **b** lateral wandering

$$+\begin{bmatrix} 0 \\ \frac{2C_{af}}{m} \\ 0 \\ \frac{2C_{af}l_f}{I_z} \end{bmatrix}\delta + \begin{bmatrix} 0 \\ 0 \\ 0 \\ 0 \\ \sigma_{\mathrm{AV}} \end{bmatrix}\mathrm{dW} \tag{8.10}$$

The stochasticity of the system is determined by the parameter σ_AV. It should be smaller than that of HDV due to several factors: (1) AVs have more precise sensing capabilities, (2) negligible reaction delay (as low as milliseconds), and (3) exact application of control inputs. According to Zhou et al., the noise level is around 10% of HDV. Figure 8.3 illustrates the lateral behavior of AVs with $\sigma_{\mathrm{AV}} = 0.01$, utilizing the parameter set defined as follows: total vehicle mass m is 1.22510^3 kg; moment of inertia mass about z axis I_z is $1.538*10^3$ kgm^2; distance of center of gravity to front axle l_f is 0.883 m; distance of center of gravity to rear axle l_r is 1.508 m; Corning stiffness coefficient (front) C_{af} is 20.89/radius; corning stiffness coefficient (rear) C_{ar} is 20.89/radius. Figure 8.3a displays the lateral paths, with a magnitude ranging from −0.1 m to 0.1 m. The lateral wandering distribution, presented in Fig. 8.3b, shows a deviation of approximately 0.06 m. At a speed of 20 mph (about 9 m/s, as used in Zhou et al.), the deviation is approximately 0.057 m. This observation aligns with the reported deviation range of [0.03 m, 0.07 m] in Zhou et al., indicating consistency between the theoretical model and empirical findings.

8.3.3.2 Acceleration

Numerous ACC/CACC models have been proposed in the literature, including those by Jiang et al. [10], Li [12], Adebisi et al. [2], and Wang et al. (2018). In this study, we consider the model presented by Jiang et al. [10] and extend it to incorporate longitudinal noise. The resulting formulation of the model is as follows:

$$\begin{cases} dx_i = v_i dt & (a) \\ dv_i = k_2(d - t_{hw}v_i - L) + k_3(v_{i-1} - v_i) + Z_{AV} & (b) \end{cases} \tag{8.11}$$

In the aforementioned ACC mode, the parameters k_2 and k_3 are assigned values of $0.23s^{-2}$ and $0.07s^{-1}$ respectively. The variable v_{i-1} represents the speed of the leading vehicle, while L denotes the length of the leader vehicle. This mode specifically applies when the leader vehicle is a HDV. Alternatively, in situations of free driving, when the leader vehicle exceeds the predefined time gap of 2 s, the acceleration is given by $dv_i = k_1(v_f - v_i)$, where the value of k_1 is set to $0.4s^{-1}$. However, if the leader vehicle is an AV, the follower vehicle switches to the CACC mode. The CACC model can be expressed as follows:

$$\begin{cases} dx_i = v_i dt & (a) \\ dv_i = \frac{k_p \Delta x(t-\Delta t) - 1.1 v_i(t-\Delta t) + k_d(\Delta v(t-\Delta t) - t_l v_i(t-\Delta t))}{\Delta t} + Z_{AV} & (b) \end{cases} \tag{8.12}$$

k_p is $0.45s^{-1}$ and k_d is set to 0.0125.

8.3.4 Parameters of the Model

The model parameters can be estimated using maximum likelihood estimation methods based on observed data. By discretizing the system described in Eq. 8.8 using the Euler-Maruyama scheme, we obtain the form $S(k+1) = S(k) + F(S(k))\delta t + L(S(k))\delta W$, where δt represents the time step and δW is a Gaussian variable with variance δt. Consequently, the state transition from $S(k)$ to $S(k+1)$ follows a normal distribution, with a mean given by $\mu_{S(k+1)} = S(k) + F(S(k))\delta t$ and a variance of $Var_{S(k+1)} = L^T(S(k))L(S(k)) \cdot \delta t$. Given a set of observations $\mathbb{S} = \{S_i\}$ representing individual vehicle trajectories, the likelihood function can be formulated as follows:

$$L_\Theta(\mathbb{S}) = \prod_{k=1}^{N-1} \mathcal{N}\big(S(i); \mu_{S(i)}, Var_{S(i)}\big) \tag{8.13}$$

The parameters can be estimated by maximizing the likelihood function $L_{\Theta}(\mathbb{S})$, assuming a normal distribution $\mathcal{N}$. The parameter optimization process can be achieved using different algorithms, including particle swarm optimization. Hence, the optimal parameter values Θ^* can be obtained through this optimization procedure.

8.3.5 Example of Simulation Results

To showcase the model's capabilities, we conduct a simulation on a two-lane road with lane widths of 3.6 m for both the left and right lanes. Each lane is populated with a density of 40 veh/km, resulting in a total of 20 vehicles in each lane. The road length is set at 500 m. Initially, the vehicles are evenly distributed in each lane, with equal distances between leader and follower vehicles. The initial speeds are determined based on the equilibrium relationship between density and speed. Subsequently, the vehicles operate according to the proposed two-dimensional microscopic traffic flow model.

Figure 8.4 presents the two-dimensional trajectory, time-space diagram, and lateral coordinate distribution for each lane. The speed is represented by the color in subplots (a)–(c), with the black lines indicating the road boundary and the yellow line representing the lane mark in Fig. 8.4a. The trajectory exhibits stochastic two-dimensional movements, while the time-space diagrams (Fig. 8.4b and c) depict the formation of stochastic traffic jams. To assess the model's consistency with empirical findings, we calculate the lateral distributions of trajectories at selected longitudinal locations (Fig. 8.4d). Notably, there is no significant distinction among the distributions at different longitudinal locations, and thus they are not separately depicted in Fig. 8.4d. The average standard deviation for both lanes across all longitudinal locations is 0.32, aligning with the recommendations in the Mechanistic-Empirical Pavement Design Guide (MEPDG), which suggests a standard deviation of 0.305 for lane widths exceeding 3.6 m. A four-lane road simulation and the trajectories are presented in Fig. 8.5. Similar results can be observed.

Figure 8.6 presents the Flow-Density (FD) relationships for different Mixed Percentage Ratios (MPRs). In Fig. 8.6a, the FD curve for a two-lane road is plotted, while Fig. 8.6b illustrates the FD curve for a three-lane road. In Fig. 8.6a, when the MPR is 0.1, the capacity reaches approximately 2000 veh/h, and the optimal density for achieving this capacity is around 50 veh/km. Conversely, when the MPR is 0.9, the capacity increases to approximately 4000 veh/h, nearly doubling the capacity of the pure HDV case, with an optimal density of 80 veh/h. Generally, as the MPR increases, both the capacity and the optimal density show an upward trend. In Fig. 8.6b, it is observed that the capacity at MPR $= 0.9$ is approximately 3000 veh/h, which is lower than that of the two-lane road. This discrepancy is due to the mutual influence

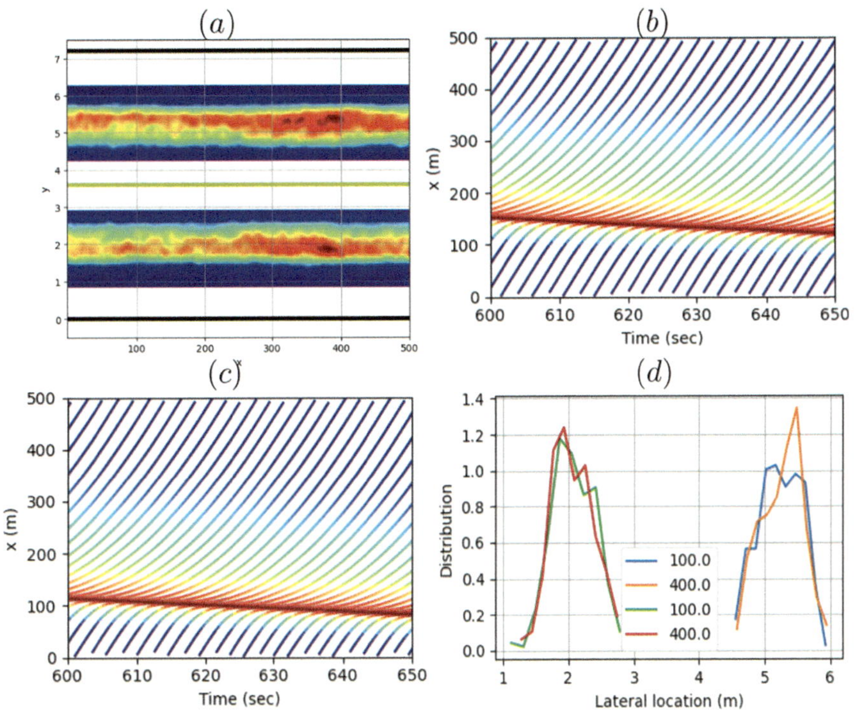

Fig. 8.4 Simulation results for two-lanes road. **a** two dimensional trajectory; **b** time-space diagram for left lane; **c** time-space diagram for right lane; **d** lateral distribution

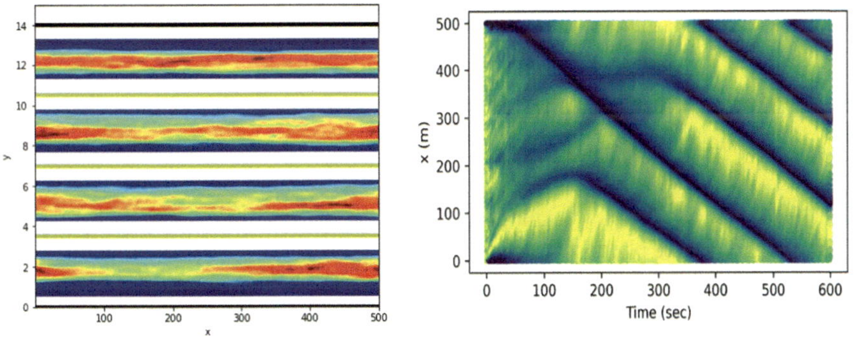

Fig. 8.5 Left: four-lane road; right: trajectories

between vehicles in neighboring lanes, as indicated by the longitudinal and lateral forces in Eq. 8.6.

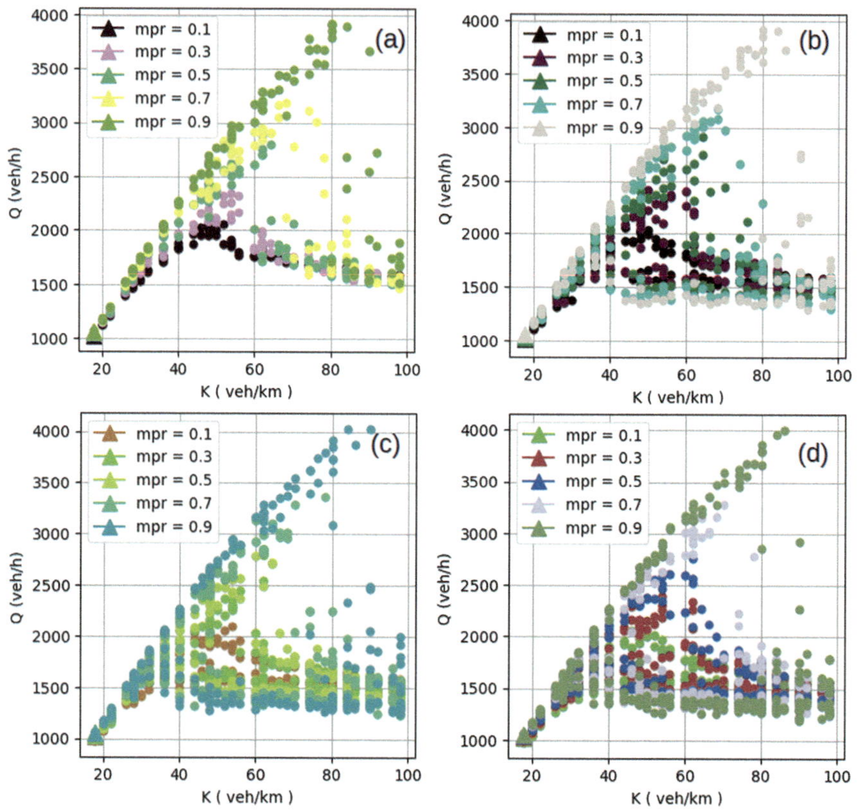

Fig. 8.6 Fundamental diagram for the case of a two-lane, three-lane, four-lane and five-lane road with different MPRs; **a** two-lane road; **b** three-lane road; **c** four-lane; **d** five-lane

8.4 Model Validation by Replicating of the Lateral Friction Phenomenon

To validate the consistency of our proposed model with existing research, we specifically investigate the phenomenon known as lateral friction. This phenomenon occurs between two adjacent lanes, where one lane serves as a general lane (GL), while the other functions as a managed lane (ML), such as an HOV lane [7, 14]. Typically, MLs are physically separated from GLs by painted strips or narrow buffers. As a result, congestion on the GL has a detrimental impact on the capacity of the ML, as vehicles on the ML experience discomfort and decelerate due to the GL congestion. Liu et al. [14] demonstrated that even when lane changing is prohibited, the GL still influences the ML through lateral interactions. The fundamental diagram illustrating the existence of lateral friction is presented in Delpiano et al. [7], Fig. 8.4b. Conventional one-dimensional microscopic models fail to capture such phenomena, which led Delpiano et al. [7] to develop a two-dimensional traffic flow model to address this

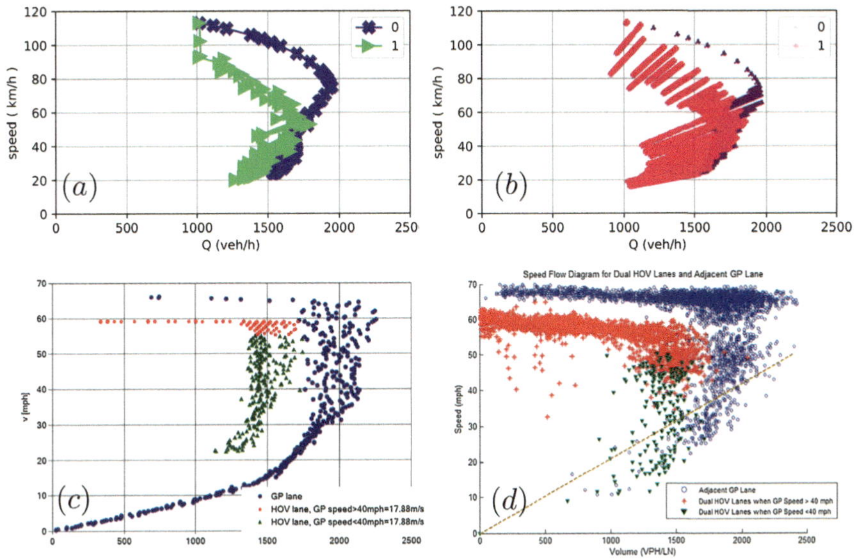

Fig. 8.7 Fundamental diagram mimicking the lateral friction; **a**: average over time horizon; **b**: distribution of the FD. Reprint from ref [16], copyright 2024, with permissions from American Society of Civil Engineers

issue. The resulting FD from their two-dimensional traffic flow model is depicted in Liu et al. [14], Figure 8.7a. However, this model does not consider stochasticity, resulting in a less dispersed FD.

To showcase the predictive capability of our proposed model in replicating the friction phenomenon, we employ a stochastic two-dimensional model to simulate a two-lane road, mimicking the dynamics between the ML and GL, where the GL exerts an influence on the ML. The lane widths of the two lanes are both 3.5 m. Figure 8.7a and b present the speed-volume relationships. Figure 8.7a represents the average of the stochastic flow rates depicted in Fig. 8.7b. It can be observed that the scatter plot in Fig. 8.7b closely aligns with the empirical observation depicted in [14], Fig. 8.7a. The empirical observation indicates a significant reduction in flow rate at the ML when speeds fall below 40 mph (or 64 km/h) due to friction. The simulation results in Fig. 8.7b support this observation, with the ML capacity under the influence of the GL being approximately 1500 veh/h.

8.5 Experiment Design and CAF Determination Methods

8.5.1 Experiment Design

The vehicles under consideration consist of two types: HDV and Autonomous Vehicles (AV). The ratios of HDV and AV are denoted as ρ_{HDV} and ρ_{AV}, respectively. The Market Penetration Rate (MPR) is used to represent ρ_{AV}. The MPR is gradually increased from 0 to 100% in equal increments. To cover a wide range of Flow-Density (FD) relationships, the density is increased from a relatively low value of 10 veh/km to the jam density of approximately 130 veh/km. The maximum flow rate obtained from the FD is determined as the capacity, while the corresponding density is identified as the optimal density. The simulation duration is set to 1 h (3600 s), with the initial 15 min for warming up and the remaining 45 min for data collection. Each combination of attributes, such as lane width, number of lanes, and MPR, is simulated to obtain the FD and determine the capacity. Additionally, the presence of a dedicated lane (DL), exclusively designated for AVs, is also taken into account. The consideration of the DL is important due to the different lateral interaction between the DL and other lanes, referred to as the general lane (GL) hereafter. The specific lane reserved as the DL can influence the lateral interaction and, consequently, impact the overall capacity.

8.5.2 CAF Determination Methods

8.5.2.1 Base Capacity Determination

The base capacity refers to the capacity of a single lane when all vehicles are HDVs. It is determined by applying the two-dimensional intelligent driver model with calibrated parameters to describe vehicle movements. As the system is stochastic, multiple simulations of a single-lane road with periodic conditions are conducted, considering a desired density, k_{desire}. The capacity is then calculated as the average across these iterations. The road has a length, L, of 1 km, and a time horizon, T. At each time step, t_i, the average speed, v_{avg}, is determined by the equation $v_{avg} = \frac{\sum_i v_i}{(L \times k_{desire})}$, where $L \times k_{desire}$ represents the number of vehicles on the road. Subsequently, the flow rate, Q, is obtained by multiplying v_{avg} by k_{desire}. Due to the stochastic nature of the simulation, Q varies with the same value of k_{desire}. To establish a result, the average value, $\overline{Q}$, is calculated over a given time horizon.

8.5.2.2 Capacity Adjustment Calculation (CAF)

The Capacity Adjustment Factor (CAF) represents the ratio of the capacity of the analyzed scenario to the capacity of the base scenario. This relationship can be mathematically expressed as $C_{adj} = CAF \times C_{base}$, where C_{adj} is the adjusted capacity. CAF is a dimensionless coefficient. Furthermore, the CAF can be determined by calculating $CAF = \frac{C_{adj}}{C_{base}}$.

8.5.3 *Results*

8.5.3.1 Capacity Adjustment of Lanes Number for Mixed AVs Flow

The subsequent analysis explores the relationship between lane number and capacity as the Market Penetration Rate (MPR) varies. We consider lane numbers ranging from 2 to 6, with MPRs changing incrementally from 0 to 100% in steps of 20%. The corresponding results are illustrated in Fig. 8.8. Notable findings include:

- Capacities increase alongside the MPR, aligning with expectations. However, the relationship between capacity and MPR is nonlinear.
- Holding the MPR constant, an increase in lane number leads to a decrease in average capacity. This phenomenon is attributable to lateral friction, where stochasticity propagates in a chain-like manner from one lane to another. Thus, on a two-lane road, stochasticity circulates between the two lanes, while on a six-lane road, stochasticity from one lane affects the remaining five lanes. This chain-style propagation results in reduced capacity as lane number increases.

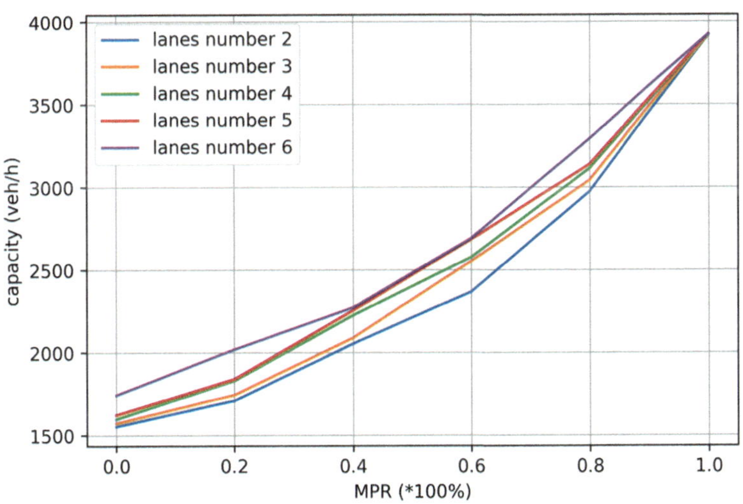

Fig. 8.8 Capacity for different MPRs and lanes numbers. Reprint from ref [16], copyright 2024, with permissions from American Society of Civil Engineers

- At a 100% MPR, capacity remains consistent. Pure Autonomous Vehicle (AV) flow experiences no lateral interaction, thus yielding the same capacity regardless of lane number.

The capacity adjustments can be derived from the base capacity of 1952 veh/h for a single lane road. These adjustments highlight the significant impact of lateral friction when vehicles are not exclusively Autonomous Vehicles (AVs) in terms of flow.

8.5.3.2 FD for Different Lanes Number and MPR

The following analysis focuses on the flow-density (FD) results of multi-lane roads under varying Market Penetration Rates (MPRs). Specifically, Fig. 8.9 presents the FD for three-lane and four-lane roads when the MPR is 50%. Each lane is labeled numerically, from 0 (leftmost) to N−1 (rightmost). In Fig. 8.9a, it is observed that the capacity is highest in the side lanes (lane 0 and lane 2), reaching 3000 veh/h due to the penetration of Autonomous Vehicles (AVs). Conversely, the capacity of the middle lane (lane 1 in Fig. 8.9a) is reduced to approximately 1800 veh/h, even though there are AVs present. This reduction is attributed to the lateral influences experienced by the middle lane from both sides, whereas the side lanes are unaffected. This observation is further supported by the four-lane case in Fig. 8.9b, where the middle lane experiences a decrease in both capacity and optimal density (i.e., the density at which the flow rate reaches capacity). Comparing these results with the single-lane scenario, where lateral interaction is absent, the optimal density for lane 1 (middle lane) in Fig. 8.9a and b is reduced to 40 veh/km from the typical 50 veh/km. Consequently, it is deduced that the FD characteristics vary lane by lane, underscoring the importance of evaluating capacity holistically.

Figure 8.10 depicts the flow-density (FD) outcomes for a higher Market Penetration Rate (MPR) of 70%. In Fig. 8.10a, the capacities of the side lanes (lane 0

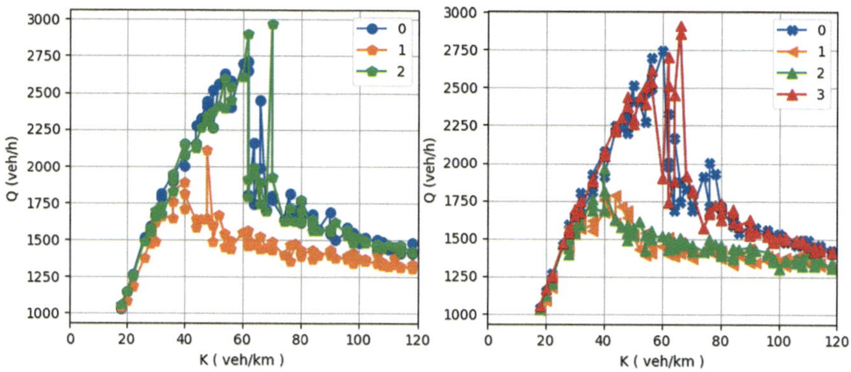

Fig. 8.9 Fundamental diagram for three-lanes (**a**) and four-lanes road (**b**), MPR = 50%

and lane 2) increase to 3500 veh/h, while the capacity of the middle lane remains compromised. Comparing Fig. 8.9a and Fig. 8.10a, it is noted that the capacity for an MPR of 70% improves to 3500 veh/h, an increase of about 500 veh/h compared to the MPR of 50%. However, even with the benefits of MPR, the capacity still falls short of the base capacity for a single lane, which is 1950 veh/h. These findings highlight how lateral interaction affects the FD pattern and diminishes capacity on multi-lane roads. Furthermore, although AVs yield certain capacity advantages, these improvements cannot outweigh the negative impacts of lateral interactions when the MPR is insufficiently high (Similar results can be found for the case when MPR is 90% in Fig. 8.11. Its discussion is omitted here).

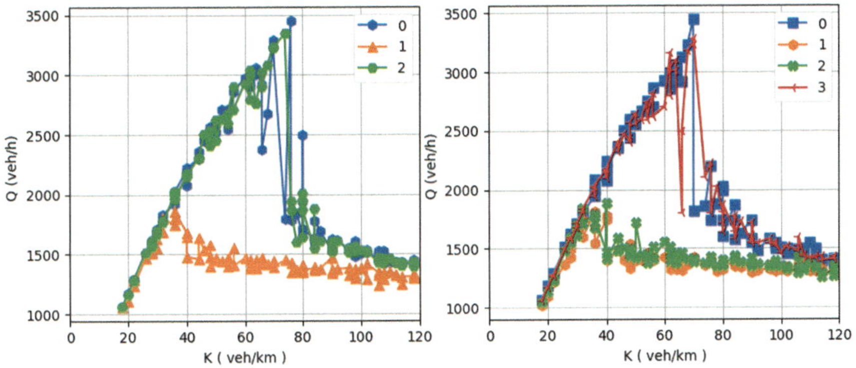

Fig. 8.10 Fundamental diagram for three-lanes (**a**) and four-lanes road (**b**), MPR = 70%

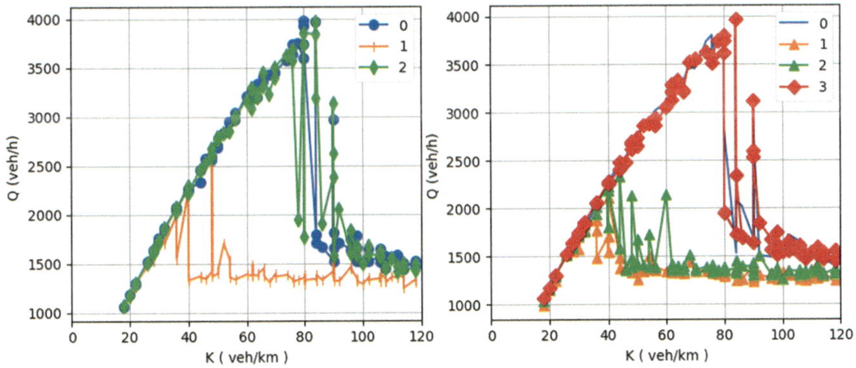

Fig. 8.11 Fundamental diagram for three-lanes (**a**) and four-lanes road (**b**), MPR = 90%

8.5.3.3 Influence of the Dedicated Lane Considering the Lateral Influence

In our analysis, we focus on a four-lane road and investigate three distinct scenarios: (1) all lanes function as general purpose lanes (GLs); (2) the leftmost lane is designated as a dedicated lane (DL); (3) one of the middle lanes serves as a DL. To ensure comparability, the lane widths for the three configurations are uniformly set at 3.5 m. Regarding demand, we assume that the density, k_{AV}, on the DL lane is fixed at 80 veh/km. For HDVs, we incrementally increase the HDVs flow from a small initial value to examine flow rate variations. In the latter two cases, these HDVs flows are evenly distributed among the remaining three GLs (since there are four lanes). Conversely, in the first case, both HDVs and AVs are intermingled and equally assigned to all four lanes. Supposing K represents density (measured in veh/km/lane), the density of each lane in the first case is calculated as $\frac{3K+k_{AV}}{4}$. Consequently, the Market Penetration Rate (MPR) is derived as $\frac{k_{AV}}{3K+k_{AV}}$. As k_AV remains constant, the MPR decreases proportionally with increasing K.

Based on the provided demand settings, it can be inferred that the total demand remains constant across the three scenarios. Figure 8.12 illustrates the flow rates for all four lanes corresponding to different values of K. From the analysis, the following conclusions can be drawn:

(1) The operational efficiency benefit of the dedicated lane (DL) setting becomes evident when the density exceeds a certain threshold, approximately 65 veh/km. This corresponds to an MPR of approximately 30%, calculated as $\frac{k_{AV}}{3K+k_{AV}} = \frac{80}{3*65+80} \approx 30\%$. When K is below 65 veh/km, the MPR exceeds 30%. As the density of HDVs increases from 10 veh/km, the MPR becomes approximately 72%, calculated as 80/(3*10 + 80). In case 1, the average density on each lane is 27.5 veh/km. The decrease in flow rate for DL configurations (cases 2 and 3) can be attributed to the higher volume resulting from the mixed AVs flow (case 1) across all four lanes compared to the flow consisting of pure HDVs

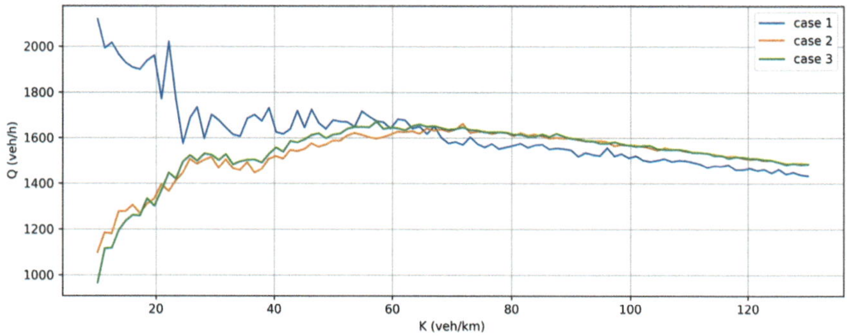

Fig. 8.12 FD considering the DL. Reprint from ref Qi [16], copyright 2024, with permissions from American Society of Civil Engineers

plus one-lane AVs flow. These findings indicate that the benefit of the DL may not always be positive. The capacity variation should be carefully evaluated by considering the mixed AVs flow and HDVs flow. The former is influenced by the MPR, as demonstrated in Fig. 8.6, highlighting the importance of a cautious evaluation of the capacity benefit provided by the DL.

(2) A comparison between case 2 and case 3 reveals that the lane assigned as a dedicated lane also affects the capacity. It is observed that after reaching congestion (K exceeds 65 veh/km), the capacity difference becomes negligible. Below this density, the flow rate difference is discernible. However, there is a density threshold (approximately 20 veh/km) at which case 3 attains a higher flow rate, while below this density, case 2 exhibits a higher flow rate.

The observations can be attributed to the following factors: (i) Although AVs exhibit less wandering, there is still a slight influence of AVs on HDVs, as HDVs may encounter AVs during their wandering process. (ii) At lower densities, the influence of AVs on HDVs is more significant. However, as the density increases, the mutual interaction among HDVs becomes more dominant, surpassing the influence of AVs on HDVs. Consequently, for K values exceeding 20 veh/km, the mutual interaction among HDVs in the three right-hand lanes (case 2) leads to a reduction in flow rate.

8.6 Notes

AVs revolutionize the transportation system with notable impacts on capacity. Current studies primarily focus on the longitudinal dimension, where AVs can operate with smaller headways, consequently increasing capacity. However, the lateral dimension, which accounts for approximately ten percent of the wandering observed in traditional heavy-duty vehicles (HDVs), is often overlooked. AVs demonstrate superior ability in predicting HDVs' lane-changing behavior with high accuracy, leading to reduced lateral interactions within traffic flow on neighboring lanes. As a result, the MPR suggests an incremental decrease in lateral interaction strength, potentially enhancing capacity. Given the presence of two-dimensional stochasticity, a re-evaluation of the capacity adjustment factor is necessary to account for both longitudinal and lateral benefits. While most research focuses on capacity analysis from a longitudinal perspective, neglecting the lateral effects may lead to biased capacity evaluations. To address this gap, a two-dimensional stochastic microscopic model is developed to investigate the capacity adjustment factor for varying lane numbers. Notably, the model's effectiveness is demonstrated by replicating the lateral friction phenomenon and examining the influence of dedicated lanes on total capacity.

Future research directions include investigating inhomogeneous road facilities such as merging/weaving areas in addition to the current focus on homogeneous road scenarios. The analysis of urban multilane streets requires consideration of how two-dimensional noise propagation differs due to turning direction influence.

Furthermore, future studies can incorporate different vehicle types, particularly heavy vehicles, to further enhance the comprehensiveness of the findings.

References

1. Adebisi A, Guo Y, Schroeder B et al (2022) Highway capacity manual capacity adjustment factor development for connected and automated traffic at signalized intersections. J Transp Eng Part Syst 148:04021121. https://doi.org/10.1061/JTEPBS.0000631
2. Adebisi A, Liu Y, Schroeder B et al (2020) Developing highway capacity manual capacity adjustment factors for connected and automated traffic on freeway segments. Transp Res Rec J Transp Res Board 2674:401–415. https://doi.org/10.1177/0361198120934797
3. Algomaiah M, Li Z (2022) Enhancing work zone capacity by a cooperative late merge system using decentralized and centralized control strategies. J Transp Eng Part Syst 148:04021107. https://doi.org/10.1061/JTEPBS.0000632
4. Carrone AP, Rich J, Vandet CA, An K (2021) Autonomous vehicles in mixed motorway traffic: capacity utilisation, impact and policy implications. Transportation 48:2907–2938. https://doi.org/10.1007/s11116-020-10154-4
5. Chen F, Song M, Ma X, Zhu X (2019) Assess the impacts of different autonomous trucks' lateral control modes on asphalt pavement performance. Transp Res Part C Emerg Technol 103:17–29. https://doi.org/10.1016/j.trc.2019.04.001
6. Chen S, Wang H, Xiao L, Meng Q (2022) Random capacity for a single lane with mixed autonomous and human-driven vehicles: bounds, mean gaps and probability distributions. Transp Res Part E Logist Transp Rev 160:102650. https://doi.org/10.1016/j.tre.2022.102650
7. Delpiano R, Herrera JC, Laval J, Coeymans JE (2020) A two-dimensional car-following model for two-dimensional traffic flow problems. Transp Res Part C Emerg Technol 114:504–516. https://doi.org/10.1016/j.trc.2020.02.025
8. Ghiasi A, Hussain O, Qian Z (Sean), Li X (2017) A mixed traffic capacity analysis and lane management model for connected automated vehicles: a Markov chain method. Transp Res Part B Methodol 106:266–292. https://doi.org/10.1016/j.trb.2017.09.022
9. Hu C, Wang J (2022) Trust-based and individualizable adaptive cruise control using control barrier function approach with prescribed performance. IEEE Trans Intell Transp Syst 23:6974–6984. https://doi.org/10.1109/TITS.2021.3066154
10. Jiang Q, Schroeder B, Ma J et al (2022) Developing highway capacity manual capacity adjustment factors for connected and automated traffic on roundabouts. J Transp Eng Part Syst 148:04022014. https://doi.org/10.1061/JTEPBS.0000674
11. Lewis M, Albright D (2007) Evaluating the highway capacity manual's adjustment factor for annual weekday to annual average daily traffic: applying a consistent traffic data methodology. Transp Res Rec J Transp Res Board 1993:117–123. https://doi.org/10.3141/1993-16
12. Li X (2022) Trade-off between safety, mobility and stability in automated vehicle following control: an analytical method. Transp Res Part B: Methodol 166:1–18. https://doi.org/10.1016/j.trb.2022.09.003
13. Liu H, Kan X (David), Shladover SE et al (2018) Impact of cooperative adaptive cruise control on multilane freeway merge capacity. J Intell Transp Syst 22:263–275. https://doi.org/10.1080/15472450.2018.1438275
14. Liu X, Schroeder BJ, Thomson T et al (2011) Analysis of operational interactions between freeway managed lanes and parallel, general purpose lanes. Transp Res Rec J Transp Res Board 2262:62–73. https://doi.org/10.3141/2262-07
15. Makridis M, Mattas K, Anesiadou A, Ciuffo B (2021) OpenACC. An open database of car-following experiments to study the properties of commercial ACC systems. Transp Res Part C Emerg Technol 125:103047. https://doi.org/10.1016/j.trc.2021.103047

16. Qi H (2024) Capacity adjustment of lane number for mixed autonomous vehicles flow considering stochastic lateral interactions. J Transp Eng, Part A: Systems 150:04023134. https://doi.org/10.1061/JTEPBS.TEENG-8142
17. Qi H, Chen C, Hu X, Zhang J (2022) Online inference of lane changing events for connected and automated vehicle applications with analytical logistic diffusion stochastic differential equation. Transp Res Part C Emerg Technol 144:103874. https://doi.org/10.1016/j.trc.2022.103874
18. Rahman MM, Hasan T, Nakamura F (2008) Development of professional driver adjustment factors for the capacity analysis of signalized intersections. J Transp Eng 134:532–536. https://doi.org/10.1061/(ASCE)0733-947X(2008)134:12(532)
19. Razmi Rad S, Farah H, Taale H et al (2020) Design and operation of dedicated lanes for connected and automated vehicles on motorways: a conceptual framework and research agenda. Transp Res Part C Emerg Technol 117:102664. https://doi.org/10.1016/j.trc.2020.102664
20. Sala M, Soriguera F (2021) Capacity of a freeway lane with platoons of autonomous vehicles mixed with regular traffic. Transp Res Part B Methodol 147:116–131. https://doi.org/10.1016/j.trb.2021.03.010
21. Sharath MN, Velaga NR, Quddus MA (2020) 2-dimensional human-like driver model for autonomous vehicles in mixed traffic. IET Intell Transp Syst 14:1913–1922. https://doi.org/10.1049/iet-its.2020.0297
22. Tian J, Zhu C, Chen D et al (2021) Car following behavioral stochasticity analysis and modelling: perspective from wave travel time. Transp Res Part B Methodol 143:160–176. https://doi.org/10.1016/j.trb.2020.11.008

Chapter 9
Stochastic Lateral Wandering Patterns of Mixed Traffic Flow

9.1 Lateral Wandering and Its Implications

The stochastic nature of lateral vehicle movement across lanes, known as lateral wandering, is an important factor in pavement design and sustainability analysis [4, 27, 28]. Normative guides such as the Mechanistic-Empirical Pavement Design Guide (MPEDG) [15] and NCHRP [1] document this phenomenon. Figure 9.1 illustrates the stochastic lateral distribution of human-driven vehicles during off-peak and rush hours, revealing that HDVs rarely strictly follow the center of the lane. Numerous studies have been conducted to evaluate this wandering behavior, with some researchers suggesting a normal distribution approximation with varying observations of standard deviation [4, 16].

In recent years, the gradual integration of AVs into the transportation system has brought both benefits and potential negative impacts. One such negative impact is the shortened lifespan of pavements [27]. This is attributed to the disparate lateral wandering behaviors exhibited by AVs and HDVs, stemming from their behavioral differences. AVs are equipped with high-precision sensors and controllers, enabling them to accurately follow the desired track, typically the center of the lane [27]. Consequently, the stress imposed on the pavement by AVs is concentrated on the lane centerline. Conversely, HDVs exhibit stochastic and fluctuating behavior around the lane centerline. This differential behavior contributes to the deterioration of pavement performance, including rutting and fatigue, under AV flow conditions [4]. Empirical observations have validated the aforementioned impact of AVs, prompting the recognition of the need to consider AV integration in the pavement design process [7, 8, 17, 26]. Correspondingly, efforts have been made on the AV control side [3, 27], proposing alternative lateral location distributions for trajectory generation, or designing deterministic trajectories deviating from the lane centerline for AV feedback controllers.

However, an additional distinction in the lateral behavior of HDVs and AVs remains unexplored and has not yet been incorporated into the wandering pattern

© The Author(s), under exclusive license to Springer Nature Singapore Pte Ltd. 2024

H. Qi, *Stochastic Two-Dimensional Microscopic Traffic Model*, Lecture Notes in Intelligent Transportation and Infrastructure,
https://doi.org/10.1007/978-981-97-3597-6_9

Fig. 9.1 Real-world stochastic lateral wandering for congested state and free flow

model. This distinction pertains to the HDV's tendency to react to approaching vehicles in the adjacent lane and maintain a safe lateral gap, a behavior referred to as lateral friction [6] or collateral anomaly [5]. In contrast, AVs, enabled by advanced technologies such as deep learning [2, 20], possess the capability to accurately predict the behavior of HDVs in neighboring lanes and are less impacted by the stochastic lateral wandering exhibited by these counterparts. Consequently, the presence of AVs diminishes the mutual interaction among neighboring lanes.

9.2 Real World Observations and the Work in This Chapter

The highD[1] dataset is utilized to illustrate the interaction and resulting collateral wandering between vehicles on neighboring lanes. Figure 9.2 presents the collective lateral distribution, with the upper plot representing three-lane road segments and the lower plot depicting two-lane road segments. The x-axis corresponds to the relative lateral location, calculated as y/l_w, where y denotes the deviation from the ego-lane middle line and l_w represents the lane width. The range of y/l_w on the x-axis is $[-0.5, 0.5]$, with $x = 0$ indicating the lane middle line. To compute the distributions, several longitudinal locations are selected, and the deviation frequencies are calculated. The results consistently demonstrate an asymmetric pattern. In Fig. 9.2a, the leftmost lane trajectory displays positive skewness, while the rightmost lane trajectory reveals a negative mean value. The lateral distribution in the middle lane is relatively symmetrical, but it generally skews to the left. Similarly, in Fig. 9.2b, the two-lane road exhibits a comparable asymmetric pattern.

[1] www.highd-dataset.com/.

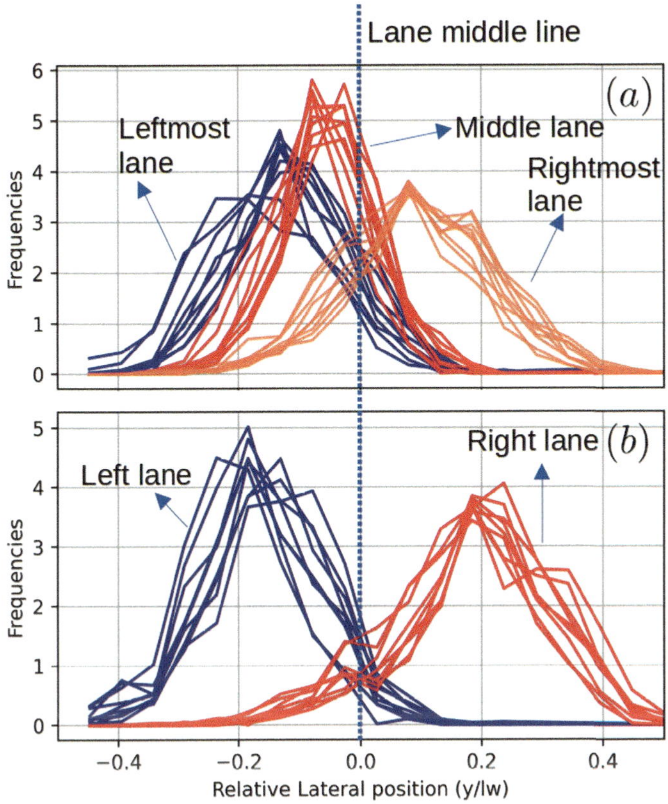

Fig. 9.2 Distribution of lateral position in highD dataset; **a** three-lane road segments; **b** two-lane road segments. Reprint from Ref. [21], copyright 2023, with permissions from Elsevier

The highD dataset also provides insights into the dynamic interactions between two neighboring vehicles on adjacent lanes, referred to as the lateral repulsive effect, which may help explain the observed asymmetric pattern. Figure 9.3 serves as an illustrative example. In Fig. 9.3a, two distances are defined: Δy_1, the distance between the ego vehicle (on the left lane) and the lane mark, and Δy_2, the distance of its neighboring vehicle to the same lane mark. Both distances fall within the range $[0, l_w]$, where l_w represents the lane width. Figure 9.3b presents the trajectory of $(\Delta y_1(t), \Delta y_2(t))$, with each line representing the continuous evolution of a vehicle pair in the four quadrants, at a time resolution of 0.04. The horizontal and vertical thick black lines represent the location of the lane middle line for the left and right lanes, respectively. The trajectory points are predominantly concentrated in Quadrant I, indicating that both vehicles maintain a safe distance from the lane mark. Quadrants II and IV, where one vehicle approaches the lane mark while the other moves away to maintain a safe distance, also contain a significant number of trajectory points. In

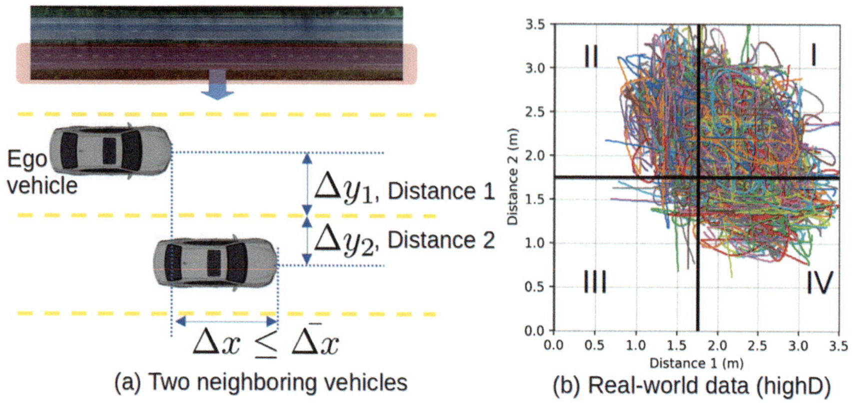

Fig. 9.3 Illustration of lateral repulsive effect in highD dataset. Reprint from Ref. [21], copyright 2023, with permissions from Elsevier

contrast, Quadrant III, where both vehicles remain close to the lane mark and consequently drive with a short lateral distance between them, displays the fewest data points. It is noticeable that once a vehicle pair enters Quadrant III, they adjust their trajectory to move towards the other quadrants. The distribution of trajectory points among these four quadrants reveals the presence of the lateral repulsive effect, which ensures that vehicles either maintain a distance from their neighbors or both vehicles move away from the common lane mark. This effect avoids the risky Quadrant III, where accidents are more likely to occur.

In light of these observations, the objective of this chapter is to investigate lateral behavior and establish a theoretically robust framework for generating consistent patterns of wandering. These patterns can be further utilized for the analysis of pavement sustainability. Our approach involves developing models for the lateral behavior of both human-driven vehicles on single-lane roads and autonomous vehicles on multiple-lane roads. A stochastic differential equation (SDE) formulation is utilized to capture the stochastic lateral dynamics, taking into account factors such as the road boundary, lane marks, vehicles on neighboring lanes, and stochastic noise. To generate the desired lateral wandering pattern, we establish the Fokker–Planck equation, followed by constructing a Markov chain model to solve this problem efficiently.

9.3 Current Practive and Research Gaps

Vehicles typically exhibit both longitudinal and lateral movement during operation. While the longitudinal aspect has been extensively explored in microscopic traffic flow models, particularly within the car following model, the lateral behavioral component has received little attention. Numerous studies have documented the tendency of vehicles to deviate from the exact lane middle line [9, 27]. In a given road section, passing vehicles exhibit varying lateral positions along their wheel paths. The collective lateral characteristic is commonly referred to as lateral wandering or lateral positioning. Lateral wandering is influenced by several factors, including road width, driver characteristics, road pavement [9], lateral disturbances [12], road marks [19], pavement edge lines [22], and others.

Lateral wandering has been examined by researchers using various data collection methods, such as specially designed cable systems [16] and video recordings [14]. Findings from these studies consistently demonstrate a bell-shaped distribution, which can be reasonably approximated by a normal distribution [4, 16]. While the mathematical accuracy of the normal distribution is limited due to its requirement for infinite support, it remains acceptable and widely utilized as the majority of wheel tracks fall within the lane width [17]. However, Gunay [9] highlights that shoulder lanes can lead to different distributions, where a single vehicle may occupy two lanes, deviating from the conventional normal distribution. Porter et al. [19] identify three categories of drivers: those inclined towards the lane middle line, those favoring the edge line, and those preferring a position near the center line. In the context of the normal approximation for lateral wandering, the standard deviation (STD) is commonly employed and varies depending on specific circumstances. The Mechanistic-Empirical Pavement Design Guide (MEPDG) [15] recommends deviations of 0.203, 0.254, and 0.305 for different lane widths. Additionally, the National Cooperative Highway Research Program (NCHRP) [1] suggests a deviation of 10 inches (approximately 0.254 m).

The aforementioned studies primarily focus on HDVs. However, in recent years, autonomous vehicles (AVs) have gradually been integrated into the transportation system. AVs exhibit different behavioral characteristics compared to HDVs, which impact the wandering patterns of mixed AV traffic. Firstly, AVs are equipped with precise sensors and controllers [27], enabling them to accurately position themselves along the desired track, often the lane middle line. Consequently, the long-term stress exerted by AVs on the pavement is concentrated around the lane middle line. In contrast, HDVs perceive imperfect information and exhibit stochastic behavior, resulting in fluctuations in their lateral positioning around the lane middle line. Secondly, HDVs react to approaching vehicles in neighboring lanes and maintain a safe lateral gap [5], a behavior known as lateral friction [6] or collateral anomaly [5]. Conversely, AVs can accurately predict the behavior of HDVs in neighboring lanes, owing to advancements in technologies such as deep learning [2], and are less influenced by their counterparts in adjacent lanes.

The research on modeling the lateral wandering of autonomous vehicles is limited in its versatility. Many researchers assume a default lateral wandering pattern, such as uniform-wander, zero-wander, or normal-wander [25]. Field tests conducted by Zhou et al. [27] revealed that the standard deviation of AVs' lateral wandering ranges from 0.03 to 0.075 m. This indicates that the lateral wandering of AVs is approximately 10–20% of that of HDVs, which is about 0.2 to 0.3 m. In the study by Zhou et al. [27], a deviation of 0.25 m (for HDVs) and 0.075 m (for AVs) was utilized. Yeganeh et al. [25] employed simulation to explore rutting under different AV wandering modes. The disparity between HDVs and AVs suggests that the lateral stress of AV traffic is more concentrated, thereby affecting pavement performance, including rutting and fatigue [4, 27]. Field observations have validated this influence of AVs [27], emphasizing the need to consider the introduction of AVs in the pavement design process [7, 8, 17, 26]. Efforts have also been made on the control side of AVs [3, 27]. These efforts propose specific lateral location distributions (rather than the fixed lane middle point) for trajectory generation or design deterministic trajectories that deviate from the lane middle line for AVs' feedback controllers. However, considering that HDVs are influenced by vehicles in neighboring lanes (including AVs), further investigation is needed for these new AV control methods. This is because the lateral location changes of AVs may alter the behavior of HDVs in adjacent lanes, subsequently affecting pavement performance in a chain manner on other lanes.

The empirical observations mentioned above provide crucial input for the road pavement design process, including the Mechanistic-Empirical Pavement Design Guide. However, there is limited attention given to the generation mechanism of the stochastic lateral wandering pattern. The driving process involves numerous factors that influence the lateral dynamics of vehicles, resulting in stochastic trajectories that form the lateral wandering pattern. In the case of multi-lane roads, there are lateral interactions among vehicles in neighboring lanes, leading to collateral wandering. To develop a wandering model and understand the generation mechanism of this wandering, it is necessary to investigate vehicle movement within the context of these influencing factors. Taking a behavioral viewpoint allows for the consideration of the influence of autonomous vehicles (AVs) as well.

9.4 Stochastic Lateral Behavioral Model

l_w: the lane width.

$S = [y, u, z]^T$: state variable for single-lane road. y is lateral deviation, u is lateral speed and z is the lateral noise.

$[y_i, u_i, z_i, y_j, u_j, z_j]^T$: state variable for two-lanes road. Lane i is the first lane and j is the second lane. The lane order is from left to right. The state variable for the multi-lane road can be defined similarly.

y_{MAX} and u_{MAX}: the threshold of the value of y and u.

$f_{mk}(\Delta y)$, $f_{bnd}(\Delta y)$, $f_{ne}(\Delta y)$, $f_u(y)$: the force functions of the lane mark, road boundary, neighboring lanes, and lane middle line. Note that in the former three definitions, the independent variables are defined as the distance from the ego vehicle to the lane mark, road boundary, and vehicles at neighboring lanes. And in $f_u(y)$, the independent variable is the deviation from the lane middle line.

α and β: the parameters used in the lane boundary and lane mark force expressions.

γ: the scaling parameter corresponding to the lane middle line attraction force.

erf(y): the conventional error function;

η_u, η_{bnd} η_{ne} and η_{mk}: the adjusting factor for the lane middle attraction force, lane bound force, intra-lane fore, and lane mark force.

b(·), L(·): the drift and diffusion term for single-lane system state equation. $b = [b_1, b_2, b_3]^T$ and $L = [l_1, l_2, l_3]$.

$\mathcal{N}(\mu, \Sigma)$: normal distribution with expectation μ and the covariance matrix Σ.

Dirac(·): the Dirac function;

O: the observations.

$L_\Theta(O)$: the likelihood function for the observations, which depends on parameters set Θ.

$[e_1, \dot{e}_1, e_2, \dot{e}_2]^T$: the state variable for a lateral-tracking control AV. The four components are lateral deviation (from the desired track), first-order differential of deviation, angular deviation from the desired track, and first-order differential of the angular deviation.

z_{AV}: the noise source of the AV.

σ_{AV}: the noise parameter for AV's lateral dynamics.

Θ_{CAV} and Θ_{HDV}: the parameters set for CAV and HDV respectively. $\Theta \overset{\text{def}}{=} \{\alpha, \beta, \gamma, \theta, \sigma, \eta_u, \eta_{bnd}, \eta_{ne}, \eta_{mk}\}$.

δ: the control input of the AV, which is the steering angle of the wheel.

Q and R: the matrix for the AV's control objective.

N_y and N_u: the number of discretization of the state y and u.

$[y]_m$, $[u]_n$: the m-th and n-th interval for y and u respectively.

$\mathcal{P}_{N_yN_u \times N_yN_u}$: transition matrix of the single-lane road.

ρ: the MPR of the CAV. ρ_i is the MPR at the i-th lane.

9.4.1 Lateral Behavior Model for HDVs

Vehicle movement is governed by several influencing factors, including the right of way defined on roads (such as lane boundaries and lane markings), the states of surrounding vehicles, and noise. The phenomenon of lateral wandering is a result of these factors. In this study, we consider these components simultaneously by incorporating (1) a static force defined by lane boundaries and markings, (2) a dynamic force generated by surrounding vehicles, and (3) noise represented by a stochastic differential equation. To represent the state of a single vehicle within a lane, we use the vector $S = [y, u, z]^T$, where y represents the lateral location, u represents the lateral speed, and z represents the noise. The distribution of y characterizes the lateral wandering within the lane. For a two-lane road, the state variables are denoted as $S = [y_i, u_i, z_i, y_j, u_j, z_j]$, representing lanes i and j. The evolution of the vehicle state within a general lane i is governed by the following stochastic differential equation:

$$\begin{cases} \mathrm{dy_i} = u_i dt & (a) \\ \mathrm{du_i} = forces([y_i, u_i, z_i], neighbores, noise, lane\ boundary\ and\ mark)dt & (b) \\ (\mathrm{dz_i} = f_{\#}([y_i, u_i, z_i])dt + L_{\#}([y_i, u_i, z_i])dW & (c) \end{cases} \quad (9.1)$$

We employ functions $f_{\#}$ and $L_{\#}$ to represent the noise, with $f_{\#}$ denoting the drift term and $L_{\#}$ representing the diffusion term. Subsequently, we proceed to construct models for the forces exerted by lane boundaries and lane marks, as well as the forces originating from neighboring lanes, and the noise factor.

9.4.1.1 Lateral Forces and Lateral Noise

We consider the following cources: lateral repulsive force of the road boundary and lane marks, Lateral attraction force of the lane middle line, Lateral repulsive force of the vehicles at neighboring lanes, and finally Lateral noise. Regarding lateral repulsive force of the road boundary and lane marks, when the vehicle is at the left or rightmost lane, where a road boundary exists, it must remain within the lane boundary. However, within the lane itself, there are soft constraints imposed by the lane mark. Consequently, at the boundary, the potential becomes infinitely high, while at the lane mark, it remains bounded. Commonly utilized potential expressions include arctangent or inverse proportional functions. In this case, we employ the inverse proportional function to represent the lane boundary. The general formula for the force exerted by the road boundary is defined as follows, with the distance to the boundary (Δy) serving as an input parameter:

$$\mathrm{f_{bnd}}(\Delta y) = \frac{1}{\left(\Delta y * \left(\frac{2}{l_w}\right)\right)^{\alpha}} - 1 \quad (9.2)$$

To modulate the rate at which the potential changes, a parameter α is introduced. If the driver does not consider the boundary until the vehicle is on the verge of touching it, a small value close to zero can be assigned to α (e.g., $\alpha = 0.1$). When the vehicle is positioned in the middle of the lane $\left(\Delta y = \frac{l_w}{2}\right)$, the force exerted is exactly zero.

In contrast to the lane boundary, the influence of lane marks on the vehicle is more flexible, allowing the vehicle to cross the lane mark if necessary to avoid a collision with other vehicles. The force exerted by the lane mark is described by Eq. 9.3, where Δy represents the distance between the vehicle and the lane mark. Deviations of the vehicle from the lane centerline elicit a force from the lane marks, prompting a realignment of the vehicle back to the middle of the lane. The direction of the force can be easily discerned.

$$\mathrm{f_{mk}}(\Delta y) = e^{-(\beta \cdot \Delta y)^2} \quad (9.3)$$

Regarding the lateral attraction force of the lane middle line, the primary objective of lateral movement is to align the vehicle with the center of the lane. When there is a certain deviation from the lane's centerline, the vehicle will steer either left or right to mitigate the deviation. The lateral speed and acceleration are represented by dy/dt and $\frac{d^2y}{dt^2}$, respectively. In order to create a model for the lane's centerline lateral movement, we observe that it exhibits an S-shaped trajectory, which can be described by various well-known mathematical functions such as the logistic curve and the error function. Given the simplicity of its derivative and its successful application in studying microscopic traffic behavior [18], we choose the error function as the basis for our model. The error function is formulated as $\mathrm{erf}(y) = \int_{-y}^{y} e^{-t^2} dt$. The derivative of the error function is given by e^{-y^2}. When a deviation y occurs, the vehicle will make adjustments to mitigate the deviation. In this process, the lateral speed should be positive when the deviation is negative, and vice versa. By scaling the derivative with γ and shifting the scaled function by $\frac{l_w}{2}$, we express the lateral speed u as follows:

$$u = \frac{dy}{dt} = \left(e^{-\gamma^2\left(y+\frac{l_w}{2}\right)^2} - e^{-\gamma^2\left(y-\frac{l_w}{2}\right)^2}\right)dt \overset{def}{=} f_y(y)dt \tag{9.4}$$

Acceleration can be derived further:

$$\frac{du}{dt} = \frac{df_u(y)}{dy}\frac{dy}{dt} = u\left(\gamma^2(2y - l_w)e^{-\gamma^2\left(y-\frac{l_w}{2}\right)^2} - \gamma^2(2y + l_w)e^{-\gamma^2\left(y+\frac{l_w}{2}\right)^2}\right) \overset{def}{=} uf_u(y) \tag{9.5}$$

By combining the lateral speed and acceleration, we obtain the following ordinary differential equation (ODE) as depicted in Eq. 9.6. The parameter γ signifies the extent of lateral regression for the deterministic scenario.

$$\begin{cases} dy = udt & (a) \\ du = u\left(\gamma^2(2y - l_w)e^{-\gamma^2\left(y-\frac{l_w}{2}\right)^2} - \gamma^2(2y + l_w)e^{-\gamma^2\left(y+\frac{l_w}{2}\right)^2}\right)dt & (b) \end{cases} \tag{9.6}$$

Regarding the lateral repulsive force of the vehicles at neighboring lanes, the lateral movement of a vehicle is influenced by the flow in neighboring lanes [6]. This stochastic movement may cause the vehicle to approach nearby vehicles. To prevent collisions, the vehicle adjusts its trajectory accordingly. When two vehicles are positioned precisely at the centerline of their respective lanes (with a lateral distance equal to the lane width), there is no mutual influence between them. However, if one vehicle is situated at the ego-lane's centerline while the other approaches the lane boundary, a significant repulsive force is required to avoid collision. Hence, we utilize the following formula to express the repulsive force $f_{ne}(\Delta y)$ (where 'ne' denotes neighboring), assuming that the lateral distance between the two vehicles is Δy:

$$f_{ne}(\Delta y) = \begin{cases} \frac{1}{l_w} - \frac{1}{\Delta y}, & \Delta y \leq l_w \\ 0, & else \end{cases} \tag{9.7}$$

Regarding the lateral random noise force, the vehicle's lateral movement is subject to disturbances caused by various sources of noise, such as driving behavior and stochastic road conditions. In microscopic traffic flow modeling, the Ornstein–Uhlenbeck process, also known as the OU process or Vasicek model in certain domains like financial modeling, is a commonly used noise model. The OU process has been successfully applied in modeling microscopic traffic behavior [23]. Consequently, we employ the Ornstein–Uhlenbeck process to characterize the noise. This type of noise possesses the following characteristics: (1) It exhibits fluctuations around a desired or expected value, and (2) it has finite variance (which is necessary for traffic flow noise) as compared to conventional Brownian noise. The basic formulation of the OU process is as follows:

$$\mathrm{dz} = \theta(\mu - z)dt + \sigma dW \tag{9.8}$$

The parameters in question hold distinct physical interpretations: θ represents the noise's reversal rate, μ signifies the equilibrium state, and σ denotes the noise level. In the case of the noise, the equilibrium state is set at zero, implying $\mu = 0$, resulting in the noise formulation $\mathrm{dz} = -\theta\,\mathrm{zdt} + \sigma\,\mathrm{dW}$. Here, θ and σ function as parameters. The process's mean and variance are expressed as:

$$E(z(t)) = z(0)e^{-\theta t};\ Var(z(t)) = \frac{\sigma^2}{2\theta}\left(1 - e^{-2\theta t}\right) \tag{9.9}$$

9.4.1.2 Lateral Behavior Model for HDV Flow on Single-Lane and Two-Lane Roads

Systhesis the model components and use state variable $[y, u, z]^T$, we have the following equation:

$$\begin{cases} dy = udt & (a) \\ du = \left[\eta_u u f_u(y) + \eta_{bnd} f_{bnd}(y) + \eta_{mk} f_{mk}(y) + z\right]dt & (b) \\ dz = -\theta zdt + \sigma dW & (c) \end{cases}$$

$$\Rightarrow \mathrm{d}\begin{bmatrix} y \\ u \\ z \end{bmatrix} = \begin{bmatrix} u \\ \eta_u u f_u(y) + \eta_{bnd} f_{bnd}(y) + \eta_{mk} f_{mk}(y) + z \\ -\theta z \end{bmatrix} dt + \begin{bmatrix} 0 \\ 0 \\ \sigma \end{bmatrix} dW$$

$$\stackrel{\mathrm{def}}{=} \mathrm{d}\begin{bmatrix} y \\ u \\ z \end{bmatrix} = b\left(\begin{bmatrix} y \\ u \\ z \end{bmatrix}\right)dt + L\left(\begin{bmatrix} y \\ u \\ z \end{bmatrix}\right)dw$$

$$\overset{\text{def}}{=} d\begin{bmatrix} y \\ u \\ z \end{bmatrix} = \begin{bmatrix} b_1 \\ b_2 \\ b_3 \end{bmatrix} dt + \begin{bmatrix} l_1 \\ l_2 \\ l_3 \end{bmatrix} dw \quad (9.10)$$

In the matrix formulation provided above, b and L are vectors, with b_i and l_i representing their respective components. Adjusting factors for the forces are denoted by η_u, η_{bnd} and η_{mk}. The Wiener process W is utilized.

In the case of multi-lane roads, we simplify the analysis by focusing on the two-lane road model, which can be naturally extended to roads with three or more lanes. Without loss of generality, we label the two lanes as i and j. The state of each lane is represented by $[y_i, u_i, z_i]^T$ and $[y_j, u_j, z_j]^T$, respectively, and these states are combined into a 6-dimensional vector to define the state variable of the two-lane road. By integrating the state equations of each lane, the system state equation can be expressed as follows:

$$\begin{cases} dy_i = u_i dt & (a) \\ du_i = \left[\eta_u u_i f_u(y_i) + \eta_{bnd} f_{bnd}(y_i) + \eta_{mk} f_{mk}(y_i) + \eta_{ne} f_{ne}(y_i - y_j) + z_i\right] dt & (b) \\ dz_i = -\theta z_i dt + \sigma\, dW & (c) \\ dy_j = u_j dt & (d) \\ du_j = \left[\eta_u u_j f_u(y_j) + \eta_{bnd} f_{bnd}(y_j) + \eta_{mk} f_{mk}(y_j) + \eta_{ne} f_{ne}(y_j - y_i) + z_j\right] dt & (e) \\ dz_j = -\theta z_j dt + \sigma dW & (f) \end{cases} \quad (9.11)$$

Upon comparing Eqs. 9.10b and 9.11b, it is evident that Eq. 9.11b incorporates an additional term, $\eta_{ne} f_{ne}(y_i - y_j)$, representing the interaction between vehicles. A similar term is present in Eq. 9.11e with the reversed sign for the distance, i.e., $y_j - y_i$ instead of $y_i - y_j$. In the case of a multiple-lane road comprising N lanes, the state variable dimension is 3*N. The system state equation follows a similar development. In scenarios involving more than two lanes, the intra-lane interaction ($f_{ne}(\cdot)$) exists solely between adjacent lanes.

9.4.2 *Lateral Model for AV*

AVs, in general, are designed to adhere to a predetermined or dynamically optimized trajectory. Despite being equipped with highly precise sensors and capable of swift reactions, they are still subject to noise disturbances. For instance, the interaction between the tire and road surface is stochastic and can be influenced by factors like rain or rutting. In this study, we introduce noise input to the feedback controller and investigate the lateral wandering pattern of AVs. The default plant model, as proposed by Xu et al. [24], is given in Eq. 9.12. The state variable $[e_1, \dot{e_1}, e_2, \dot{e_2}]^T$ represents the lateral deviation from the desired track, the first-order derivative of the deviation, the angular deviation from the desired track, and the first-order derivative of the angular deviation, respectively. The control input is denoted as δ, representing the steering angle of the wheel.

$$\frac{\mathrm{d}}{\mathrm{dt}}\begin{bmatrix} e_1 \\ \dot{e}_1 \\ \mathrm{e}_2 \\ \dot{e}_2 \end{bmatrix} = \begin{bmatrix} 0 & 1 & 0 & 0 \\ 0 & -\frac{2C_{af}+2C_{ar}}{mV_x} & \frac{2C_{af}+2C_{ar}}{m} & \frac{-2C_{af}l_f+2C_{ar}l_r}{mV_x} \\ 0 & 0 & 0 & 1 \\ 0 & -\frac{2C_{af}l_f-2C_{ar}l_r}{I_zV_x} & \frac{2C_{af}l_f-2C_{ar}l_r}{I_z} & -\frac{2C_{af}l_f^2+2C_{ar}l_r}{I_zV_x} \end{bmatrix}\begin{bmatrix} e_1 \\ \dot{e}_1 \\ \mathrm{e}_2 \\ \dot{e}_2 \end{bmatrix} dt + \begin{bmatrix} 0 \\ \frac{2C_{af}}{m} \\ 0 \\ \frac{2C_{af}l_f}{I_z} \end{bmatrix}\delta$$

$$\stackrel{\mathrm{def}}{=} \frac{dS}{dt} = A \cdot S + \mathrm{B} \cdot \delta \tag{9.12}$$

Assuming that the control gain is represented by the matrix G, the system can be reformulated as $\frac{\mathrm{dS}}{\mathrm{dt}} = (\mathrm{A} - \mathrm{BG})\delta$. If the control objective is defined as $\mathrm{J} = \int(\mathrm{S^TQS} + \mathrm{R} \cdot \delta^2)\mathrm{dt}$, the control gain G can be computed as $\mathrm{G} = (\mathrm{B^TXB} + \mathrm{R})^{-1}\mathrm{B^TXA}$, where X is determined by solving the Riccati equation $\mathrm{X} = \mathrm{Q} + \mathrm{A^T}(\mathrm{I} + \mathrm{XBR^{-1}B^T})^{-1}\mathrm{XA}$. For a detailed derivation, reference can be made to Xu et al. [24]. Given the stochastic nature of the system, we introduce a noise term $\mathrm{z_{AV}}$ to the system (now represented as $\frac{\mathrm{dS}}{\mathrm{dt}} = (\mathrm{A} - \mathrm{BG})\delta$. This noise term is sourced from the same origin as that for HDV but with a distinct parameter σ_{AV}. Consequently, the system state is expanded to $[\mathrm{e}_1, \dot{\mathrm{e}}_1, \mathrm{e}_2, \dot{\mathrm{e}}_2, \mathrm{z_{AV}}]^{\mathrm{T}}$, and the system equation can be expressed as follows:

$$\frac{\mathrm{d}}{\mathrm{dt}}\begin{bmatrix} e_1 \\ \dot{e}_1 \\ \mathrm{e}_2 \\ \dot{e}_2 \\ z_{av} \end{bmatrix} = \begin{bmatrix} 0 & 1 & 0 & 0 & 0 \\ 0 & -\frac{2C_{af}+2C_{ar}}{mV_x} & \frac{2C_{af}+2C_{ar}}{m} & \frac{-2C_{af}l_f+2C_{ar}l_r}{mV_x} & 0 \\ 0 & 0 & 0 & 1 & 0 \\ 0 & -\frac{2C_{af}l_f-2C_{ar}l_r}{I_zV_x} & \frac{2C_{af}l_f-2C_{ar}l_r}{I_z} & -\frac{2C_{af}l_f^2+2C_{ar}l_f^2}{I_zV_x} & 0 \\ 0 & 0 & 0 & 0 & -\theta \end{bmatrix}\begin{bmatrix} e_1 \\ \dot{e}_1 \\ \mathrm{e}_2 \\ \dot{e}_2 \\ z_{Av} \end{bmatrix} dt$$

$$+ \begin{bmatrix} 0 \\ \frac{2C_{af}}{m} \\ 0 \\ \frac{2C_{af}l_f}{I_z} \end{bmatrix}\delta + \begin{bmatrix} 0 \\ 0 \\ 0 \\ 0 \\ \sigma_{\mathrm{AV}} \end{bmatrix}\mathrm{dW} \tag{9.13}$$

The parameter σ_{AV} determining the system's stochasticity. The value of σ_{AV} should be smaller than that of HDVs due to several reasons: (1) the AV has more precise sensing capabilities, (2) the reaction delay can be negligible (in milliseconds), and (3) the control input can be precisely applied. According to Zhou et al. [27], the AV noise is approximately ten percent of that of HDVs. Therefore, we can set σ_{AV} to be ten percent of the HDV value. Figure 9.4 illustrates the AV's lateral behavior with $\sigma_{\mathrm{AV}} = 0.01$. The parameter set defined as follows: total vehicle mass m is 1.22510^3 kg; moment of inertia mass about z axis I_z is $1.538*10^3$ kgm^2; distance of center of gravity to front axle l_f is 0.883 m; distance of center of gravity to rear axle l_r is

1.508 m; Corning stiffness coefficient (front) C_{af} is 20.89/radius; corning stiffness coefficient (rear) C_{ar} is 20.89/radius. In Fig. 9.4a, the lateral paths are shown using the vehicle parameters from Table 9.1. Comparing them with the trajectories of HDV (Fig. 9.5b in subsequent simulations), we observe that the AV tracks are more centered with minimal oscillations, ranging approximately from −0.1 to 0.1 m. Figure 9.4b displays the standard deviation for various longitudinal speeds. At low speeds (3 m/s, or around 10 km/h), the deviation is approximately 0.08 m. At a speed of 20 mph (around 9 m/s, as used in Zhou et al. [27]), the deviation is roughly 0.057 m. The reported deviations in Zhou et al. [27] fall within the range of [0.03 m, 0.07 m]. Thus, the results obtained from the theoretical model align with empirical observations. Figure 9.4c presents the distribution of lateral wandering.

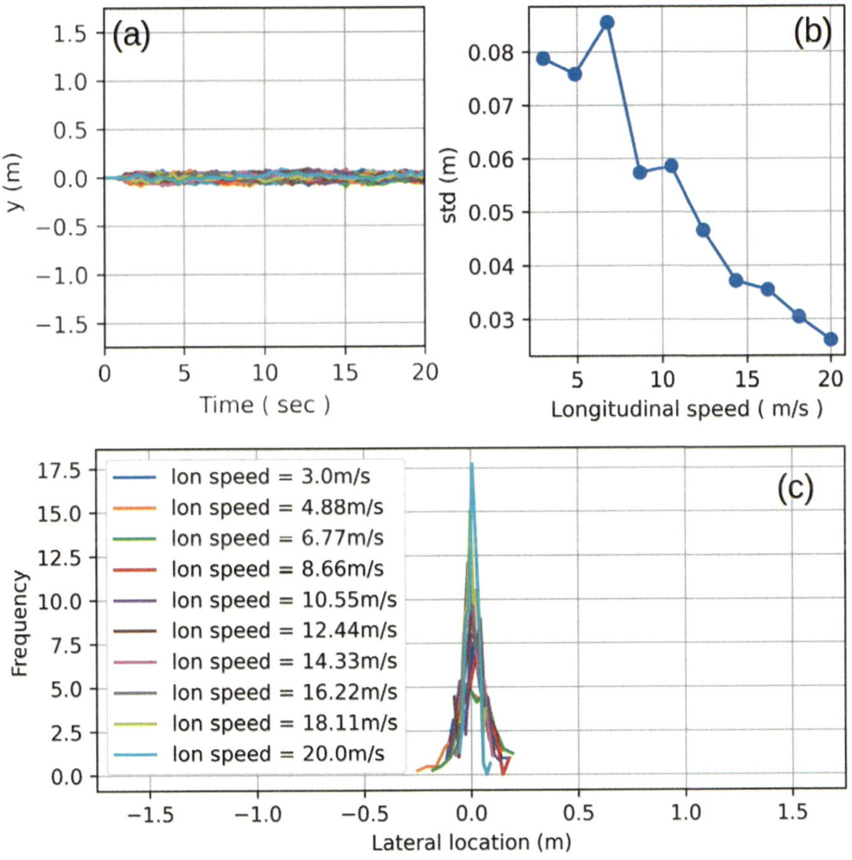

Fig. 9.4 Behaviors of the AV's lateral wandering; **a** tracks; **b** standard deviation; **c** lateral wandering. Reprint from Ref. [21], copyright 2023, with permissions from Elsevier

Table 9.1 Comparison of the standard deviation for different lane width

Lane width l_w (m)	3	3.12	3.37	3.62
Theoretical standard deviation (m)	0.206	0.222	0.257	0.302
MEPDG [15]	0.203 ($l_w < 3m$)	0.254 ($3 \le l_w < 3.6$)		0.305 ($l_w > 3.6m$)
Existing research	0.24 ($l_w \in [2.88, 3.12]$)		0.26 ($l_w \in [3.13, 3.37]$)	0.29 ($l_w \in [3.38, 3.62]$)

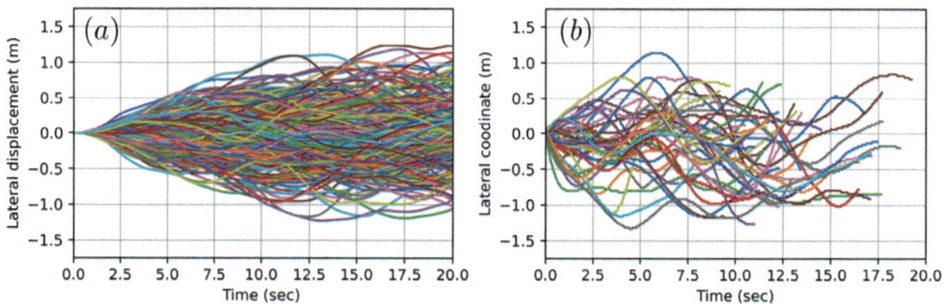

Fig. 9.5 Simulation (**a**) and the real-world data (**b**). Reprint from Ref. [21], copyright 2023, with permissions from Elsevier

9.5 Stochastic Lateral Behavioral Model Calibration

The model parameters encompass α, β, γ, θ, σ, and $\eta_{\#}$, where $\# \in \{mk, bnd, ne\}$. Collectively, these parameters are denoted as $\Theta \overset{def}{=} \{\alpha, \beta, \gamma, \theta, \sigma, \eta_{\#}\}$, representing the characteristics of the road-lane-driver-noise system. The known data, used for parameter determination, consists of $\left(y_i(k), u_i(k), y_j(k), u_j(k)\right)$ observations from two-lane roads. Here, 'k' denotes the k-th observation for the two vehicles in neighboring lanes. We express these observations as $O = \left\{\left(y_i(k), u_i(k), y_j(k), u_j(k)\right)\right\}, k = 1,2\ldots N$, with k ranging from 1 to N. The system's transition equation is developed as follows:

$$\begin{bmatrix} y_i(k+1) \\ u_i(k+1) \\ z_i(k+1) \\ y_j(k+1) \\ u_j(k+1) \\ z_j(k+1) \end{bmatrix} = \begin{bmatrix} y_i(k) \\ u_i(k) \\ z_i(k) \\ y_j(k) \\ u_j(k) \\ z_j(k) \end{bmatrix} + F\left(\begin{bmatrix} y_i(k) \\ u_i(k) \\ z_i(k) \\ y_j(k) \\ u_j(k) \\ z_j(k) \end{bmatrix}\right)\Delta t + L\left(\begin{bmatrix} y_i(k) \\ u_i(k) \\ z_i(k) \\ y_j(k) \\ u_j(k) \\ z_j(k) \end{bmatrix}\right)\Delta W_{2\times 1} \quad (9.14)$$

In the given expression for a single step, $\Delta W_{2\times 1}$ represents a two-dimensional Gaussian variable with a mean of zero and a variance matrix Q_L given by diag(Δt,

Δt). Equation 9.14 defines the matrices F(·) and L(·) as follows:

$$\begin{cases} F\left(\begin{bmatrix} y_i(k) \\ u_i(k) \\ z_i(k) \\ y_j(k) \\ u_j(k) \\ z_j(k) \end{bmatrix}\right) = \begin{bmatrix} u_i(k) \\ \eta_u u_i(k) f_u(y_i(k)) + \eta_{bnd} f_{bnd}(y_i(k)) \\ +\eta_{mk} f_{mk}(y_i(k)) + \eta_{ne} f_{ne}\left(y_i(k) - y_j(k)\right) + z_i(k) \\ -\theta z_i(k) \\ u_j(k) \\ \eta_u u_j(k) f_u\left(y_j(k)\right) + \eta_{bnd} f_{bnd}\left(y_j(k)\right) + \eta_{mk} f_{mk}\left(y_j(k)\right) \\ +\eta_{ne} f_{ne}\left(y_j(k) - y_i(k)\right) + z_j(k) \\ -\theta z_j(k) \end{bmatrix} \\ L\left(\begin{bmatrix} y_i(k) \\ u_i(k) \\ z_i(k) \\ y_j(k) \\ u_j(k) \\ z_j(k) \end{bmatrix}\right) = \begin{bmatrix} 0 & 0 \\ 0 & 0 \\ \sigma & 0 \\ 0 & 0 \\ 0 & 0 \\ 0 & \sigma \end{bmatrix} \end{cases} \tag{9.15}$$

Therefore, the one-step conditional distribution can be expressed as:

$$p_\Theta\left(\begin{bmatrix} y_i(k+1) \\ u_i(k+1) \\ z_i(k+1) \\ y_j(k+1) \\ u_j(k+1) \\ z_i(k+1) \end{bmatrix} \begin{bmatrix} y_i(k) \\ u_i(k) \\ z_i(k) \\ y_j(k) \\ u_j(k) \\ z_i(k) \end{bmatrix}\right) \sim \text{Dirac}\left(\begin{bmatrix} y_i(k+1) \\ u_i(k+1) \\ z_i(k+1) \\ y_j(k+1) \\ u_j(k+1) \\ z_j(k+1) \end{bmatrix} - \begin{bmatrix} y_i(k) \\ u_i(k) \\ z_i(k) \\ y_j(k) \\ u_j(k) \\ z_j(k) \end{bmatrix} - F\left(\begin{bmatrix} y_i(k) \\ u_i(k) \\ z_i(k) \\ y_j(k) \\ u_j(k) \\ z_j(k) \end{bmatrix}\right)\Delta t\right) \cdot \mathcal{N}\left(\begin{bmatrix} -\theta z_i(k) \\ -\theta z_j(k) \end{bmatrix}, \begin{bmatrix} \sigma & 0 \\ 0 & \sigma \end{bmatrix}\right) \tag{9.16}$$

The Dirac function, denoted as Dirac(·), is used in the context. $\mathcal{N}\left(\begin{bmatrix} -\theta z_i(k) \\ -\theta z_j(k) \end{bmatrix}, \begin{bmatrix} \sigma & 0 \\ 0 & \sigma \end{bmatrix}\right)$ represents a bivariate Gaussian distribution. Using the one-step transition probability from Eq. 9.16, the likelihood function for the observation O can be expressed as follows:

$$\mathrm{L}_{\Theta}(O)=\prod_{k=1}^{N-1}\mathrm{p}_{\Theta}\left(\begin{bmatrix} y_i(k+1)\\ u_i(k+1)\\ z_i(k+1)\\ y_j(k+1)\\ u_j(k+1)\\ z_j(k+1)\end{bmatrix}\begin{bmatrix} y_i(k)\\ u_i(k)\\ z_i(k)\\ y_j(k)\\ u_j(k)\\ z_j(k)\end{bmatrix}\right) \tag{9.17}$$

Therefore, the optimization of the parameters can be achieved by finding $\Theta^* = \mathrm{argmax}\, \mathrm{L}_{\Theta}(\mathrm{O})$. This parameter optimization can be accomplished using the particle swarm optimization (PSO) algorithm. The same procedure can be applied to both single and multi-lane roads.

9.6 Stochastic Lateral Wandering Model

The lateral displacement distribution known as lateral wandering is caused by stochastic lateral movement. The aforementioned section has elaborated on the stochastic lateral movement for HDVs and AVs. The distribution p(y) is employed to depict the lateral wandering. Several methods can be utilized to derive the distribution p(y) from the behavioral model defined in Eqs. 9.10 and 9.11. In the subsequent sections, we outline three approaches for the wandering pattern of HDVs. Initially, we present the Euler–Maruyama scheme, followed by the derivation of the Fokker–Planck equation (FPE), and ultimately, the development of a Markov chain based approach.

9.6.1 Pure HDV Case

9.6.1.1 Euler–Maruyama Scheme

One way to obtain the lateral wandering distribution p(y) is through the utilization of numerical methods. In this regard, we employ the Euler–Maruyama scheme on a single lane road. Assuming the system state at time k is $[y(k), u(k), z(k)]^T$, the state at time k + 1 can be represented as:

$$\begin{bmatrix} y(k+1)\\ u(k+1)\\ z(k+1)\end{bmatrix}=\begin{bmatrix} y(k)\\ u(k)\\ z(k)\end{bmatrix} + \begin{bmatrix} u(k)\\ \eta_u u(k) f_u(y(k)) + \eta_{bnd} f_{bnd}(y(k)) + \eta_{mk} f_{mk}(y(k))\\ -\theta z(k)\end{bmatrix}\Delta t + \begin{bmatrix} 0\\ 0\\ \sigma\end{bmatrix}\Delta W \tag{9.18}$$

The time step is denoted as Δt, while ΔW represents a Gaussian variable with zero mean and a variance of Δt. Consequently, given the state vector $\left[y(k), u(k), z(k)\right]^T$, the state vector $\left[y(k+1), u(k+1), z(k+1)\right]^T$ follows a normal distribution. Its mean, denoted as $\mu_{\begin{bmatrix} y(k+1) \\ u(k+1) \\ z(k+1) \end{bmatrix}}$, and covariance matrix Σ can be formulated as:

$$\begin{cases} \mu_{\begin{bmatrix} y(k+1) \\ u(k+1) \\ z(k+1) \end{bmatrix}} = \begin{bmatrix} y(k) \\ u(k) \\ z(k) \end{bmatrix} + \begin{bmatrix} u(k) \\ \eta_u u(k) f_u(y(k)) + \eta_{\text{bnd}} f_{\text{bnd}}(y(k)) \\ + \eta_{mk} f_{mk}(y(k)) + z(k) \\ -\theta z(k) \end{bmatrix} \delta t \quad (a) \\ \Sigma = \begin{bmatrix} 0 & 0 & 0 \\ 0 & 0 & 0 \\ 0 & 0 & \sigma^2 \end{bmatrix} \quad (b) \end{cases} \tag{9.19}$$

In Fig. 9.5, a comparison is shown between simulation data and real-world data obtained from the same initial location $y = 0$, using the parameters specified in the case study section. The evolution of both the simulation and real-world data exhibits stochastic behavior. Approximately 20 s into the experiment, both the real-world data and simulation deviate from the lane middle line ($y = 0$) by about 1 m. Based on the trajectories, the distribution $p(y)$ can be easily obtained for any given moment.

For the two-lane road behavioral model in Eq. 9.11, the Euler–Maruyama scheme can be obtained as follows:

$$\begin{bmatrix} y_i(k+1) \\ u_i(k+1) \\ y_j(k+1) \\ u_j(k+1) \end{bmatrix} = \begin{bmatrix} y_i(k) \\ u_i(k) \\ y_j(k) \\ u_j(k) \end{bmatrix} + \begin{bmatrix} u_i(k) \\ \eta_u u_i(k) f_u(y_i(k)) + \eta_{bnd} f_{bnd}(y_i(k)) \\ + \eta_{mk} f_{mk}(y_i(k)) + \eta_{ne} f_{ne}\left(y_i(k) - y_j(k)\right) \\ u_j(k) \\ \eta_u u_j(k) f_u\left(y_j(k)\right) + \eta_{bnd} f_{bnd}\left(y_j(k)\right) \\ + \eta_{mk} f_{mk}\left(y_j(k)\right) + \eta_{ne} f_{ne}\left(y_j(k) - y_i(k)\right) \end{bmatrix} \Delta t + \begin{bmatrix} 0 \\ z_i(k) \\ 0 \\ z_j(k) \end{bmatrix} \Delta t \tag{9.20}$$

The numerical scheme for multi-lanes road can be derived in a similar way, which is omitted here.

9.6.1.2 Fokker–Planck Equation

The state distribution of a stochastic differential equation can be described by a partial differential equation, known as the Fokker–Planck equation or FPE. In this case, we are interested in solving Eq. 9.10, which represents an evolving distribution P(y, u, z, t). Due to the challenges associated with directly solving the stochastic differential equation, we instead derive the FPE for the system described in Eq. 9.10.

$$\frac{\partial P}{\partial t} = -\left[\frac{\partial(b_1 \cdot P)}{\partial y} + \frac{\partial(b_2 \cdot P)}{\partial u} + \frac{\partial(b_3 \cdot P)}{\partial z}\right] + \frac{1}{2}\left[\frac{\partial^2(\sigma^2 p)}{\partial z^2}\right] \tag{9.21}$$

The functions are defined as follows: $b_1([y, u, z]^T) = u$, $b_2([y, u, z]^T) = [\eta_u u f_u(y) + \eta_{bnd} f_{bnd}(y) + \eta_{mk} f_{mk}(y) + z]$, and $b_3 = -\theta z$. And hence:

$$\frac{\partial(b_1 \cdot P)}{\partial y} = b_1\frac{\partial P}{\partial y} + P\frac{\partial b_1}{\partial y} = b_1\frac{\partial P}{\partial y} \tag{9.22}$$

The partial diferential $\frac{\partial(b_2 \cdot P)}{\partial u}$ can be expressed:

$$\begin{aligned}\frac{\partial(b_2 \cdot P)}{\partial u} &= b_2\frac{\partial P}{\partial u} + P\frac{\partial b_2}{\partial u} \\ &= \left[u f_u(y) + f_{bnd}(y) + f_{mk}(y) + z\right]\frac{\partial P}{\partial u} + f_u(y)P\end{aligned} \tag{9.23}$$

Further:

$$\frac{\partial(b_3 \cdot P)}{\partial z} = b_3\frac{\partial P}{\partial z} + P\frac{\partial b_3}{\partial z} = -\theta z\frac{\partial P}{\partial z} - \theta P \tag{9.24}$$

Combining the aforementioned components, we can express Eq. 9.21 in the following manner:

$$\frac{\partial P}{\partial t} = -\left\{\begin{aligned} & b_1\frac{\partial P}{\partial y} + \left[\eta_u u f_u(y) + \eta_{bnd} f_{bnd}(y) + \eta_{mk} f_{mk}(y) + z\right]\frac{\partial P}{\partial y} \\ & + f_u(y)P - \theta z\frac{\partial P}{\partial z} - \theta P \end{aligned}\right\} + \frac{\sigma^2}{2}\left[\frac{\partial^2 p}{\partial z^2}\right] \tag{9.25}$$

For stable distribution, we can directly set $\frac{\partial P}{\partial t}$ to zero, and hence we have the final expression for the stable distribution P:

$$0 = (\theta - f_u(y))P - \overbrace{\left\{b_1 \frac{\partial P}{\partial y} + \left[\eta_u u f_u(y) + \eta_{bnd} f_{bnd}(y) + \eta_{mk} f_{mk}(y) + z\right] \frac{\partial P}{\partial u} - \theta z \frac{\partial P}{\partial z}\right\}}^{firstorderterm}$$

$$+ \overbrace{\frac{\sigma^2}{2}\left[\frac{\partial^2 p}{\partial z^2}\right]}^{2ndorderterm} \tag{9.26}$$

By solving the equation mentioned above, we obtain the distribution P(y, u, z). To obtain the desired marginal distribution p(y), we integrate out the variables u and z. However, solving the aforementioned equation is challenging, and even if solvable, it becomes even more complex in multi-lane scenarios with additional independent variables. Therefore, we address this issue by employing the Markov chain model.

9.6.1.3 Markov Chain Solution

While we have derived the Fokker–Planck equation (FPE) for the distribution p(y, u, z) that allows us to obtain the desired distribution p(y), it cannot be directly applied. Initially, we focus on the one-lane road model described in Eq. 9.10. We observe that the noise z follows an Ornstein–Uhlenbeck (OU) process with a known distribution, thus enabling us to exclude the third equation from Eq. 9.18. This simplification results in the following expression:

$$\begin{bmatrix} y(k+1) \\ u(k+1) \end{bmatrix} = \begin{bmatrix} y(k) \\ u(k) \end{bmatrix} + \begin{bmatrix} u(k) \\ \eta_u u f_u(y) + \eta_{bnd} f_{bnd}(y) + \eta_{mk} f_{mk}(y) \end{bmatrix} \delta t + \begin{bmatrix} 0 \\ z(k) \end{bmatrix} \delta t \tag{9.27}$$

We denote z(k) separately in Eq. 9.27 as an independent random variable. Referring to Iacus [10], we can express the mean and variance of z(k) as follows:

$$E(z(k)) = z(0)e^{-\theta k \cdot \delta t}; \; Var(z(k)) = \frac{\sigma^2}{2\theta}\left(1 - e^{-2\theta \cdot k \cdot \delta t}\right) \tag{9.28}$$

Therefore, when given $\left[y(k), u(k)\right]^T$, the distribution of $\left[y(k+1), u(k+1)\right]^T$ can be represented by a normal distribution.

$$\left[y(k+1), u(k+1)\right]^T \sim \mathcal{N}\left(\mu\left|\begin{bmatrix} y(k) \\ u(k) \end{bmatrix}\right., \Sigma\left|\begin{bmatrix} y(k) \\ u(k) \end{bmatrix}\right.\right) \tag{9.29}$$

If we assume zero noise at the initial state, the mean and variance of this distribution can be expressed as follows:

$$\begin{cases} \mu_{\left\|\begin{bmatrix} y(k) \\ u(k) \end{bmatrix}\right.} = \begin{bmatrix} y(k) + u(k)\Delta t \\ u(k) + \eta_u u(k) f_u(y(k)) \cdot \Delta t + \eta_{\text{bnd}} f_{\text{bnd}}(y(k)) \cdot \Delta t + \eta_{mk} f_{mk}(y(k)) \cdot \Delta t \end{bmatrix} & (a) \\ \Sigma_{\left\|\begin{bmatrix} y(k) \\ u(k) \end{bmatrix}\right.} = \Delta t^2 \begin{bmatrix} 0 & \sigma^2 \\ 0 & \frac{0}{2\theta}\left(1 - e^{-2\theta k \delta t}\right) \end{bmatrix} & (b) \end{cases} \tag{9.30}$$

The equation above represents the one-step transition of the state variable $[y, u]^T$. As k approaches a sufficiently large value, the term $e^{-2\theta k \delta t}$ in $\Sigma_{\left\|\begin{bmatrix} y(k) \\ u(k) \end{bmatrix}\right.}$ tends to zero. Consequently

$$\Sigma_{\left\|\begin{bmatrix} y(k) \\ u(k) \end{bmatrix}\right.} \sim \delta t^2 \begin{bmatrix} 0 & 0 \\ 0 & \frac{\sigma^2}{2\theta} \end{bmatrix} \tag{9.31}$$

To establish a Markov chain, we discretize the lateral position (y) and speed (u) into intervals. The domain of y, denoted as $\left[-\frac{l_w}{2}, \frac{l_w}{2}\right]$ or equivalently $\left[-y_{MAX}, y_{MAX}\right]$, and the domain of u, $[-u_{MAX}, u_{MAX}]$, are discretized into N_y and N_u intervals respectively. The step sizes, Δy and Δu, are derived as $\frac{l_w}{N_y}$ and $\frac{2u_{MAX}}{N_u}$ respectively. Each interval of y, denoted as $[y]_i$, has a domain of $\left[-\frac{l_w}{2} + \frac{(i-1)l_w}{N_y}, -\frac{l_w}{2} + \frac{il_w}{N_y}\right]$. Similarly, each interval of u, denoted as $[u]_j$, has a domain of $\left[-u_{MAX} + \frac{2(n-1)u_{MAX}}{N_u}, -u_{MAX} + \frac{2n \cdot u_{MAX}}{N_u}\right]$. Suppose the current system state is $([y]_i, [u]_j)$, and the state at the next time step is $([y]_m, [u]_n)$. For a specific point $[y, u]^T$ within the space $([y]_i, [u]_j j)$, the transition probability to the space $([y]_m, [u]_n)$ can be calculated as follows:

$$p([y]_m, [u]_n | y, u) = \int_{-u_{MAX} + \frac{2(n-1)u_{MAX}}{N_u}}^{-u_{MAX} + \frac{2n \cdot u_{MAX}}{N_u}} \left[\int_{-\frac{l_w}{2} + \frac{(m-1)l_w}{N_y}}^{-\frac{l_w}{2} + \frac{ml_w}{N_y}} Dirac\left(y - \mu_{\left\|\begin{bmatrix} y \\ u \end{bmatrix}\right.}\right) - \mathcal{N}\left(\begin{bmatrix} y_\# \\ u_\# \end{bmatrix}; \mu_{\left\|\begin{bmatrix} y \\ u \end{bmatrix}\right.}, \Sigma_{\left\|\begin{bmatrix} y \\ u \end{bmatrix}\right.}\right) dy_\# \right] du_\# \tag{9.32}$$

By utilizing the Dirac function $\mathrm{Dirac}\left(y - \mu\left(\begin{bmatrix} y \\ u \end{bmatrix}\right)\right)$, we can integrate the values of y that do not equal $\mu\left(\begin{bmatrix} y \\ u \end{bmatrix}\right)$ using Eq. 9.30. In the case where $[y]_m$ does not contain the value $\mu\left(\begin{bmatrix} y \\ u \end{bmatrix}\right)$, the probability $p([y]_m, [u]_n|y, u)$ becomes zero. However, if there exists a $y_{\#} \in \left[-\frac{l_w}{2} + \frac{(m-1)l_w}{N_y}, -\frac{l_w}{2} + \frac{ml_w}{N_y}\right]$ that is equal to $\mu\left(\begin{bmatrix} y \\ u \end{bmatrix}\right)$, the probability is calculated as follows:

$$p([y]_m, [u]_n|y, u) = \int_{-u_{MAX} + \frac{2(n-1)u_{MAX}}{N_u}}^{-u_{MAX} + \frac{2n \cdot u_{MAX}}{N_u}} \mathcal{N}\left(\begin{array}{c} u_{\#}; u + \eta_u u f_u(y) \cdot \Delta t + \eta_{bnd} f_{bnd}(y) \cdot \Delta t \\ + \eta_{mk} f_{mk}(y) \cdot \Delta t, \frac{\sigma^2}{2\theta} \end{array}\right) du_{\#} \tag{9.33}$$

The integration described above is a standard integration over a normal distribution and cannot be further simplified. It can be computed using the error function:

$$\begin{aligned} p([y]_m, [u]_n|y, u) = {} & \mathrm{erf}\left(\frac{\sqrt{\theta}\left(-u_{MAX} + \frac{2n \cdot u_{MAX}}{N_u} - \left(\begin{array}{c} u + \eta_u u f_u(y) \cdot \Delta t \\ + \eta_{bnd} f_{bnd}(y) \cdot \Delta t + \eta_{mk} f_{mk}(y) \cdot \Delta t \end{array}\right)\right)}{\sigma}\right) \\ & - \mathrm{erf}\left(\frac{\sqrt{\theta}\left(-u_{MAX} + \frac{2(n-1)u_{MAX}}{N_u} - \left(\begin{array}{c} u + \eta_u u f_u(y) \cdot \Delta t + \\ \eta_{bnd} f_{bnd}(y) \cdot \Delta t + \eta_{mk} f_{mk}(y) \cdot \Delta t \end{array}\right)\right)}{\sigma}\right) \end{aligned} \tag{9.34}$$

The transition probability from $([y]_i, [u]_j)$ to $([y]_m, [u]_n)$ is the integration over the interval $([y]_i, [u]_j)$:

$$p([y]_m, [u]_n|[y]_i, [u]_j) = \int_{[u]_j} \int_{[y]_i} p([y]_m, [u]_n|y, u) dy du \tag{9.35}$$

This probability cannot be concisely expressed and must be numerically calculated. By obtaining $p([y]_m, [u]_n|[y]_i, [u]_j)$, for all i, j, m, n, the transition matrix of the system states can be constructed. The transition matrix is denoted as $\mathcal{P}_{N_y N_u \times N_y N_u}$. The equilibrium distribution, π, can be obtained by solving the following linear equation:

$$\pi_{1\times N_yN_u} = \pi_{1\times N_yN_u}\mathcal{P}_{N_yN_u\times N_yN_u} \tag{9.36}$$

The discretized state interval $([y]_m, [u]_n)$ can be indexed as $(m-1)\cdot N_u + n$, representing the row and column index. Consequently, the marginal distribution of the state interval $([y]_m, [u]_n)$ corresponds to the appropriate element in $\pi_{1\times N_yN_u}$.

Similarly, the Markov chain model for a two-lane road given in Eq. 9.11 can be derived. The Euler–Maruyama scheme, shown in Eq. 9.20, already provides an expression. The variables $z_i(k)$ and $z_j(k)$ in Eq. 9.20 follow a normal distribution, and their mean and variance can be expressed as follows, in accordance with the theory of the OU process:

$$\begin{cases} E(z_i(k)) = z_i(0)e^{-\theta_i k\cdot\Delta t};\ Var(z_i(k)) = \frac{\sigma_i^2}{2\theta_i}\left(1-e^{-2\theta_i\cdot k\cdot\Delta t}\right) \\ E(z_j(k)) = z_j(0)e^{-\theta_j k\cdot\Delta t};\ Var(z_j(k)) = \frac{\sigma_j^2}{2\theta_j}\left(1-e^{-2\theta_j\cdot k\cdot\Delta t}\right) \end{cases} \tag{9.37}$$

Considering $[y_i(k), u_i(k), y_j(k), u_j(k)]^T$, the distribution $[y_i(k+1), u_i(k+1), y_j(k+1), u_j(k+1)]^T$ follows a Multivariate Gaussian distribution with certain mean and variance:

$$\begin{cases} \mu\left(\begin{bmatrix} y_i(k) \\ u_i(k) \\ y_j(k) \\ u_j(k) \end{bmatrix}\right) = \begin{bmatrix} y_i(k) + \Delta t u_i(k) \\ \left(\begin{matrix} \eta_u u_i(k) f_u(y_i(k)) + \eta_{bnd} f_{bnd}(y_i(k)) + \\ \eta_{mk} f_{mk}(y_i(k)) + \eta_{ne} f_{ne}(y_i(k) - y_j(k)) \end{matrix}\right)\Delta t \\ y_j(k) + \Delta t u_j(k) \\ \left(\begin{matrix} \eta_u u_j(k) f_u(y_j(k)) + \eta_{bnd} f_{bnd}(y_j(k)) + \\ \eta_{mk} f_{mk}(y_j(k)) + \eta_{ne} f_{ne}(y_j(k) - y_i(k)) \end{matrix}\right)\Delta t \end{bmatrix} & \text{(a)} \\ \Sigma\left(\begin{bmatrix} y_i(k) \\ u_i(k) \\ y_j(k) \\ u_j(k) \end{bmatrix}\right) = \Delta t^2 \begin{bmatrix} 0 & 0 & & \\ 0 & \frac{\sigma_i^2}{2\theta_i}\left(1-e^{-2\theta_i k\Delta t}\right) & 0 & 0 \\ & & 0 & 0 \\ & 0 & & \frac{\sigma_j^2}{2\theta_j}\left(1-e^{-2\theta_j k\Delta t}\right) \end{bmatrix} & \text{(b)} \end{cases} \tag{9.38}$$

The covariance matrix is singular and diagonal, implying the elements of $[y_i(k+1), u_i(k+1), y_j(k+1), u_j(k+1)]^T$ can be treated as independent. Hence, the distribution can be expressed as:

$$p\left(\begin{bmatrix} y_i(k+1) \\ u_i(k+1) \\ y_j(k+1) \\ u_j(k+1) \end{bmatrix}\middle|\begin{bmatrix} y_i(k) \\ u_i(k) \\ y_j(k) \\ u_j(k) \end{bmatrix}\right) = \mathrm{Dirac}\left(\begin{bmatrix} y_i(k+1) \\ y_j(k+1) \end{bmatrix} - \begin{bmatrix} y_i(k) + \Delta t u_i(k) \\ y_j(k) + \Delta t u_j(k) \end{bmatrix}\right)$$

$$\times \mathcal{N}\left(\mu_{\left[\begin{array}{c} y_i(k) \\ u_i(k) \\ y_j(k) \\ u_j(k) \end{array}\right]}, \Sigma_{\left[\begin{array}{c} y_i(k) \\ u_i(k) \\ y_j(k) \\ u_j(k) \end{array}\right]}\right) \tag{9.39}$$

Utilizing the Dirac function, similar to the single-lane road case, we can construct the Markov chain model for a two-lane road. The process is essentially identical to that of a single-lane road.

9.6.2 *Mixed AVs Flow Case*

The mixed AV flow's wandering can be understood as the combination of the wandering observed in pure HDV traffic and pure AV traffic, with the AV's Market Penetration Rate (MPR) denoted by ρ within the range of [0, 1]. The behavior disparity between HDV and AV can be categorized into two aspects: (1) AV exhibits minimal lateral fluctuation compared to HDV; (2) AV is less affected by neighboring lane vehicles' lateral deviations due to its accurate inference capability of HDV lane-changing behavior. In the case of single-lane roads, the second effect can be disregarded. Therefore, when AV's MPR is ρ, the overall lateral wandering is simply the weighted sum of the AV's lateral wandering and that of HDVs.

$$p(y) = \rho \cdot p(y|\Theta_{AV}) + (1 - \rho)p(y|\Theta_{HDV}) \tag{9.40}$$

The wandering of the AV, denoted as $p(y|\Theta_{AV})$, and the wandering of HDV, denoted as $p(y|\Theta_{HDV})$, characterize the respective behaviors. The distribution $p(y)$ can be obtained through Euler–Maruyama simulation. In the context of multi-lane roads with mixed AV and HDV flow, each lane may have its own Market Penetration Rate, denoted as ρ_i. For a two-lane case, we represent the MPR of AV in each lane (lane 0 and lane 1) as a vector $\mathbb{P} = [\rho_0, \rho_1]$, where $[1 - \rho_0, 1 - \rho_1]$ represents the ratio of HDV. The interaction modes between neighboring lanes encompass four combinations: HDV-AV, HDV-HDV, AV-HDV, and AV-AV. These modes are characterized by the parameters $[\Theta_{HDV}, \Theta_{AV}]$, $[\Theta_{HDV}, \Theta_{HDV}]$, $[\Theta_{AV}, \Theta_{HDV}]$ and $[\Theta_{AV}, \Theta_{AV}]$, respectively. For example, $[\Theta_{HDV}, \Theta_{AV}]$ indicates that the first lane consists of HDV while the second lane consists of AV. The probabilities for these four modes are $(1 - \rho_0)\rho_1$, $(1 - \rho_0)(1 - \rho_1)$, $\rho_0(1 - \rho_1)$ and $\rho_0\rho_1$, respectively. Consequently, the final lateral wandering is the weighted sum of these four modes. The calculation of the wandering $p_0(y)$ for the lane with ID 0 is as follows:

$$\begin{aligned} p_0(y) = & (1-\rho_0)\rho_1 \cdot p_0(y|[\Theta_{HDV}, \Theta_{AV}]) \\ & + (1-\rho_0)(1-\rho_1) \cdot p_0(y|[\Theta_{HDV}, \Theta_{HDV}]) \\ & + \rho_0(1-\rho_1) \cdot p_0(y|[\Theta_{AV}, \Theta_{HDV}]) \\ & + \rho_0\rho_1 \cdot p_0(y|[\Theta_{AV}, \Theta_{AV}]) \end{aligned} \quad (9.41)$$

And the wandering for lane 1, $p_1(y)$ is expressed as:

$$\begin{aligned} p_1(y) = & (1-\rho_0)\rho_1 \cdot p_1(y|[\Theta_{HDV}, \Theta_{AV}]) \\ & + (1-\rho_0)(1-\rho_1) \cdot p_1(y|[\Theta_{HDV}, \Theta_{HDV}]) \\ & + \rho_0(1-\rho_1) \cdot p_1(y|[\Theta_{AV}, \Theta_{HDV}]) \\ & + \rho_0\rho_1 \cdot p_1(y|[\Theta_{AV}, \Theta_{AV}]) \end{aligned} \quad (9.42)$$

Consequently, by aggregating $p_0(y)$ and $p_1(y)$, the overall road's wandering can be obtained. This principle can be extended to multi-lane roads, accommodating different MPR definitions for each lane.

9.7 Stochastic Wandering Validation

We begin by examining scenarios where the vehicle operates on a single-lane road, thereby lacking neighboring vehicles in the left or right lanes. Figure 9.6 displays the system paths (representing the stochastic evolution of lateral location) and lateral wandering in this context. Figure 9.6a illustrates a subset of the system paths from 1000 simulations, while Fig. 9.6b presents marginal distributions at moments beyond 10 s (each curve corresponds to a specific moment). Notably, the distributions tend to stabilize. When it comes to single-lane roads, the lateral wandering demonstrates symmetry due to the symmetric forces exerted by boundaries and noise. This behavior closely resembles the characteristics of a normal distribution.

Figure 9.7 displays the lateral wandering for varying lane widths, revealing that wider lanes correspond to larger lateral wandering. Consequently, standard deviations

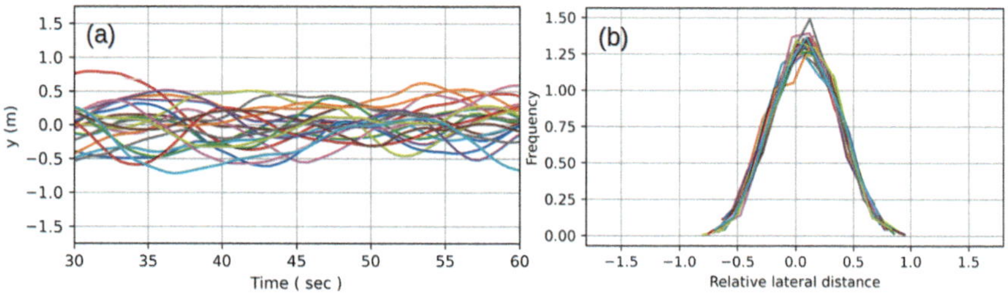

Fig. 9.6 The system paths and the marginal distribution. Reprint from Ref. [21], copyright 2023, with permissions from Elsevier

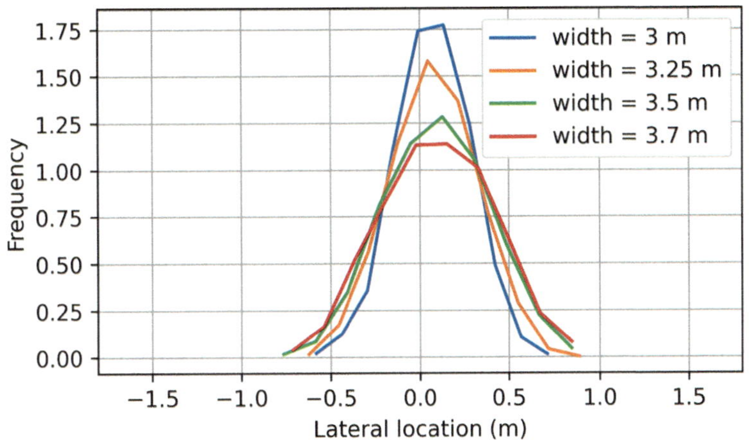

Fig. 9.7 The lateral wandering for different lane width. Reprint from Ref. [21], copyright 2023, with permissions from Elsevier

can be computed for different lane widths. For a lane width of $l_w = 3$ m, the standard deviation is approximately 0.2 m, aligning with empirical observations reported in previous studies. For instance, the MEPDG [15] suggests deviations of 0.203, 0.254, and 0.305 for $l_w < 3$ m, $3 \le l_w < 3.6$, and $l_w > 3.6$ m, respectively. By comparing the proposed model with these references, it can be calculated that the error of the proposed model is 1.4% (0.003/0.203) for $l_w = 3$ m when compared to MEPDG, a 7.4% error for $l_w = 3.12$ m when compared to Buiter et al. (1989), and a 1.1% error for $l_w = 3.37$ m. Similar error calculations can be performed for other lane widths. These findings confirm the consistency of the proposed model with existing research regarding standard deviations. Figure 9.8 depicts the plotted standard deviations for various lane widths, ranging from 0.2 to 0.35.

9.8 Case Study

9.8.1 Scenarios Settings

In this section, we present the results of the validation process and analyze the wandering patterns in various scenarios. The validation includes single-lane, two-lane, and three-lane road configurations, comparing them against existing wandering conclusions and real-world data from highD trajectories. Following the validation, we investigate the impact of MPRs. Additionally, considering that the lateral wandering pattern in multi-lane situations is interconnected, modifying the usage of one lane can have a cascading effect on adjacent lanes. Therefore, apart from the general lanes (GL) scenario, we also examine the dedicated lane (DL) scenario as proposed

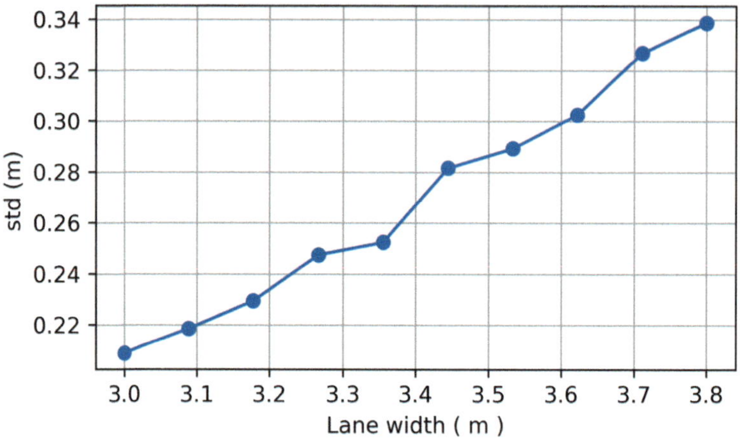

Fig. 9.8 The standard deviation for different lane width. Reprint from Ref. [21], copyright 2023, with permissions from Elsevier

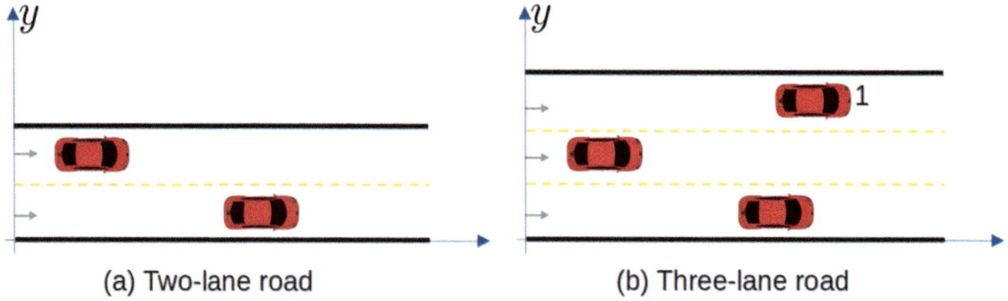

Fig. 9.9 Two-lane road and three-lane road. Reprint from Ref. [21], copyright 2023, with permissions from Elsevier

by Liu and Song [13]. In all examples, unless otherwise specified, the lane width is set to 3.5 m. The model's parameter values used for simulation are as follows: $\alpha = 0.1$, $\beta = 3.6$, $\gamma = 0.56$, $\sigma_{HDV} = 0.1$, $\sigma_{AV} = 0.01$, $\theta = 0.1$, $\eta_{ne} = 20$, $\eta_{\#} = 1.0$, $\forall \# \in \{mk, u, bnd\}$. Figure 9.9 illustrates the coordinate system for two-lane and three-lane roads, where the lateral origin is positioned at the boundary of the rightmost lane.

9.8.1.1 Two-Lane and Three-Lane Road

In a two-lane road configuration (as depicted in Fig. 9.9a), a key distinction from a single-lane road is the interaction between vehicles in adjacent lanes. Both lanes have a width of 3.5 m in our setup. The system paths, represented by lateral trajectories, and the corresponding lateral distribution are visually presented in Fig. 9.10. In Fig. 9.10a, 50 paths are shown for each lane, with stochastic lateral tracks. The stable distribution is displayed in Fig. 9.10b, with lane IDs indicated in Fig. 9.10a (Lane 0 refers to the left lane, and Lane 1 corresponds to the right lane). Notably, the x-axis in Fig. 9.10b is normalized by dividing the lane width, allowing for relative lateral location representation within the range of $[-0.5, 0.5]$. Location 0 corresponds to the midpoint of each lane. In the left lane, vehicles tend to deviate away from the middle line of the road, while in the right lane, path tracks are also shifted towards the road boundary.

To assess the disparity between the theoretical model and real-world data (specifically the highD dataset), we calculate the standard deviation for the relative lateral distance. The lateral distance is determined by $\frac{y-y_{middle}}{l_w}$, where y_{middle} represents the lane midpoint and l_w denotes the lane width. Employing the relative lateral distance is necessary due to the varied lane widths in the real-world data, ranging from 2.62 to 3.34 m. The resulting standard deviations for the two neighboring lanes in the highD dataset are 0.09 and 0.11 in Table 9.2, respectively. In comparison, the theoretical

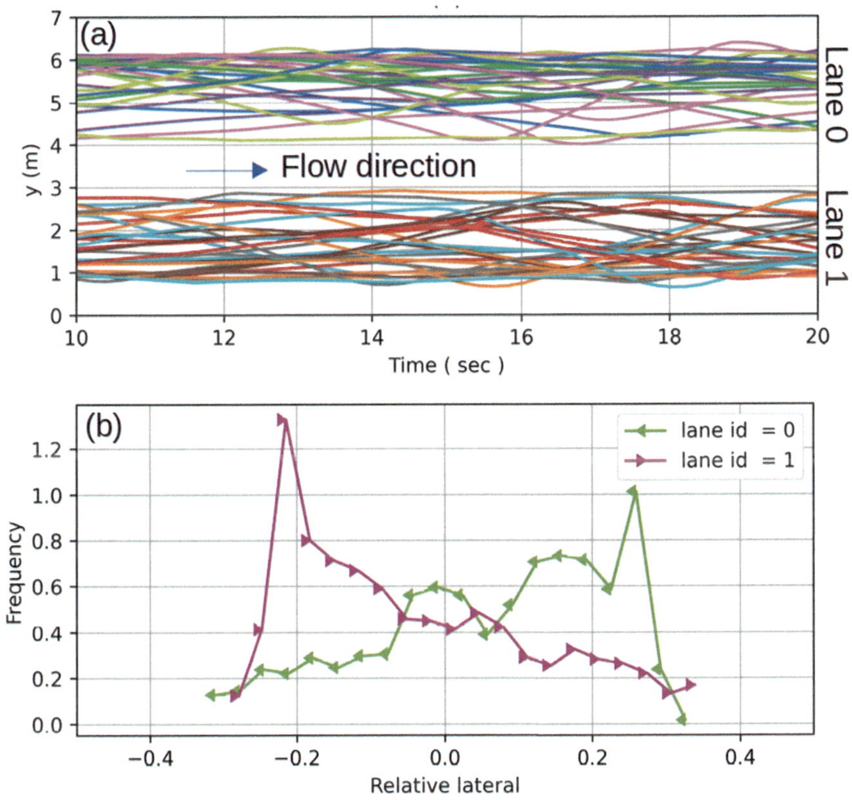

Fig. 9.10 Lateral wandering for a two-lane road. Reprint from Ref. [21], copyright 2023, with permissions from Elsevier

results are 0.125 and 0.129. These findings indicate a close alignment between the model predictions and the observed real-world data.

Figure 9.11 showcases the coupled distance between neighboring vehicles, utilizing $\left(\Delta y_1(t), \Delta y_2(t)\right)$ as the x-axis and y-axis. The phase plot in Fig. 9.11a exhibits trajectories from the highD dataset, while Fig. 9.11b displays trajectories generated by the proposed model. Both sets of trajectories demonstrate similar distributions across the four quadrants, indicating that the proposed model effectively replicates the microscopic-level behavior of vehicle pairs while capturing the lateral wandering dynamics of the overall traffic flow. Moving on to the three-lane road configuration, Fig. 9.12 illustrates the relative lateral wandering. Lane ID 0 corresponds to the leftmost lane, ID 2 corresponds to the rightmost lane, and ID 1 represents the middle lane. It is noteworthy that the lateral wandering patterns at the leftmost and rightmost lanes exhibit skewness, while the wandering at the middle lane is symmetric. This aligns with empirical observations in Fig. 9.2a, attributed to intra-lane interactions. The standard deviations, detailed in Table 9.3, indicate a notable resemblance between the theoretical model's standard deviation and real-world observations.

Table 9.2 The standard deviation comparison

The standard deviation of relative lateral distance	highD	Theoretical model
Lane 0	0.09	0.125
Lane 1	0.11	0.129

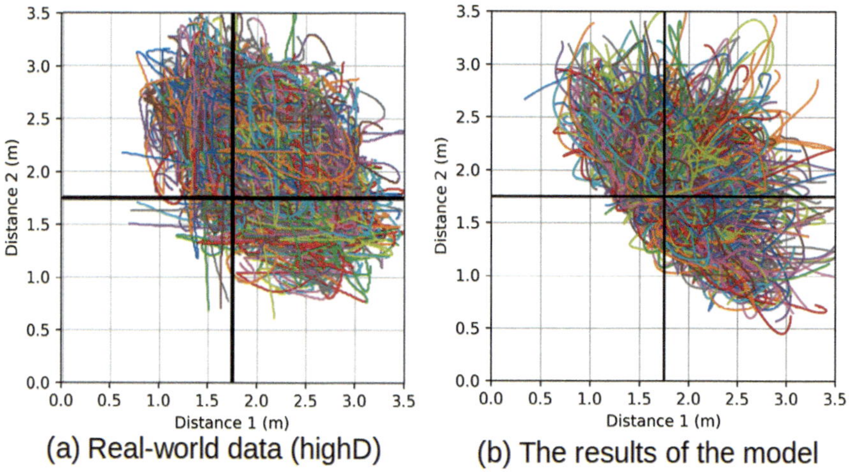

Fig. 9.11 The coupled distance of neighboring vehicles. Reprint from Ref. [21], copyright 2023, with permissions from Elsevier

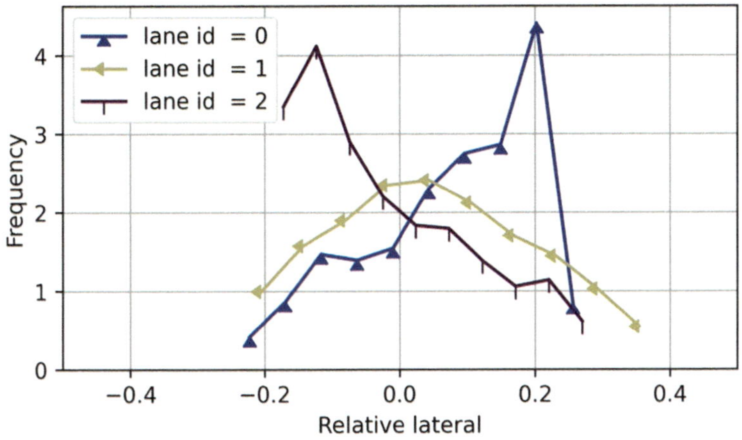

Fig. 9.12 Lateral wandering for a three-lane road. Reprint from Ref. [21], copyright 2023, with permissions from Elsevier

Table 9.3 The standard deviation comparison for the three-lane scenario

Std of relative lateral distance	highD	Theoretical model
Lane 0	0.12	0.124
Lane 1	0.08	0.129
Lane 2	0.115	0.126

9.8.2 *The Influence of the MPRs*

9.8.2.1 One-Lane Road

To investigate the impact of the MPR on lateral wandering, the total wandering is calculated as the weighted sum of HDV's wandering and AV's wandering. Figure 9.13 displays the total wandering and the corresponding standard deviation. In Fig. 9.13a, the lateral distribution of wandering is depicted for various MPR values. The results reveal that as the MPR increases, the distribution becomes more centered. Furthermore, the standard deviation steadily decreases with an increase in MPR. Specifically, when $\rho = 1$, the deviation reaches 0.08. In contrast, for $\rho = 0$, the deviation stands at 0.26. Notably, this ratio indicates that the HDV case is approximately three times wider than the AV case, aligning with empirical observations reported by Zhou et al. [28], which state that AVs are three times narrower.

9.8.2.2 Two-Lanes Road

We make use of two MPR values, denoted as ρ_0 and ρ_1, assigned to the left and right lanes, respectively. Specifically, ρ_0 represents the MPR for the left lane, while ρ_1

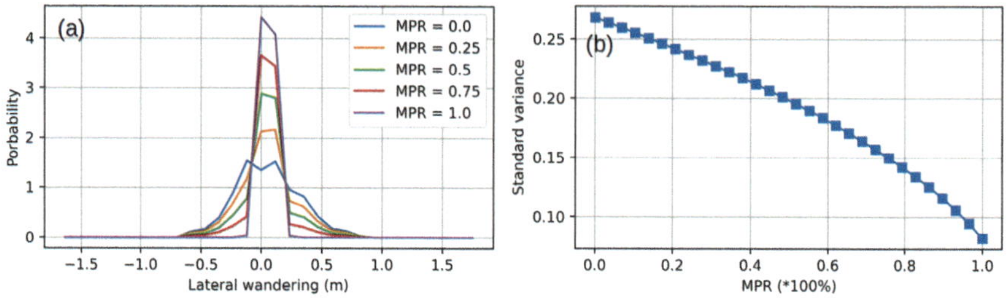

Fig. 9.13 Lateral wandering for different AV MPR; **a** wandering; **b** standard deviation. Reprint from Ref. [21], copyright 2023, with permissions from Elsevier

represents the MPR for the right lane. Our investigation includes two categories of combinations: (1) $\rho_0 = 0$, $\rho_1 > 0$; (2) $\rho_0 > 0$, $\rho_1 > 0$. Within each category, we vary the MPR to analyze its impact on wandering behavior.

Analyzing the first case ($\rho_0 = 0$, $\rho_1 > 0$), Fig. 9.14 presents the results. The lateral range of [0, 3.5 m] represents lane 1 (right lane), while the range of [3.5 m, 7 m] represents lane 0 (left lane). In Fig. 9.14, the bold black line signifies the lane mark and boundary, while the dashed black lines depict the lane middle line for reference. When the MPR is zero (indicated as the 'pure HDV' curve), the wandering in the left lane is shifted to the left, and the wandering in the right lane is shifted to the right, which aligns with the observations in real world dataset. As the MPR for lane 1 increases to 0.09, the wandering becomes more centered in comparison to the pure HDV case. Moreover, it is evident that as the MPR increases in the right lane, the wandering becomes increasingly centered. Additionally, the figure reveals that the MPR in the right lane influences the wandering in the left lane: alterations in the lateral wandering pattern in the right lane lead to changes in the lateral distribution in the left lane. This can be attributed to the decrease in lateral interactions among neighboring vehicles when more AVs are present in the right lane, as AVs have a tendency to drive closer to the lane middle line. Consequently, the wandering in the left lane also becomes more centered. This conclusion can be extended to scenarios with multiple lanes. Consequently, evaluating the lateral wandering in mixed AV flows necessitates an assessment of the entire road segment, rather than focusing solely on a single lane.

In the second case ($\rho_0 > 0$, $\rho_1 > 0$), we consider $\rho_0 = 0.5$ and vary ρ_1. The outcomes are illustrated in Fig. 9.15. Similar findings are observed, indicating that an increasing MPR results in a more centered wandering behavior. Comparing the wandering in the left lane between Fig. 9.14 (with $\rho_0 = 0$) and Fig. 9.15 (with $\rho_0 = 0.5$), it is evident that the wandering is also more centered in the latter case.

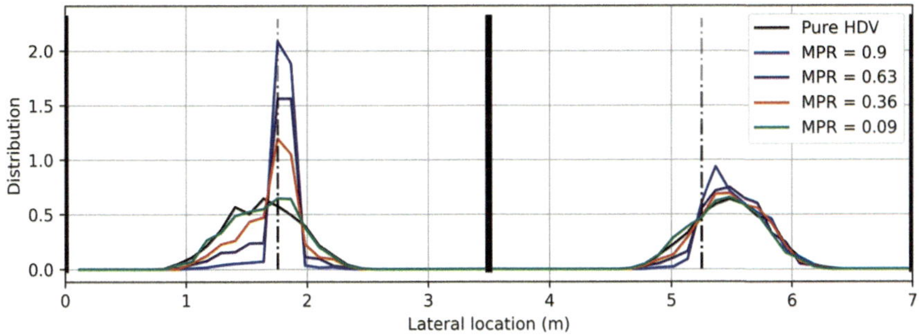

Fig. 9.14 Case 1: $\rho_0 = 0$, $\rho_1 = 0.09$–0.9. Note that the MPR in the legend corresponds to MPR at the right lane (or ρ_1). Reprint from Ref. [21], copyright 2023, with permissions from Elsevier

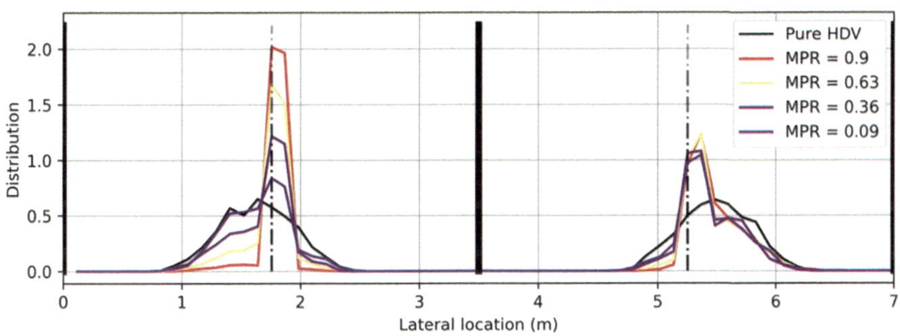

Fig. 9.15 Case 1: $\rho_0 = 0.5$, $\rho_1 = 0.09$–0.9. Note that the MPR in legend corresponds to MPR at the right lane. Reprint from Ref. [21], copyright 2023, with permissions from Elsevier

9.8.3 Dedicated Lane Case

We analyze the lateral wandering behavior in the context of AV-dedicated lanes. The concept of dedicated lanes for AVs is not novel and brings several advantages, including increased efficiency. In terms of lateral wandering, the introduction of dedicated lanes has a dual impact: (1) AVs within the dedicated lanes do not affect neighboring HDVs, as AVs tend to drive near the lane middle line while HDVs are aware of AV behavior, and (2) AVs within the dedicated lanes are not influenced by neighboring HDVs, assuming accurate perception of HDV behavior by the AVs [20]. Consequently, the dedicated lanes obstruct the lateral interaction between vehicles, thereby altering the wandering pattern. We consider two configurations of dedicated lanes on a three-lane road, taking the MPR into account.

Configuration (1), the MPR assignment for the three lanes follows the sequence [1, 0, 0], starting from the leftmost to the rightmost lane. In Configuration (2), the MPR is defined as [0, 1, 0]. Considering that AVs may require narrower lane widths [11], we examine two scenarios: Case (a) involves lane widths of [3.5 m, 3.5 m,

3.5 m] for the three lanes; Case (b) entails a dedicated lane (DL) width of 2.8 m, with the remaining width (i.e., 3.5 m * 3 − 2.8 m) divided equally between the remaining two lanes.

The width of the dedicated lane (DL), which is set at 2.8 m based on Jehanfo et al.' s study [11], gives rise to four distinct configurations: 1-a, 1-b, 2-a, and 2-b. The corresponding results are depicted in Figs. 9.16, 9.17, 9.18 and 9.19. Figure 9.16 reveals that the leftmost lane (with an x-axis interval of [7 m, 10.5 m]) exclusively comprises CAVs, exhibiting a narrower wandering behavior compared to HDVs (represented by the GL case curve). In the remaining two lanes, intra-lane interactions persist, causing the original symmetrical lateral distribution pattern to become asymmetric in the middle lane due to the presence of intra-lane forces only from the right-hand side. Figure 9.17 showcases a notable change in lateral wandering when the DL width (i.e., the leftmost lane) is reduced from 3.5 to 2.8 m. This alteration in lane configuration yields a significant shift in the wandering pattern across all three lanes, resulting from the change in the lane middle line or the reference line for HDV driving. A similar pattern to that observed in Fig. 9.16 is discernible.

Figures 9.18 and 9.19 depict the relocation of the dedicated lane (DL) to the center of the road, similar to one of the cases studied by Jehanfo et al. [11]. Consequently, the lateral wandering patterns of the remaining two lanes undergo corresponding adjustments. In Fig. 9.18, the leftmost and rightmost lanes are unaffected by the intra-lane force originating from the middle line, resulting in a shift from an asymmetric wandering pattern (as observed in the GL case curve) to a relatively symmetric one. In Fig. 9.19, the reduced width of the middle DL lane leads to a translation of the wandering patterns in the leftmost and rightmost lanes. These results highlight the interconnection between lateral wandering patterns across the entire road section, where changes in one lane impact the wandering behavior in other lanes. Such considerations are crucial for pavement sustainability analysis and design.

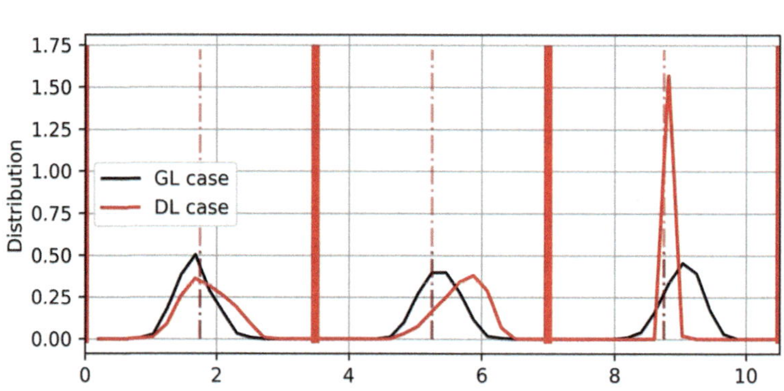

Fig. 9.16 Lateral wandering for DL case 1-a. Reprint from Ref. [21], copyright 2023, with permissions from Elsevier

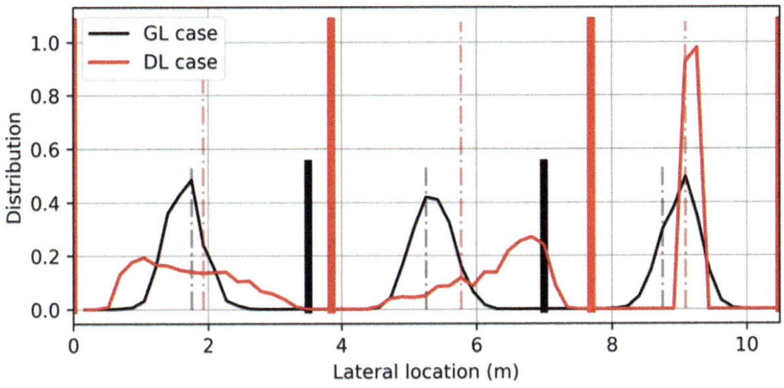

Fig. 9.17 Lateral wandering for DL case 1-b. Reprint from Ref. [21], copyright 2023, with permissions from Elsevier

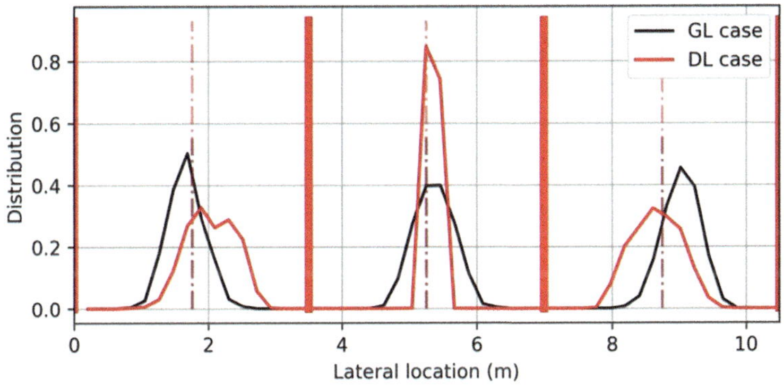

Fig. 9.18 Lateral wandering for DL case 2-a. Reprint from Ref. [21], copyright 2023, with permissions from Elsevier

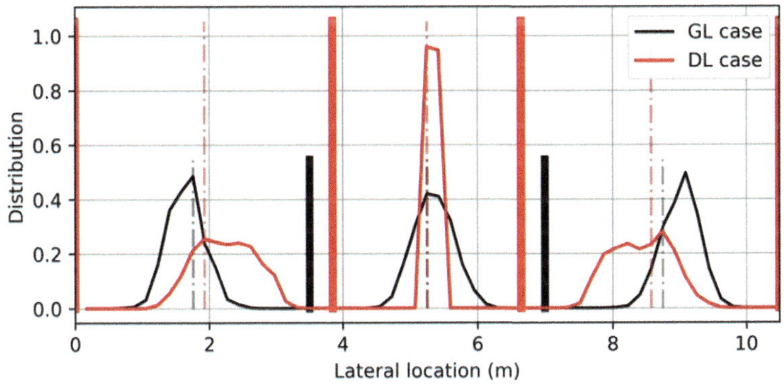

Fig. 9.19 Lateral wandering for DL case 2-b. Reprint from Ref. [21], copyright 2023, with permissions from Elsevier

9.9 Notes

The stochastic nature of vehicle lateral movement, known as lateral wandering, significantly impacts pavement sustainability and serves as a crucial determinant for pavement design. While empirical observations of lateral wandering exist, a theoretical investigation into this phenomenon under mixed AV flow scenarios is lacking, posing challenges in analysis. To bridge this gap, this study presents a theoretical model that depicts stochastic lateral movement influenced by road boundaries, lane markings, neighboring vehicles, and stochastic noise. By utilizing the Fokker–Planck equation and a Markov chain method, the model calculates lateral wandering. The results demonstrate that the model accurately replicates empirical observations of the wandering pattern. Additionally, the study examines the impact of AV market penetration rates on one-lane and multi-lane roads, offering insights for analyzing pavement sustainability under mixed AV flow conditions, including scenarios involving dedicated lanes. Potential extensions of this research include: (1) Considering platoon formations and the influence of control algorithms on the lateral wandering of the entire platoon in the forthcoming era of connected and autonomous vehicles; (2) Investigating complex road structures, such as on–off ramps and highly curved roads, which were not considered within the homogeneous road structure analyzed in this study.

References

1. ARA Inc. (2004) Guide for mechanistic empirical design of new and rehabilitated pavement structures. Washington, DC
2. Ali Y, Zheng Z, Haque MdM et al (2021) CLACD: a complete LAne-changing decision modeling framework for the connected and traditional environments. Transp Res Part C: Emerg Technol 128:103162. https://doi.org/10.1016/j.trc.2021.103162
3. Chen F, Song M, Ma X, Zhu X (2019) Assess the impacts of different autonomous trucks' lateral control modes on asphalt pavement performance. Transp Res Part C: Emerg Technol 103:17–29. https://doi.org/10.1016/j.trc.2019.04.001
4. Chen F, Song M, Ma X (2020) A lateral control scheme of autonomous vehicles considering pavement sustainability. J Clean Prod 256:120669. https://doi.org/10.1016/j.jclepro.2020.120669
5. Delpiano R, Herrera MJC, Coeymans AJE (2015) Characteristics of lateral vehicle interaction. Transp A: Transp Sci 11:636–647. https://doi.org/10.1080/23249935.2015.1059377
6. Delpiano R, Herrera JC, Laval J, Coeymans JE (2020) A two-dimensional car-following model for two-dimensional traffic flow problems. Transp Res Part C: Emerg Technol 114:504–516. https://doi.org/10.1016/j.trc.2020.02.025
7. Georgouli K, Plati C, Loizos A (2021) Autonomous vehicles wheel wander: structural impact on flexible pavements. J Traffic Transp Eng (English Edition) 8:388–398. https://doi.org/10.1016/j.jtte.2021.04.002
8. Georgouli K, Plati C (2023) Autonomous trucks' (ATs) lateral distribution and asphalt pavement performance. Int J Pavement Eng 24:2046274. https://doi.org/10.1080/10298436.2022.2046274
9. Gunay B (2004) An investigation of lane utilisation on turkish highways. Proc Inst Civ Eng-Transp 157:43–49. https://doi.org/10.1680/tran.2004.157.1.43

10. Iacus SM (2008) Simulation and inference for stochastic differential equations: with R examples. Springer, New York, New York, NY
11. Jehanfo H, Hu S, Kaparias I et al (2022) Redesigning highway infrastructure systems for connected autonomous truck lanes. J Transp Eng Part A: Systems 148:04022104. https://doi.org/10.1061/JTEPBS.0000762
12. Liu Y, Alsaleh R, Sayed T (2022) Modeling lateral interactions between motorized vehicles and non-motorized vehicles in mixed traffic using accelerated failure duration model. Transp A: Transp Sci 18:910–933. https://doi.org/10.1080/23249935.2021.1908443
13. Liu Z, Song Z (2019) Strategic planning of dedicated autonomous vehicle lanes and autonomous vehicle/toll lanes in transportation networks. Transp Res Part C: Emerg Technol 106:381–403. https://doi.org/10.1016/j.trc.2019.07.022
14. Luo W, Wang KCP (2013) Wheel path wandering based on field data. In: Airfield and highway pavement 2013. American society of civil engineers, Los Angeles, California, United States, pp 506–515. https://doi.org/10.1061/9780784413005.040
15. McCullah J, Gray D (2005) Cooperative highway program
16. McGarvey T (2016) Vehicle lateral position depending on road type and lane width: vehicle position surveys carried out on the swedish road network. Statens väg-och transportforskningsinstitut
17. Okte E, Al-Qadi IL (2022) Impact of autonomous and human-driven trucks on flexible pavement design. Transp Res Rec: J Transp Res Board 2676:144–160. https://doi.org/10.1177/03611981221077083
18. Papageorgiou M, Mountakis K-S, Karafyllis I et al (2021) Lane-free artificial-fluid concept for vehicular traffic. Proc IEEE 109:114–121. https://doi.org/10.1109/JPROC.2020.3042681
19. Porter RJ, Donnell ET, Mahoney KM (2004) Evaluation of effects of centerline rumble strips on lateral vehicle placement and speed. Transp Res Rec 1862:10–16. https://doi.org/10.3141/1862-02
20. Qi H, Chen C, Hu X, Zhang J (2022) Online inference of lane changing events for connected and automated vehicle applications with analytical logistic diffusion stochastic differential equation. Transp Res Part C: Emerg Technol 144:103874. https://doi.org/10.1016/j.trc.2022.103874
21. Qi H, Hu X (2023) Behavioral investigation of stochastic lateral wandering patterns in mixed traffic flow. Transp Res Part C: Emerg Technol 155:104310
22. Sun X, Park J, Tekell VO "Dean", Ludington NS (2007) Evaluation of pavement edge lines on narrow highways with low traffic volume in Louisiana. Transp Res Rec 1989–1:115–122https://doi.org/10.3141/1989-13
23. Tian J, Zhu C, Chen D et al (2021) Car following behavioral stochasticity analysis and modeling: perspective from wave travel time. Transp Res Part B: Methodol 143:160–176
24. Xu S, Peng H, Lu P et al (2020) Design and experiments of safeguard protected preview lane keeping control for autonomous vehicles. IEEE Access 8:29944–29953. https://doi.org/10.1109/ACCESS.2020.2972329
25. Yeganeh A, Vandoren B, Pirdavani A (2022) Impacts of load distribution and lane width on pavement rutting performance for automated vehicles. Int J Pavement Eng 23:4125–4135. https://doi.org/10.1080/10298436.2021.1935938
26. Zhao G, Liu L, Li S, Tighe S (2021) Assessing pavement friction need for safe integration of autonomous vehicles into current road system. J Infrastruct Syst 27:04021007. https://doi.org/10.1061/(ASCE)IS.1943-555X.0000615
27. Zhou F, Hu S, Chrysler ST et al (2019) Optimization of lateral wandering of automated vehicles to reduce hydroplaning potential and to improve pavement life. Transp Res Rec 2673:81–89. https://doi.org/10.1177/0361198119853560
28. Zhou F, Hu S, Xue W, Flintsch GW (2019) Optimizing the lateral wandering of automated vehicles to improve roadway safety and pavement life

Chapter 10
Lateral Stochasticity in Lane Changing via Logistic Diffusion Process

10.1 Introduction

The emergence of connected and automated vehicle technology has the potential to revolutionize the future transportation system, particularly in terms of traffic safety and efficiency. The CAVs rely on sensing the surrounding traffic flow, particularly when human-driven vehicles are present, to inform their planning, decision-making, and control processes. Therefore, understanding the behavioral attributes of HDVs and inferring their intentions is paramount for ensuring the benefits and safety of CAVs.

Accurately predicting the trajectories of surrounding vehicles and taking preemptive actions to avoid collisions are essential for the safe operation of connected and autonomous vehicles. In general, microscopic traffic behavior can be categorized into two main types: car-following (CF) and lane-changing (LC). CF pertains to the kinematic characteristics of an HDV following another vehicle, while LC involves the decision-making and execution process of changing lanes. Both behaviors exhibit stochastic characteristics in both longitudinal and lateral directions. However, an efficient and accurate prediction module is crucial for CAV operations. One important task is inferring the lane-changing behavior, i.e., determining whether a nearby vehicle's lateral deviation is indicative of an upcoming lane change or is simply normal noise or a result of driving error.

To address this issue, numerous methods have been proposed to determine whether a driver is in car-following (CF) or lane-changing mode. For example, Yang et al. [18] utilized the zero slope of the trajectory to identify the switching moment, while Chen et al. [4–6] determined the time window for lane changing by applying a predefined spatial range to the smoothed trajectory. Ali et al. [1] employed the wavelet method to detect the onset of lane changing. Other approaches include logistic regression and discrete choice models. Furthermore, once the initiation of a lane-changing maneuver occurs, the vehicle requires a specific duration to complete the process. This duration varies based on factors such as driver preference [4–6], surrounding

© The Author(s), under exclusive license to Springer Nature Singapore Pte Ltd. 2024

H. Qi, *Stochastic Two-Dimensional Microscopic Traffic Model*, Lecture Notes in Intelligent Transportation and Infrastructure,
https://doi.org/10.1007/978-981-97-3597-6_10

vehicle conditions [16], and even road surfaces [7]. Two specific moments of interest to CAVs are the crossing of the lane mark moment (CLMM) and the lane changing duration (LCD) [10]. However, to the best of our knowledge, real-time inference of these two metrics (CLMM and LCD) remains unresolved.

Despite the availability of various methods to assess lane changing intentions, we assert that the decision-making process for lane changing is latent and not directly observable or detectable. Additionally, the initiation of lane changing (referred to as lane changing initiation moment or LCIM) can be consistently defined alongside CLMM and LCD. Therefore, instead of attempting to estimate drivers' decision processes, a unified framework can be employed to model the detection of LCIM and infer CLMM and LCD. In this chapter, we propose the use of a logistic diffusion model to uncover the switching moment (LCIM) and derive the distribution of CLMM and LCD.

10.2 Current Practice and Drawbacks

The interaction between connected and automated vehicles (CAVs) and human-driven vehicles (HDVs) necessitates the inference of HDV behavior as an important input for CAV planning and control algorithms [15, 19]. In recent years, various methods and models have been utilized to infer HDV behavior. These methods encompass time series analysis [20], deep learning [17, 21], queuing network-based models [3], dynamic Bayesian networks [11], static Bayesian networks, ensemble learning methods [8], hidden Markov chain models [16], logit-based models [13], among others. The inferred HDV behavior includes the estimation of lane changing intention [16], lane keeping or lane changing behavior [13], failed lane changing attempts [2], and the probability of lane changing [13]. However, these models typically focus on specific aspects of lane changing and cannot provide a comprehensive set of lane changing information, such as LCIM, LCD, and CLMM.

Furthermore, data-driven models mentioned above exhibit various limitations, including: (1) the requirement for training prior to deployment, which encounters difficulties in achieving global optimization of system parameters under specific conditions; (2) uncertainties in transferability of trained models to different locations, as the models are trained using data from specific places and impractical to retrain for every new application.

Although a variety of methods exist for inferring HDV behavior, several challenges remain to be addressed: (1) In addition to the stochastic nature of traffic flow, HDVs themselves exhibit stochastic behavior. Consequently, real-time vehicle-level inference is necessary. (2) Apart from common lane changing intentions, obtaining information on lane changing duration and the moment of crossing lane markings would contribute additional knowledge for safer trajectory planning. Hence, real-time implementation of inference is crucial. Currently, no existing methods can achieve online inference of these three metrics. (3) Given the expectation of an online model, it is desirable to keep the parameters as simple as possible. However, data-driven approaches such as Bayesian networks, ensemble learning, and deep learning suffer from this limitation and cannot offer effective online inference tools.

To address the aforementioned requirements, we investigate the concurrent online inference of lane changing intention, cross-lane-mark moment, and lane changing duration. For this purpose, we develop a lane changing model rooted in logistic diffusion theory. A stochastic differential equation is formulated to capture the logistic diffusion process. Notably, the model comprises only two adjustable parameters, which possess physical interpretations and can be dynamically tuned in real-time.

10.3 Modeling

10.3.1 Notation and Abbreviation List

For easy reference, we list the notations and abbreviations as follows.

Y: the lateral displacement.

Y_C: the reference line when vehicle changes lane.

Z: A variable which is the translation of Y in the derivation of the analytical expression of Y.

k, σ: the parameters in the logistic diffusion model;

l_w: the lane width;

ρ: used in the LCIM identification. It takes the values in the set of $\{-1, 1\}$.

Δt: the time interval in the discretization of SDE.

10.3.2 Deterministic Logistic Lane Changing Model

The logistic function, also known as the logistic curve, is an S-shaped function extensively employed in various studies related to lane changing, including the calculation of lane changing probability [12]. The standard deterministic formulation of the logistic function is as follows:

$$Y = \frac{L}{1 + e^{-k(x - x_0)}} \tag{10.1}$$

The logistic curve is characterized by three parameters: L, k, and x_0. L denotes the threshold value at which the curve reaches its maximum, x_0 represents the midpoint along the x-axis, and k governs the rate of change. The plot below illustrates the logistic curve for various parameter values, with x_0 set to 0. The graph exhibits an initial exponential growth phase, followed by a gradual deceleration as it approaches the saturation point (L). Eventually, the growth ceases, reaching maturity.

The curve's shape bears resemblance to the trajectory observed during lane changing, whether in terms of temporal lateral trajectory or the two-dimensional spatial trajectory. The parameter L governs the upper threshold of the curve, while k regulates its steepness. A higher value of k results in a steeper curve. In order to simulate the lane changing process, we replace L with the lane width, which denotes the lateral displacement before and after the lane change. Consequently, the deterministic formula is as follows, where Y represents the lateral displacement.

$$\mathrm{Y} = \frac{l_w}{1 + \mathrm{e}^{-k(x - x_0)}} \tag{10.2}$$

10.3.3 Logistic Diffusion Formulation of the Lane Changing Process

10.3.3.1 Model Development

To incorporate the lane changing process into a logistic diffusion model, we begin by taking the derivative of Y. This yields the following expression:

$$\mathrm{Y}' = \frac{l_w \mathrm{k}\mathrm{e}^{-k(x-x_0)}}{\left(1 + \mathrm{e}^{-k(x-x_0)}\right)^2} = \frac{l_w}{1 + \mathrm{e}^{-k(x-x_0)}} \frac{\mathrm{k}\mathrm{e}^{-k(x-x_0)}}{1 + \mathrm{e}^{-k(x-x_0)}} = \frac{k}{l_w}\mathrm{Y}(l_w - Y) \tag{10.3}$$

Subsequently, the logistic curve can be defined as:

$$dY = \frac{k}{l_w}\mathrm{Y}(l_w - Y)\mathrm{dt} \tag{10.4}$$

To account for stochasticity, we introduce noise, resulting in a typical logistic diffusion formula:

$$dY = \frac{k}{l_w}\mathrm{Y}(l_w - Y)dt + \sigma Y dW \tag{10.5}$$

In the above equations, σ represents a parameter, and W denotes a Wiener process. Since the lateral coordinate of a particular lane may not start from 0, we introduce a base lateral coordinate, $\mathrm{Y_C}$, which represents the lateral coordinate of the lane's middle line where the lane changing occurs. By normalizing Y in Eq. 10.5 with $\mathrm{Y_C}$, the final equation is derived.

$$dY = \frac{k}{l_w}(Y - Y_C)(l_w - (Y - Y_C)) + \sigma(Y - Y_C)dW \tag{10.6}$$

The shape of the stochastic differential equation (SDE) is governed by two parameters. Larger values of σ correspond to more stochastic paths, while greater values of k lead to faster lane changing processes. The parameter σ originates from stochastic factors such as driver behavior, road surface conditions, or vehicle states. The randomness in parameter k arises from influences such as surrounding vehicles and driver attributes. We perform random sampling of parameter k within the range of [0.5, 2] and generate simulated trajectories. Subsequently, these trajectories are translated both temporally and spatially, based on the point where they intersect the lane mark ($\frac{l_w}{2}$). As a result of the translation, all paths align with the origin.

10.3.3.2 Solution of the Proposed Model

Equation 10.6 represents a nonlinear stochastic differential equation (SDE). In general, deriving an analytical solution for a nonlinear SDE is a challenging task. Therefore, alternative approaches must be employed. One common technique involves transforming the nonlinear SDE into a linear counterpart, using specific transformation rules under certain conditions [9]. To assess the transformability of Eq. 10.6, we introduce three functions: a(Y), b(Y), and A(y).

$$\begin{cases} a(\mathrm{Y}) = \frac{k}{l_w}(Y - Y_C)(l_w - (Y - Y_C)) \\ b(Y) = \sigma(Y - Y_C) \\ A(y) = \frac{a(y)}{b(y)} - \frac{1}{2}\frac{db(y)}{dy} = \frac{k(l_w - (Y - Y_C))}{l_w \sigma} - \frac{\sigma}{2} \end{cases} \tag{10.7}$$

If either of the two conditions mentioned below is satisfied, Eq. 10.6 can be transformed into a linear stochastic differential equation (SDE).

$$\begin{cases} \frac{\mathrm{d(A(y))}}{\mathrm{dy}} = 0 \\ \frac{\mathrm{d}}{\mathrm{dy}}\left(\frac{\frac{\mathrm{d}\left(\mathrm{b(y)}\frac{\mathrm{d(A(y))}}{\mathrm{dy}}\right)}{\mathrm{dy}}}{\frac{\mathrm{d(A(y))}}{\mathrm{dy}}}\right) = 0 \end{cases} \tag{10.8}$$

To verify the transformability, we analyze the conditions:

$$\begin{cases} \frac{\mathrm{d(A(y))}}{\mathrm{dy}} = -\frac{\mathrm{k}}{\mathrm{l_w}\sigma} \\ \frac{\mathrm{d}}{\mathrm{dy}}\left(\frac{\frac{\mathrm{d}}{\mathrm{dy}}\left(\mathrm{b(y)}\frac{\mathrm{d(A(y))}}{\mathrm{dy}}\right)}{\frac{\mathrm{d(A(y))}}{\mathrm{dy}}}\right) = \frac{\mathrm{d}\sigma}{\mathrm{dy}} = 0 \end{cases} \tag{10.9}$$

Upon verifying that the second condition of Eq. 10.8 holds true, it is concluded that Eq. 10.6 can indeed be transformed into a linear SDE. Let us denote the transformed SDE as:

$$dZ = (a_1 Z + a_2)dt + (b_1 Z + b_2)dW \tag{10.10}$$

The argument in Eq. 10.10 encompasses a_1, a_2, b_1 and b_2. Notably, b_1 can be directly determined as:

$$b_1 = -\frac{\frac{d}{dy}\left(b(y)\frac{d(A(y))}{dy}\right)}{\frac{d(A(y))}{dy}} = -\sigma \tag{10.11}$$

To establish an appropriate transformation from Y in Eq. 10.6 to Z in Eq. 10.10:

$$\begin{aligned} Z &= U(Y) = \exp\{b_1 \cdot B(Y)\} = e^{b_1 \int_{Y_0}^{y} \frac{ds}{b(s)}} = e^{b_1 \int_{Y_0}^{y} \frac{ds}{\sigma(s - Y_C)}} \\ &= e^{-\int_{Y_0}^{y} \frac{ds}{(s - Y_C)}} = e^{\ln(Y_0 - Y_C) - \ln(Y - Y_C)} = \frac{(Y_0 - Y_C)}{(Y - Y_C)} \end{aligned} \tag{10.12}$$

we select arguments a_1, a_2 and b_2 such that the following equation is satisfied:

$$\begin{cases} a(Y)\frac{dU(Y)}{dY} + \frac{1}{2}b_2(Y)\frac{d^2U(Y)}{dY^2} = a_1 U(Y) + a_2 (1) \\ b(Y)\frac{dU(Y)}{dY} = b_1 U(Y) + b_2 = -\sigma U(Y) + b_2 (2) \end{cases} \tag{10.13}$$

By substituting $b(Y) = \sigma(Y - Y_C)$ into the second equation in Eq. 10.13, we obtain:

$$b(Y)\frac{dU(Y)}{dY} = \sigma(Y - Y_C) \cdot \left(-\frac{(Y_0 - Y_C)}{(Y - Y_C)^2}\right) = -\sigma\frac{(Y_0 - Y_C)}{(Y - Y_C)} \tag{10.14}$$

Comparing this expression with the right-hand side (RHS) of Eq. 10.13.2, we deduce that $b_2 = 0$. Consequently, the left-hand side (LHS) of the first equation in Eq. 10.13 becomes:

$$\begin{aligned} & a(Y)\frac{dU(Y)}{dY} + \frac{1}{2}b_2(Y)\frac{d^2U(Y)}{dY^2} \\ & \overset{b_2=0}{\Rightarrow} \frac{k}{l_w}(Y - Y_C)(l_w - (Y - Y_C))\left(-\frac{(Y_0 - Y_C)}{(Y - Y_C)^2}\right) \\ & = -\frac{k}{l_w}(l_w - (Y - Y_C))\frac{(Y_0 - Y_C)}{(Y - Y_C)} = \frac{k}{l_w}((Y_0 - Y_c) - U(Y)) \end{aligned} \tag{10.15}$$

Comparing this LHS with the RHS of Eq. 10.13.1, we determine that:

$$a_1 = -\frac{k}{l_w};\ a_2 = \frac{k(Y_0 - Y_c)}{l_w} \tag{10.16}$$

Thus, the transformed SDE can be expressed as:

$$dZ = \left(-\frac{k \cdot Z}{l_w} + \frac{k(Y_0 - Y_c)}{l_w}\right)dt + (-\sigma)ZdW \tag{10.17}$$

The analytical solution to Eq. 10.17 is:

$$Z = \Phi(t)\left(Z_0 + \frac{k(Y_0 - Y_c)}{l_w}\int_0^t \frac{ds}{\Phi(s)}\right) \tag{10.18}$$

where:

$$\Phi(t) = e^{\left(\left(-\frac{k}{l_w} - \frac{1}{2}\left(\frac{k(Y_0 - Y_c)}{l_w}\right)^2\right)t + \frac{k(Y_0 - Y_c)}{l_w}W(t)\right)} \tag{10.19}$$

Hence, the solution to Eq. 10.6 can be presented as:

$$Z = \frac{(Y_0 - Y_C)}{(Y - Y_C)} \Rightarrow Y = Y_C + \frac{(Y_0 - Y_C)}{Z} \overset{Z_0=1}{\Rightarrow} Y = Y_C + \frac{(Y_0 - Y_C)}{\Phi(t)\left(1 + \frac{k(Y_0 - Y_c)}{l_w}\int_0^t \frac{ds}{\Phi(s)}\right)} \tag{10.20}$$

10.3.4 *Fokker–Planck Equation (FP Equation) of the Lane Changing Trajectories*

Equation 10.6 represents the fundamental principle governing the lateral movement observed during lane changing. Its solution, denoted as P(Y, t), corresponds to a stochastic process. This process can be constructed using either Monte Carlo simulation or determined by solving a specialized partial differential equation known as the Fokker–Planck (FP) equation. In this section, we first derive the FP equation and subsequently present the Monte Carlo simulation, which enables the analysis of the distribution.

10.3.4.1 Theoretical Distribution P(Z, t)

To obtain the Fokker–Planck (FP) equation, we express Eq. 10.6 in a modified form for compact derivation and adherence to the conventions of stochastic differential equation (SDE) symbols.

$$dY = \frac{k}{l_w}(Y - Y_C)(l_w - (Y - Y_C))dt + \sigma(Y - Y_C)dW = f(Y, t)dt + L(Y, t)dW \tag{10.21}$$

The FP equation can be defined as:

$$\frac{\partial P(Y, t)}{\partial t} = -\frac{\partial (f(Y, t) \cdot P(Y, t))}{\partial Y} + \frac{1}{2}\left[\frac{\partial^2\left((L(Y,\ t))^2 P(Y, t)\right)}{\partial Y^2}\right] \tag{10.22}$$

In Eq. 10.22, the first term on the right-hand side (RHS) can be calculated as:

$$\frac{\partial (f(Y, t) \cdot P(Y, t))}{\partial Y} = P(Y, t)\frac{\partial (f(Y, t))}{\partial Y} + f(Y, t)\frac{\partial P(Y, t)}{\partial Y} \tag{10.23}$$

The calculation of $\frac{\partial (f(Y,t))}{\partial Z}$ in the above equation yields:

$$\frac{\partial (f(Y, t))}{\partial Y} = \frac{k}{l_w}(1 - 2Y + 2Y_C) \tag{10.24}$$

Substituting Eq. 10.24 into Eq. 10.23, we have:

$$\begin{aligned} &\frac{\partial (f(Z, t) \cdot P(Z, t))}{\partial Z} \\ &= P(Z, t)\left(\frac{k}{l_w}(1 - 2Y + 2Y_C)\right) + \frac{k}{l_w}(Y - Y_C)(1 - (Y - Y_C))\frac{\partial P(Y, t)}{\partial Y} \end{aligned} \tag{10.25}$$

To formulate $\frac{\partial^2\left((L(Y,\ t))^2 P(Y,t)\right)}{\partial Y^2}$ in Eq. 10.22, we first compute the first partial derivative as:

$$\frac{\partial\left((L(Y,\ t))^2 P(Y, t)\right)}{\partial Y} = P(Y, t)\frac{\partial\left((L(Y,\ t))^2\right)}{\partial Y} + (L(Y,\ t))^2\frac{\partial P(Y, t)}{\partial Y} \tag{10.26}$$

Additionally, the second partial derivative can be calculated as:

$$\begin{aligned} &\frac{\partial^2\left((L(Y,\ t))^2 P(Y, t)\right)}{\partial Y^2} \\ &= \left[P(Y, t)\frac{\partial^2\left((L(Y,\ t))^2\right)}{\partial Y^2} + \frac{\partial\left((L(Y,\ t))^2\right)}{\partial Y}\frac{\partial P(Y, t)}{\partial Y}\right] \end{aligned}$$

$$+\left[\frac{\partial\left((L(Y,\ t))^2\right)}{\partial Y}\frac{\partial P(Y,t)}{\partial Y}+(L(Y,\ t))^2\frac{\partial^2 P(Y,t)}{\partial Y^2}\right]$$

$$= P(Y,t)\frac{\partial^2\left((L(Y,\ t))^2\right)}{\partial Y^2}+2\frac{\partial\left((L(Y,\ t))^2\right)}{\partial Y}\frac{\partial P(Y,t)}{\partial Y}$$

$$+(L(Y,\ t))^2\frac{\partial^2 P(Y,t)}{\partial Y^2} \tag{10.27}$$

Expanding upon $\frac{\partial\left((L(Y,\ t))^2\right)}{\partial Y}$ in Eq. 10.27, we find:

$$\frac{\partial\left((L(Y,\ t))^2\right)}{\partial Y}=2\sigma^2(Y-Y_C) \tag{10.28}$$

For the $\frac{\partial^2\left((L(Y,\ t))^2\right)}{\partial Y^2}$, we have:

$$\frac{\partial^2\left((L(Y,\ t))^2\right)}{\partial Y^2}=2\sigma^2 \tag{10.29}$$

Substituting Eq. 10.28 and Eq. 10.29 into Eq. 10.27, we obtain:

$$\begin{aligned}\frac{\partial^2\left((L(Y,\ t))^2 P(Y,t)\right)}{\partial Y^2} &= 2\sigma^2\cdot P(Y,t)\\ &\quad+4\sigma^2(Y-Y_C)\frac{\partial P(Y,t)}{\partial Y}\\ &\quad+\sigma^2(Y-Y_C)^2\frac{\partial^2 P(Y,t)}{\partial Y^2}\end{aligned} \tag{10.30}$$

By incorporating Eq. 10.23 and Eq. 10.27 into Eq. 10.22, the distribution of Y can be derived, as presented below (detailed calculations omitted):

$$\begin{aligned}\frac{\partial P(Y,t)}{\partial t} &= -\left[P(Z,t)\left(\frac{k}{l_W}(1-2Y+2Y_C)\right)+\frac{k}{l_W}(Y-Y_C)(1-(Y-Y_C))\frac{\partial P(Y,t)}{\partial Y}\right]\\ &\quad+\frac{1}{2}\left[2\sigma^2\cdot P(Y,t)+4\sigma^2(Y-Y_C)\frac{\partial P(Y,t)}{\partial Y}+\sigma^2(Y-Y_C)^2\frac{\partial^2 P(Y,t)}{\partial Y^2}\right]\\ &=\left[\sigma^2-\left(\frac{k}{l_W}(1-2Y+2Y_C)\right)\right]P(Y,t)+\left[\begin{array}{l}\frac{k}{l_W}(Y-Y_C)(1-(Y-Y_C))\\ +2\sigma^2(Y-Y_C)\end{array}\right]\frac{\partial P(Y,t)}{\partial Y}\\ &\quad+\sigma^2(Y-Y_C)^2\frac{\partial^2 P(Y,t)}{\partial Y^2}\end{aligned} \tag{10.31}$$

Furthermore, the boundary condition for the FP equation, given an initial lateral location Y_0, is as follows:

$$\begin{cases} P(Y = Y_0, t = 0) = 1(a) \\ P(Y \neq Y_0, t = 0) = 0(b) \end{cases} \tag{10.32}$$

10.3.4.2 Monte Carlo Simulation of the P(Y, t)

In this section, we present the Monte Carlo solution for the distribution P(Y, t) using varying parameters of k and sigma. The simulation process is as follows. We discretize Eq. 10.6 using the Euler method:

$$Y_{i+1} - Y_i = \left(\frac{k}{l_w}(Y_i - Y_C)(l_w - (Y_i - Y_C))\right)\Delta t + \sigma(Y_i - Y_C)\Delta W_i$$

$$\Rightarrow Y_{i+1} = Y_i + \Delta t\left(\frac{k}{l_w}(Y_i - Y_C)(l_w - (Y_i - Y_C))\right) + \sigma(Y_i - Y_C)\Delta W_i \tag{10.33}$$

Here, ΔW_i represents independent Gaussian variables with zero mean and a variance of Δt (we assume a fixed time step without loss of generality). The system path simulation follows these steps: (1) Initialize the lateral coordinate y_0 and set the counter $j = 1$; (2) Generate a Gaussian random number $N(0, \Delta t)$; (3) Calculate y_{j+1} using Eq. 10.33 from y_j. Repeat steps 2 and 3 to obtain a system path.

10.3.5 CLMM and LCD Distribution

The completion of a lane changing behavior occurs when the vehicle's lateral displacement reaches the lane width, assuming a constant lane width. The cross-lane-mark moment refers to the point in time when the trajectory crosses the lane mark, and the duration from the onset of lane changing is denoted as CLMM (Cross-Lane-Mark Moment). These definitions are stated as follows:

$$\begin{cases} \textit{LC duration distribution} : P(Y = l_w, t) \ (a) \\ \textit{CLMM distribution} : P\left(Y = \frac{l_w}{2}, t\right) \quad (b) \end{cases} \tag{10.34}$$

To determine the distributions of these variables, the solution to the Fokker–Planck (FP) equation is required. However, since an analytical solution for the FP equation is not available, Monte Carlo methods are employed as an alternative approach.

10.4 Online Parameters Calibration and Inference of LCIM, CLMM and LCD

The model follows a general workflow consisting of the following stages: (1) data collection, where the input consists of lateral observations of the HDV; (2) online parameter calibration; (3) inference of the Cross-Lane-Mark Moment (CLMM) and Lane Changing Duration (LCD).

10.4.1 Parameter Calibration

For online parameter calibration, the assumed input is a series of lateral coordinates, $y_0, y_1, \ldots y_N$. The likelihood is formulated as:

$$l(k,\sigma) = p_{\sigma,k}(y_1)\prod_{2}^{N} p_{\sigma,k}(y_i|y_{i-1}) \overset{def}{=} p(y_1)\prod_{2}^{N} p(y_i|y_{i-1}) \tag{10.35}$$

To incorporate the probability in the likelihood, the Euler discretization method is employed using Eq. 10.6 in Eq. 10.33. Thus, given Y_i, the distribution of Y_{i+1} follows a normal distribution with mean and variance:

$$\begin{cases} \mu_{Y_{i+1}} = Y_i + \Delta t\left(\frac{k}{l_w}(Y_i - Y_C)(l_w - (Y_i - Y_C))\right) \\ \sigma^2_{Y_{i+1}} = (\sigma(Y_i - Y_C))^2 \Delta t = \Delta t\sigma^2 (Y_i - Y_C)^2 \end{cases} \tag{10.36}$$

In other words,

$$p(y_i|y_{i-1}) = \frac{1}{\sqrt{2\pi\,\Delta t\sigma^2(y_{i-1} - Y_C)^2}} e^{-\frac{\left(y_i - \left(y_{i-1} + \Delta t\left(\frac{k}{l_w}(y_{i-1} - Y_C)(l_w - (y_{i-1} - Y_C))\right)\right)\right)^2}{\Delta t\sigma^2(y_{i-1} - Y_C)^2}} \tag{10.37}$$

Taking the logarithm yields

$$\ln p(y_i|y_{i-1}) = -\pi\,\Delta t\sigma^2(y_{i-1} - Y_C)^2 - \frac{\left(y_i - \left(y_{i-1} + \Delta t\left(\frac{k}{l_w}(y_{i-1} - Y_C)(l_w - (y_{i-1} - Y_C))\right)\right)\right)^2}{\Delta t\sigma^2(y_{i-1} - Y_C)^2} \tag{10.38}$$

Thus, the likelihood can be expressed as:

$$
\begin{aligned}
L(k,\sigma) &= \ln(l(k,\sigma)) \\
&= -\pi\Delta t\sigma^2\sum_i\left((y_{i-1}-Y_C)^2\right) \\
&\quad -\frac{1}{\Delta t\sigma^2}\sum_i\left(\frac{\left(y_i-\left(y_{i-1}+\Delta t\left(\frac{k}{l_w}(y_{i-1}-Y_C)(l_w-(y_{i-1}-Y_C))\right)\right)\right)^2}{(y_{i-1}-Y_C)^2}\right)
\end{aligned}
\tag{10.39}
$$

To adjust the parameters, the partial derivative is set to zero:

$$
\begin{cases}\frac{\partial L}{\partial k}=0\\ \frac{\partial L}{\partial \sigma}=0\end{cases}
\tag{10.40}
$$

The resulting derivation for k is as follows:

$$
\begin{aligned}
&\sum_i -2\Delta t\frac{(l_w-(y_{i-1}-Y_C))}{l_w}\left(\frac{y_i-y_{i-1}-\Delta t\left(\frac{k}{l_w}(y_{i-1}-Y_C)(l_w-(y_{i-1}-Y_C))\right)}{y_{i-1}-Y_C}\right)=0\\
&\Rightarrow \sum_i 2\Delta t\frac{(l_w-(y_{i-1}-Y_C))}{l_w}\left(\frac{y_i-y_{i-1}}{y_{i-1}-Y_C}-\left(\frac{k\Delta t}{l_w}(l_w-(y_{i-1}-Y_C))\right)\right)=0\\
&\Rightarrow \sum_i 2\Delta t\frac{(l_w-(y_{i-1}-Y_C))}{l_w}\frac{y_i-y_{i-1}}{y_{i-1}-Y_C}=\sum_i\frac{2k\Delta t^2}{l_w^2}(l_w-(y_{i-1}-Y_C))^2\\
&\Rightarrow k=\frac{\sum_i\Delta t\frac{(l_w-(y_{i-1}-Y_C))}{l_w}\frac{y_i-y_{i-1}}{y_{i-1}-Y_C}}{\sum_i\frac{\Delta t^2}{l_w^2}(l_w-(y_{i-1}-Y_C))^2}
\end{aligned}
\tag{10.41}
$$

Similarly, σ is calibrated as:

$$
\begin{aligned}
&-\pi\Delta t\sigma\sum_i\left((y_{i-1}-Y_C)^2\right)+\frac{\sum_i\left(\frac{\left(y_i-\left(y_{i-1}+\Delta t\left(\frac{k}{\rho\cdot l_w}(y_{i-1}-Y_C)(l_w-(y_{i-1}-Y_C))\right)\right)\right)^2}{(y_{i-1}-Y_C)^2}\right)}{\Delta t\sigma^3}=0\\
&\Rightarrow \sigma=\sqrt[4]{\frac{\sum_i\left(\frac{\left(y_i-\left(y_{i-1}+\Delta t\left(\frac{k}{\rho\cdot l_w}(y_{i-1}-Y_C)(l_w-(y_{i-1}-Y_C))\right)\right)\right)^2}{(y_{i-1}-Y_C)^2}\right)}{\Delta t^2\pi\sum_i\left((y_{i-1}-Y_C)^2\right)}}
\end{aligned}
\tag{10.42}
$$

10.4.2 Online LCIM Identification

During the initiation of a lane change, the vehicle gradually aligns its lateral position with the centerline of the intended lane. Subsequently, when another lane change is initiated, the lateral position once again aligns with the new selected lane's centerline. In the case of a single lane change process as described by Eq. 10.6, the centerline coordinate of the current lane is denoted as Y_c, while the centerline of the target lane can be determined as $Y_c + l_w$. Following the completion of the lane change, the vehicle's lateral position fluctuates around the $Y_c + l_w$ coordinate. Considering the possibility of the vehicle changing lanes to either the left or right, the direction of the lane change is denoted by the variable ρ, where $\rho \in \{1, -1\}$. Consequently, Eq. 10.6 can be reformulated as follows:

$$dY = \frac{k}{\rho \cdot l_w}(Y - Y_C)(\rho \cdot l_w - (Y - Y_C))dt + \sigma(Y - Y_C)dW \tag{10.43}$$

The LCIM's inference can be translated as detecting changes in Y_C and ρ. An example is that he vehicle is on a lane with its middle line on Y_2 at time t. Assuming a historic dummy lane change from Y_1 to Y_2, Y_C and ρ default as Y_1 and 1, respectively. To determine the vehicle's behavior, we observe Y_C and ρ within a specific time window. The values for Y_C and ρ for lane keeping, lane changing to the left, and lane changing to the right are presented below.

$$\begin{cases} Y_C = Y_2, \rho = 1 & \textit{change lane to left} \\ Y_C = Y_2, \rho = -1 & \textit{change lane to right} \\ Y_C = Y_1, \rho = 1 & \textit{lane keeping} \end{cases} \tag{10.44}$$

Hence, the behavior can be determined by selecting the highest probability among the three cases. To generalize this principle, let Y_C' and ρ' represent the cases to be examined, and ρ_{last} indicate the direction of the previous lane change. Then, the three outcomes become:

$$\begin{cases} case1 : \rho' = 1, Y_C' = Y_c & (a) \\ case2 : \rho' = 1, Y_C' = Y_c + \rho_{last} \cdot l_w & (b) \\ case3 : \rho' = -1, Y_C' = Y_c + \rho_{last} \cdot l_w & (c) \end{cases} \tag{10.45}$$

Given an observed series of $y_1, \ldots y_N$ within a specific time window (y_N is typically considered the lateral coordinate at time t), the inference of the LCIM aims to identify which case has the highest probability. Assuming that y_0 represents the data just before the time window, the likelihood can be expressed as follows:

$$\mathfrak{L}(Y_C', \rho') = \ln\left(p_{Y_C',\rho'}(y_0)\prod_1^N p_{Y_C',\rho'}(y_i|y_{i-1})\right) \tag{10.46}$$

The optimal Y'_C and ρ' are:

$$\rho^*, Y_C^* = \operatorname{argmax} \mathfrak{L}(Y'_C, \rho') \tag{10.47}$$

Once ρ^* and Y_C^* are obtained, it becomes straightforward to ascertain whether the vehicle intends to maintain its current lane or initiate a lane change to the left/right.

The aforementioned reasoning process operates in real-time. Sometimes, it becomes necessary to determine the LCIM when a historical trajectory is provided. Without loss of generality, let's assume the given trajectory is y_1 to y_N. To establish the LCIM, we formulate the likelihood based on the switching moment k (which is considered the LCIM):

$$\mathfrak{L}_k = \ln\left(p_{\rho,Y_c}(y_1)\prod_2^k p_{\rho,Y_c}(y_i|y_{i-1})\prod_{k+1}^N p_{\rho',Y'_c}(y_i|y_{i-1})\right) \tag{10.48}$$

Then:

$$\begin{aligned}\mathfrak{L}_k &= \ln\left(p_{\rho,Y_c}(y_1)\prod_2^k p_{\rho,Y_c}(y_i|y_{i-1})\prod_{k+1}^N p_{\rho',Y'_c}(y_i|y_{i-1})\right)\\ &= -\pi\,\Delta t\sigma^2\sum_{i=1}^{k}\left(\left(y_{i-1}-Y_C\right)^2\right)\\ &-\frac{1}{\Delta t\sigma^2}\sum_{i=1}^{k}\left(\frac{\left(y_i-\left(y_{i-1}+\Delta t\left(\frac{k}{\rho\cdot l_w}\left(y_{i-1}-Y_C\right)\left(\rho\cdot l_w-\left(y_{i-1}-Y_C\right)\right)\right)\right)\right)^2}{\left(y_{i-1}-Y_C\right)^2}\right)\\ &-\pi\,\Delta t\sigma^2\sum_{i=k+1}^{N}\left(\left(y_{i-1}-Y'_C\right)^2\right)\\ &-\frac{1}{\Delta t\sigma^2}\sum_{i=k+1}^{N}\left(\frac{\left(y_i-\left(y_{i-1}+\Delta t\left(\frac{k}{\rho\cdot l_w}\left(y_{i-1}-Y'_C\right)\left(\rho'\cdot l_w-\left(y_{i-1}-Y'_C\right)\right)\right)\right)\right)^2}{\left(y_{i-1}-Y'_C\right)^2}\right)\end{aligned} \tag{10.49}$$

Similar to the online version, we have:

$$\rho^*, k^* = \underset{\substack{k\in 2\sim N;\\ \rho', Y'_C\in\{(1, Y_C), (1, Y_c+\rho_{last}\cdot l_w), (-1, Y_c+\rho_{last}\cdot l_w)\}}}{\operatorname{argmax}} \mathfrak{L}_k \tag{10.50}$$

Initialization at moment $k = 2$ allows for the first term (i.e. $p(y_1)$) to be ignored during optimization due to its constant value across different k values.

$$\rho^*, k^* = \operatorname{argmax} \prod_{2}^{k} p_{\rho, Y_c}(y_i | y_{i-1}) \prod_{k+1}^{N} p_{\rho', Y'_c}(y_i | y_{i-1}) \tag{10.51}$$

This is facilitated by the normal distribution of each term on the right-hand side of the equation. However, to solve the optimization problem, it is necessary to enumerate all possibilities and combinations of k, and (ρ', Y'_C).

10.4.3 Mean of CLMM and LCD

Once a lane change maneuver is detected, the involved vehicle requires a specific duration to complete the lane change behavior. In accordance with Eq. 10.34, this duration is determined by the solution to the FP equation for the Lane Change Duration (LCD) and the CLMM. By employing the Monte Carlo (MC) method, the distribution can be simulated in real time. The obtained results from the simulation are then utilized to infer the values of CLMM and LCD. The desired outcome is derived by calculating the average of the collected samples. The process involved is relatively straightforward and will not be further discussed in this context.

10.5 Case Study

10.5.1 Data Description and Preprocessing

To assess the efficacy of our proposed algorithm, we have conducted validation tests on two distinct datasets, namely the NGSIM and highD datasets. Whereas NGSIM is an older dataset with low-quality data, highD is a newer dataset with smoother trajectory data. The time resolution rate of NGSIM is 0.1 s, compared to 1/25 s in highD. We began processing both datasets in the same way, first by extracting the trajectory of each vehicle, then by extracting the data concerning lane changes. Due to differences in their coordinate systems, each trajectory was translated both temporally and spatially to ensure they shared a common point of origin - the middle line of the lane with a Y value of zero.

10.5.2 LCIM Inference

10.5.2.1 LCIM Inference on Observations with Lane Changing

An instance of the LCIM using highD data is illustrated for analysis in this section. Figure 10.1 showcases the selected observation and the corresponding inference results. In Fig. 10.1a, the trajectory of a vehicle is depicted, highlighting a lane change to the left. The lane width is measured at 3.7 m, and the x-axis represents the frame, with one frame corresponding to 1/25 s. The trajectory spans approximately 12 s. Figure 10.1b visualizes the logarithm of the lane keeping probability, defined as $\rho = 0$ in the figure (equivalent to case 1 in Eq. 10.45), alongside the logarithm of the probabilities for left lane changing ($\rho = 1$) and right lane changing ($\rho = -1$). To enhance clarity, smaller logarithm probability values are omitted from the visualization. The graph demonstrates that the lane keeping probability is highest from the beginning until the 170th moment. After that point, the probability of left lane changing becomes the highest among the three options, indicating the determination of the LCIM at the 170th moment. The corresponding lateral location is approximately 0.2 m. When comparing the probabilities of $\rho = 1$ and $\rho = -1$, it is evident that in the initial period (around moment 0 to moment 40), the probability of $\rho = 1$ is higher due to a tendency towards leftward trajectory shift. However, after around moment 70, the trajectory begins to regress towards the middle line of the lane ($Y = 0$). By moment 140, the vehicle trajectory converges to the middle line of the target lane ($Y = 3.7$). Consequently, the probability of $\rho = 1$ continues to rise and surpasses the probability of $\rho = 0$. This finding leads to the conclusion that the LCIM occurs at approximately moment 170.

A total of 317 highD dataset trajectories exhibiting left lane changing behavior were analyzed. For each trajectory, the LCIM was determined along with the corresponding lateral coordinate within the range of −1.85 to 1.85 m, considering a lane width of 3.7 m.

The NGSIM dataset is known to exhibit a higher level of noise. Figure 10.2 illustrates the LCIM inference results for NGSIM observations. A comparison between Fig. 10.2a and 10.1a demonstrates noticeable fluctuations in the trajectory. Starting from a lateral coordinate of 0.8 m, the trajectory initially turns to the right. Figure 10.2b presents the logarithms of the probabilities, showing similar trends. Initially, lane keeping is preferred, but at moment 500, the probability of $\rho = 1$, indicating a left lane change, becomes the highest among the three possibilities. Due to the considerable noise in the trajectory, the logarithms also display fluctuations when compared to Fig. 10.1b.

10.5.2.2 The False Negative and False Positive Error of LCIM

An effective LCIM inference should possess robustness to noise present in the data. In general, there exist two types of errors associated with LCIM inference: (1) the algorithm fails to identify a lane changing event, resulting in a false negative; (2) the algorithm falsely indicates a lane changing event, resulting in a false positive. To evaluate these error types, both lane changing data and lane keeping data are utilized. To determine the false positive error rate, 393 lane keeping trajectories from NGSIM are employed for LCIM inference. Among these trajectories, only two are identified as having a LCIM, with maximum lateral differences of 2.11 m and 2.29 m respectively. The resulting false positive rate is calculated as $2/393 \approx 0.005$,

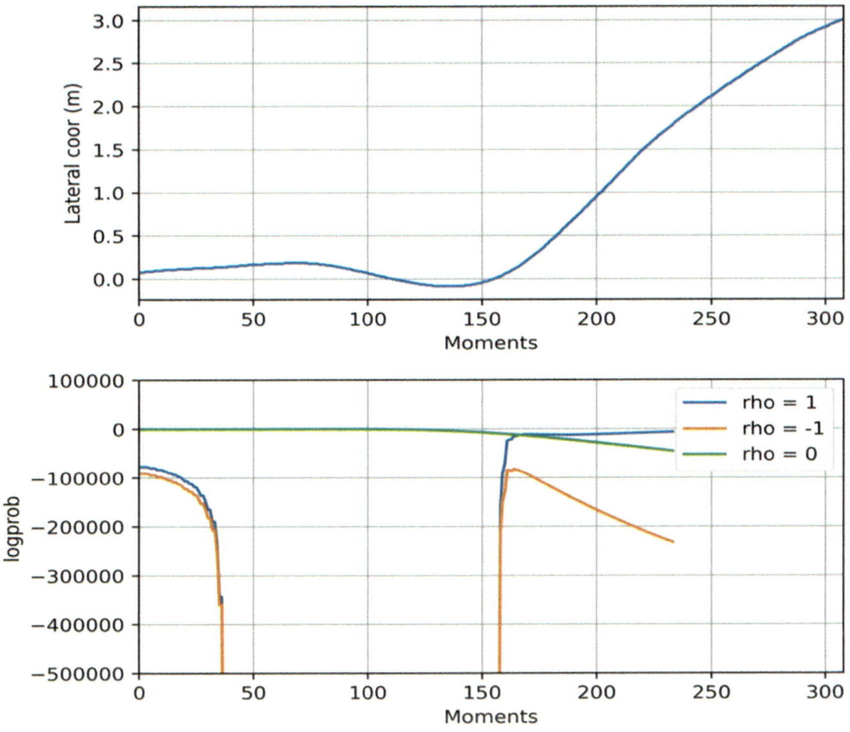

Fig. 10.1 LCIM inference. Reprint from Ref. Qi et al. [14], copyright 2024, with permissions from Elsevier

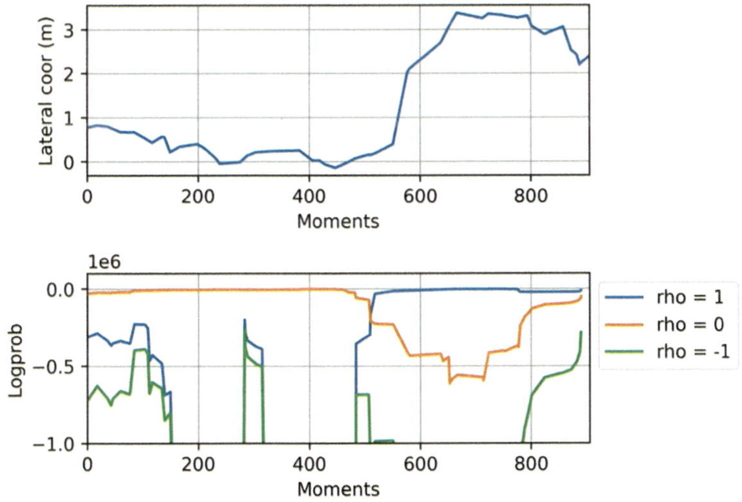

Fig. 10.2 LCIM inference for NGSIM observation example. Reprint from Ref. Qi et al. [14], copyright 2024, with permissions from Elsevier

indicating robustness of the LCIM against stochastic noise in terms of false positives. For the false negative error, 317 lane changing trajectories are employed. Out of the total trajectories, 312 are successfully identified as having a LCIM. Consequently, the false negative error is computed as $5/317 \approx 0.015$, implying a relatively low occurrence of false negatives.

10.5.2.3 Leading Time of LCIM

The timely identification of LCIM in CAVs is crucial for effective trajectory planning of surrounding vehicles in the target lane. A longer leading time, which refers to the duration between LCIM identification and lane traversal, signifies a superior algorithm performance. The average time is approximately 2.24 s and the maximum one is about 4.5 s. In comparison to existing models like [17], our proposed method demonstrates superior performance.

10.5.2.4 Lane Changing Duration Inference Results

The performance of the CLMM and LCD inference is evaluated based on their ability to predict accurate ground truth durations. To calculate the ground truth duration for CLMM, the lateral coordinate (Y_{CLMM}) corresponding to the identified LCIM is recorded, and the duration between this moment and the moment the vehicle traverses the target lateral coordinate (i.e., the lane mark $Y = \frac{l_w}{2}$) is defined as the ground truth CLMM. The identified Y_{CLMM} is used as the initial state of Eq. 10.6, and the state path is simulated 5,000 times using the Monte Carlo method for the CLMM inference. The same process is repeated for the LCD inference, with the only difference being that the target lateral coordinate is $Y = l_w$. The results are presented in Fig. 10.3, where the x-axis shows the inference and the y-axis shows the ground truth. The green line at $y = x$ serves as the reference line, and deviations from this line indicate errors. The inference and ground truth for CLMM generally follow the reference line with minimal deviation, whereas there is a certain deviation in the LCD inference. The error distribution for CLMM is controlled within [−2 s, 2 s], suggesting a good fit. The LCD inference error is distributed slightly wider, with most of the error falling within [−4 s, 2 s], as shown in Fig. 10.4.

10.5.3 Sensitivity of LCIM to Lane Width Argument

To assess the sensitivity of the model's parameters on the inference, we examined the impact of lane width on LCIM results. In instances where lanes marks are difficult for sensors to identify, or high-resolution maps do not provide assistance, lane width data may be missing, which is required for the model's input. Utilizing highD and NGSIM data, we assessed LCIM results using different lane widths ranging from 2.7

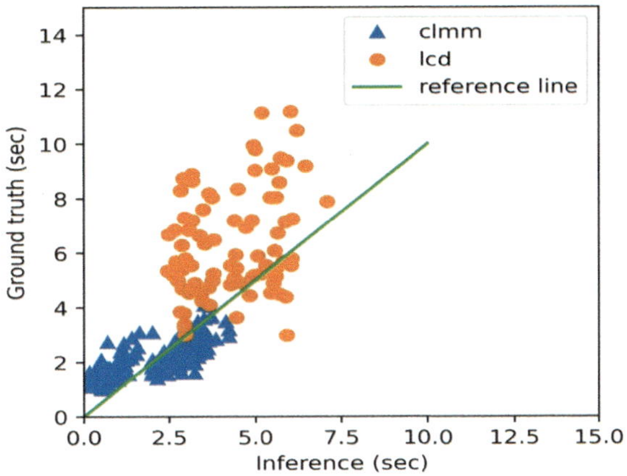

Fig. 10.3 Comparison between inference and ground truth for CLMM and LCD. Reprint from Ref. Qi et al. [14], copyright 2024, with permissions from Elsevier

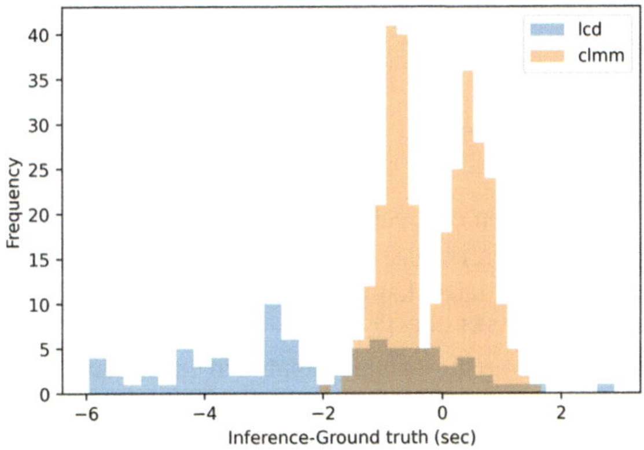

Fig. 10.4 Errors of CLMM and LCD inference result. Reprint from Ref. Qi et al. [14], copyright 2024, with permissions from Elsevier

to 3.7 m. The figures presented in Fig. 10.5 illustrate our findings, with Fig. 10.5a, b reflecting highD observation results and Fig. 10.5c, d NGSIM observation results. While minor differences were observed for highD observations, ranging from 0.56 to 14 s between LCIMs at moments 276, 283, and 290, NGSIM observations for all LCIMs were consistent at 110. Therefore, our findings suggest minimal influence of lane width on LCIM results.

10.5.3.1 Sensitivity to the Speed When Lane Changing is Initiated

The LCIM is affected by the surrounding traffic conditions, specifically the behavior of nearby vehicles. We conducted tests to evaluate the sensitivity of the LCIM inference to these surrounding vehicle states. We investigate the leading time for various speeds when the LCIM is identified. The leading time is defined according to the method described in previous section. The results indicate that the distribution of leading time remains relatively consistent across different speeds, with the average

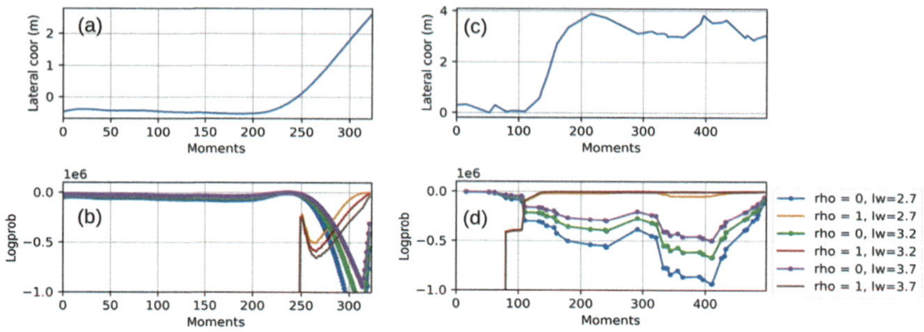

Fig. 10.5 Sensitivity to the lane width argument. Reprint from Ref. Qi et al. [14], copyright 2024, with permissions from Elsevier

values being very similar. Consequently, our proposed method is not significantly affected by variations in speed or traffic flow conditions.

10.6 Notes

The operation of a vehicle is characterized by stochastic driver behavior, resulting in random variations in vehicle states. A comprehensive framework for real-time inference of lane changing initiation moment (LCIM), combined with the estimation of CLMM and LCD, is lacking in current literature. To address this gap, we propose a logistic diffusion approach that models lateral movement using a stochastic logistic diffusion type of stochastic differential equation (SDE). We derive an analytical solution and implement parameter calibration, LCIM inference, CLMM estimation, and LCD estimation within a unified framework. Results demonstrate that our proposed method successfully provides real-time output for the desired information. The CLMM error is controlled within 2 s, while the majority of LCD errors are less than 4 s. This framework holds promise for informing the design of trajectory planning for connected and autonomous vehicles.

References

1. Ali Y, Bliemer MC, Zheng Z, Haque MM (2020) Cooperate or not? Exploring drivers' interactions and response times to a lane-changing request in a connected environment. Transp Res Part C: Emerg Technol 120:102816
2. Ali Y, Zheng Z, Haque MM, Yildirimoglu M, Washington S (2020) Understanding the discretionary lane-changing behaviour in the connected environment. Accid Anal Prev 137:105463
3. Bi L, Wang C, Yang X, Wang M, Liu Y (2015) Detecting driver normal and emergency lane-changing intentions with queuing network-based driver models. Int J Hum-Comput Interact 31(2):139–145. https://doi.org/10.1080/10447318.2014.986638

4. Chen Q, Gu R, Huang H, Lee J, Zhai X, Li Y (2021) Using vehicular trajectory data to explore risky factors and unobserved heterogeneity during lane-changing. Accid Anal Prev 151:105871
5. Chen Q, Huang H, Li Y, Lee J, Long K, Gu R, Zhai X (2021) Modeling accident risks in different lane-changing behavioral patterns. Anal Methods Accid Res 30:100159
6. Chen Y, Li G, Li S, Wang W, Li SE, Cheng B (2021) Exploring behavioral patterns of lane change maneuvers for human-like autonomous driving. IEEE Trans Intell Transp Syst 23(9):14322–14335
7. Ghasemzadeh A, Ahmed MM (2018) Utilizing naturalistic driving data for in-depth analysis of driver lane-keeping behavior in rain: non-parametric MARS and parametric logistic regression modeling approaches. Transp Res Part C: Emerg Technol 90:379–392
8. Hou Y, Edara P, Sun C (2015) Situation assessment and decision making for lane change assistance using ensemble learning methods. Expert Syst Appl 42(8):3875–3882
9. Kloeden PE, Platen E (1992) Numerical solution of stochastic differential equations, 1st Corrected ed. 1992, Corr. 4th printing 2011 edition. Springer, Berlin, New York
10. Li T, Guo F, Krishnan R, Sivakumar A, Polak J (2020) Right-of-way reallocation for mixed flow of autonomous vehicles and human driven vehicles. Transp Res Part C: Emerg Technol 115:102630
11. Liu J, Xiong H, Wang T, Huang H, Zhong Z, Luo Y (2020) Probabilistic vehicle trajectory prediction via driver characteristic and intention estimation model under uncertainty. Ind Robot: Int J Robot Res Appl 48(6):778–791. https://doi.org/10.1108/IR-06-2020-0114
12. Ng C, Susilawati S, Kamal MAS, Chew IML (2020) Development of a binary logistic lane change model and its validation using empirical freeway data. Transp B: Transp Dyn 8(1):49–71. https://doi.org/10.1080/21680566.2020.1715309
13. Pang M-Y, Jia B, Xie D-F, Li X-G (2020) A probability lane-changing model considering memory effect and driver heterogeneity. Transp B: Transp Dyn 8(1):72–89. https://doi.org/10.1080/21680566.2020.1715310
14. Qi H, Chen C, Hu X, Zhang J (2022) Online inference of lane changing events for connected and automated vehicle applications with analytical logistic diffusion stochastic differential equation. Transp Res Part C: Emerg Technol 144:103874. Schubert R, Schulze K, Wanielik G (2010) Situation assessment for automatic lane-change maneuvers. IEEE Trans Intell Transp Syst 11(3):607–616. https://doi.org/10.1109/TITS.2010.2049353
15. Wang C, Sun Q, Fu R, Li Z, Zhang Q (2018) Lane change warning threshold based on driver perception characteristics. Accid Anal Prev 117:164–174. https://doi.org/10.1016/j.aap.2018.04.013
16. Xia Y, Qu Z, Sun Z, Li Z (2021) A human-like model to understand surrounding vehicles' lane changing intentions for autonomous driving. IEEE Trans Veh Technol 70(5):4178–4189
17. Xing Y, Lv C, Wang H, Cao D, Velenis E (2020) An ensemble deep learning approach for driver lane change intention inference. Transp Res Part C: Emerg Technol 115:102615
18. Yang D, Zhu L, Ran B, Pu Y, Hui P (2016) Modeling and analysis of the lane-changing execution in longitudinal direction. IEEE Trans Intell Transp Syst 17(10):2984–2992. https://doi.org/10.1109/TITS.2016.2542109
19. Yoo J, Langari R (2019) A predictive perception model and control strategy for collision-free autonomous driving. IEEE Trans Intell Transp Syst 20(11):4078–4091. https://doi.org/10.1109/TITS.2018.2880409
20. Zhang H, Fu R (2020) A hybrid approach for turning intention prediction based on time series forecasting and deep learning. Sensors 20(17):4887. https://doi.org/10.3390/s20174887
21. Zyner A, Worrall S, Nebot E (2020) Naturalistic driver intention and path prediction using recurrent neural networks. IEEE Trans Intell Transp Syst 21(4):1584–1594. https://doi.org/10.1109/TITS.2019.2913166

Chapter 11
Game Behavior Within the Intersection

11.1 Introduction

Intersection plays a vital role in the urban road system. Most congestions originate from intersections. The operational efficiency of the entire road network is basically determined by the traffic efficiency of the intersections [1]. Therefore, the correct assessment of intersection capacity is of great significance for traffic planning, operational assessment and intersection management [2].

According to the Highway Capacity Manual, the basic method for calculating intersection capacity is based on the saturation flow rate [3]. The traditional way to calculate the saturation flow rate is the headway method. However, the depature headways are unstable due to many reasons including the driver behavior, vehicles performances etc. [4]. In order to describe the instability of the headway, a log-normal distribution model was introduced to model the headway with good accuracy [5]. Jin et al. [6] further found that departure headways followed a position-dependent log-normal distributions [6].

The headways above are defined at stop-line. After the onset of green, vehicles depart from the stop-line, drive through the intersection and exit from the intersection outlet. In rush hours, the vehicles may compete at the intersection outlet, and thus some vehicles get stuck within the intersection, and finally the queue influence the headways at stop line. The theoretical capacity using the SFR at stop-line cannot be achieved and the true capacity is overestimated. As a result, the intersection planning and control optimization performances may be compromised.

It is clear that SFR (saturation flow rate) defined at the stop line cannot reflect this mechanism. As the vehicle that enter the intersection must exit from the corresponding outlet lanes, the SFR defined at outlet lanes would be more accurate compared with that defined at stop-line during competing scenarios. Thus we propose a concept, OSFR (outlet saturation flow rate), to describe the operational efficiency at the intersection outlet. The OSFR is influenced by the competing behavior among vehicles, such as the case in Fig. 11.1. In this paper, the method of game theory is

© The Author(s), under exclusive license to Springer Nature Singapore Pte Ltd. 2024
H. Qi, *Stochastic Two-Dimensional Microscopic Traffic Model*, Lecture Notes
in Intelligent Transportation and Infrastructure,
https://doi.org/10.1007/978-981-97-3597-6_11

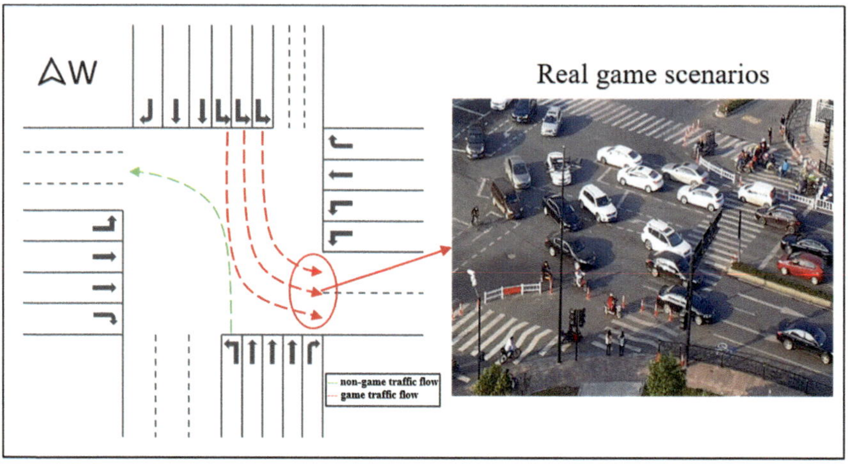

Fig. 11.1 The channelization of the intersection and the trajectory of the two kinds of traffic flow

used to model the competing behavior. The advantage of game theory over traditional rule-based models is that it can reflect dynamic interactions among drivers [7]. In recent years, game theory has been successfully used in human-like decision-making of autonomous vehicles [8–10] and lane changing behavior modeling in a connected environment (19; 18).

11.2 Data Source and Processing

The vehicle trajectory dataset used in this paper was collected by camera from an intersection (YuHangTang RD & JiangDun RD) in Hangzhou, China on October 30, 2020. The recording time is from 7:30 to 8:30 AM (morning peak). The main reason for choosing this intersection is that for the left-turn vehicles at the west entrance, the number of incoming lanes is greater than the number of exit lanes (outlet lanes).As such, the three incoming movements (each incoming lane corresponds to a movement) compete with each other for the two outlet lanes. While the incoming lanes number equals the outlet lanes number for left-turn flow at east approach, and there is no competing behavior. In this paper, we defines the left-turn traffic flow at the east approach as non-game traffic flow, and the left-turn traffic flow at the west approach as game traffic flow. The following analysis will also focus on these two kinds of traffic flow. The channelization status of the intersection is shown in Fig. 11.1.

The original video of the vehicle trajectory was taken from a high position on the 11 floor of a building near the intersection. The video resolution is 1080P (1920*1080), and 30 frames per second. Through the image recognition method based on deep

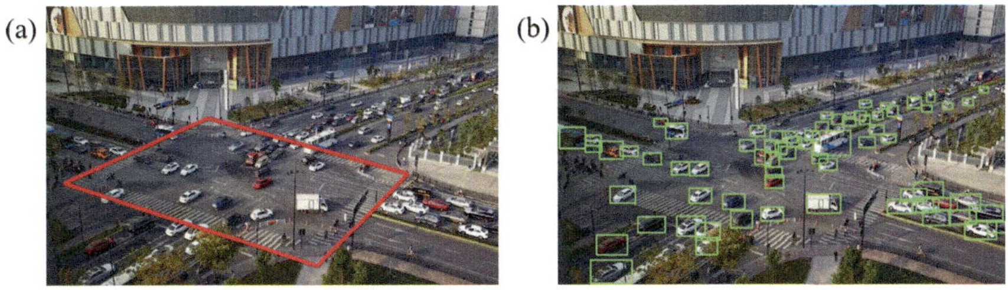

Fig. 11.2 **a** The original video and the area within the red rectangle is the study area. **b** The recognition result

learning, the trajectory data of the vehicle is obtained, and the recognition result is shown in Fig. 11.2.

The output trajectory of the vehicles is pixel based and not meter based. We use homogeneous coordinate transformation method to convert the image coordinates to real world coordinates (i.e. meter based) [11]. The transformation formula is as follows:

$$sm = pM \tag{11.1}$$

where s is a non-zero scale factor; M is a three-dimensional world homogeneous coordinate, $M = [X\ Y\ Z\ 1]^T$; m is a two-dimensional image homogeneous coordinate, $m = [u\ v\ 1]^T$; p is the mapping transformation matrix.

Since the three-dimensional world coordinates are not considered in this paper, the movement of vehicles at the intersection is regarded as two dimension trajectory. Therefore, it can be assumed that the vehicles movement plane is located on the plane of the world coordinate system $Z = 0$, and the conversion equation can be rewritten as:

$$s \times \begin{bmatrix} u \\ v \\ 1 \end{bmatrix} = \begin{bmatrix} p_1\ p_2\ p_3 \\ p_4\ p_5\ p_6 \\ p_7\ p_8\ p_9 \end{bmatrix} \times \begin{bmatrix} X \\ Y \\ 1 \end{bmatrix} \tag{11.2}$$

After the conversion, the unit of the vehicle trajectory data is changed from pixels to meters, and the trajectory error caused by the shooting angle is greatly reduced, making it possible to calculate the speed. The conversion result is shown in Fig. 11.3.

After transformation, the speed can be calculated from the distance of the trajectory points and the corresponding time stamps differences. However, due to the random factors (the image capturing error, the vehicle identification error, etc.), the speed will be very large or even negative [12]. To overcome this, we smooth the data using similar method in [12]. The average of the coordinates of the vehicle within a predefined time horizon is regarded as the current position of the vehicle. In other similar studies, Wan et al. chose 1 s as the time horizon [12]. In this article, we selects 5 frames (0.17 s), 15 frames (0.5 s) and 30 frames (1 s) as the time horizon

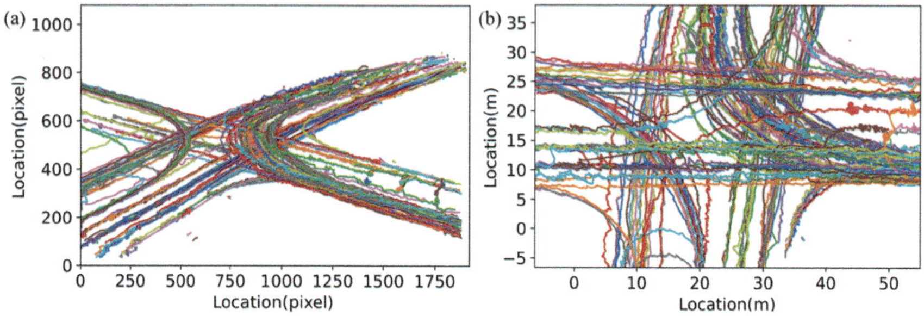

Fig. 11.3 **a** The original vehicle trajectory in pixel. **b** The converted trajectory of the vehicle in the intersection

to compare the results after smoothing. Acceleration is used to determine the best horizon. Under normal driving conditions, the range of vehicle acceleration should be from −6 to 5 m/s2 [12, 13, 20]. The influence of different smooth horizon is shown in Fig. 11.4. The error rates (percentage of acceleration outliers) of 5, 15 and 30 frames are 61.4%, 1.0% and 0.1%, respectively. Since the speed curve is too smooth when the 30 frame range is selected, the driver's decision is difficult to be recognized, 15 frames are selected as the final time range.

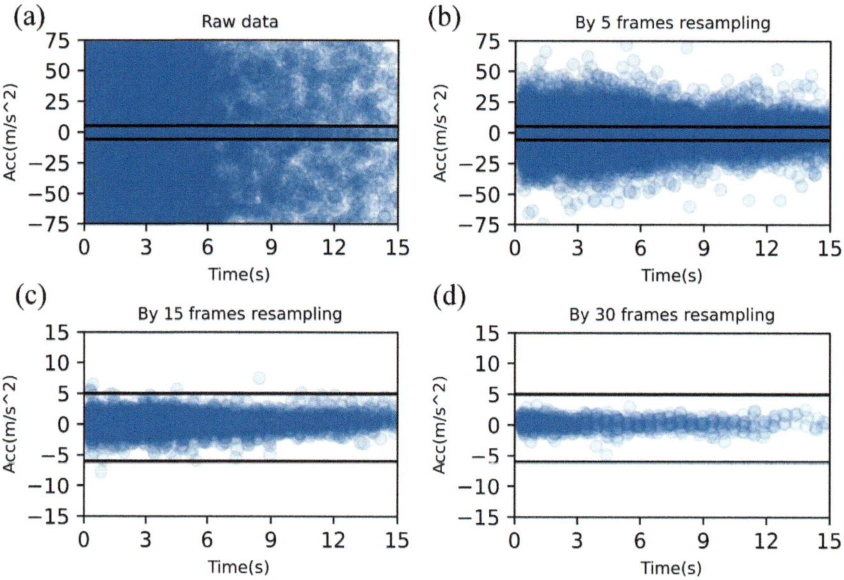

Fig. 11.4 The acceleration distribution scatter plots of raw and each resampling data (time refers to the driving time of vehicles in the intersection)

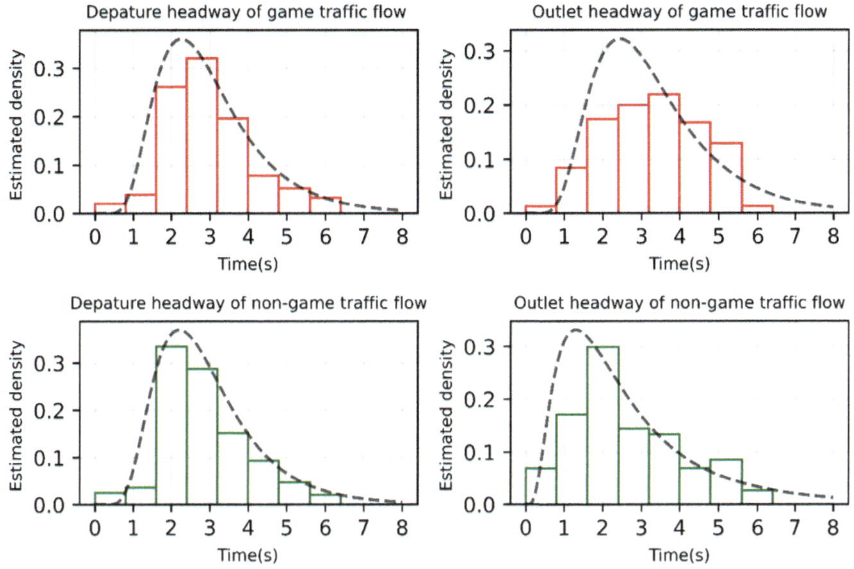

Fig. 11.5 The distribution of headway and the fitting results of log-normal distribution

11.3 Model Development

11.3.1 Framework

The trajectory of a vehicle inside an intersection can be divided into three parts: entering, accelerating, and exiting. Vehicles enter the intersection from the stop-line according to a certain headway distribution (SFR model). After that, the vehicles will drive along its trajectory based on certain speed profiles. When the vehicle is ready to exit the intersection (enter the outlet), the vehicle needs to play games with other vehicles (if there is any) to get the opportunity to exit. The models that describe the above three parts will be presented in this section.

11.3.2 SFR Model at Stopline

Many models have been used to describe the distribution of headway, such as Gaussian distribution [13], Shifted exponential distribution [14] and Gamma distribution [15]. In this section, we follow Jin's approach [6] and use a log-normal distribution to fit both the departure headway and the empirical outlet headway, the result is shown in the Fig. 11.5. The probability density function of the lognormal distribution:

Table 11.1 The K-S test results and estimated parameters of different log-normal distributions

Type	Flow	Sample size	p-value (α = 0.02)	μ	σ	E(X)
depature	game	597	0.3617	1.0113	0.4442	3.0344
outlet	game	597	0.8143	1.0991	0.4580	3.3329
depature	non-game	600	0.6541	1.0117	0.3709	2.9461
outlet	non-game	600	0.8802	0.8090	0.4569	2.4928

$$f(x, \mu, \sigma) = \begin{cases} \frac{1}{x\sqrt{2\pi}\sigma} exp\left[-\frac{1}{2\sigma^2}(lnx - \mu)^2\right], x > 0 \\ 0, x <= 0 \end{cases} \tag{11.3}$$

where x: Continuous random variable with positive value; μ: Expected value; σ: Standard deviation. The estimated parameters of log-normal distribution and corresponding Kolmogorov–Smirnov hypothesis testing (K-S test) results are given in Table 11.1.

It can be observed that (1) the departure headways are very alike. Although the departure headway is greater for game scenario, the difference is trivial; (2) the outlet headways of game scenario increase substantially about 33.7% (calculated as (3.33–2.49)/2.49). This is caused by the game behavior of vehicles in the intersection. Such influences are never reported in the literature. The specific game behavior modeling and how the model affects the headway will be described below.

11.3.3 Speed Curve Model

Traditionally it is assumed that vehicles depart from the stopline and accelerate until the speed limit is reached. It is not true when the game behavior of vehicles exist, as the vehicles that fail in the competing need to decelerate to give way to other vehicles. The speed profile of the game traffic flow and the non-game traffic flow thus are quite different. This conclusion can be confirmed in Fig. 11.6. Figure 11.6a and b are speed curves for non-game scenario while Fig. 11.6c and d are speed curves for game scenario. In Fig. 11.6, the first few cars before the stop line in the same phase are selected for comparison. For non-game vehicles, their speed tends to increase and they can pass the intersection in less than 8 s. For the game vehicles, their speed is very unstable, most of the vehicles experience more than one deceleration process.

By analyzing the data of 18 intersections in Melbourne and Sydney, Akcelik obtained queue discharge speed model for signalized intersections [16]. From this model, the relationship between speed and the time since the start of green can be established:

$$v = v_m[1 - e^{-m(t-t_0)}] \tag{11.4}$$

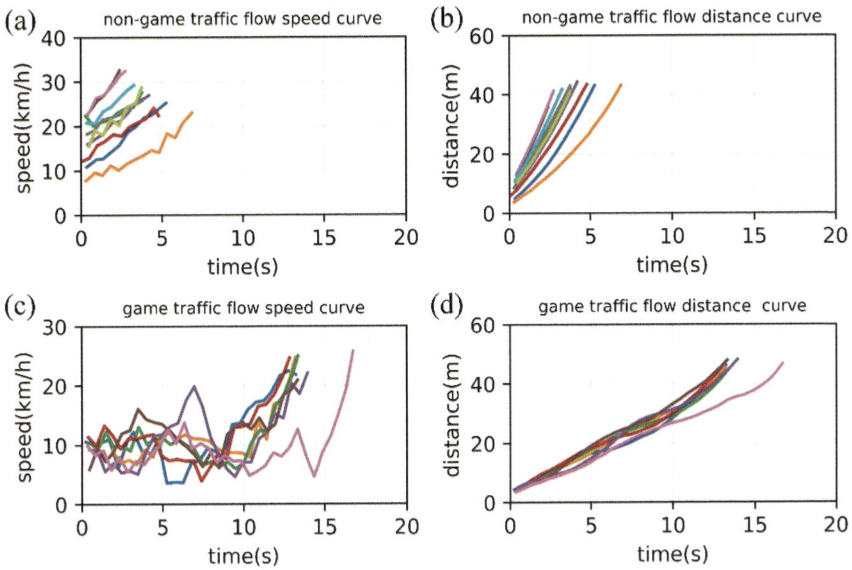

Fig. 11.6 The speed and distance curve of the two kinds of traffic flow ($t = 0$ is the time when the vehicle leaves the stop line)

where v is the queue discharge speed (km/h) at time t; vm is the maxinum queue discharge speed; t is the time since the start of green (s); m and t0 are the parameters to be calibrated.

The advantage of this model is that it uses time since the green start as the independent variable, rather than the time since the vehicles leave the stop line. In this way, the speed of all vehicles can be considered in a single model. We employ this model and fit using the speed data, and get the following result (the result is shown in Fig. 11.7):

$$v = 26.671 \times [1 - e^{-0.268(t+1.36)}] \quad (11.5)$$

11.3.4 Game Behavior Modeling at the Intersection Outlet

11.3.4.1 Game Environment Setting

Applying game theory to the microscopic analysis of traffic flow is not new [7–10, 17, 18]. Some researches use game theory to analyze how drivers make decisions when changing lanes [7]. The problem studied in this paper is more like lane-choosing, which is similar to lane-changing or merging. Figure 11.8 shows the lane-choosing

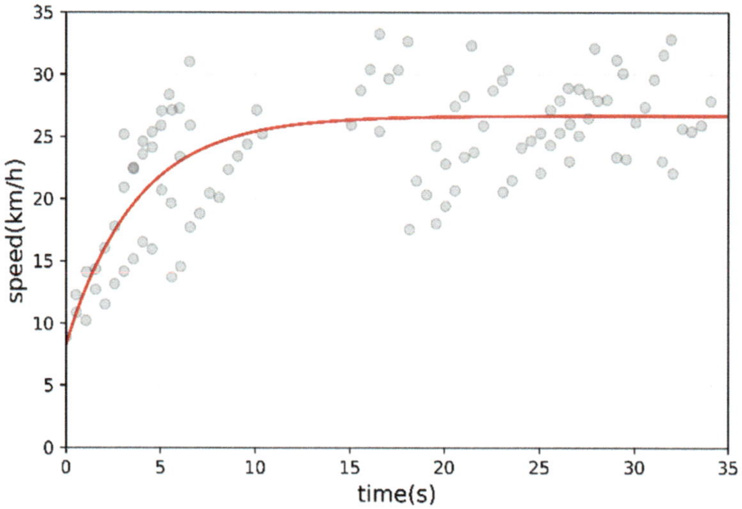

Fig. 11.7 The speed model of non-game vehicles at the intersection ($t = 0$ is the time when the green start)

scenario of the game traffic flow. The movements of the vehicles are divided into three zones: free zone, game zone and outlet.

In the free zone, the vehicle can accelerate freely without the influence of competing behavior from other movements while obeying the speed profiles in Eq. 11.5. In the game zone, the vehicle needs to play with each other when competing for outlet lane. Note that the game may only occur when the vehicles are close enough, or equivalently more than one vehicles are presented simultaneously. For example, if the middle vehicle is in front and far away from the left and right vehicle, the middle vehicle can go straight into the outlet without playing games.

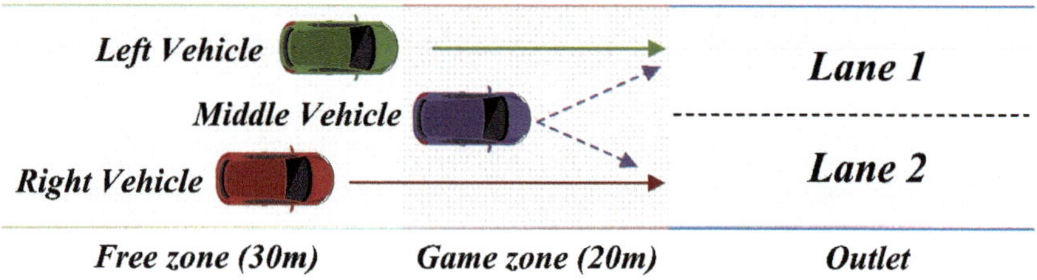

Fig. 11.8 The schematic of the lane-choosing of the game traffic flow

11.3.4.2 Game Model Formulation

Different types of drivers have different driving habits, which are related to their gender, age and personality [19–21]. In this research, three types of drivers are considered: polite, neutral, and rude. For polite drivers, they tend to give way to competing vehicles. The reasonable explanation is that these drivers want to minimize the possibility of collision and they are not in a hurry. For rude drivers, they tend to be aggressive. For neutral drivers, they have no obvious inclinations and will make judgments based on specific circumstances.

The following assumptions are made:

- The middle vehicle is called the ego vehicle, and the left or right vehicle is called the target vehicle;
- The ego vehicle will only play with one target vehicle at a time;
- The drivers are aware of each other's speed and distance;

In this paper, the target vehicle does not know whether the driver in the ego vehicle is rude or polite. So the lane-choosing behavior is modeled as a two-person non-zero-sum non-cooperative game under incomplete information. The payoff matrix is shown in Table 11.2. In this study, all types of drivers have the same action set (give way or don't give way to other drivers, or equivalently, accelerate and decelerate). The difference is that different types of drivers will have different payoff under the same action. For example, P11 will be less than R11, because rude drivers are less worried about the impact of a collision than polite drivers.

Because the game of incomplete information is difficult to solve, we converts the game of incomplete information into the game of imperfect information by the Hysanyi transformation [22]. This approach introduces a virtual participant "nature". Nature moves first and determine the type of ego driver. And every driver knows the specific distribution of the probability. As the extensive form of the game was very complex, we made a schematic diagram under a simplified scenario in Fig. 11.9 (assume that the driver in the target vehicle is only neutral, while the driver in the ego vehicle is a mixture of rude and polite).

Table 11.2 The payoff matrix corresponding to the different strategic combinations of the two drivers (*N*, *P* and *R* correspond to the payoff for neutral, polite, and rude driver, respectively)

Action	*Ego vehicle*	
	A_1 (Accelerate)	A_2 (Decelerate)
Target vehicle		
B_1 (Accelerate)	(N_{11} or P_{11} or R_{11}, N_{11} or P_{11} or R_{11})	(N_{12} or P_{12} or R_{12}, N_{12} or P_{12} or R_{12})
B_2 (Decelerate)	(N_{21} or P_{21} or R_{21}, N_{21} or P_{21} or R_{21})	(N_{22} or P_{22} or R_{22}, N_{22} or P_{22} or R_{22})

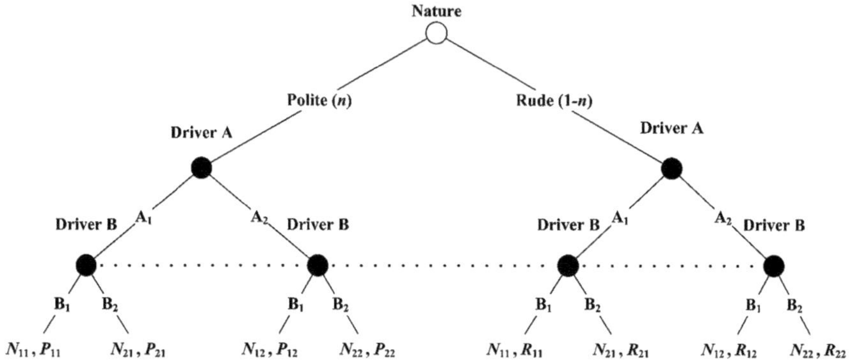

Fig. 11.9 Lane-choosing game in extensive form

11.3.4.3 Payoff Function

In the game of lane-changing, [7] construct the payoff function based on the speed changes before and after the lane change and the acceleration required to avoid collisions. In this study, in addition to the speed and distance benefits of the vehicle, the payment function will also consider the psychological benefits of different types of drivers. Rude drivers will get greater psychological benefits when accelerating (more inclined to accelerate). The payoff function formulas for different drivers types are as follows. For neutral drivers:

$$\begin{cases} N_{11} = a_{brake} + D/d + b^n_{acc} \\ N_{12} = a_{acc} + \frac{D}{d} + b^n_{acc} \\ N_{21} = a_{dec} - \frac{D}{d} + b^n_{dec} \\ N_{22} = a_{dec} - D/d + b^n_{dec} \end{cases} \tag{11.6}$$

where a_{brake}: Both drivers choose to accelerate, the deceleration corresponding to the sudden braking of the vehicle in order to avoid collision; a_{acc}: When the other driver chooses to decelerate, the acceleration that the vehicle can obtain; a_{dec}: Deceleration when the vehicle actively chooses to decelerate; D: The total distance of the intersection; *d*: The distance between the vehicle and the lane entrance; $\mathrm{b}^{\mathrm{n}}_{\mathrm{acc}}$, $\mathrm{b}^{\mathrm{n}}_{\mathrm{dcc}}$: Psychological benefits of neutral drivers when they choose to accelerate and decelerate, respectively. For polite drivers:

$$\begin{cases} P_{11} = a_{brake} + D/d + b^p_{acc} \\ P_{12} = a_{acc} + \frac{D}{d} + b^p_{acc} \\ P_{21} = a_{dec} - \frac{D}{d} + b^p_{dec} \\ P_{22} = a_{dec} - D/d + b^p_{dec} \end{cases} \tag{11.7}$$

where b^{p}_{acc}, b^{p}_{dec}: Psychological benefits of polite drivers when they choose to accelerate and decelerate, respectively. For rude drivers:

$$\begin{cases} R_{11} = a_{brake} + D/d + b^{r}_{acc} \\ R_{12} = a_{acc} + \frac{D}{d} + b^{r}_{acc} \\ R_{21} = a_{dec} - \frac{D}{d} + b^{r}_{dec} \\ R_{22} = a_{dec} - D/d + b^{r}_{dec} \end{cases} \tag{11.8}$$

In above equations, b^{r}_{acc}, b^{r}_{dec}: Psychological benefits of rude drivers when they choose to accelerate and decelerate, respectively.

By analyzing the data, the values of the above parameters are obtained. Under general circumstances, $a_{brake} = -5$ m/s2, $a_{acc} = 2$ m/s2 and $a_{dec} = -2$ m/s2. The same type of drivers is assumed to have the same psychological benefits, $b^{n}_{acc} = b^{n}_{dec} = 0$, $b^{r}_{acc} = 6$, $b^{r}_{dec} = -2$, $b^{p}_{acc} = -1$ and $b^{p}_{dec} = 2$. Note that the setting of these values has taken into account the range of data values.

11.3.4.4 Nash Equilibrium Solution

In this section, the details of solving Nash equilibrium are described. The first step is to decompose the incomplete information game into three complete information games, that is, to play with the neutral, polite and rude drivers respectively. For the complete information game, the driver's strategy can be uniquely determined.

$$s(action, type) = \begin{cases} acc \ if \ u^{type}_{acc} \geq u^{type}_{dec}, \\ dec \qquad else. \end{cases} \tag{11.9}$$

where s is the strategy, which is determined by the action of the driver and the type of another driver; action include accelerate and decelerate; type include neutral, polite and rude; u is the payoff, which is determined by the action and the type of the driver. After that, by comparing the expected payoff of acceleration and deceleration, the driver's action can be determined:

$$\begin{cases} E(u^{type}_{acc}) = n_1 u^{type}_{s(acc,netural)} + n_2 u^{type}_{s(acc,polite)} + n_3 u^{type}_{s(acc,rude)} \\ E(u^{type}_{dec}) = n_1 u^{type}_{s(dec,netural)} + n_2 u^{type}_{s(dec,polite)} + n_3 u^{type}_{s(dec,rude)} \end{cases} \tag{11.10}$$

where n1, n2 and n3 are the proportions of neutral, polite and rude drivers respectively (n1 + n2 + n3 = 1). If $E(u^{type}_{acc}) > E(u^{type}_{dec})$, then the driver will choose to accelerate, otherwise it will choose to decelerate (when equal, it is assumed that the vehicle will accelerate). Since the game is symmetric, the actions of the other driver can be determined in the same way. In fact, Nash equilibrium is directly related to the distribution of driver types. Under different proportions of drivers, there will be different Nash equilibriums.

11.4 Simulation

11.4.1 Simulation Framework

To investigate the OSFR under the influence of game behavior within the intersection, we resort to the simulation method. The simulation use the models above and get the moment when vehicles depart from the outlet, given the vehicles that enters the intersection from the stop-line. Once the vehicles exiting moments are obtained, the OSFR can be calculated. The simulation framework is shown in Fig. 11.10 (the corresponding game scenario is shown in Fig. 11.8). First of all, vehicles are generated at the stop line, and the depature headways distribution of vehicles follows the log normal distribution. After that, the vehicles drive according to the proposed speed model. When a vehicle enters the game zone, if there are other vehicles in the game zone that choose the same target lane, then the game needs to be played, otherwise it will exit from the outlet lane directly.

The details of the implementation of the proposed game theory-based OSFR model are illustrated in Algorithm 1. Note that the distribution of vehicle headway

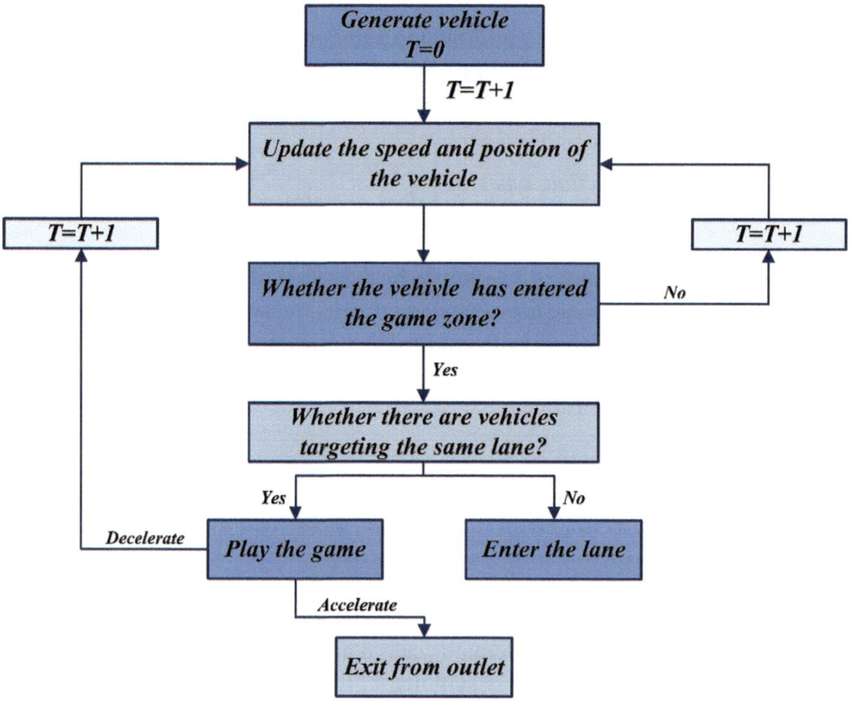

Fig. 11.10 The simulation framework of game model between vehicles (the simulation step size is 0.3 s)

in the simulation follows the log-normal distribution. And the details of speed model and game model are illustrated in Sect. 11.3.2 and Sect. 11.3.4, respectively.

Algorithm 1 Proposed game theory-based OSFR model.

Input	• Time when the vehicles leave the stop line, *depature_set*; Speed model, $v(t)$; • Driver type ratio, n_1, n_2, n_3; • Game model, *game*; • Number of simulated vehicles, n;
output	• Time when the vehicles enter the outlet line, *outlet_set*; • $t = 0, i = 0$;

```
while len(outlet_set) < n do
    t = t + 1;
    update vehicle.distance = ∫ v(t)dt ;
    if depature_set[i] > t then
        i = i + 1;
        generate vehicle[i];
    end
    if vehicle.distance > 30 and vehicle[id_1].target_lane ==
      vehicle[id_2].target_lane then
        game(vehicle[id_1], vehicle[id_2]);
        if vehicle[id_1].action == acc and vehicle[id_1].action == acc then
            update vehicle[id_1].speed = vehicle[id_1].speed – 2.5m/s;
            update vehicle[id_2].speed = vehicle[id_2].speed – 2.5m/s;
        end
        if vehicle[id_1].action == dec and vehicle[id_1].action == dec then
            update vehicle[id_1].speed = vehicle[id_1].speed – 0.5m/s;
            update vehicle[id_2].speed = vehicle[id_2].speed – 0.5m/s;
        end
        if vehicle[id_1].action == acc and vehicle[id_1].action == dec then
            update vehicle[id_1].speed = vehicle[id_1].speed + 1m/s;
            update vehicle[id_2].speed = vehicle[id_2].speed – 0.5m/s;
        end
    end
    if vehicle[id].distance > 50 then
        delete vehicle[id];
        update outlet_set.append(t);
    end
end
```

11.4.2 Simulation Results

In the simulation, we assume that 20 cars are generated in each cycle, and a total of 10 cycles are simulated. As there are three incoming lanes, we have total 600 vehicles. After all of these vehicle leave the intersection, we gather the headways that the vehicles leave the intersection through the outlet (or equivalently the two outgoing lanes of the intersection). As there are three types of drivers, we repeat the simulation 9 times to accommodate different types ratios. We set three levels of maximal ratio for each type of drivers, i.e. 100, 80 and 60%. The remaining ratios are assigned equally to other two types of drivers. The simulation results are shown in Fig. 11.11. It is shown that the distribution of outlet headway is different under different proportions of types of drivers. The results show that when there is only one type of driver in the road network, the headway does not increase significantly (the first row in Fig. 11.11). This is because in this situation, the driver has an accurate expectation of what the other driver is going to do and will not brake sharply. When the flow are a mixture of different types of drivers, the OSFR deteriorates. The minimal headway is the case when all drivers are of rude type, as each vehicle want to exit the intersection as early as possible; the maximal headway is the case when major vehicles (80%) are rude drivers, and its mean headway is 3. 29 s.

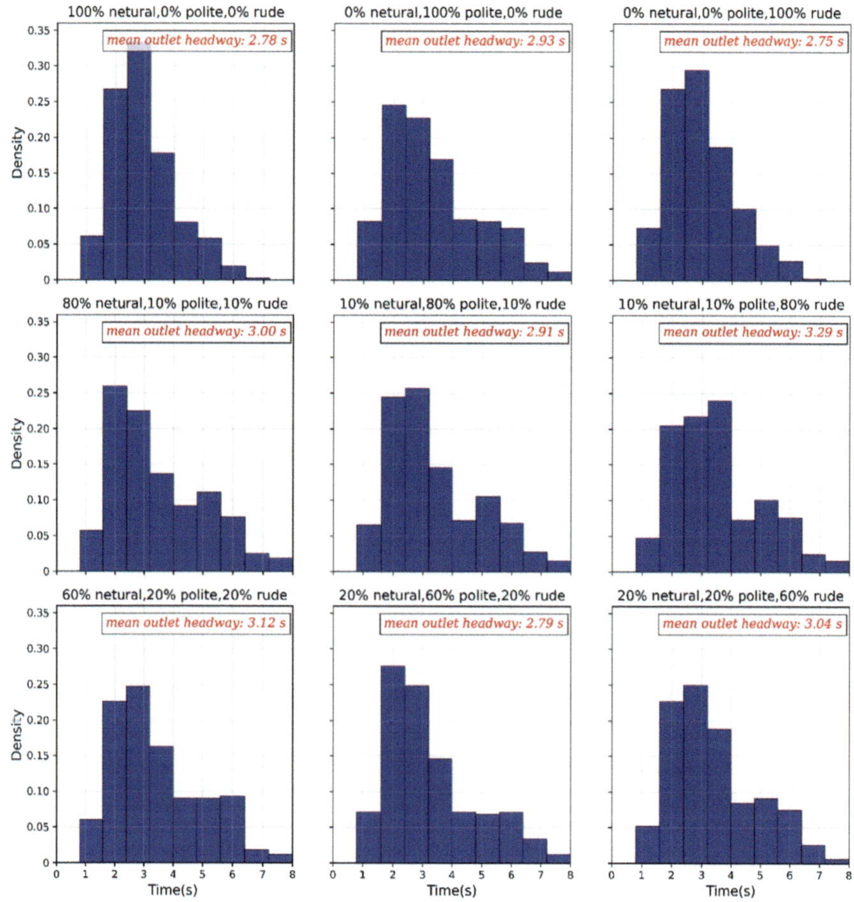

Fig. 11.11 The simulation results of outlet headway under different driver proportions

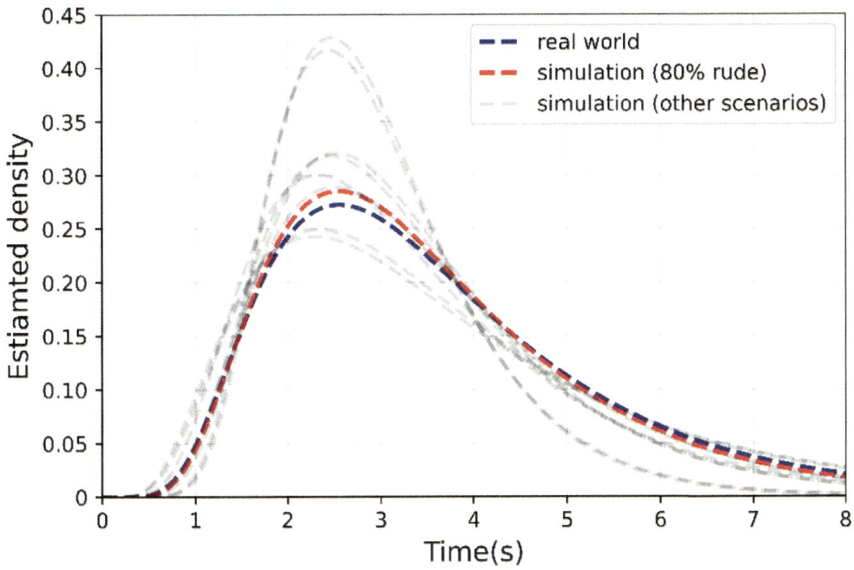

Fig. 11.12 Comparison of headway distribution between real world and simulation scenarios

In essence, the reason for the increase in headway is that both drivers choose to accelerate, which causes the driver to brake sharply in order to avoid collision. This extreme situation is mainly related to two factors, one is that drivers misjudge the intentions of other drivers (the driver thinks the other driver will decelerate, so he chooses to accelerate, but the other drive also chooses to accelerate), and the other is that rude drivers choose to accelerate when they should yield. In Fig. 11.12, the log-normal distribution is used to fit the headway time under simulation scenarios and the real scene respectively. The results show that the closest approximation to the real situation is the scenario which rude drivers accounted for 80%.

In order to further analyze the influence of the game model on the traffic flow at the intersection, we compare the outlet headway of different vehicles number under the same driver proportion (10% neutral, 10% polite, 80% rude). The vehicles number can be interpreted as the duration of the green phase, as longer green duration would feed more vehicles into the intersection. The result is shown in Fig. 11.13. With the increase of the number of simulated vehicles, the mean outlet headway increases, and the growth rate increases first and then stabilize. When the number of simulation vehicles is more than 35, the mean outlet headway is basically stable at 3.43 s. The reason for the increase of headway is that with the increase of vehicles, the game between vehicles becomes more and more frequent, and the overall speed will slow down as results of fail in the competing. As the vehicles number increase to certain level, each vehicle would experience a game, and thus the outlet headway becomes stable, which can be interpreted as the saturated headway at the outlet (OSFR). This headway is the output headways of the intersection, which is different from the SFR at the stop line.

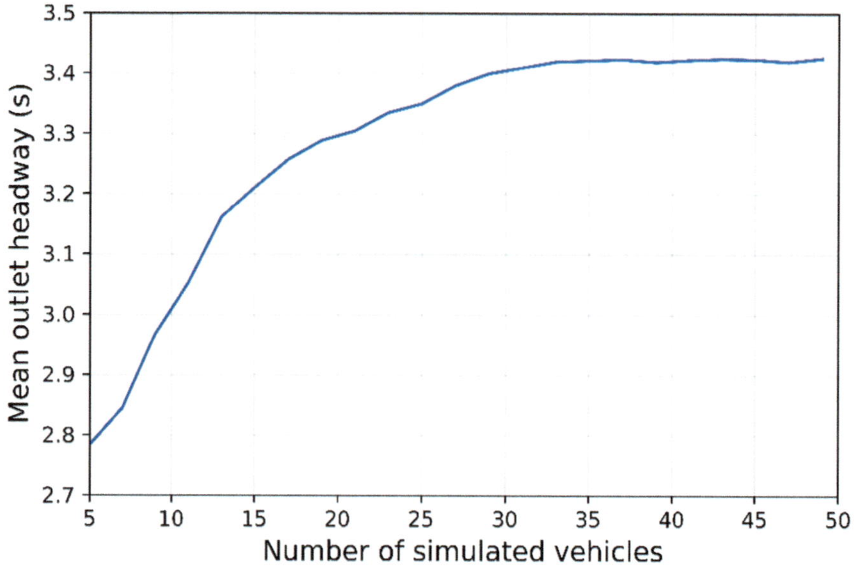

Fig. 11.13 The simulation results of outlet headway under different number of simulated vehicles

11.4.3 Application in Calculating SFR Adjustment Factor

Based on the above conclusions, the formula for calculating intersection capacity can be modified. According to HCM 2010 [3], the capacity of each lane group at the intersection can be calculated as follow:

$$c_i = s_i \frac{g_i}{C} \tag{11.11}$$

where c_i: Capacity of lane group i (veh/h); s_i: Saturation flow rate for lane group i (veh/h); g_i: Effective green time for lane group i (s); C: Intersection cycle length (s). And the saturation flow rate s_i is related to the basic saturation flow rate and the adjustmengt factors:

$$s_i = s_{out} f_w f_{HV} f_g f_p f_{bb} f_a f_{LU} f_{LT} f_{RT} f_{Lpb} f_{Rpb} \tag{11.12}$$

where s_0: Basic saturation flow rate (veh/h); f: the adjustment factors. Note that the saturation flow rate is calculated from the departure headway, which is suitable for most intersections. As shown in Fig. 11.5, when there is no game between vehicles, departure headway is greater than outlet headway. In this case, the above calculation can be applied. However, when the number of incoming lanes is greater than the number of exit lanes, the effective saturation flow rate should be determined by the outlet headway rather than the departure headway based on the above results. Therefore, for intersections with mismatched lanes, the saturation flow rate calculation formula needs to be adjusted:

$$s_i = s_{out} f_w f_{HV} f_g f_p f_{bb} f_a f_{LU} f_{LT} f_{RT} f_{Lpb} f_{Rpb} \quad (11.13)$$

where s_{out} is basic saturation flow rate based on outlet headway(veh/h).

$$s_{out} = 1/\bar{h}_{out} \quad (11.14)$$

where $\bar{h}_{out}$: Mean outlet headway(s), which can be calculated by the model present in this paper.

11.5 Notes

This chapter is a detailed version of Zhang and Qi (2023b). Solving the congestion that occurs in the intersection is a challenge. The prerequisite to solve this problem is that we have to understand how the vehicles operate in the intersection. The existing results show that scholars focus on the study of SFR, but ignore the OSFR. The game behavior of vehicles in the intersection will affect the operation of the traffic flow.

Based on empirical data, this paper proposes an OSFR model to explain the impact of game behavior between vehicles on SFR. And define different types of drivers to make the model more in line with the real scene. The experimental results show that the game behavior between vehicles will increase the headway. And under different proportions of drivers, there will be different effects. These results help to re-understand how the congestion in the intersection is generated and how to calculate the OSFR when the number of entrances and exits does not match. Besides, it can be used to adjust the calculation method of intersection capacity.

Due to the lack of information such as heading Angle, accelerator aperture and specific size of the vehicle, the model presented in this paper does not consider the vehicle dynamics model. In the future work, vehicle dynamics model can be introduced to make the game model more accurate. Furthermore, it needs to be realized that drivers in different regions have different driving habits, and the payoff function needs to be redefined when the model is applied elsewhere.

References

1. Arshi AN, Alhajyaseen WKM, Nakamura H, Zhang X (2018) A comparative study on the operational performance of four-leg intersections by control type. Transp Res Part A: Policy Pract 118:52–67
2. Yang CC, Wang JW, Dong JS (2020) Capacity model of exclusive right-turn lane at signalized intersection considering pedestrian-vehicle interaction. J Adv Transp 2020:19
3. Transportation Research Board (2010) Highway capacity manual

4. Li L, Chen XM (2017) Vehicle headway modeling and its inferences in macroscopic/microscopic traffic flow theory: A survey. Transp Res Part C: Emerg Technol 76:170–188
5. Shao CQ, Liu XM (2012) Estimation of saturation flow rates at signalized intersections. Discret Dyn Nat Soc 2012:1–9
6. Jin X, Zhang Y, Wang F, Li L, Yao D, Su Y, Wei Z (2009) Departure headways at signalized intersections: a log-normal distribution model approach. Transp Res Part C: Emerg Technol 17(3):318–327
7. Talebpour A, Mahmassani HS, Hamdar SH (2015) Modeling lane-changing behavior in a connected environment: A game theory approach. Transp Res Part C: Emerg Technol 59:216–232
8. Hang P, Lv C, Xing Y, Huang C, Hu Z (2021) Human-like decision making for autonomous driving: a noncooperative game theoretic approach. IEEE Trans Intell Transp Syst 22(4):2076–2087
9. Ji K, Orsag M, Han K (2021) Lane-merging strategy for a self-driving car in dense traffic using the stackelberg game approach. Electronics 10(8):894
10. Sankar GS, Han K (2020) Adaptive robust game-theoretic decision making strategy for autonomous vehicles in highway. IEEE Trans Veh Technol 69(12):14484–14493
11. Qu Z, Chen H, Li Z, Hu H, Wei W (2007) 2D view reconstruction method based on single calibration pattern. J Jilin Univ (Eng Technol Ed) 37(5):1159–1163
12. Wan Q, Peng G, Li Z, Inomata FHT (2020) Spatiotemporal trajectory characteristic analysis for traffic state transition prediction near expressway merge bottleneck. Transp Res Part C: Emerg Technol 117:102682
13. Adams WF (1936) Road traffic considered as a random series. J Inst Civ Eng 4(1):121–130
14. Gerlough DL, Schuhl A (1955) Use of poisson distribution in highway traffic. Eno foundation for highway traffic control saugatuck, conn.
15. Buckley DJ (1968) A semi-poisson model of traffic flow. Transp Sci 2(2):107–133
16. Akcelik R, Besley M (2002) Queue discharge flow and speed models for signalised intersections. Transportation and Traffic Theory in the 21st Century, 2002: 99–118
17. Na X, Cole DJ, Li G (2021) Two nash-equilibrium-based steering control models for representing a driver's interaction with vehicle automated steering. Veh Syst Dyn 2021:1–35
18. Smirnov N, Liu Y, Validi A, Morales-Alvarez W, Olaverri-Monreal C (2021) A game theory-based approach for modeling autonomous vehicle behavior in congested. Urban Lane-Chang Scenarios Sens 21(4):1523
19. Dey KC, Yan L, Wang X, Wang Y, Shen H, Chowdhury M, Yu L, Qiu C, Soundararaj V (2016) A review of communication, driver characteristics, and controls aspects of cooperative adaptive cruise control (CACC). IEEE Trans Intell Transp Syst 17(2):491–509
20. Wu YP, Zhao XH, Yao Y, Ma JM, Rong J (2017) Optimization method to improve ecodriving acceptance and effectiveness based on driver type classification. Transp Res Rec 2665:21–29
21. Zheng Y, Chase RT, Elefteriadou L, Sisiopiku V, Schroeder B (2017) Driver types and their behaviors within a high level of pedestrian activity environment. Transp Lett 9(1):1–11
22. Harsanyi JC (1967) Games with incomplete information played by "bayesian" players, I-III Part I. Basic Model Manag Sci 14(3):159–182

Chapter 12
Intersection Deadlock Probability: Stochastic Petri Net Approach

12.1 Deadlock Definition and How It Forms

Deadlock refers to a severe congestion state that occurs at an intersection, where multiple traffic movements are locked in a standstill due to each movement waiting for others to clear the intersection and create space. As a result, traffic flow from all directions comes to a complete halt. Figuratively speaking, deadlock scenarios can be observed at both a four-leg intersection (depicted in sub-figures (a) and (c) of Fig. 12.1) and a T-intersection (depicted in sub-figures (b) and (d) of Fig. 12.1). The inherent loop structure prevents any movements from being executed, resulting in all vehicles becoming trapped in the queue.

Efficiently preventing deadlock is an essential task, as it differs from typical traffic congestion. In instances of regular congestion, vehicles heading in the direction of high demand can still proceed despite queue length. Conventional traffic management techniques like signal control and route guidance are effective in increasing vehicle queue capacity and discharge during normal congestion. However, in deadlock situations, vehicles from all directions are completely immobilized, rendering these methods ineffective. Furthermore, due to limited visibility of nearby traffic conditions, drivers may take substantial time to realize the occurrence of a deadlock. Even if drivers become aware of the deadlock, the trapped vehicles cannot immediately free themselves due to constrained space for movement. Consequently, research endeavors to model and predict deadlock formation are crucial in preventing such situations.

The occurrence of deadlock hinges upon the presence of a loop structure within vehicular trajectories. In its simplest form, deadlock manifests as a loop comprising multiple vehicle queues, with each queue obstructed by its downstream counterpart. An illustrative example can be observed in Fig. 12.2a, where four vehicle queues form a counterclockwise loop. Notably, multiple loops can combine to create a more extensive deadlock, as depicted in Fig. 12.2b. In the case of a conventional four-leg intersection, numerous conflicting points exist, giving rise to multiple trajectory loop

© The Author(s), under exclusive license to Springer Nature Singapore Pte Ltd. 2024

H. Qi, *Stochastic Two-Dimensional Microscopic Traffic Model*, Lecture Notes in Intelligent Transportation and Infrastructure,
https://doi.org/10.1007/978-981-97-3597-6_12

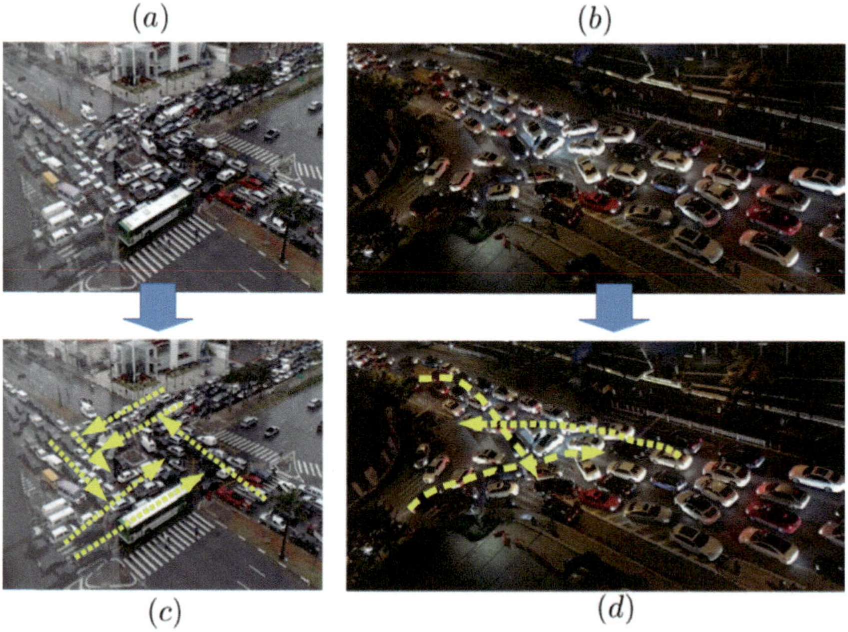

Fig. 12.1 Deadlock scenarios (**a**) and (**c**), four-leg intersections (**b**) and (**d**) at T-intersection

configurations. As evidenced in Fig. 12.3a, a typical four-leg intersection is shown, while Fig. 12.3b illustrates the points of conflict for all trajectories, which conform to an elliptical curve. From these trajectories, various loops can be derived. Figure 12.3c demonstrates a representative loop, highlighting the potential for extracting numerous similar loops from the trajectory data. Crucially, the presence or absence of a loop structure serves as the determining factor for the occurrence or prevention of deadlock, respectively. For instance, a deadlock scenario would not arise in a single intersection governed by two one-way streets.

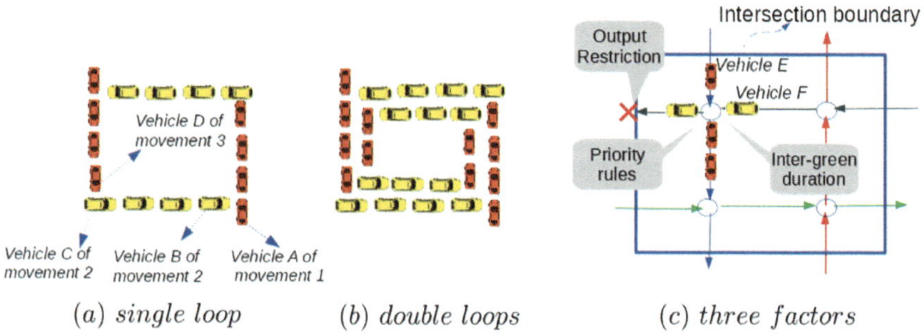

Fig. 12.2 Single loop, double loops, and three factors for deadlock

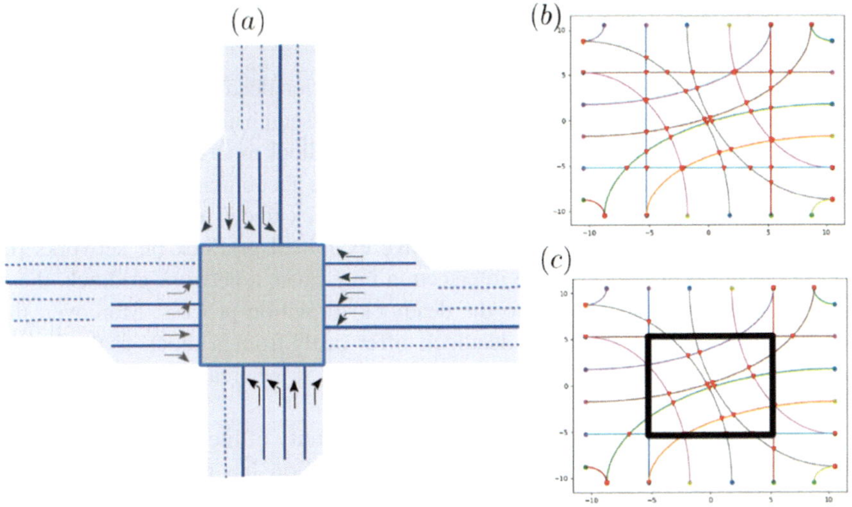

Fig. 12.3 **a** An example of an intersection; **b** The trajectories; **c** Trajectory loop

The occurrence of a deadlock depends on various factors. It is crucial that vehicles entering an intersection do so only after conflicting vehicles have exited the intersection boundaries, thus preventing a deadlock. Three factors that contribute to a deadlock at a common intersection can be summarized as follows: Firstly, outflow restriction occurs when a downstream queue obstructs the intersection outlet, causing vehicles to wait within the intersection. Secondly, insufficient inter-green time results in conflicting movements at the intersection point impeding one another. For example, heavy rain can degrade the saturation flow rate (SFR), leading to a substantial decrease in the maximum flow rate depicted by the fundamental diagram. Consider the example of vehicles E and F in Fig. 12.2c, where the distance from the stop-line of vehicle E to the confliction point is represented by l_E and that of vehicle F by l_F. Assuming an average speed of v, the clearance time (i.e., all-red time) can be calculated as $\frac{|l_E - l_F|}{v}$. When the SFR degrades, the speed decreases to v', resulting in an increased needed clearance time of $\frac{|l_E - l_F|}{v'}$. As a consequence, the clearance time designed for normal conditions becomes insufficient, leading to a confliction. Lastly, the third factor involves the implementation of priority rules for confliction points. For example, in Fig. 12.2a, if vehicle A from movement 1 yields to vehicle B from movement 2, movement 2 can proceed ahead, avoiding a blockage of vehicle D (the first vehicle of the south-bound movement). Therefore, even when factors 1 (outlet restriction) and 2 (insufficient inter-green time) coexist, an appropriate priority arrangement can prevent a deadlock from occurring.

12.2 Current Practice

Investigating the formation and dynamics of intersection deadlock is essential due to its detrimental effects. However, there exists limited research in this area. Prior studies have focused on block queues, with [3] describing gridlock as an exponential increase in vehicle accumulation leading to low exit and more accumulation. Reference [12] proposed a model to replicate hysteresis and gridlock under non-steady-state and homogeneous conditions. Other studies have examined gridlock on networks [6, 11, 14]. While a deadlock at one intersection may cause a network gridlock, these findings offer limited insight into the deadlock formation process. Moreover, the mechanisms causing intersection deadlock differ vastly from gridlock, as the former involves vehicle conflicts from opposing directions [5], whereas the latter is a result of vehicle queue propagation and overflow [6].

Further research opportunities arise from the scarcity of literature on the subject of deadlock. Deadlock is not limited to transportation but manifests in various domains including software and hardware development [4], manufacturing systems [9], distributed systems [15], robots and automated guided vehicles [10, 21], queuing networks [16], among others. The essential conditions for deadlock, as outlined by [2] and Silberschatz et al. [20], can be summarized into four components: (1) mutual exclusion, which prohibits multiple tasks from simultaneously occupying a resource; (2) no pre-emption, implying that resources cannot be released until they have been fully utilized; (3) hold and wait, wherein tasks retain resources while awaiting others; and (4) circular wait, characterized by a loop of tasks requesting resources from one another. Notably, these four conditions align with the intersection deadlock depicted in Fig. 12.2a as follows: (1) the confliction point can only be occupied by a single vehicle; (2) vehicle B cannot compel vehicle A to vacate the space; (3) the vehicles must wait for downstream vehicles to advance; and (4) the vehicle queues form a loop, exhibiting circular wait.

Deadlock research in various fields is documented in the academic literature, primarily conducted within the discrete event system (DES) framework. Petri Net (PN) is the predominant tool used within the DES framework, offering a robust approach for simulating and analyzing features such as concurrency, asynchrony, synchrony conflict, mutual exclusion, precedence relations, and deadlocks. In the domain of transportation, PN models have been successfully applied to signal control, co-piloting systems, and other related areas. However, past PN models in transportation studies often represented intersections with only a few places, neglecting the consideration of vehicle trajectories. This simplification hinders the investigation of intersection deadlocks.

In this chapter, we present a PN-based modeling approach to investigate deadlock formation in transportation systems. Our approach utilizes Petri Nets to model and analyze deadlock dynamics, including formation probability and duration. To achieve this, we design four sub-PN models: DemandPN for demand input, ConflictPN for vehicles conflictions, TrafficLightPN for traffic signal plan, and CellularPN for vehicles movements. These models are integrated into an IntersectionPN model, allowing us to study the dynamics of intersection deadlocks. We define three system states: deadlock state, trap state, and live state, and formulate the deadlock occurrence as an integer programming problem. Using a reachability graph, we analyze the formation probability and duration of the deadlock, resulting in a Markov chain representation of the system. We illustrate the effectiveness of our proposed model through a case study.

The objective of this chapter is to address the following objectives: Apply a Petri Net (PN) approach to model the flow of traffic at intersections and analyze the occurrence of deadlocks; Define the deadlock state and establish a methodology for investigating the metrics associated with deadlock formation, such as probability and duration; Examine the connection between factors contributing to deadlock formation, including priority rules, outlet restrictions, and SFR degradation.

12.3 Preliminaries on Petri Net

12.3.1 Elements of Petri Net (PN)

Petri Nets (PN) are an established formal model for analyzing distributed and concurrent systems, with widespread use across multiple fields, including computer science. Mathematically, a PN can be represented as a directed bipartite graph, described by a 4-tuple (P, T, F, W), where P and T are finite and non-empty, disjoint sets of places and transitions, respectively. The flow relationship of the net is represented by arcs between places and transitions, and transitions and places, but not between places or transitions. A mapping function W: $(P \times T) \cup (T \times P) \rightarrow \mathbb{N}$ assigns integer weights to each arc, where $W(x, y) > 0$ if and only if $(x, y) \in F$, and $W(x, y) = 0$ otherwise.

Petri Nets, a formal model for analyzing concurrent and distributed systems, consist of places and transitions. Places contain tokens, representing the system's state, with the distribution of tokens known as a marking. Deadlock is a specific type of marking. The initial marking is the system's starting state. In the PN diagram, a transition can fire if it is enabled, meaning it has enough tokens in all its input places. When a transition fires, tokens are removed from input places according to arc weights. Simultaneously, new tokens are added to output places based on arrow weights.

The preset of a node x, which can be either a place or transition, is a subset consisting of all elements y in P or T respectively, such that the element (y, x) exists in the PN diagram's set of arcs F. Conversely, the postset of x is defined as a subset of all elements y in P or T, such that the element (x, y) exists in F. For a given PN, a transition t within T is only enabled at a marking M if and only if for all p in t, M(p) is greater than or equal to the weight of the arc (p, t), referred to as W(p, t). Firing enabled transition t changes a marking M to a new marking M′, where for all places p in P, $M'(p) = M(p) - W(p, t) + W(t, p)$. M′ is referred to as an immediately reachable marking from M. The set of markings reachable from M in PN (N, M) is designated as the reachability set R(N, M). A PN is bounded if there exists k in $\mathbb{N}^+$, such that for all M in R(N, M_0) and all places p in P, M(p) is less than or equal to k. The rectangular matrix $\mathbb{A}_{|P|\times|T|}$ is known as the incidence matrix and is defined as $\mathbb{A} = \left[\mathcal{A}_{ij} | \mathcal{A}_{ij} = W(p_i, t_j) - W(t_j, p_i)\right]$. Each element $\mathcal{A}_{ij}$ of the incidence matrix is the weight of the arc, refering to the number of tokens that change in place p_i after firing transition t_j.

In an extended Petri net, there are specific arcs known as inhibitor arcs. These arcs go from a place p to a transition t. When place p is marked (contains tokens), transition t is not enabled. As a result, tokens in the input places to transition t remain on hold until the tokens in place p are cleared. A timed Petri net is a network where certain durations are associated with certain places or transitions. These durations can be either deterministic or stochastic. In this paper, we focus on timed Petri nets, where a specific duration is assigned to certain transitions. To illustrate this concept, we present an example in Fig. 12.4. Each circle in the figure represents a place, and a thin bar represents an instantaneous transition, while a bold bar represents a timed transition with a specified duration.

The system under consideration exhibits three transitions and three places. Each system status at different times can be represented by a row. The nomenclature used for the places and transitions follow their appearance order. The system at the onset moment, t = 0 (initial row), displays one token in the 1st place making the 1st transition enabled. The 1st transition is instantaneous and hence fired immediately. The enabled 1st transition deposits one token each into the 1st and 2nd places (second row at t = 0^+s) and the 1st place is subsequently always left with one token. An inactivated 1st transition is created by the dashed arc from the 2nd place to the 1st

Fig. 12.4 An example of a Petri Net

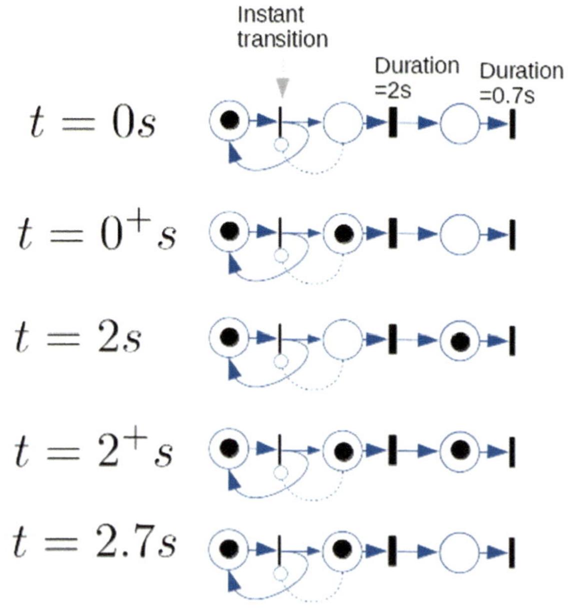

transition. Moreover, the 2nd transition is timed, and within 2 s from its onset moment (t = 2s), it is fired, and the token is shifted to the 3rd place. The 3rd transition is also timed, and at moment t = 2.7s, it is activated, and the token is eradicated from the system.

The PN model structure is a useful tool for modeling concurrent systems, such as the intersection traffic flow state that involves deadlock phenomena. To encode the target system into the PN framework, it is important to consider the operational characteristics. We decompose the intersection into various components and model them using the appropriate PN framework. These components are then combined to form the IntersectionPN, which can be used for deadlock analysis.

12.3.2 Reachability Graph

In the context of Petri Nets, the evolution of states can be represented through reachability and reachability graphs (RG). A state, also known as a marking, undergoes a change from one state to another. The reachability problem examines whether a specific state can be reached from an initial state M_0. Typically, this problem is highly complex, exhibiting exponential difficulty [13]. The RG, denoted as RG(·, ·) in this study, is a directed graph where each node represents a state. The directed edges in RG(·, ·) capture the state evolution, signifying that if a state M' can be reached from state M'', an arc exists from node M'' to node M'. In the case of a bounded PN with finite tokens, the graph is also finite. By gradually simulating the PN, the RG(·, ·) can be constructed, enabling the capture of system dynamics. This manuscript formulates the problem of reaching deadlock states as an integer program, while simultaneously constructing the RG(·, ·) using a simulation-based approach.

12.3.3 Symbols in the Model Development

The symbols in the following model construction include the following:

v_{free}, k_{jam}, q_{max}: free-flow speed, jam density and saturation flow rate respectively.

τ and $\delta\tau$: general time reference and time step;

δx: The cell size after the discretization. Once cell correspond to a place in PN.

l_{veh} and w_{veh}: the length and width of the vehicle;

$N_{cells,conflict}$: The number of cells that the conflict area is discretized.

β: the saturation flow degradation coefficient;

h_1, h_2, h_3..: the time headways of the vehicles after the green onset;

$\mathcal{L} = \{\mathbb{L}_k\}\mathbb{L}_k = \left\{\mathbb{R}_{i_1j_1}, \mathbb{R}_{i_2j_2} \ldots \mathbb{R}_{i_{|\mathbb{L}_k|}j_{|\mathbb{L}_k|}}\right\}$: the loops which are extracted from the trajectories; $\mathbb{L}_k$ are the k-the loop; $\mathbb{R}_{i_1j_1}$ is a leg (a set of successive places and transitions) along the trajectory of movement $i_1 \rightarrow j_1$; $|\mathbb{L}_k|$ is the number of legs in the loop;

$\mathbb{D}_{\mathbb{L}_k}$: the deadlock which is formed at loop $\mathbb{L}_k$. It is a collection of special places or cells. These cells together constitute a deadlock state. Note that deadlock $\mathbb{D}_{\mathbb{L}_k}$ and loop $\mathbb{L}_k$ both represent a set, their meaning are different: $\mathbb{L}_k$ convers all the places and transitions along the loop k, while $\mathbb{D}_{\mathbb{L}_k}$ is only set of places that will guarantee deadlock occurrence, and thus $\mathbb{D}_{\mathbb{L}_k} \subset \mathbb{L}_k$;

$\mathcal{D} = \{\mathbb{D}_{\mathbb{L}_k}\}$: the set of all deadlock;

$\mathbb{C}_{i_{m-1}j_{m-1}\#i_m j_m}$: Conflicted area formed by trajectory $i_{m-1}j_{m-1}$ and $i_m j_m$.

ξ: formation probability of certain deadlock;

ω: Average formation time (number of time steps) of certain deadlock;

C: Cycle length.

ϕ: general reference to the signal stages; ϕ_k is the duration of stage k.

$\mathcal{S}$: signal setting considered in the simulation.

(P, T, F, W): Basic PN notations. P and T are place and transitions, and F are arcs in PN; W is the weight of each arc in F. $|P|$ and $|T|$ are the number of places and transitions respectively.

M_0: initial marking.

$\mathbb{M}_D, \mathbb{M}_L, \mathbb{M}_T$, $\mathbb{M}$ and $\overrightarrow{\mathbb{M}}$:$\mathbb{M}_D, \mathbb{M}_L, \mathbb{M}_T$ are deadlock states, live states and trap states. They are three subsets of $\mathbb{M}$. $|\mathbb{M}|$ is the cardinal of $\mathbb{M}$; $\overrightarrow{\mathbb{M}}$ is the vector representations of the markings, where M_0 is the first element.

$\mathbb{P} = \left[\wp_{M_i M_j}\right]_{|\mathbb{M}|\times|\mathbb{M}|}$: transition probability matrix of the underlying markov chain of IntersectionPN.

$\mathbb{P} = \begin{bmatrix} \mathbb{P}_{\mathbf{LL}} & \mathbb{P}_{\mathbf{LT}} & \mathbb{P}_{\mathbf{LD}} \\ 0 & \mathbb{P}_{\mathbf{TT}} & \mathbb{P}_{\mathbf{TD}} \\ 0 & 0 & \mathbb{P}_{\mathbf{DD}} \end{bmatrix}$: Block representation of $\mathbb{P}$. $\mathbb{P}_{\mathbf{LL}}$ size is $|\mathbb{M}_L|$; $\mathbb{P}_{\mathbf{TT}}$ size is $|\mathbb{M}_T|$; $\mathbb{P}_{\mathbf{DD}}$ size is $|\mathbb{M}_D|$. 'L', 'T' and 'D' indicate live, trap and deadlock respectively.

π: The initial distribution vector.

$\mu_{1\times(|\mathbb{M}_T|+|\mathbb{M}_L|)}$: The probability that the system would evolve to deadlock from each trap state and live state.

RG$(\cdot,\cdot)$: The reachability graph of the system states.

$\mathfrak{U}$ and $\mathfrak{B}$:$\mathfrak{U}$ is the visited states during the construction of RG$(\cdot,\cdot)$;$\mathfrak{B}$ is the branches that are cached during the construction of RG$(\cdot,\cdot)$.

$\mathbb{A}_{|P|\times|T|} = \begin{bmatrix} A_1 \\ A_{\mathbb{L}_k} \\ A_3 \end{bmatrix}$: Incident matrix of the IntersectionPN.

$t_{h1}, t_{h2}, t_{h3}\ldots$: the transitions of the input DemandPN. They correspond to the discharging headways at stop line;

$p_{i,h_m}{}'$, p_{i,h_m}, $t_{i,h_m}{}'$, t_{i,h_m}, $\forall m \in \{1, 2, 3, 4, 5\}$: places and transitions for the DemandPN;

$t_{i,input}$, $p_{i,input}$: the input transition and place for inflow i;

$t_{i,reset}$: reset transition of the DemandPN, which means that the green signal terminates.

$t_{i,absorb}$: the absorb transition of the DemandPN. It works when the red signal begins, and $t_{i,absorb}$ will absorb the remaining tokens in the DemandPN.

V: an integer which describes the maximum cells number that vehicles can traverse during one tome-step, i.e.$\delta\tau$.

$p_{i,j,m}$, $t_{i,j,m}$: m-th place and transition of inflow i to outflow j.

$p^{k}_{i,j,m}$ and $t^{k+}_{i,j,m}$, $\forall m, \forall k \in \{1 \sim V\}$: the forward places and transitions of place.

$p_{SIGNAL,Approach,Turn,Color}$: place of certain turning direction of certain approach. For instance, $Approach \in \{W, N, S, E\}$, TURN $\in$ {LT, TH}, and Color $\in$ {Green, Red, Amber}. The proposed model can take right-turn flow into account;

$t_{SIGNAL,Approach,Turn,Color}$: transition of certain turning direction of certain approach. The subscripts are interpreted the same as above;

p_{SIGNAL,ϕ_2} and $p_{SIGNAL,\phi_{2y}}$: the place for phase ϕ_2 and its yellow duration.

$p_{RESTRICT,i,j,green}$ and $p_{RESTRICT,i,j,red}$: the PN place of periodic outlet restriction of movement i $\rightarrow$ j. The two places correspond to 'virtual green' and 'virtual red' respectively.

$t_{RESTRICT,i,j,green}$ and $t_{RESTRICT,i,j,red}$: the PN transition of periodic outlet restriction of movement i $\rightarrow$ j. The two transitions correspond to 'virtual green' and 'virtual red' respectively.

12.4 Formulation of Intersection Traffic Dynamics as a Petri Net (PN)

12.4.1 Intersection Petri Net Framework

We examine a generalized intersection, as depicted in Fig. 12.3, with multiple lanes available for each turning maneuver. Upon entering the intersection, a vehicle follows a specific path, which intersects with other trajectories, creating points of conflict. These conflict points delineate segments within a vehicle's trajectory. The collection of trajectories and conflict points forms a graph, where nodes represent conflict points and edges represent trajectory segments. Within this graph, trajectory loops can be identified, characterized by a consistent movement direction (either clockwise or counterclockwise). Each of these loops has the potential to result in a deadlock situation.

In our modeling framework, we make several assumptions. Firstly, we assume that the vehicle arrivals at each lane approaching the intersection are known and are affected by the signal settings. Secondly, we assume that once a vehicle enters the intersection, it follows a predetermined trajectory within the intersection. Furthermore, when two vehicles arrive at a conflict point simultaneously, a priority rule is applied to resolve the conflict. We will discuss two priority rules. Based on these assumptions, we develop the following Petri nets (PNs) to represent the traffic dynamics within the intersection: (1) a CellularPN to describe a vehicle's expected movement along the predetermined trajectory curves, (2) a TrafficLightPN to regulate the vehicle's movements according to specific traffic light settings, (3) a DemandPN to represent the vehicle arrivals at each lane, and (4) a ConflictPN to handle the

Fig. 12.5 Model framework

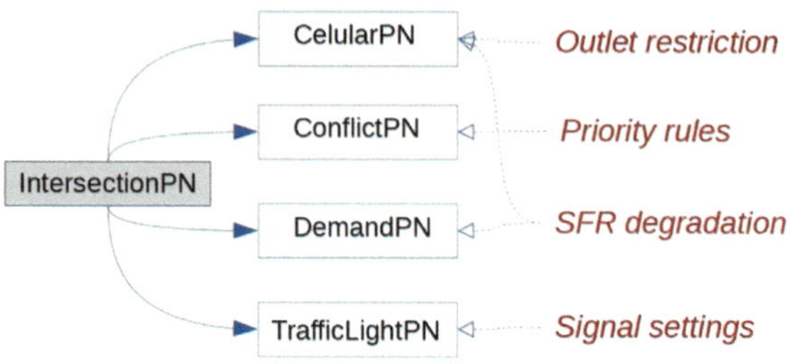

conflicts among trajectories at the intersection points. Finally, these four sub-models are combined to form the IntersectionPN. If we want to investigate the influence of new signal settings on deadlock formation, we can make slight revisions to specific sub-PN-models. For example, the TrafficLightPN can be modified to accommodate the new signal setting. The model framework is illustrated in Fig. 12.5.

12.4.2 IntersectionPN Components

12.4.2.1 CellularPN: Cellular Automata Equipped PN Sub-Model

This section details the development of a CellularPN to depict a vehicle's projected movement along predetermined trajectory curves. Initially, the CellularPN is established solely upon the vehicle's velocity in order to facilitate comprehension. Subsequently, guidance on encoding acceleration is provided. The trajectory is regarded as a series of cells, visualized in Fig. 12.6a, with a length of δx, and each vehicle covering 2 cells. Figure 12.6b exhibits a sample CellularPN formation. The maximum velocity for the featured vehicle is 5 cells per time step, with the model's parameters, including the vehicle's maximum speed, cell size, and number of occupied cells, being adjustable as needed. Each cell showcases a place in the CellularPN. The trajectory's cells are presented in the first row of Fig. 12.6b, with neighboring cells connected by a timed transition with a duration of $\delta\tau$, also the model's time step. In traditional PNs, tokens move forward from one place to the next in a places-transition-places sequence.

To model various speeds, an additional PN structure is required, expanding the places and transitions. Figure 12.6b includes multiple rows representing different speeds. Let V represent the number of cells a vehicle can traverse during a single time step, $\delta\tau$. Accordingly, $V\frac{\delta x}{\delta\tau}$ denotes the free flow speed, v_{free}. In the figure, the superscript of $p^2_{i,j,m} \sim p^5_{i,j,m}$ corresponds to speeds ranging from 2 cells/per time step to 5 cells/per time step (i.e., V) for the tokens in place, $p_{i,j,m}$. These are organized vertically. Transitions $t^{2,+}_{i,j,m}$ to $t^{5,+}_{i,j,m}$ are connected to places p $p^2_{i,j,m} \sim p^5_{i,j,m}$, with the '+ ' indicating forward token movement. When firing is enabled, transition $t^{v,+}_{i,j,m}$ will

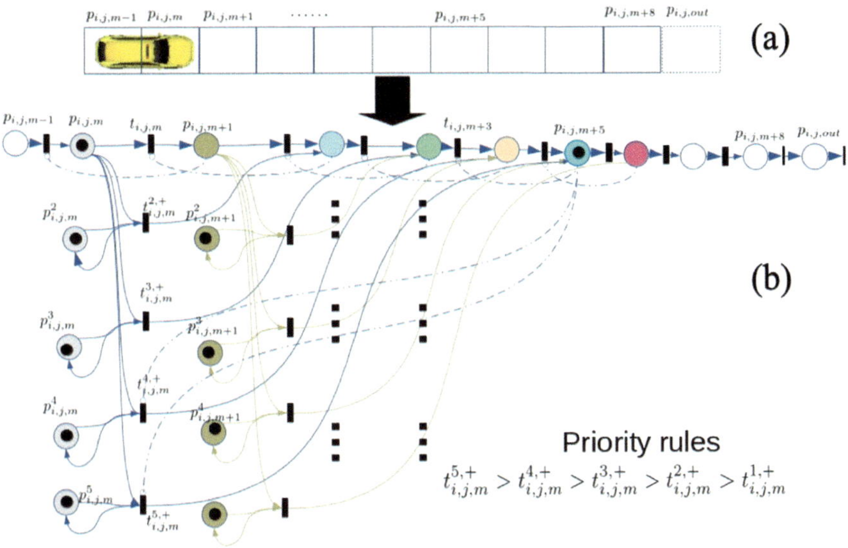

Fig. 12.6 CellularPN structure

deposit one token back to place $p^{v}_{i,j,m}$, as there is also an arc from $t^{v,+}_{i,j,m}$ to $p^{v}_{i,j,m}$. Hence, places $p^{2}_{i,j,m} \sim p^{5}_{i,j,m}$ are continuously occupied by one token.

In Fig. 12.6b, each place, $p_{i,j,m}$ and $p^{2}_{i,j,m} \sim p^{5}_{i,j,m}$, contains a single token. Consequently, transitions $t^{2,+}_{i,j,m} \sim t^{5,+}_{i,j,m}$ are all eligible for firing. However, due to conflicting options, only one transition can be executed. This situation, known as confliction in Petri net terminology, is addressed by implementing priority rules. In Fig. 12.6b, the priority rule at the bottom right stipulates that transition $t^{v,+}_{i,j,m}$ always takes precedence over $t^{v-1,+}_{i,j,m}$. Consequently, transition $t^{5,+}_{i,j,m}$ is favored over the other transitions. When $t^{5,+}_{i,j,m}$ is fired, it places one token in $p_{i,j,m+5}$, effectively enabling the vehicle to progress by four places or cells. This concept aligns with the scenario depicted in Fig. 12.7.

The deployment of dashed lines indicative of inhibitors emanating from the location $p_{i,j,m+5}$ prohibits any transitions from being executed when a token exists in that

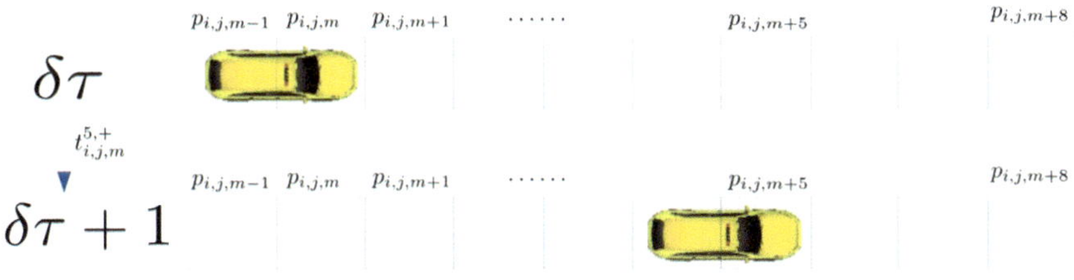

Fig. 12.7 Vehicle movement example1 in CellularPN

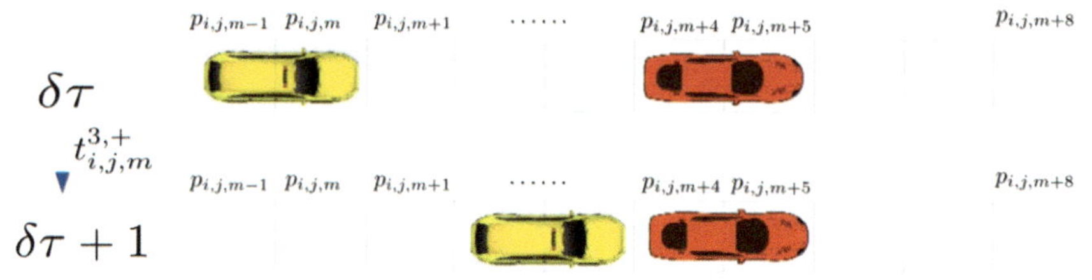

Fig. 12.8 Vehicle movement example 2 in CellularPN

position. This mechanism is applied to minimize the gaps between vehicles. Based on the hypothesis that every vehicle assumes two spaces, the inhibiting terminus comprises the transitions $t_{i,j,m+3}$, $t^{4,+}_{i,j,m}$ and $t^{5,+}_{i,j,m}$. Any attempt to trigger $t_{i,j,m+3}$ or $t^{4,+}_{i,j,m}$ interjects a token into $p^{4}_{i,j,m}$, while firing $t^{5,+}_{i,j,m}$ places a token in $p_{i,j,m+5}$, leading to a fatal run-in. The use of inhibitors circumvents this issue by generating a constraint that only allows $t^{3,+}_{i,j,m}$ to activate if there is a vehicle present in $p_{i,j,m}$. The vehicle located at $p_{i,j,m}$ can exclusively shift to $p_{i,j,m+3}$ via $t^{3,+}_{i,j,m}$ in an instance where $p_{i,j,m+5}$ already contains a single vehicle, as presented in Fig. 12.8.

To summarize, the model uses places $p_{i,j,m}$, $\forall m$ to represent vehicle storage cells, $p^{k}_{i,j,m}$, $\forall k = 2, ..V$ to model vehicle speeds and limit their movement per time step, and transitions $t^{k,+}_{i,j,m}$, $\forall k = 1, 2, ..V$ to prioritize vehicle movement. The parameters V, $\delta\tau$ and δx may be modified to support the v_{free} speed limit.

Figure 12.6 demonstrates the immediate acceleration of the vehicle to speed V = 5, which may lack realism. Therefore, Fig. 12.6's basic CellularPN is expanded to Fig. 12.9's CellularPN. In addition to the existing places $p_{i,j,m}$ and transitions t_(i,j,m), the extended model incorporates additional places $p'^{k}_{i,j,m}$, k and transitions $t'^{k,+}_{i,j,m}$, where k ranges from 2 to V. This revised model accounts for acceleration and follows the following workflow.

Assuming the black dots within each place in Fig. 12.9 represent the initial tokens, there are two vehicles: one located in $p_{i,j,m}$ and the other in $p_{i,j,m+5}$. The vehicle in $p_{i,j,m}$ is currently traveling at a speed of 2, indicated by the presence of one token in $p'^{2}_{i,j,m}$. Due to inhibitors, the maximum speed of the vehicle in $p_{i,j,m}$ from $p_{i,j,m+5}$ is 3, but since there is no token in $p'^{3}_{i,j,m}$, transition $t^{3,+}_{i,j,m}$ cannot be enabled. As a result, transition $t^{2,+}_{i,j,m}$ is enabled instead, and firing it will shift the vehicle from $p_{i,j,m}$ to $p_{i,j,m+2}$. Firing transition $t^{2,+}_{i,j,m}$ will also result in one token being placed in both $p'^{2}_{i,j,m+2}$ and $p'^{3}_{i,j,m+2}$. This implies that if the present speed is 2 (i.e., transition $t^{2,+}_{i,j,m}$), then the possible speed for the next step could either be 2 or 3, exhibiting a maximal acceleration of one cell per time step. The absence of tokens in $p'^{4}_{i,j,m+2}$ and $p'^{5}_{i,j,m+2}$ renders transitions $t^{4,+}_{i,j,m+2}$ and $t^{5,+}_{i,j,m+2}$ infeasible. Determining which speed to adopt depends on other factors such as inhibitors from other vehicles.

Using the described approach, we make the assumption that if the current speed in a given time interval is 2, then the potential speeds for the subsequent time interval

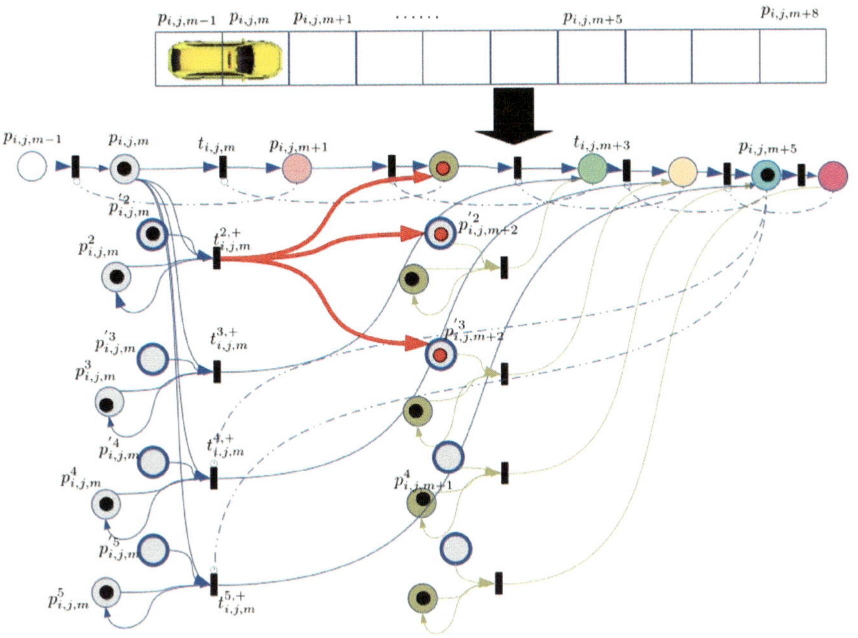

Fig. 12.9 CellularPN with acceleration consideration

are either 2 or 3. The mapping between the current speed and the possible speeds for the next interval can be determined by adjusting the relationship according to the maximum acceleration value, denoted as a_{MAX}.

12.4.2.2 TrafficLightPN: Traffic Signal Sub-Model

Assuming protected left-turn (LT) movements, Fig. 12.10b illustrates a signaling system consisting of eight signals and four phases. These signal settings can be modified to accommodate different timing plans. Each approach has two traffic lights, namely for through (TH) and left-turn movements. Modeling these traffic lights using Petri nets (PN) is shown in Fig. 12.10. The eight pink squares represent the displayed signals, where symbol $p_{SIGNAL,N,TH,Red}$ in Fig. 12.10a corresponds to the place associated with the red display signal for through traffic at the north approach. Each pink square comprises three places, with the color indicating the function of the place. A marked place indicates a specific signal color. The shallow blue square represents the phase settings, with symbol p_{SIGNAL,ϕ_2} denoting the second phase at the bottom of the figure. Each signal has eight inhibitors originating from p_{SIGNAL,ϕ_2}. For example, the inhibitor for the north left-turn sub-PN connects to $t_{SIGNAL,N,TH,Red}$, preventing tokens from moving from $p_{SIGNAL,N,TH,Red}$ to move to $p_{SIGNAL,N,TH,Green}$. Consequently, as long as p_{SIGNAL,ϕ_2} is marked (i.e., ϕ_2 has not completed), the through signal at the

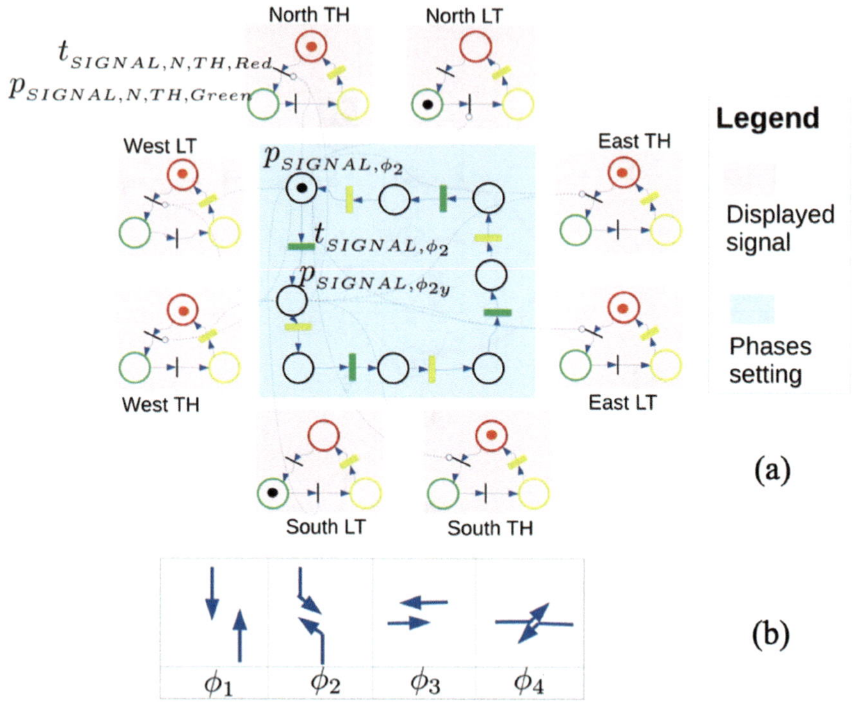

Fig. 12.10 TrafficLightPN for a common four phases plan

north approach remains red. Similar interpretations apply to other inhibitors. The timed transition t_{SIGNAL,ϕ_2} in the figure has a duration identical to ϕ_2. At the end of this phase, the token in p_{SIGNAL,ϕ_2} is removed and placed in $p_{SIGNAL,\phi_{2y}}$, representing the yellow state of ϕ_2. There are six inhibitors from $p_{SIGNAL,\phi_{2y}}$ mirroring those from p_{SIGNAL,ϕ_2}. The inhibitors to $t_{SIGNAL,N,LT,Green}$ and $t_{SIGNAL,S,LT,Green}$ are removed, indicating that the left-turn signals for north and south directions display yellow. For the sake of clarity, the other inhibitors are omitted from the figure.

12.4.2.3 DemandPN: Input Demand Sub-Model

DemandPN stores data related to vehicles entering an intersection. Upon the initiation of a green signal, vehicles enter the intersection in a sequential manner. The time difference between the first vehicle crossing the stop-line and the onset of the green signal is referred to as the initial time headway, denoted as h_1. Subsequent time headways are designated as h_2, h_3, and so forth. When the green signal ends, the demand input is reset to zero. Figure 12.11 presents the DemandPN model that depicts the aforementioned demand input process along with its initial markings. In the figure, the thick bars represent timed transitions associated with the headway

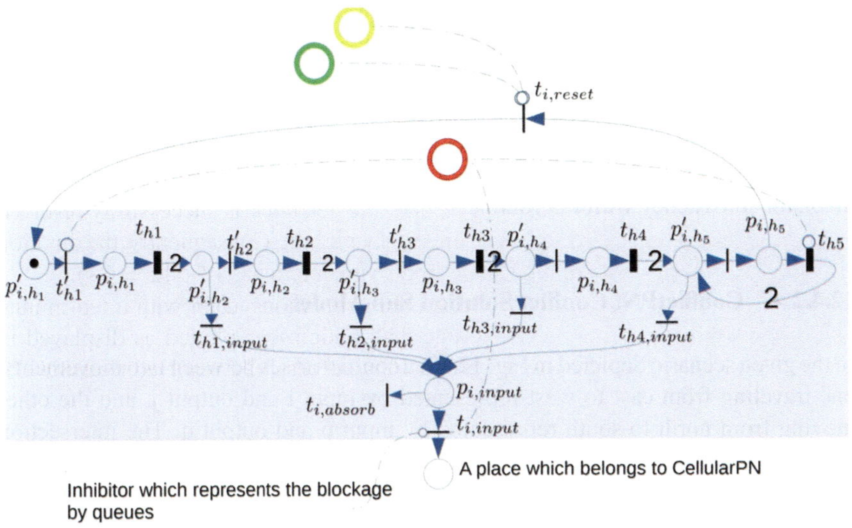

Fig. 12.11 DemandPN for movement i

series, whereas the thin bars represent instantaneous transitions. Positioned at the top of Fig. 12.11 are three exogenous places (in relation to DemandPN) that indicate the color of the signal for this particular movement.

The DemandPN operates in the following manner:

1. Initialization: Initially, there is a token in the left-most place, denoted as $p_{i,h_1}{}'$.
2. Green signal starts: If the signal is green, there is no token in the red signal place. Consequently, transition t'_{h_1} can be activated. The token in $p_{i,h_1}{}'$ is then immediately moved to p_{i,h_1}.
3. One vehicle is outputted: After a duration of time, t_{h1}, (1) transition t_{h1} is activated, and two tokens are placed in $p_{i,h_2}{}'$ (for simplicity, the duration is considered as the transition itself). (2) As a result, both $t_{h_1,input}$ and t'_{h_2} can be fired. (3) Consequently, one token is added to place $p_{i,input}$. This token represents one vehicle entering the intersection through the $t_{i,input}$ transition. However, the vehicle's entry depends on the state of the external inhibitor connected to $t_{i,input}$, which represents queue blockage.
4. Other vehicles are outputted: At the third stage, step 2, a token is also placed in p_{i,h_2}. After a duration of t_{h2}, the t_{h2} transition can be activated, similar to t_{h1}. Consequently, a series of vehicles is generated.
5. Once the red signal is initiated, a token will be placed in the red place to represent the active red signal. Simultaneously, the tokens in the green and yellow places are removed. This allows for the activation of the $t_{i,reset}$ transition, causing the token in p_{i,h_5} to revert back to p'_{i,h_1}. Additionally, the remaining tokens are retrieved through the "$t_{i,absorb}$" transition.

The time headways for discharging flow are represented by durations t_{h1}, t_{h2}, and t_{h3}. These durations can be extended or truncated to model various arrival vehicle patterns, such as saturated or unsaturated flow. In DemandPN, conflict arises between $t_{i,absorb}$ and $t_{i,input}$, which compete for a single token in $p_{i,input}$. These transitions cannot fire simultaneously due to a priority rule, which specifies that $t_{i,input}$ takes precedence over $t_{i,absorb}$. This rule is expressed as $t_{i,input} > t_{i,absorb}$ and serves to maintain consistency with PN routine.

12.4.2.4 ConflictPN: Conflict Solution Sub-Model

In the given scenario depicted in Fig. 12.12, a conflict arises between two movements: one traveling from east to west represented by input i and output j, and the other moving from north to south represented by input p and output q. The intersection of their trajectories results in a conflict. The discretization around the conflict point is shown in the diagram, where four cells, labeled a-d, represent the conflicting sections. It is important to note that the provided discretization in Fig. 12.12 serves for illustrative purposes only, as a higher resolution would make the figure too complex. Occupancy of cells a or c prevents movement pq from crossing the conflict point, while occupancy of cells b or d obstructs movement ij from crossing the same point. Two priority rules are under consideration in this context: (1) First Enter First Service (FEFS), and (2) Stochastic rule. FEFS dictates that if a vehicle enters the conflict area before other vehicles, the latter must wait. Figure 12.13 illustrates the scenario where two vehicles are expected to collide at the interval $\tau + \delta\tau$ in Fig. 12.13.b. The red vehicle is projected to enter at $\tau + 2/3\delta\tau$ (with cells a and b outside the conflict area, and cell c within it), while the yellow vehicle is projected to enter at $\tau + 1/2\delta\tau$. Consequently, the yellow vehicle is granted priority as it enters earlier. The stochastic rule, on the other hand, randomly determines the vehicle that enters the conflict area first if multiple vehicles from each conflict are expected to enter simultaneously.

Figure 12.12b illustrates the Petri nets (PNs) containing the vehicles, with each cell representing a PN, and the confliction cells are $p_{i,j,m+3}$, $p_{i,j,m+4}$ (or cells a and c), $p_{p,q,r+4}$ and $p_{p,q,r+5}$ (or cells b and d). Auxiliary places and transitions are required to resolve the conflict, and the ConflictPN in Fig. 12.14 shows the consistent colors for the confliction area places in Fig. 12.12b. The ConflictPN is symmetric, with the upper and lower halves corresponding to movements pq and ij, respectively. The colors of the arcs in Fig. 12.14, as compared with Fig. 12.12b, include two additional types: red arcs and green arcs. Green arcs indicate priority, while red arcs signal that the corresponding vehicles do not have priority to cross the confliction point. The confliction is resolved with the aid of the middle place colored yellow, i.e., place $p_{ij\#pq}$.

Assuming the initial marking depicted in Fig. 12.12b, vehicles are present at positions $p_{i,j,m+2}$ and $p_{p,q,n+2}$. Based on the state of $p^2_{i,j,m+2}$ (unmarked) and $p^2_{p,q,n+2}$ (marked) in Fig. 12.14, the token in $p_{p,q,n+2}$ would advance to $p_{p,q,n+4}$ at a speed of 2, and the token in $p_{i,j,m+2}$ would advance to $p_{i,j,m+3}$ at a speed of 1, if the conflict is disregarded. However, due to the conflicting nature of the expected positions, only

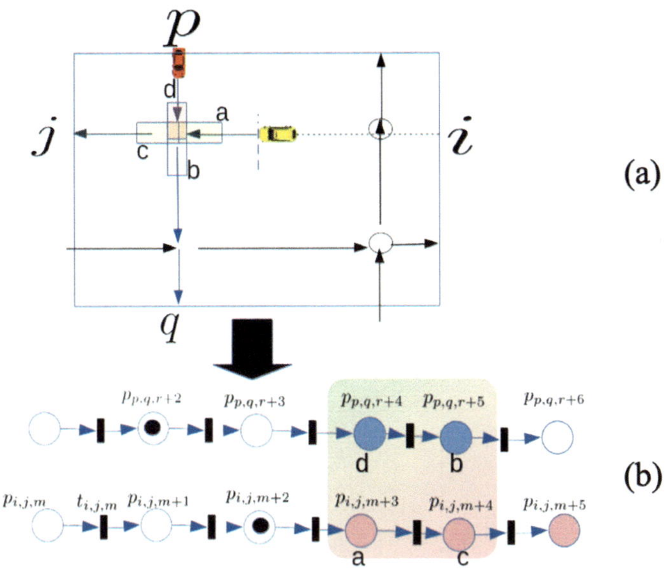

Fig. 12.12 Discretization of the area of conflict

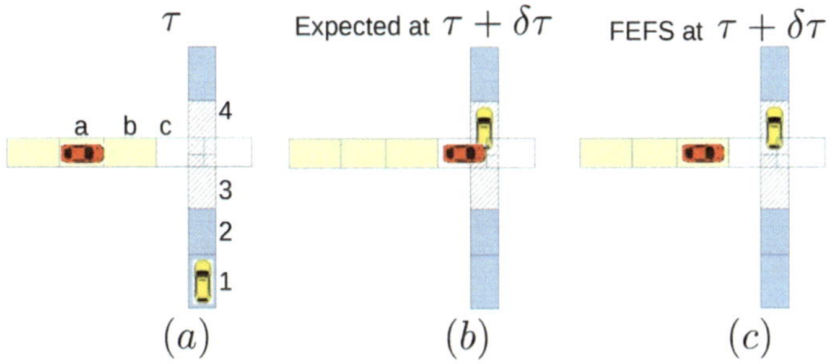

Fig. 12.13 Illustration of First Enter First Service (FEFS)

one vehicle can access the conflicted areas. The procedures of the ConflictPN can be casted:

(1) Initially, both transition $t^{2,+}_{p,q,n+2}$ and $t_{i,j,m+2}$ can be activated, resulting in the deposit of one token in places $p_{ij,pq}$ and $p_{pq,ij}$, respectively. Additionally, one token each is placed in $p^{BKP}_{ij\to m+3,pq}$ and $p^{BKP}_{pq\to n+4,ij}$, which serve as records for successful and unsuccessful vehicle movements during the confliction, respectively. The term 'BKP' signifies backup.
(2) Priority is given to the activation of transitions $t^{+}_{ij,pq}$ and $t^{+}_{pq,ij}$ over $t^{-}_{ij,pq}$ and $t^{-}_{pq,ij}$ (upper-left annotations in Fig. 12.14). The symbol '+' denotes a forward movement, while '−' denotes a backward movement.

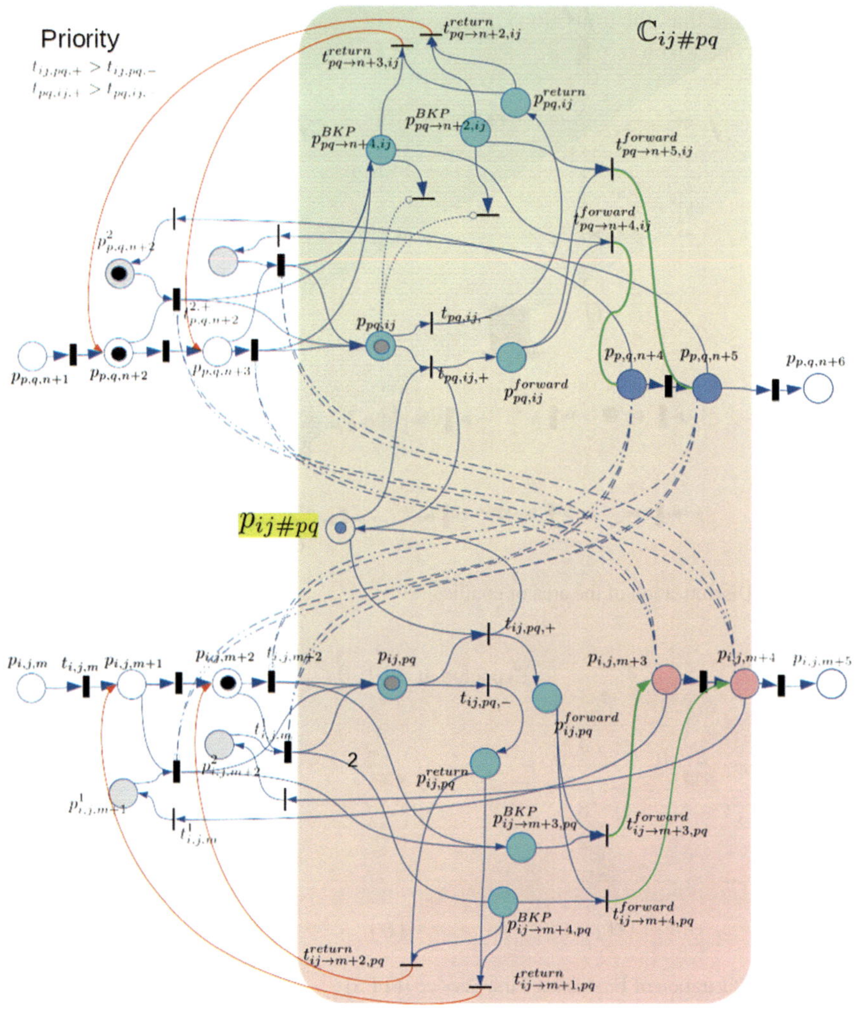

Fig. 12.14 ConflictPN

(3) A conflict arises between $t^{+}_{ij,pq}$ and $t^{+}_{pq,ij}$, which means only one of them can be activated. If one of the transitions is randomly selected, the second rule (stochastic rule) is employed.

(4) Let's randomly select pq as the privileged movement, resulting in the firing of $t^{+}_{pq,ij}$. Although $t^{+}_{pq,ij}$ cannot be fired, $t^{-}_{ij,pq}$ remains enabled. Consequently, both $t^{+}_{pq,ij}$ and $t^{-}_{ij,pq}$ are fired. The '+' indicates the movement of the token (i.e., vehicle in movement pq) forward, occupying the confliction area. Meanwhile, the vehicle in movement ij must wait outside the confliction area. After the firing, one token is placed in $p^{\text{forward}}_{pq,ij}$ and $p^{\text{return}}_{ij,pq}$.

(5) Following that, the transition $t^{forward}_{pq,n+4,ij}$ and $t^{return}_{ij,m+3,pq}$ can be activated. The former places one token in $p_{p,q,n+4}$, while the latter places one token back in $p_{i,j,m+2}$.

12.4.2.5 Other Petri Net (PN) Settings in the Deadlock Formation Analysis

The deadlock arises from a combination of multiple factors, which need to be encoded in the Petri net (PN) model. Three factors are considered: output flow restriction, saturation flow degradation, and insufficient clearance time. These factors may operate simultaneously, requiring corresponding revisions to the PN model.

We examine the recurring limitation on outlet access, which occurs when a queue from a downstream direction overflows and obstructs the intersection. This constraint can affect one or multiple outlets of the intersection. When a spillback is caused by a signal downstream, the restriction follows a periodic pattern resembling a "virtual signal." During the "virtual red signal," the outlet is blocked, while during the "virtual green signal," vehicle movement resumes as usual. The periodic restriction can be realized using the method depicted in Fig. 12.15. An external signal comprises two places and two transitions: $p_{RESTRICT,i,j,green}$, $p_{RESTRICT,i,j,red}$, $t_{RESTRICT,i,j,green}$ and $t_{RESTRICT,i,j,red}$. Offsets for the signal timing can be implemented through the use of places $p_{RESTRICT,i,j,offset}$. and $t_{RESTRICT,i,j,offset}$. When the virtual red signal place ($p_{RESTRICT,i,j,red}$) contains a token, any transition leading to $p_{i,j,out}$ is inhibited, preventing vehicle departure from the intersection. This establishment ensures a periodic restriction.

Regarding the SFR, β is utilized to implement the degradation of the saturation flow rate (SFR) at an intersection during adverse weather conditions like rain or snow. In such circumstances, the duration of timed transitions is extended by β,

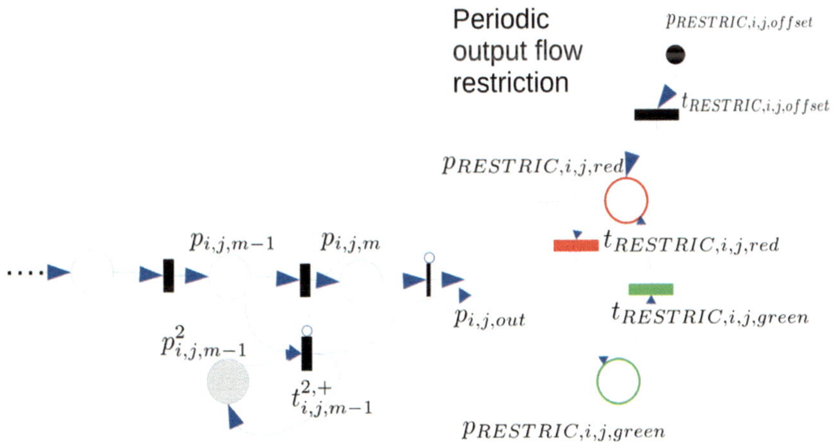

Fig. 12.15 A periodic restriction

effectively decreasing the passing rate of vehicles. For instance, if $\beta = 2$, it means that the vehicle passing rate is halved. In addition, the clearance time, also known as the all-red time, is employed to ensure the smooth clearance of vehicles from the intersection. Inadequate clearance time can result in conflicts between competing traffic streams. To establish an appropriate clearance time, adjustments to the all-red signal are made.

12.4.3 Intersection Discretization and IntersectionPN Assemble

In the case of an common intersection, the trajectories of traffic movements are calculated via a functional curve, such as a cubic spline, and then discretized using CellularPN. To achieve this, the discrete parameters, which consist of the cell length δx, time step $\delta \tau$, and the maximum number of cells V that a vehicle can traverse during a particular time step, must be determined. The values of these parameters are generated through a multistage process. Firstly, based on the confliction area illustrated in Fig. 12.16, the length of the conflict area is approximated to be $l_{veh}+w_{veh}$, where l_{veh} and w_{veh} are the length and width of the vehicle, respectively. Taking into account the minimal space requirements between vehicles, the number of cells $N_{cells,conflict}$ required to divide the confliction area is identified. Secondly, δx is set as $\frac{l_{veh}+w_{veh}}{N_{cells,conflict}}$, following which the free flow speed is computed as $\frac{V\delta x}{\delta \tau} \leq v_{free}$. Here, V is established by taking the ceil of the result of $\frac{v_{free}\delta \tau}{\delta x}$. These actions allow determination of the number of cells in a given trajectory. It is worth noting that actions is notably smaller than the trajectory length. This formalism uses $v_{free} = 50\text{km/h}$, $l_{veh} = 5\text{m}$, $w_{veh} = 2\text{m}$, and if $N_{cells,conflict} = 7$, then $\delta x = 1\text{m}$, $\delta \tau = 0.084\text{sec}$. The resultant values of these parameters could be adjusted for customization based on particular requirements. Given a scenario wherein there are 12 movements, each approach having three movements, and 16 confliction points, the assembly of 12 DemandPNs, 12 CellularPNs, 16 ConflictPNs, and 1 TrafficLightPN collectively forms the IntersectionPN. The purpose of this configuration is to facilitate an analysis of the deadlock dynamics that may arise within the intersection.

12.5 Reachability and Reachability Graph-Based Deadlock Modeling

After constructing the IntersectionPN and configuring the essential settings, such as the signals, it becomes possible to conduct an analysis. The initial step is to establish three distinct system states: deadlock state, trap state, and live state. Subsequently, we will proceed to introduce reachability analysis techniques that can be employed to investigate deadlocks.

Fig. 12.16 Size of the conflict area

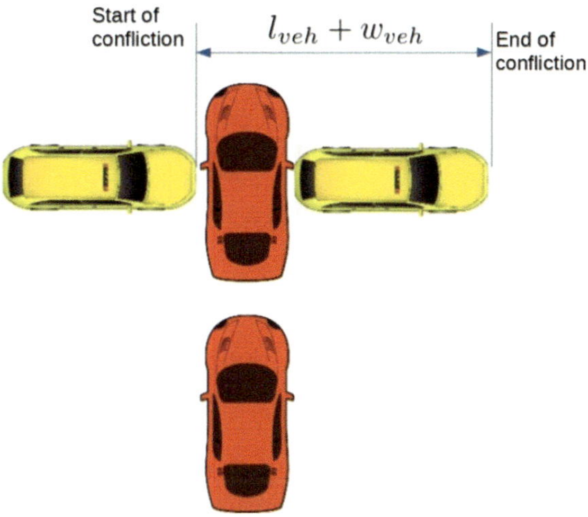

12.5.1 System States Definition and Deadlock Metrics

12.5.1.1 Deadlock State Definition and Metrics

The deadlock state is characterized by a specific arrangement of vehicles forming a queue along a particular loop. It should be noted that each deadlock occurs within a designated loop. For illustrative purposes, let us consider loop $\mathbb{L}_k = \left\{\mathbb{R}_{i_1j_1}, \mathbb{R}_{i_2j_2} \ldots \mathbb{R}_{i_{|\mathbb{L}_k|}j_{|\mathbb{L}_k|}}\right\}$ as an example to define the deadlock state. $\mathbb{L}_k$ represents the k-th loop and consists of a total of $|\mathbb{L}_k|$ legs. The first leg of $\mathbb{L}_k$ is denoted as $\mathbb{R}_{i_1j_1}$. Figure 12.17 illustrates the loop arrangement. To exemplify the definition, we can select any two trajectories $i_m \rightarrow j_m$ and $i_{m-1} \rightarrow j_{m-1}$ (where i_{m-1} represents the movement entry and j_{m-1} represents the movement exit) within the loop. It should be noted that $\mathbb{R}_{i_{m-1}j_{m-1}}$ only represents a section of the trajectory $i_{m-1} \rightarrow j_{m-1}$, specifically six cells (cell 2~cell 7) as depicted in the figure. The direction of the loop is indicated in the center of Fig. 12.17 subfigure (a).

In the context of deadlock analysis, it is worth noting that a deadlock occurs when there is a blockage of the queue along leg $\mathbb{R}_{i_{m-1}j_{m-1}}$ caused by a queue along leg $\mathbb{R}_{i_mj_m}$. This blockage is represented by the ConflictPN denoted as $\mathbb{C}_{i_{m-1}j_{m-1}\#i_mj_m}$. In Fig. 12.17, $\mathbb{C}_{i_{m-1}j_{m-1}\#i_mj_m}$ encompasses cells 6 and 7 on leg $\mathbb{R}_{i_{m-1}j_{m-1}}$) and cells 2 and 3 on leg $\mathbb{R}_{i_mj_m}$. In the case of a deadlock within this loop, the last vehicle in the queue along $\mathbb{R}_{i_mj_m}$ (located at the queue tail) will occupy cells 2 and 3, thus preventing the first vehicle in the queue along $\mathbb{R}_{i_{m-1}j_{m-1}}$ from entering the conflict area. Specifically, only cell 3 is considered in this deadlock definition as it is the minimal requirement. Occupying cell 3 by the queue on $\mathbb{R}_{i_mj_m}$ will effectively block the vehicles from $\mathbb{R}_{i_{m-1}j_{m-1}}$, regardless of whether cell 2 is occupied or not. Alternatively, within the deadlock analysis framework, it is imperative to consider the placement of the initial vehicle along leg

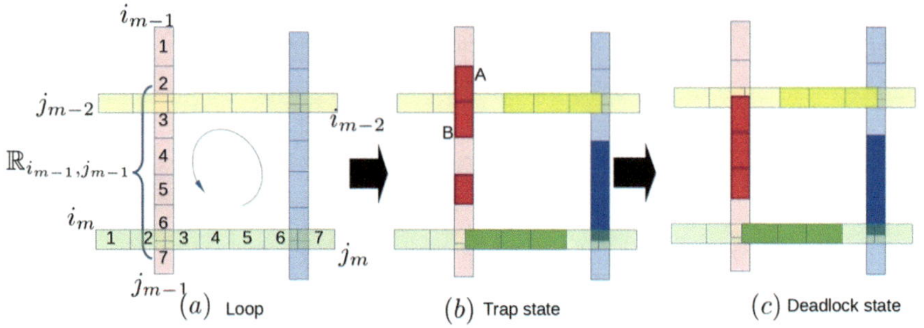

Fig. 12.17 Deadlock definition: **a** loop without vehicles; **b** trap state; and **c** deadlock state. Note that, in (**a**), the direction of the deadlock is specified by the arrow in the center

$\mathbb{R}_{i_{m-1}j_{m-1}}$ with regards to the conflict area. Consequently, for an accurate characterization of deadlock, it is necessary for cell 6 on $\mathbb{R}_{i_{m-1}j_{m-1}}$ to remain unoccupied due to its inclusion within the conflict area. Moreover, compliance with the deadlock definition requires the final vehicle to be located in $\mathbb{C}_{i_{m-2}j_{m-2}\#i_{m-1}j_{m-1}}$, denoting the occupancy of cell 3 on $\mathbb{R}_{i_{m-1}j_{m-1}}$. These stipulations are visually represented in Fig. 12.17c to depict the aforementioned deadlock scenario.

The mathematical formulation of the deadlock definition is the subsequent step to be taken. It is imperative to identify the indices of the locations, or cells along the loop, to accurately define a deadlock scenario. The jam density, k_{jam}, is fundamental in this regard as it denotes the average space occupied by a vehicle, accounting for its length and intra-vehicle gap, which is equivalent to $\frac{1}{k_{jam}}$. The difference between neighboring vehicle indexes is determined as $\frac{1/k_{jam}}{\delta x} = \frac{1}{k_{jam}\delta x}$, where the operator $\lfloor\cdot\rfloor$ rounds down towards the nearest integer. Assuming n represents the index of the cell on loop $\mathbb{R}_{i_{m-1}j_{m-1}}$ immediately outside conflict area $\mathbb{C}_{i_{m-1}j_{m-1}\#i_m j_m}$, such that $p_{i_{m-1}j_{m-1},n+1}$ is present in $\mathbb{C}_{i_{m-1}j_{m-1}\#i_m j_m}$, the occupied cell indexes along $\mathbb{R}_{i_{m-1}j_{m-1}}$ satisfying the deadlock definition are stated as $\left\{p_{i_{m-1}j_{m-1},n-\mu\frac{1}{k_{jam}\delta x}}\right\}$, where μ ranges from 0 to $\mathfrak{N}$.

$$\begin{cases} p_{i_{m-1}j_{m-1},n-\mathfrak{N}\frac{1}{k_{jam}\delta x}+1} \notin \mathbb{C}_{i_{m-2}j_{m-2}\#i_{m-1}j_{m-1}} & \text{(a)} \\ p_{i_{m-1}j_{m-1},n-\mathfrak{N}\frac{1}{k_{jam}\delta x}} \in \mathbb{C}_{i_{m-2}j_{m-2}\#i_{m-1}j_{m-1}} & \text{(b)} \end{cases}, \forall m \in \{0 \sim |\mathbb{L}_k|\} \tag{12.1}$$

In condition (b), the index $n-\mathfrak{N}\frac{1}{k_{jam}\delta x}$ refers to the cell index of the final vehicle on leg $\mathbb{R}_{i_{m-1}j_{m-1}}$ within the deadlock definition. This vehicle is responsible for obstructing the queue of vehicles upstream (on leg $\mathbb{R}_{i_{m-2}j_{m-2}}$), thereby necessitating their presence in the conflict area $\mathbb{C}_{i_{m-2}j_{m-2}\#i_{m-1}j_{m-1}}$. Conversely, in condition (a), the index $n-\mathfrak{N}\frac{1}{k_{jam}\delta x}+1$ identifies the vehicle immediately downstream of the aforementioned conflict area $\mathbb{C}_{i_{m-2}j_{m-2}\#i_{m-1}j_{m-1}}$. Since condition (b) guarantees the obstruction of $\mathbb{R}_{i_{m-2}j_{m-2}}$ by $\mathbb{R}_{i_{m-1}j_{m-1}}$, condition (a) ensures the minimal definition of a deadlock. Here, "minimal" refers to excluding any vehicle that satisfies the conditions but does not contribute to a deadlock, taking into account that additional vehicles do not alter the deadlock state. For each leg within the loop $\mathbb{L}_k$, we define the occupied cell indexes as described above, resulting in a deadlock definition as depicted in Fig. 12.17c. Similar definitions can be established for other loops apart from $\mathbb{L}_k$, thereby constituting the set of deadlock states denoted as $\mathbb{M}_D$.

12.5.1.2 Trap State Definition

In order to study the formation process of deadlocks, a complementary state known as the trap state is introduced. The trap state signifies the point at which an intersection enters a state where the occurrence of a deadlock is inevitable and certain. The time period between the trap state and the deadlock state can be considered as buffer time or lead time. For example, in Fig. 12.17b, all three sections of the loop are already occupied by the queue. Vehicles A and B, depicted in the figure, still possess the ability to move forward, and thus, technically speaking, the traffic flow is not yet completely "dead". Nevertheless, it is evident that a deadlock cannot be circumvented. All trap states are identified and denoted by the set $\mathbb{M}_T$. States that are free from deadlock (termed as live states and that neither belong to $\mathbb{M}_D$ nor $\mathbb{M}_T$) are represented by $\mathbb{M}_L$. The comprehensive collection of all possible states of the system is encompassed by $\mathbb{M}$. The method for identifying $\mathbb{M}_T$ is elaborated upon below, as the identification process involves the transition matrix of the state Markov chain.

12.5.1.3 Deadlock Metrics

Within our framework, we establish a set of metrics to characterize the occurrence of deadlocks, namely the formation time ω and the formation probability ξ. The formation time ω represents the point at which a deadlock emerges, measured from the initial state M_0. It should be noted that ω can be represented as a vector when referring to multiple initial states. On the other hand, the formation probability ξ captures the likelihood of a specific deadlock transpiring. Similarly, ξ may also be represented as a vector when addressing the probabilities of multiple deadlocks.

12.5.2 *Reachability as an Integer Programming Problem*

When conducting an analysis of deadlocks, the primary inquiry entails the determination of whether a deadlock event is likely to occur. Within the context of Petri nets (PN), this pertains to establishing whether or not a particular state within the set $\mathbb{M}_D$ can be reached from the initial state M_0. If it is not reachable, then the intersection is deemed free from deadlocks. To simplify the analysis, our focus is exclusively on a specific loop denoted as $\mathbb{L}_k$, wherein the deadlock is anticipated to manifest. For a given general state denoted as M, if it can be reached from the initial state M_0, then the state equation outlined by Bourdeaud'huy et al. [1] is applicable:

$$\mathrm{M}_{|\mathrm{P}|\times 1} = M_0 + \mathbb{A}_{|\mathrm{P}|\times|\mathrm{T}|}\sigma_{|\mathrm{T}|\times 1} \tag{12.2}$$

The incidence matrix of PN, denoted as $\mathbb{A}_{|\mathrm{P}|\times|\mathrm{T}|}$, and the non-negative integer vector $\sigma_{|\mathrm{T}|\times 1}$, which is indexed by all transitions T, are defined. In this vector, the entry corresponding to a certain transition j represents the number of times that transition j is fired, denoted as n_j. If the system state M in Eq. 12.2 belongs to the set $\mathbb{M}_\mathrm{D}$, then there exists an evolution path that leads to a deadlock state. Thus, determining whether a deadlock will occur can be transformed into finding a non-negative integer solution for Eq. 12.2. To address this, we can organize the state M in Eq. 12.2 as $\mathrm{M} = \begin{bmatrix} \mathrm{m}_1 \\ \mathrm{m}_{\mathbb{L}_k} \\ \mathrm{m}_3 \end{bmatrix}$ without loss of generality. Here, the vector $\mathrm{m}_{\mathbb{L}_k}$ represents the deadlock state in loop $\mathbb{L}_\mathrm{k}$, and its length corresponds to the number of places (discretized cells) in the loop. The vectors m_1 and m_3 correspond to the places outside of loop $\mathbb{L}_\mathrm{k}$. The unknown variables in Eq. 12.2 include $\sigma_{|\mathrm{T}|\times 1}$, m_1, and m_3. The inclusion of m_1, and m_3 is necessary because the definition of a deadlock state only specifies the exact locations of vehicles in the loop (i.e., each entry in $\mathrm{m}_{\mathbb{L}_k}$), without indicating the state of all other places.

Consequently, we proceed to partition the incidence matrix $\mathbb{A}$ and M_0 into three distinct components.

$$\mathbb{A} = \begin{bmatrix} \mathrm{A}_1 \\ \mathrm{A}_{\mathbb{L}_k} \\ \mathrm{A}_3 \end{bmatrix}; \mathrm{M}_0 = \begin{bmatrix} \mathrm{M}_{01} \\ \mathrm{M}_{\mathbb{L}_k} \\ \mathrm{M}_{03} \end{bmatrix} \tag{12.3}$$

Thus, Eq. 12.2 can be expressed accordingly:

$$\mathbb{A}\sigma = \begin{bmatrix} \mathrm{A}_1 \\ \mathrm{A}_{\mathbb{L}_k} \\ \mathrm{A}_3 \end{bmatrix}\sigma = \begin{bmatrix} \mathrm{m}_1 \\ \mathrm{m}_{\mathbb{L}_k} \\ \mathrm{m}_3 \end{bmatrix} - \mathrm{M}_0 = \begin{bmatrix} \mathrm{m}_1 - \mathrm{M}_{01} \\ \mathrm{m}_{\mathbb{L}_k} - \mathrm{M}_{\mathbb{L}_k} \\ \mathrm{m}_3 - \mathrm{M}_{03} \end{bmatrix} \Rightarrow \begin{cases} \mathrm{A}_1\sigma = \mathrm{m}_1 - \mathrm{M}_{01} \text{ (a)} \\ \mathrm{A}_{\mathbb{L}_k}\sigma = \mathrm{m}_{\mathbb{L}_k} - \mathrm{M}_{\mathbb{L}_k} \text{ (b)} \\ \mathrm{A}_3\sigma = \mathrm{m}_3 - \mathrm{M}_{03} \text{ (c)} \end{cases} \tag{12.4}$$

Given the existence of three different unknowns (σ, m_1 and m_3), we can rearrange Eq. 12.4 accordingly, resulting in Eq. 12.4.

$$\begin{bmatrix} \mathrm{A}_1 & -\mathrm{e} & 0 \\ \mathrm{A}_{\mathbb{L}_k} & 0 & 0 \\ \mathrm{A}_3 & 0 & -\mathrm{e} \end{bmatrix} \begin{bmatrix} \sigma \\ \mathrm{m}_1 \\ \mathrm{m}_3 \end{bmatrix} = \begin{bmatrix} -\mathrm{M}_{01} \\ \mathrm{m}_{\mathbb{L}_k} - \mathrm{M}_{\mathbb{L}_k} \\ -\mathrm{M}_{03} \end{bmatrix} \tag{12.5}$$

Consequently, the occurrence of deadlock in loop $\mathbb{L}_\mathrm{k}$ can be determined by solving Eq. 12.4: if there is no feasible non-negative solution, deadlock cannot transpire. To address this, we can convert this existence problem into an integer programming formulation:

$$\min_{\sigma,m_1,m_3} Y_1 = \begin{bmatrix} A_1 & -e & 0 \\ A_{\mathbb{L}_k} & 0 & 0 \\ A_3 & 0 & -e \end{bmatrix} \begin{bmatrix} \sigma \\ m_1 \\ m_3 \end{bmatrix} - \begin{bmatrix} -M_{01} \\ m_{\mathbb{L}_k} - M_{\mathbb{L}_k} \\ -M_{03} \end{bmatrix} \tag{12.6}$$

$$\text{s.t. } \sigma,\ m_1,\ m_3 \text{ are all non negative interger vectors (b)}$$

Alternatively, the existence problem in Eq. 12.4 can be transformed into a standard integer programming problem:

$$\min_{\sigma,m_1,m_3} Y_2 = 1 \begin{bmatrix} A_1 & -e & 0 \\ A_{\mathbb{L}_k} & 0 & 0 \\ A_3 & 0 & -e \end{bmatrix} \begin{bmatrix} \sigma \\ m_1 \\ m_3 \end{bmatrix}$$

$$\text{s.t.} \begin{cases} \sigma,\ m_1,\ m_3 \text{ are all non negative interger vectors (a)} \\ \begin{bmatrix} A_1 & -e & 0 \\ A_{\mathbb{L}_k} & 0 & 0 \\ A_3 & 0 & -e \end{bmatrix} \begin{bmatrix} \sigma \\ m_1 \\ m_3 \end{bmatrix} \geq \begin{bmatrix} -M_{01} \\ m_{\mathbb{L}_k} - M_{\mathbb{L}_k} \\ -M_{03} \end{bmatrix} \text{(b)} \end{cases} \tag{12.7}$$

These aforementioned formulations can be solved using conventional optimization solvers. If the optimal values of Y_1 or Y_2 are non-zero, deadlock occurrence is impossible. However, the above programming models do not incorporate confliction rules and priority configurations. Moreover, the provided solutions do not offer insights into formation probability and time, thus providing limited information on deadlock formation. We must rely on the reachability graph approach for further analysis.

12.5.3 Reachability Graph Construction by Simulation

To generate the reachability graph, we commence at state M_0 and execute the system in an iterative manner. In the event that a deadlock state is reachable from M_0, it has the potential to result in a deadlock. Conversely, if no deadlock state can be reached, the intersection is deemed free from deadlock. At a specific state denoted as M'_τ, we assess the transitions that can be fired, remove the tokens from the incoming places, and subsequently add the tokens to the outgoing places. Consequently, a new state, denoted as M''_τ, is obtained, indicating that from state $M_{\tau'}$, it is possible to reach M''_τ. Thus, within the graph, an arc exists from state M'_τ to M''_τ. Assuming that the system behaves stochastically from state $M_{\tau'}$, i.e., it evolves to state M''_τ with probability p and to state M''_τ with probability 1-p, the reachability graph will encompass arcs from $M_{\tau'}$ to both M''_τ and M''_τ. Each arc is assigned a probability value. The construction process of the reachability graph for IntersectionPN is displayed as follows. Throughout this process, a cached set $\mathfrak{B}$ is employed to store the potential divergence of the system state for later reference. However, when the resolution of both space and time is high, the construction process may become time-consuming. In such situations, it is advisable to simulate a specific time horizon, such as 10 h, and investigate the deadlock mechanism within this horizon instead. The pseudo code for the reachability graph is given in Fig. 12.18.

Input	• Intersection topology (e.g., size, lane channelization) • Signal settings • Demand information (vehicle arrivals) • M_0 (initial with no vehicle in the intersection)
Output	• $RG(\cdot,\cdot)$

Get the trajectories of all approaches, and the intersection points;

Construct the CellularPN, DemandPN, TrafficLightPN, ConflictPN, respectively, and assemble them into IntersectionPN

Set time τ=0, and set the initial tokens, or initial state $M_\tau = M_0$; set cached branches set $\mathfrak{B} = \emptyset$; set reachability graph as $RG(\cdot,\cdot) = \emptyset$, visited states set $\mathfrak{U} = \emptyset$;

WHILE $\mathfrak{B} \neq \emptyset$ or $RG(\cdot,\cdot) == \emptyset$

1. Determine the collection of enabled transitions set $\{T_i\}, i = 1{\sim}N_{M_\tau}$ and corresponding probabilities $\{p_i\}, i = 1{\sim}N_{M_\tau}$, based on current state M_τ N_{M_τ} is the number of enabled transition sets.
2. Dump the state-transitions tuple $\{M_\tau \oplus \{T_i\}, i = 2{\sim}N_{M_\tau}\}$ into $\mathfrak{B}$, i.e. $\mathfrak{B} = \mathfrak{B} \cup \{M_\tau \oplus \{T_i\}, i = 2{\sim}N_{M_\tau}\}$.
3. Fire the transitions in T_1, and get a new marking $M_{t+\delta t},$; then assign $RG(M_\tau, M_{\tau+\delta\tau'}) = p_1$, the first probability in $\{p_i\}$;
4. if $M_{\tau+\delta\tau} \notin \mathfrak{U}$ (i.e. $M_{\tau+\delta\tau}$ is a new state), $\tau \leftarrow \tau + \delta\tau$, $\mathfrak{U} = \mathfrak{U} \cup \{M_{\tau+\delta\tau}\}$, return to step 1;
5. If $M_{\tau+\delta\tau} \in \mathfrak{U}$, the state has already been visited. Go to step 6;
6. Load the first state in $\mathfrak{B}$ as the new starting state. For instance, first state-transition tuple in the queue of $\mathfrak{B}$ is $(M_{\tau''}, \{T_i\})$, then system will roll back to moment $\tau \leftarrow \tau'$ and the state is $M_{\tau'}$. return to step 1.

WHILE LOOP END

OUTPUT $RG(\cdot,\cdot)$

Fig. 12.18 Pseudo code for the reachability graph construction

12.5.4 Markov Chain-Based Deadlock Inference

The reachability graph is inherently associated with a Markov chain representing the system states $\mathbb{M}$. Let $|\mathbb{M}|$ denote the total number of possible states. The transition matrix p of the Markov chain for the IntersectionPN can thus be represented as $\mathrm{I} = \left[\surd_{M_iM_j}\right]_{|\mathbb{M}|\times|\mathbb{M}|}$. For any given pair of states M_i and M_j within the IntersectionPN, the transition probability is denoted as $\surd_{M_iM_j}$ (It should be noted that, to avoid conflict with Petri net convention, we use the symbol $\surd$ to represent transition probability). The set $\mathbb{M}_D$ encompasses the deadlock states and is defined according to Eq. 12.1. Trap states, by definition, inevitably lead to deadlock states. The identification of trap states and live states is determined by the following condition:

$$\begin{cases} \sum\limits_{j\in\mathbb{M}_D} \sum\limits_{n=1}^{|\mathbb{M}|} (\mathbb{P}^n)_{ij} = 100\% \Rightarrow \mathrm{i} \in \mathbb{M}_T \text{ (a)} \\ \sum\limits_{j\in\mathbb{M}_D} \sum\limits_{n=1}^{|\mathbb{M}|} (\mathbb{P}^n)_{ij} < 100\% \Rightarrow \mathrm{i} \in \mathbb{M}_L \text{ (b)} \end{cases} \tag{12.8}$$

The i-j term in the matrix $\mathbb{P}^n$, denoted as $(\mathbb{P}^n)_{ij}$, is derived from Eq. 12.8, which represents the nth power of matrix p. Equation 12.8a specifies that, if state i proceeds to state j within n steps while operating in a deadlock state with a probability of 100%, then it is classified as a trap state. Another approach to identify the trap states is to determine $\mathbb{M}_T$ iteratively given $\mathbb{M}_D$. For a state $\mathbb{M}_i$ not yet classified, its membership in live states $\mathbb{M}_L$ or trap state $\mathbb{M}_T$ is established by verifying whether it remains removed from $\mathbb{M}_D$ after $|\mathbb{M}|$ steps. This allows us to classify the states into the three categories of interest. Matrices can be expressed in a block-wise fashion for improved organization, where matrix $\mathbb{P}$ is no exception.

$$\mathbb{P} = \begin{bmatrix} \mathbb{P}_{\mathbf{LL}} & \mathbb{P}_{\mathbf{LT}} & \mathbb{P}_{\mathbf{LD}} \\ 0 & \mathbb{P}_{\mathbf{TT}} & \mathbb{P}_{\mathbf{TD}} \\ 0 & 0 & \mathbb{P}_{\mathbf{DD}} \end{bmatrix} \tag{12.9}$$

The transition behavior of live states is described by the $\mathbb{P}_{LL}$ block, which has an analogous interpretation for other blocks. The probability of transitioning from a trap state to a live state is always zero ($\mathbb{P}_{TL} = 0$), as trap states are not permitted to transition back to live states. Similarly, a deadlock state is unable to transition to either a live state or a trap state, as the state definition does not permit such transitions. The initial state vector is represented by $\pi_{1\times|\mathbb{M}|}$, where M_0 corresponds to a live state, and the state vector is expressed as [1,0,0,....] with M_0 purposefully assigned as the first row and column. At time step n, the marginal distribution of all states in M is given by $\pi\mathbb{P}^n$. The probability of a state from either $\mathbb{M}_T$ or $\mathbb{M}_L$ evolving to $\mathbb{M}_D$ is obtained by solving the following equation: $\xi_{1\times(|\mathbb{M}_T|+|\mathbb{M}_L|)} = 1-\mu^*$, where μ^* is the solution of the following equation [8]:

$$\mu = \begin{bmatrix} \mathbb{P}_{LL} & \mathbb{P}_{LT} \\ 0 & \mathbb{P}_{TT} \end{bmatrix} \mu \tag{12.10}$$

The time at which the state of $\mathbb{M}_D$ is reached from the trap state is represented by $\omega_{1\times(|\mathbb{M}_T|)}$ and serves as the mean timestamp. It should be noted that $\mathbb{M}_L$, the live state, may persist without transitioning into a trap state with a positive probability, leading to an infinite average time of transition into $\mathbb{M}_D$. The equation below, as presented by Kulkarni [8], provides the smallest non-negative solution for obtaining the value of $\omega_{1\times(|\mathbb{M}_T|)}$. The above procedures of calculating deadlock metrics is illustrated in Fig. 12.19.

$$\omega = e + \mathbb{P}_{TT}\omega \tag{12.11}$$

The figure below illustrates the process for determining the metrics associated with deadlock, which encompass the probability of the deadlock's formation (ξ) and its duration (ω).

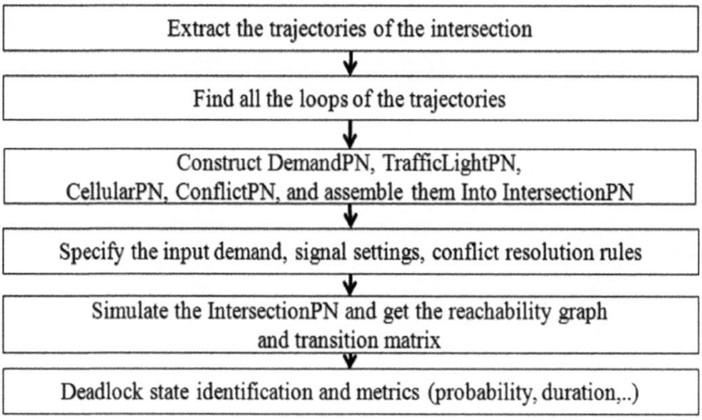

Fig. 12.19 The process of obtaining the deadlock metrics

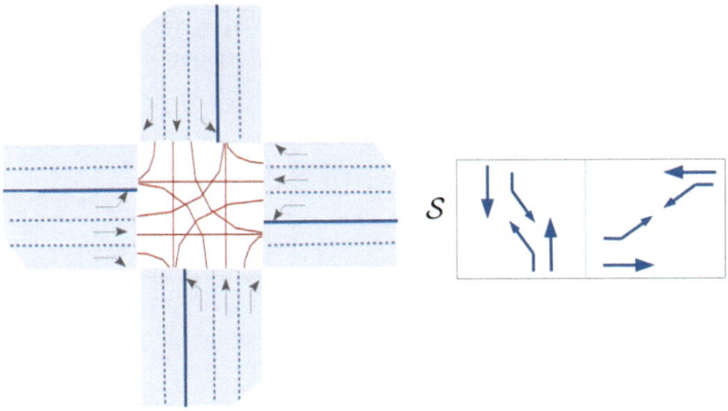

Fig. 12.20 Intersection layout, trajectory and signal setting

12.6 Case Study

12.6.1 Scenario Setup

The configuration parameters encompass the intersection layout, demand specification, signal allocation, outlet limitation, and conflict resolution rules. The respective settings are as follows: (1) The intersection layout, illustrated in Fig. 12.20, utilizes the interpolated cubic spline curve, a commonly employed technique in various simulation software, to generate the trajectories. (2) The DemandPN is determined based on the findings of [7], with the headways defined as follows: $h_1 = 4.53$, $h_2 = 3.05$, $h_3 = 2.76$, $h_4 = 2.57$, $h_5 = 2.51$, $h_{5+} = 2$. '5+', where '5+' denotes the headway after the fifth vehicle. (3) The signal configuration, displayed in Fig. 12.20, is taken into account.

12.6.2 FEFS-Based Stochastic Scenario Result Analysis

The "First Enter First Service" (FEFS) protocol dictates that the earliest entering vehicle within the conflict zone takes priority. This protocol is visually demonstrated in Fig. 12.13 and involves a stochastic element in its treatment of scenarios where two vehicles enter the area concurrently, affording equal probability to each. Both phases of the associated signal have equal durations of 50 s. The western outlet of the intersection undergoes periodic blockage. Virtual red and green signals are 50 s in duration with the opening signal beginning at 1 s. Based on the results, the earliest possible moment of deadlock is 251.58 s. The largest deadlock, consisting of four directional movements, occurs at 252 s. This information is visually depicted in the accompanying Fig. 12.21.

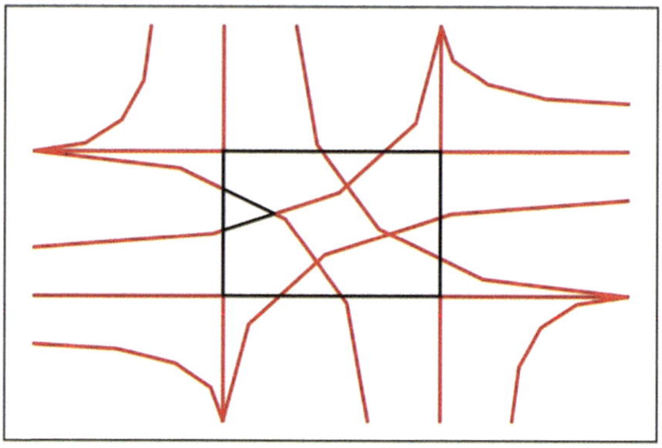

Fig. 12.21 Two deadlocks. Small one occurs at 251.58

This study examines the probability of system deadlock occurrence at each time step within the confine of each deadlock state, as represented by the corresponding entries in $\pi \mathrm{I}^n$ vector, where π refers to the initial distribution, and I the state transition matrix. Meanwhile, non-deadlock probability is represented by $1 - \pi \mathrm{I}^n$. By summing the odds of entering every state in $\mathbb{M}_D$, we determine the probability that the system will eventually enter $\mathbb{M}_D$ at each time step n, which is depicted in Fig. 12.22. Since the likelihood of deadlock preceding 600 s or following 660 s, is nearly zero or one, we limit the probability representation within [600 s before, 660 s after]. Analysis reveals a significant increase in the probability of the deadlocked system. The deadlock increasing probability is almost 50% at 625 s (10 min). Although the highest probability after 660 s is close to 99.554%, it changes insignificantly. Therefore, based on the FEFS protocol, the system is most likely to reach a deadlock within 10 min.

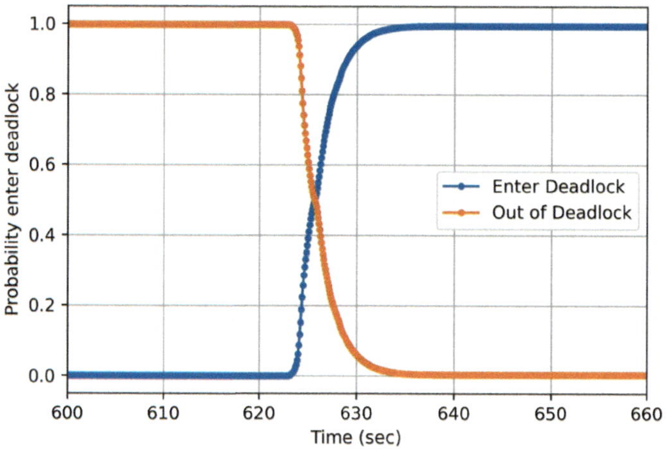

Fig. 12.22 The probability that the system would enter $\mathbb{M}_D$ for FEFS

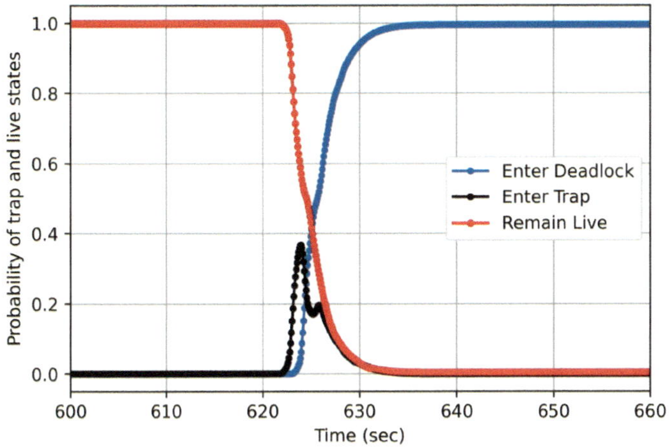

Fig. 12.23 The probability of the trap states $\mathbb{M}_T$ and the live states $\mathbb{M}_L$.

The term "trap state" refers to a condition indicating the likelihood of a deadlock. Figure 12.23 illustrates the marginal distribution of trap states at each point in time $\delta\tau$. The curves in black and red represent the probability of entering any trap state and the probability of the system remaining in a state that is not a trap state or a deadlock state, respectively. The "Out of deadlock" state in Fig. 12.22 does not exclude the possibility of a trap state, making it distinct from the red curve. The probability curve for trap states indicates that (1) it leads to a probability curve for deadlocks and that the occurrence of a trap state precedes that of a deadlock, with the lag duration representing the buffer time for the deadlock states. The distribution of the lag is presented later. (2) The probability curve for trap states increases before decreasing, owing to the fact that once a trap state is entered, the system cannot return to the live state to enter the trap again. The simulation results for the probability that the system eventually reaches the deadlock state confirm the results of Eq. 12.10, where μ^* is solved and is equal to 0, thereby making $\xi_{1\times(|\mathbb{M}_T|+|\mathbb{M}_L|)} = 1 - \mu^* = 1$.

Afterwards, the mean duration between reaching a trap state and encountering a deadlock is calculated. The minimum and maximum durations observed for all trap states to transition into deadlock states are 0.084 s and 129.78 s, respectively. This indicates that once a trap state is reached, a deadlock is expected to occur within approximately 2 min. The average duration is determined to be 38.948 s, as shown in Fig. 12.24 where a histogram showcasing the cumulative distribution of the time lag is presented. Notably, the red curve represents the cumulative distribution. The majority of trap states swiftly progress into deadlock, with more than half of them emerging within one minute (60 s). The occurrence of two minor peaks aligns with the phase duration (set to 50 s). The analytical outcomes derived from Eq. 12.11 closely resemble the findings in Fig. 12.24, consequently eliminating the need for further detailed discussion. Upon investigation of the FEFS regulation, it has been determined that should all vehicles behave in an aggressive manner, the probability of

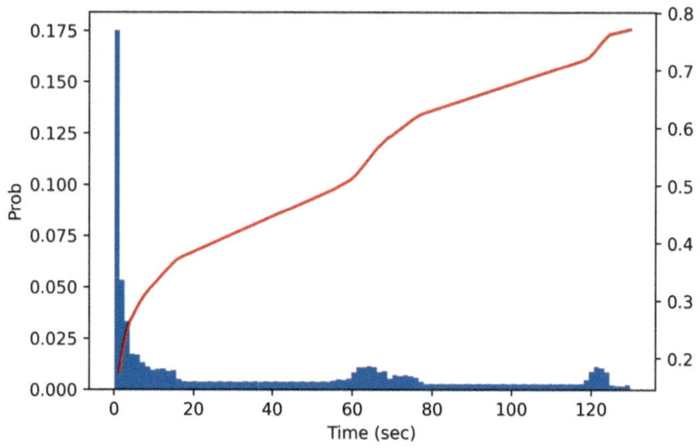

Fig. 12.24 Distribution of time used from the trap state to the deadlock state

reaching an impasse is exceedingly high. Therefore, it is imperative to adhere to the guidelines stipulating that should the intersection become clogged, vehicles ought not to attempt entry; especially during periods of high traffic when backups may ensue and obstruct the intersection. Upon identifying an obstruction, it is discovered that in approximately ten minutes, a deadlock will nearly certainly occur. As the maximum time allotted for the buffer is roughly two minutes, the eight-minute interval following the impediment is crucial for mitigating deadlock by enforcing traffic regulations at the points of convergence.

12.6.3 Pure Stochastic Scenario Result Analysis

In a hypothetical scenario governed by stochastic processes, conflicts are resolved in a probabilistic manner. Consequently, the streams within the system have equal priorities on a long-term scale. Such a mechanism may manifest in intersections that lack proper regulations and fail to adhere to right-of-way norms, or in areas where traffic regulations are not duly observed. A constraint on the outlet is applied periodically, similar to the case in a First Exclusion First Serve (FEFS) scenario, whereby the western outlet becomes obstructed. The earliest instance at which a deadlock is likely to occur is calculated to be 274.008 s, slightly later than in the FEFS scenario. Figures 12.25 and 12.26 illustrate the probabilities of entering a deadlock state, transitioning from a deadlock state to a live state, entering a trap state, and remaining in a live state, respectively. The observed trend deviates from the FEFS case where the probability of deadlock is almost certain. In the stochastic case, the probability of deadlock stabilizes at 76.56%. Therefore, in a stochastic scenario, there is always a possibility, or rather a prioritized arrangement, for the system to avoid becoming locked in a deadlock state. The shift in the probability curve occurs approximately 730 s into the simulation, which corresponds to the seventh cycle. This moment occurs around 100 s later than in the FEFS case. As for the trap state curve, most occurrences transpire around 730 s. By comparing the FEFS case with the stochastic scenario, it can be deduced that the competitive priority solution provided by FEFS leads to earlier and easier entry into deadlock situations.

The objective of this study is to examine the impact of outlet restriction offset on the likelihood of deadlock formation. The experiment considered four different offsets: 5 s, 20 s, 35 s, and 50 s. A graph of the marginal probabilities was plotted and demonstrates a correlation between the outlet restriction offset and deadlock behavior. Two key differences were identified between the cases: the probability

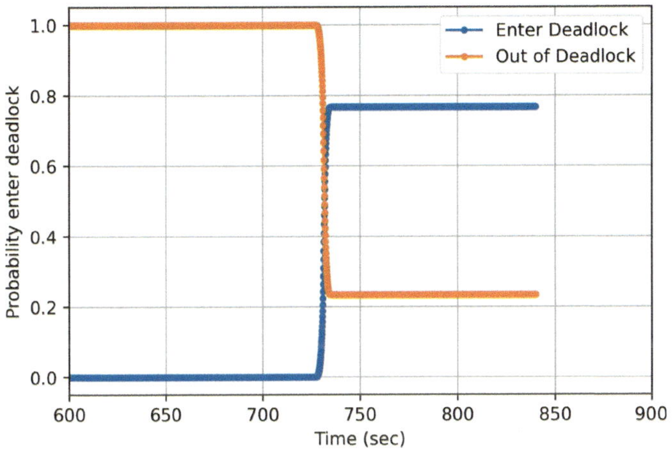

Fig. 12.25 The probability that the system would enter $\mathbb{M}_D$ for a stochastic priority

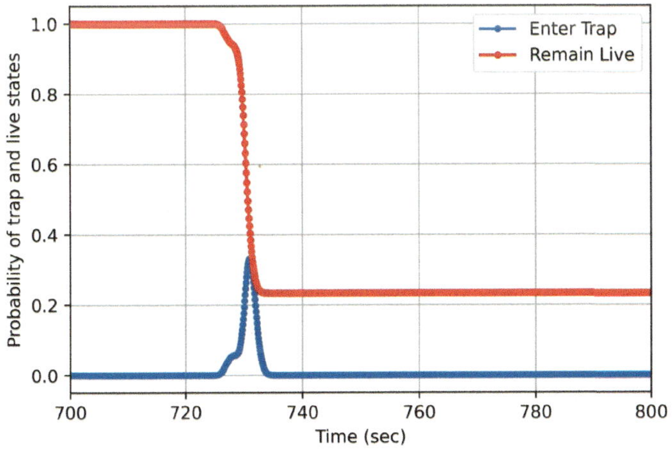

Fig. 12.26 The probability of trap states $\mathbb{M}_T$ and the live states $\mathbb{M}_L$ for stochastic priority

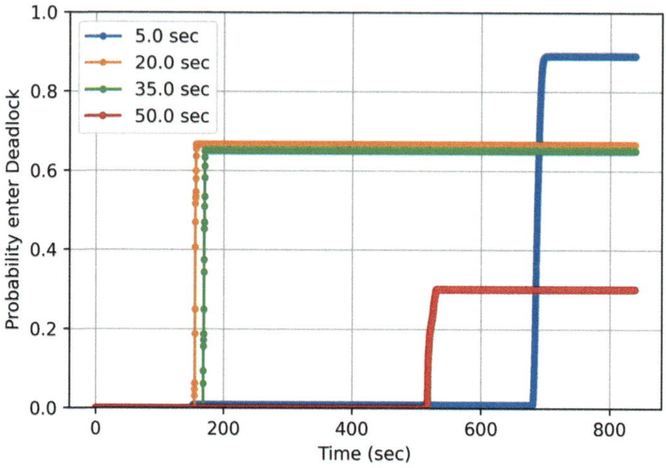

Fig. 12.27 Sensitivity analysis of the outlet restriction offset

of stability and the point at which the marginal distribution changes. The stable probability values for deadlock were determined to be 89%, 66%, 65%, and 30% for the 5 s, 20 s, 35 s, and 50 s offsets, respectively. The results are given in Fig. 12.27. The blocking mechanism of the first offset mainly affects the first cycle, while the fourth offset mainly affects the second cycle. The remaining two offsets partly affect both the first and second cycles. It should be noted that in the first phase, the east-through movement is blocked, while in the second phase, the south-left movement is blocked. Given that through movements are more likely to cause loop formation than left-turn movements, the stable probability of deadlock for the first offset is significantly higher than that of the fourth offset.

12.6.4 Extreme Weather Scenario Result Analysis

In certain situations where the intersection SFR undergoes deterioration and the vehicles present fail to exit the intersection on time, an increase in conflicts may ensue leading to a deadlock. Such incidents are typically encountered in the presence of adverse weather conditions such as rainfall [19] and snow [18]. This study simulates a scenario where heavy rain is observed, resulting in all CellularPN transitions being increased to $7{*}\delta\tau = 0.588$s, thereby causing the SFR to drop to 1/7 of its original value. The study assumes no outlet restrictions and employs the FEFS confliction rule, with signal setting being utilized once again, where each phase lasts for 50 s. Figure 12.28 displays the marginal probability of deadlock for each moment, as well as the marginal probability that the system is out of deadlock, whether it may be $\mathbb{M}_T$ or $\mathbb{M}_L$. It is noteworthy that deadlock formation occurs considerably more rapidly than in the aforementioned scenarios, with deadlock almost certainly occurring after nearly one cycle. The probability of being trapped at the tail end of the first phase is nearly 0.8 as shown in Fig. 12.29, indicating that once the trap state is entered, a deadlock is imminent. The time lag from trap state to deadlock is illustrated in Fig. 12.30, portraying that the majority of the lag times amount to less than 20 s or half a minute. These findings indicate that during instances of large-scale incidents such as rainfall and snow, where it is difficult to sustain common capacity due to substantial deterioration, timely action is essential as the system will evolve to deadlock relatively quickly.

We examine the sensitivity of the degree of SFR degradation, utilizing four levels of degradation, namely 3, 4, 5, and 6. The confliction rule undergoes a transition to

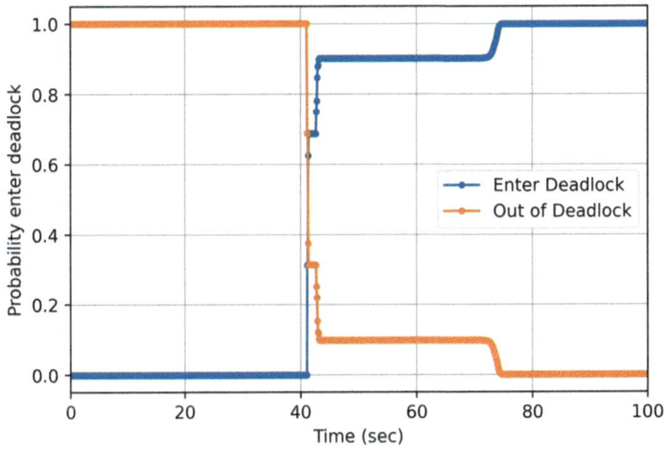

Fig. 12.28 The probability that the system would enter $\mathbb{M}_{\mathrm{D}}$.

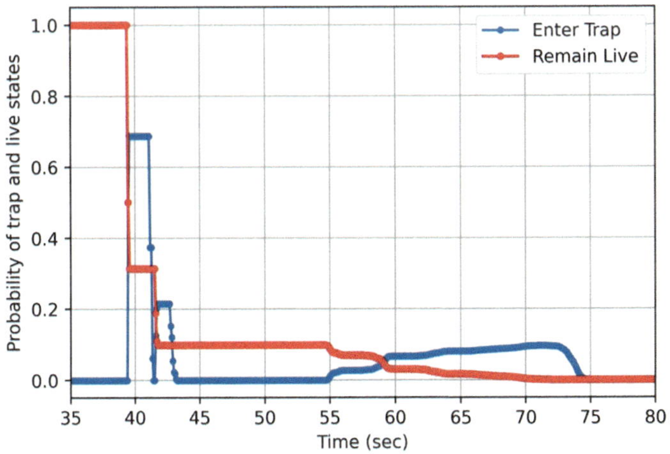

Fig. 12.29 The probability of trap states $\mathbb{M}_{\mathrm{T}}$ and the live states $\mathbb{M}_{\mathrm{L}}$.

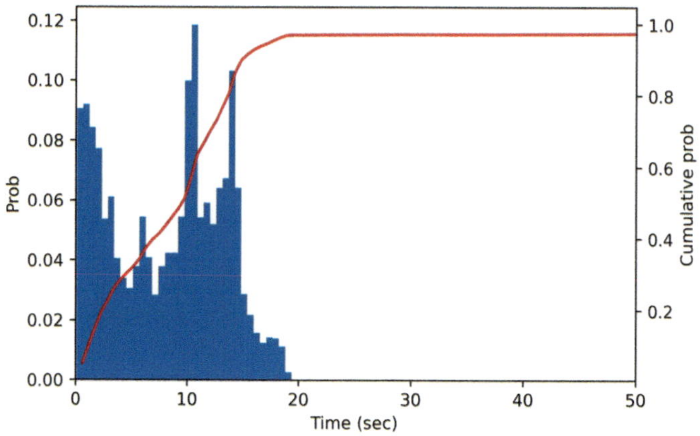

Fig. 12.30 Distribution of time used from the trap state to deadlock

the stochastic nature, while maintaining the constancy of the other parameters. The graphical representation in the subsequent figure elucidates the marginal probabilities of the trap states. Notably, it is evident that the probability remains relatively consistent, albeit the occurrence transpires earlier as the SFR degradation intensifies. This can be attributed to the fact that, under significant SFR degradation, vehicles tend to vacate the intersection at a later time, thereby triggering conflicts at an earlier stage (Fig. 12.31).

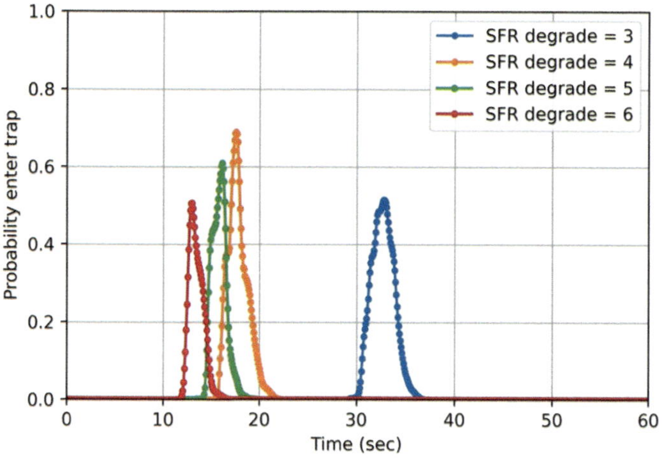

Fig. 12.31 Different SFR degradations influence on deadlock

12.7 Notes

This chapter is a revised version of [17]. In this chapter, we propose a Petri Net-based modeling approach for investigating the occurrence, likelihood, and duration of deadlock formation. We represent the key components of intersections - vehicle trajectory, traffic signals, vehicle arrivals, and conflict behavior - as separate Petri Nets named CellularPN, TrafficLightPN, DemandPN, and ConflictPN respectively. We define three system states: deadlock state, trap state, and live state. We formulate the occurrence of deadlock as an integer programming problem. To study the probability and duration of deadlock formation, we constructed a reachability graph that results in a Markov chain for the system. Through a case study, we demonstrate that driver behavior at conflict points significantly influences the probability and duration of deadlock formation. In competitive scenarios (FEFS), where driver priority is determined based on their entering order at conflict points, deadlock is almost certain to occur earlier compared to a purely stochastic priority assignment. Additionally, the degradation of SFR (Signal Failure Rate) may lead to deadlock formation, affecting the timing of its occurrence. Future research should focus on addressing deadlock effectively. First, we need to consider scenarios where only partial information is available to the intersection manager, such as signal settings, vehicle trajectories, and conflict configurations. An investigation into avoiding deadlock with incomplete vehicle information is necessary. Moreover, in our modeling, we assume that vehicles strictly follow specified trajectories. However, in congested situations within large intersections, slight deviations from the average path can occur. Thus, further analysis is needed to model deadlock in a 2-dimensional space, incorporating the kinematic and dynamic properties of vehicles (e.g., slip angle, steering angle).

References

1. Bourdeaud'huy T, Hanafi S, Yim P (2007) Mathematical programming approach to the Petri nets reachability problem. Eur J Oper Res 177(1):176–197
2. Coffman E, Elphick M (1971) System deadlocks. Comput Surv 3(2):67–78
3. Daganzo CF (2007) Urban gridlock: macroscopic modeling and mitigation approaches. Transp Res B Methodol 41(1):49–62
4. Ghaderi Z, Alqahtani A, Bagherzadeh N (2017) AROMa: aging-aware deadlock-free adaptive routing algorithm and online monitoring in 3D NoCs. Ieee Trans Parallel Distrib Syst 29(4):772–788
5. Imbsweiler J, Ruesch M, Weinreuter H, Leon FP, Deml B (2018) Cooperation behaviour of road users in T-intersections during deadlock situations. Transp Res F: Traffic Psychol Behav 58:665–677
6. Jiang R, Chen JY, Ding ZJ, Ao DC, Hu MB, Gao ZY, Jia B (2016) Network operation reliability in a manhattan-like urban system with adaptive traffic lights. Transp Res Part C Emerg 69:527–547
7. Jin X, Zhang Y, Wang F, Li L, Yao D, Su Y, Wei Z (2009) Departure headways at signalized intersections: a log-normal distribution model approach. Transp Res Part C Emerg 17(3):318–327
8. Kulkarni VG (2016) Modeling and analysis of stochastic systems. Crc Press

9. Luo J, Huang YS, Weng YS (2019) Design of variable traffic light control systems for preventing two-way grid network traffic jams using timed petri nets. IEEE Trans Intell Transp Syst 21(7):3117–3127
10. Luo J, Wan Y, Wu W, Li Z (2020) Optimal petri-net controller for avoiding collisions in a class of automated guided vehicle systems. IEEE Trans Intell Transp Syst 21(11):4526–4537
11. Lämmer S, Treiber M (2012) Self-healing networks - gridlock prevention with capacity regulating traffic lights. In: Self-adaptive and self-organizing systems workshops (SASOW), 2012 IEEE Sixth International Conference on, pp 61–65
12. Mahmassani HS, Saberi M, Zockaie A (2013) Urban network gridlock: theory, characteristics, and dynamics. Transp Res Part C Emerg 36:480–497
13. Mayr EW (1981) An algorithm for the general Petri net reachability problem. SIAM J Comput 13(3):441–460
14. Mendes GA, Da Silva LR, Herrmann HJ (2012) Traffic gridlock on complex networks. Physica A Stat Mech Appl 391(1–2):362–370
15. Mishra KN, Kumar N (2020) Voting-priority-based deadlock prevention in multi-server multi-CS distributed systems. Natl Acad Sci Lett 43(7):625–630
16. Palmer GI, Harper PR, Knight VA (2018) Modelling deadlock in open restricted queueing networksEur. J Oper Res 266(2):609–621
17. Qi H, Yu Y, Tang Q, Hu X (2022) Intersection traffic deadlock formation and its probability: a petri net-based modeling approach. IET Intell Trans Sys 16:1342–1363. https://doi.org/10.1049/itr2.12210
18. Perrin H, Martin PT, Hansen BG (2001) Modifying signal timing during inclement weather. Transp Res Rec 1748(1):66–71
19. Sun H, Yang J, Wang L, Li L, Wu B (2013) Saturation flow rate and start-up lost time of dual-left lanes at signalized intersection in rainy weather condition. Procedia Soc Behav Sci 96:270–279
20. Silberschatz A, Peterson JL, Galvin PB (1991) Operating systems concepts. Addison-Wesley, Reading, MA
21. Zhou Y, Hu H, Liu Y, Lin SW, Ding Z (2020) A distributed method to avoid higher-order deadlocks in multi-robot systems. Automatica 112:108706

Chapter 13
Intersection Deadlock Protocol on Mixed Autonomous Vehicles

13.1 Deadlock and the Autonomous Vehicles

During peak traffic periods, deadlock situations frequently arise at intersections, presenting an extreme operational state in which competing traffic movements are unable to proceed due to mutual anticipation of clearance at the intersection. This cyclic waiting behavior leads to a complete halt in traffic in all directions, constituting a deadlock. It is important to note that deadlocks differ significantly from typical traffic congestion scenarios where vehicles can still make progress albeit at reduced speeds, and where roadway capacity remains intact. In contrast, deadlocks render vehicle movements entirely immobile, resulting in the depletion of roadway capacity and a complete absence of speed and flow. Consequently, unless promptly resolved, an intersection deadlock can escalate into a widespread traffic congestion that renders conventional regulatory measures (e.g., traffic control and route guidance) ineffective. Consequently, it becomes paramount to conduct a comprehensive analysis of deadlocks, develop effective detection mechanisms, and establish preventive measures to mitigate their occurrence.

The rapid progress in vehicular technology has led to the increased popularity of connected and autonomous vehicles (CAVs), with many modern vehicles already incorporating various levels of automation. However, existing literature predominantly focuses on trajectory planning and signal control joint optimization for CAVs, overlooking the importance of cooperative path planning and deadlock detection, avoidance, and recovery. Consequently, when operating under an uncoordinated path planning strategy, CAVs lack the necessary capabilities to detect, avoid, or recover from deadlocks. As a result, once trapped in a deadlock, CAVs can only remain indefinitely stationary, awaiting human intervention to resolve the intersection blockage. Therefore, we advocate for the development of not only individual CAV capabilities but also the design of cooperative sensing and control protocols to address the multifaceted challenges presented by deadlock scenarios in the context of multiple CAVs.

© The Author(s), under exclusive license to Springer Nature Singapore Pte Ltd. 2024

H. Qi, *Stochastic Two-Dimensional Microscopic Traffic Model*, Lecture Notes in Intelligent Transportation and Infrastructure,
https://doi.org/10.1007/978-981-97-3597-6_13

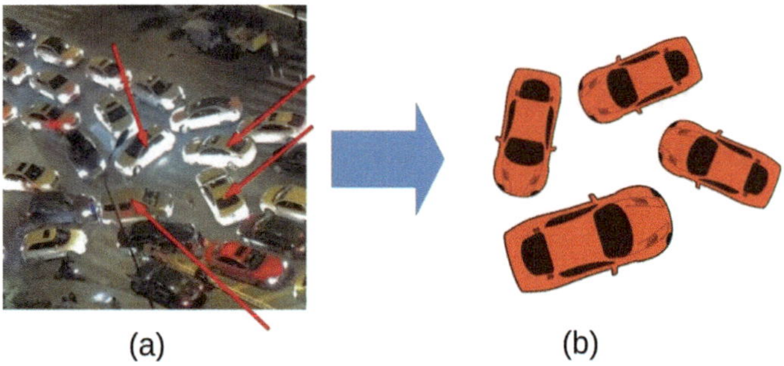

Fig. 13.1 Illustration of deadlock detection

To address this significant research gap, this chapter proposes a deadlock detection and cooperative avoidance protocol tailored for connected and autonomous vehicles (CAVs) at intersections involving a mixed flow of autonomous CAVs and human-driven vehicles (HDVs). The modeling efforts undertaken in this study primarily focus on two stages: deadlock detection, and deadlock avoidance and recovery. To illustrate these stages, two straightforward examples are presented. Firstly, concerning deadlock detection, the observed deadlock in Fig. 13.1 involves numerous vehicles. However, the crux of the issue lies in the four highlighted vehicles that contribute to the emergence of loop-style waiting behaviors within the intersection. Resolving this deadlock necessitates the identification and unlocking of these four key vehicles. To achieve this, a method is proposed that enables the identification of loops and the key vehicles within said loops. Secondly, in terms of the deadlock avoidance and recovery protocol, consider the scenario depicted in Fig. 13.2 Movements A, B, C, and D form a loop, leading to traffic obstruction. However, instead of adhering strictly to its planned trajectory (represented by movement A in Fig. 13.2b), vehicle 1 can deviate slightly to the right and follow the green dashed arrow. By doing so, it escapes from the deadlock, transforming the loop structure depicted in Fig. 13.2b into a chain-spillover structure illustrated in Fig. 13.2c.

13.2 Basic Assumptions of this Chapter

The following assumptions govern the deadlock modeling presented in this chapter:

- Connected and autonomous vehicles (CAVs) possess the capability to relay real-time information concerning their states, encompassing location, steering angle, and other relevant variables. However, human driven vehicles (HDVs) do not contribute any registration data. Nonetheless, we posit that real-time HDV locations can be acquired through a road side unit (RSU) or alternative sensors such as road side radar or video surveillance devices.

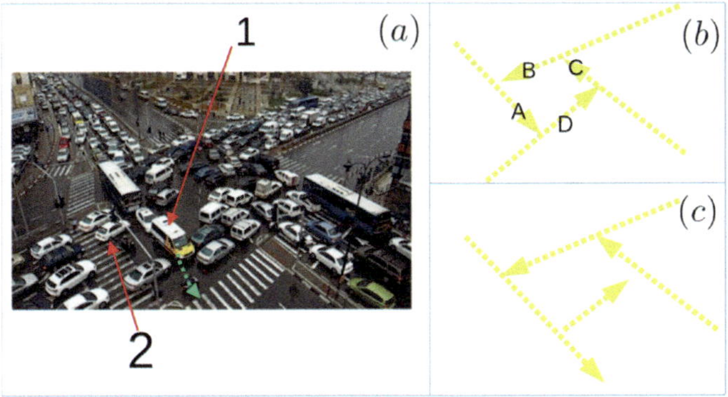

Fig. 13.2 Illustration of deadlock avoidance and recovery

- Given the typical communication delay of milliseconds, this chapter assumes that data collection and RSU sharing occur instantaneously.

Drawing upon these assumptions, the subsequent sections elucidate the methods for detecting, avoiding, and recovering from traffic deadlocks using connected and autonomous vehicle technology.

13.3 Current Practice Regarding Mixed AVs Flow

During rush hours, traffic deadlock is a frequent occurrence. However, existing literature has paid limited attention to the topic, resulting in minimal modeling efforts dedicated to understanding the formation mechanism of deadlocks, devising detection methods, and developing strategies for avoidance and recovery. Nevertheless, deadlocks as system states are observed in various domains, including software and hardware development [6], manufacturing systems [11, 12], distributed systems [14], robots and automated guided vehicles [10, 24], and queuing networks [17], among others. The necessary conditions for deadlock in these domains can be summarized into four components [4, 20]: (1) mutual exclusion, where a resource cannot be simultaneously occupied by multiple tasks,(2) no preemption, meaning resources cannot be taken away until they have been fully utilized; (3) hold and wait, indicating that tasks retain their resources while waiting for other resources; and (4) circular wait, involving the existence of a circular chain of tasks, where each task requests a resource from another task in the chain. Interestingly, when comparing these four necessary conditions with intersection deadlock, congruence can be identified in the following manner: (1) the conflict point in a traffic deadlock can only be occupied by a single vehicle; (2) a vehicle cannot compel another vehicle to vacate the space; (3) vehicles must wait until downstream vehicles advance; and (4) vehicle queues form a loop, reflecting the presence of a circular wait.

The literature has addressed the management of intersections in the presence of either fully autonomous or mixed autonomous vehicles. Several notable models have been proposed, such as the automatic intersection management approach [5], consensus-based control at signal-free intersections [15], crossroad + [8], and polling-Systems-Based Autonomous Vehicle Coordination at signal-free intersections [13], among others. These methods adopt either a centralized control approach or establish cooperative mechanisms among vehicles to facilitate the flow of fully autonomous vehicles. Researchers, including Arvin et al. [1] and Cabri et al. [3], have also investigated intersection management for mixed traffic flow, proposing negotiation or coordination mechanisms for vehicle groups, whether composed of connected and autonomous vehicles (CAVs) or a combination of CAVs and human-driven vehicles (HDVs). However, these studies assume that vehicles will either adhere to given commands (in the case of CAVs) or behave as expected (in the case of HDVs), as described by models such as the intelligent driver model [22]. Consequently, the trajectories of vehicles traversing intersections are determined through optimization models that rely on the aforementioned assumptions. However, none of the aforementioned research has explicitly modeled or provided a viable solution for intersection deadlock. We argue that, without the explicit modeling of intersection deadlock and the development of cooperative control protocols for autonomous vehicles, the occurrence of deadlock remains a possibility. Moreover, once deadlock manifests, normal traffic flow cannot spontaneously recover.

Research exploring deadlocks in these fields can be found within the existing body of literature, predominantly implemented using the discrete event system (DES) framework. Within the realm of DES tools, the Petri Net (PN) formulation approach has been extensively utilized. In the domain of transportation, PN models have been applied to various domains, including signal control [10], co-pilot systems, and other related areas. Nevertheless, due to the continuous nature of vehicle movements and the nonlinearity in vehicle dynamic behavior, employing the DES approach alone is inadequate for capturing the complex dynamics exhibited within intersections, ultimately making it an unsuitable choice for comprehensive deadlock detection, avoidance, and recovery strategies.

13.4 Notations

For easy reference, we list all the symbols and the abbreviations.

Symbols	
l_{veh}, w_{veh}	The length and the width of the vehicle
$l_f(l_r)$, $l_F(l_R)$	The distance of front (rear) wheel and front (rear)bump to the CG
$\mathbf{z} = [x, y, \psi]$	Vehicle state vector. X and Y are the coordinate of the CG, and ψ is the heading angle
$\mathbf{u} = [v, \delta^f]^T$	The control input, speed and the steering angle of front wheel

(continued)

(continued)

δ_{i}^{f}	Steering angle of front wheel for vehicle i
$\Delta\delta^{\mathrm{f}}$	The discretization length in the EDP for type II deadlock
$\delta^{\mathrm{f,MAX}}$	The maximum steering angle
$\dot{\mathbf{z}} = \boldsymbol{f}(\mathbf{z}, \mathbf{u})$	The vehicle state equation. U is the input
$\mathrm{w} = (\mathrm{x}, \mathrm{y})$	Observation output of HDV (i.e. the trajectory)
$\mathrm{x_i}, y_i, \psi_i, v_i,$	The state for i-th vehicle
A, B, C, D	The four corners of one typical vehicle. For instance, x^A, x^B, x^C and x^D are the x-coordinate of the four corners of a general vehicle
$\mathrm{d_{buffer}}$	A small number in the optimization of the steering intervals that lead to blockage
I	Identity matrix with appropriate dimension
E, F	Coefficient matrix of the dynamic system of vehicle with unknown control input
G, H	Coefficient matrix of a singular system which is the linearized edition of $\dot{\mathbf{z}} = \boldsymbol{f}(\mathbf{z}, \mathbf{u})$
M	Observation matrix of the system with unknown control input
Q, R	The covariance matrix of the vehicle state (i.e. $[\mathrm{x}, \mathrm{y}, \psi]^{\mathrm{T}}$) and the control input (i.e. $[v, \delta_f]^{\mathrm{T}}$)
P	Estimation error covariance matrix
$\mathrm{l_i} = \mathcal{L}_{\mathrm{i}\to\mathrm{j}}(\mathrm{l_j})$	The restriction function between two vehicles, i and j, where vehicle i is blocked by vehicle j. Vehicle j need to drive the distance $\mathrm{l_j}$ along its expected trajectory, if vehicle i want to move a distance of $\mathrm{l_i}$ along its expected path
$\widehat{\mathrm{l}_{0\to1}}$, $\underline{l_{1\leftarrow0}}$ and $\overline{\mathrm{l}_{0\to1}}$	$\widehat{\mathrm{l}_{0\to1}}$ is the permitted travel distance of vehicle 0; $\underline{l_{1\leftarrow0}}$ is the evasion condition of vehicle 0 and $\overline{\mathrm{l}_{0\to1}}$ is the evasion distance of vehicle 0
$\mathrm{l_i} = \mathfrak{I}_{\mathrm{i}\to\mathrm{j},\mathbb{L}_k}(\cdot)$	The restriction function for vehicle i along loop $\mathbb{L}_k$ until vehicle j
$\mathbb{C}_\mathrm{i}$	The set of the vehicles that block the vehicle i
$l_j = \mathcal{K}_{i\to j}\left(l_i, \delta_i^f, \delta_j^f\right)$	The restriction function for vehicle i which is blockd by vehicle j, when vehicle i uses the steer angle δ_i^f and j uses the steer angle δ_j^f
$l_j = \mathfrak{K}_{\mathrm{ij},\mathbb{L}_k}\left(\mathrm{l_i}, \delta_\mathrm{i}^\mathrm{f}, \boldsymbol{\delta}^f\right)$	Restriction function of vehicle i along the loop utill vehicle j; vehicle i uses the steer angle δ_i^f, and other vehicles' steer angle is given by $\boldsymbol{\delta}^f = \left\{\delta_i^f, \delta_{i+1}^f \ldots \delta_j^f\right\}$
$\mathcal{W}_{\mathrm{j}\rightsquigarrow\mathrm{s},\mathbb{L}_k}$	The maximum escaping propagated distance from node j to node s along loop $\mathbb{L}_k$
$\mathrm{BG}(\mathfrak{N}, \mathfrak{E})$	Block graph for fixed path (or fixed steering angle). $\mathfrak{N} = \{1, 2 \ldots\}$ is the nodes set. Each node represents a vehicle. $\mathfrak{E}$ is the edge set, which represents the blockage relationship between vehicles

(continued)

(continued)

$\left(\underline{\delta^f_{j,m}}, \overline{\delta^f_{j,m}},\right)$	The two bounds of the m-th steering angle interval decomposed for vehicle j
$\left]\delta^f_j\right[$	The number of decomposed intervals for vehicle j. Each interval correspond to a specific blockage relationship
$\overleftrightarrow{\mathbf{BG}}\left(\mathfrak{N}, \mathfrak{E}\|\delta^f\right)$	Blockage graphs set for flexible path (or flexible steering angle), with nodes set $\mathfrak{N}$ and edges set $\mathfrak{E}$ given the steering angles settings δ^f. Each element in $\overleftrightarrow{\mathbf{BG}}\left(\mathfrak{N}, \mathfrak{E}\|\delta^f\right)$ is a blockage graph, $\overleftrightarrow{\mathbf{BG}}(\mathfrak{N}, \mathfrak{E}\|\delta^f) = \left\{\mathbf{BG}(\mathfrak{N}, \mathfrak{E})\|\left\{\left(\underline{\delta^f_{j,m}}, \overline{\delta^f_{j,m}}\right)\right\}, \forall j, m\right\}$. Each edge in $\mathbf{BG}(\mathfrak{N}, \mathfrak{E})$ is labeled by an interval, which represent the steering angle range
$\backslash rm\mathcal{L} = \{\mathbb{L}_k\}$; $\mathbb{L}_k = \{1^{\mathbb{L}_k}, 2^{\mathbb{L}_k}, 3^{\mathbb{L}_k}\ldots\}$	The loops set in blockage graph. $\mathbb{L}_k$ are the k-the loop; $1^{\mathbb{L}_k}$ is the first vehicle in the loop; $2^{\mathbb{L}_k}$ blocks vehicle $1^{\mathbb{L}_k}$; $\|\mathbb{L}_k\|$ is the number of legs in the loop
PI^1_{DED} and PI^2_{DED}	Performances of reroute unlock strategy and backward-forward unlock strategy

13.5 Properties of Intersection Deadlock

13.5.1 Vehicle Model and the Intra-Vehicle-Blockage

13.5.1.1 Basic Vehicle Model

To maintain generality, we develop models for a generic front-wheel-driven vehicle, as depicted in Fig. 13.3a, where the physical dimensions are illustrated. The width of the vehicle is denoted byw_{veh}. The location of the vehicle j is described by the coordinates of its center of gravity (CG), given by $\left(x_j, y_j\right)$. The distances from the CG to the front and rear bumpers are represented by l_F and l_R, respectively, while the distances from the CG to the front and rear axles are denoted as l_f and l_r. The heading angle of the vehicle is represented by ψ_j. For the sake of simplicity, unless confusion arises, we may occasionally omit the vehicle index j or the time dimension, resulting in notations such as x(t), y(t) and $\psi(t)$ or x_j, y_j and ψ_j. The four corners of the generic vehicle are designated as points A, B, C, and D, with x_j^A, x_j^B, x_j^C and x_j^D denoting the respective x-coordinates. We adopt the vehicle dynamic model presented in [18]. The state of the vehicle is represented by the vector $z = \left[x, y, \psi\right]$. The control input consists of the front steering angle δ^{ϕ} and speed v. The model is provided in Eq. (13.1) (Fig. 13.3).

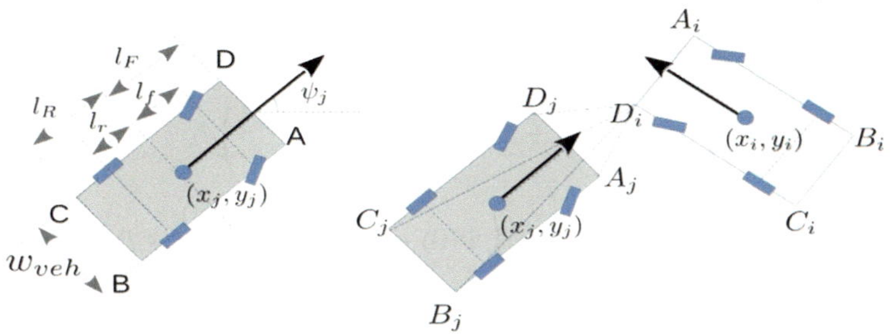

(a) Vehicle model (b) Vehicle collision check

Fig. 13.3 Vehicle illustration

$$\dot{\mathbf{z}} = \boldsymbol{f}(\mathbf{z}, \mathbf{u}) : \begin{cases} \dot{x} = v\cos(\psi + \beta) \ (a) \\ \dot{y} = v\sin(\psi + \beta) \ (b) \\ \dot{\psi} = \frac{v\cos(\beta)\tan(\delta^f)}{l_f + l_r} \ (c) \end{cases} \tag{13.1}$$

The angle β in Eq. (13.1) is defined as the arctangent of the ratio between the product of the lateral tire force factor l_r and the tangent of the front steering angle δ^f, and the sum of the front and rear tire distances l_f and l_r, respectively. The control inputs in Eq. (13.1) consist of the vehicle's speed and steering angle. In cases where direct control over the speed is not available and the acceleration a is used as input instead, Eq. (13.1) can be easily modified as follows:

$$\dot{\mathcal{Z}} = \mathcal{F}(\mathcal{Z}, \mathcal{U}) : \begin{cases} \dot{\mathrm{x}} = \mathrm{v}\cos(\psi + \beta) \ (\mathrm{a}) \\ \dot{\mathrm{y}} = \mathrm{v}\sin(\psi + \beta) \ (\mathrm{b}) \\ \dot{\Psi} = \frac{\mathrm{v}\cos(\beta)\tan(\delta^{\mathrm{f}})}{\mathrm{l_f} + \mathrm{l_r}} \ (\mathrm{c}) \\ \dot{\mathrm{v}} = \mathrm{a} \ (\mathrm{d}) \end{cases} \tag{13.2}$$

If the values of δ_f and speed v remain constant, the heading angle ψ can be determined by solving Eq. (13.1)c, which can be expressed as:

$$\psi(\mathrm{t}) = \psi(0) + \mathrm{t}\frac{\mathrm{v}\cos(\beta)\tan(\delta^{\mathrm{f}})}{\mathrm{l_f} + \mathrm{l_r}} \tag{13.3}$$

From Eq. (13.1)(a), the following relationship is obtained:

$$\begin{aligned} x(t) &= x(0) + \int_0^t v\cos(\psi(\varrho) + \beta)d\varrho \\ &= x(0) + v\int_0^t \cos\left(\psi(0) + \varrho\frac{v\cos(\beta)\tan(\delta^f)}{l_f + l_r} + \beta\right)d\varrho \\ &= x(0) + \frac{v(l_f + l_r)}{\cos(\beta)\tan(\delta^f)}\left[\sin\left(\Psi(0) + t\frac{v\cos(\beta)\tan(\delta^f)}{l_f + l_r} + \beta\right)\right. \\ &\left. - \sin(\psi(0) + \beta)\right] \end{aligned} \tag{13.4}$$

In a similar manner, the following equation is derived:

$$\begin{aligned} y(t) &= y(0) + \int_0^t v \cdot \sin(\psi(\varrho) + \beta)d\varrho \\ &= y(0) + v\int_0^t \sin\left(\psi(0) + \varrho\frac{v\cos(\beta)\tan(\delta^f)}{l_f + l_r} + \beta\right)d\varrho \\ &= y(0) + \frac{v(l_f + l_r)}{\cos(\beta)\tan(\delta^f)}\left[\cos\left(\psi(0) + t\frac{v\cos(\beta)\tan(\delta^f)}{l_f + l_r} + \beta\right)\right. \\ &\left. -\cos(\psi(0) + \beta)\right] \end{aligned} \tag{13.5}$$

The aforementioned equations allow for the real-time trajectory (represented by the center of gravity of the vehicle) to be determined. The four corners of the vehicle represent essential points on the boundary and are critical for collision detection. The coordinates of the four corners of a general vehicle can be defined as:

$$\begin{cases} x^A = x + l_F\cos(\psi) + \frac{w_{veh}}{2}\cos\left(\psi - \frac{\pi}{2}\right);\ y^A = y + l_F\sin(\psi) + \frac{w_{veh}}{2}\sin\left(\psi - \frac{\pi}{2}\right); \\ x^B = x - l_R\cos(\psi) + \frac{w_{veh}}{2}\cos\left(\psi - \frac{\pi}{2}\right);\ y^B = y - l_R\sin(\psi) + \frac{w_{veh}}{2}\sin\left(\psi - \frac{\pi}{2}\right); \\ x^C = x - l_R\cos(\psi) + \frac{w_{veh}}{2}\cos\left(\frac{\pi}{2} + \psi\right);\ y^C = y - l_R\sin(\psi) + \frac{w_{veh}}{2}\sin\left(\frac{\pi}{2} + \psi\right); \\ x^D = x + l_F\cos(\psi) + \frac{w_{veh}}{2}\cos\left(\frac{\pi}{2} + \psi\right);\ y^D = y + l_F\sin(\psi) + \frac{w_{veh}}{2}\sin\left(\frac{\pi}{2} + \psi\right); \end{cases} \tag{13.6}$$

By utilizing the aforementioned coordinates, the collision detection between vehicles (as depicted in Fig. 13.3b) can be accomplished by evaluating the following condition [21]:

$$S_{\Delta A_iA_jB_j} + S_{\Delta A_iB_jC_j} + S_{\Delta A_iC_jD_j} + S_{\Delta A_iD_jA_j} = S_{\Delta A_jB_jC_jD_j} = w_{veh}(l_F + l_R) \tag{13.7}$$

The triangle formed by connecting the corresponding three points or corners of the vehicles, denoted as $\Delta A_iA_jB_j$, is used in the evaluation. If this equation holds true, it indicates that the two vehicles have collided.

13.5.1.2 Blockage Intervals

In a scenario referred to as deadlock, vehicles experience blockage in a circular pathway. As a result, an appropriate steering angle must be determined by the respective vehicles to escape from the deadlock. The degree of blockage experienced by vehicles along the circular pathway is contingent on their respective steering angles, denoted by ψ. A graphical representation of this blockage is depicted in Fig. 13.4. The vehicle identified as j is capable of adopting different steering angles ψ_i, whereby a specific range of steering angles can be utilized by vehicle i to block vehicle j. The optimization problem is utilized to identify the interval of blockage. Assuming that the state of the two vehicles is $[x_i, y_i, \psi_i]$ and $[x_j(0), y_j(0), \psi_j(0)]$, with the omission of time, vehicle i remains stationary, and a specific range of steering angles for vehicle j is established, encapsulated by both a lower bound and an upper bound that prevents a collision between the two vehicles from occurring.

The vehicle's trajectory is represented by the trajectories of its four corners. Figure 13.4 demonstrates the concept of determining the minimum and maximum steering angle. To acquire the maximum steering angle (the leftmost angle that leads to collision), the trajectories of vehicle j's four corners traversing point D_i with the appropriate ψ_j are calculated. The highest steering angle among the four (due to the presence of four corners) is considered the desired maximum steering angle. The procedure for obtaining the minimum counterpart is similar. Therefore, the current objective is to determine the suitable ψ_j that ensures the trajectory passes a specific point D_i.

In light of the CG trajectories of vehicle j, denoted as $(x_j(t),\ y_j(t))$, the trajectory of corner A^j, given vehicle j's steer angle δ^f, can be articulated as a dependent function premised on the steering angle δ^f.

$$\begin{cases} x^{A_j}\left(t\middle|\delta^f\right) = x_j(t) + l_F\cos(\psi) + \frac{W_{veh}}{2}\cos\left(\psi_j - \frac{\pi}{2}\right)\ (a) \\ y^{A_j}\left(t\middle|\delta^f\right) = y_j(t) + l_F\sin\left(\psi_j\right) + \frac{W_{veh}}{2}\sin\left(\psi_j - \frac{\pi}{2}\right)\ (b) \end{cases} \tag{13.8}$$

Fig. 13.4 Steer interval for blockage

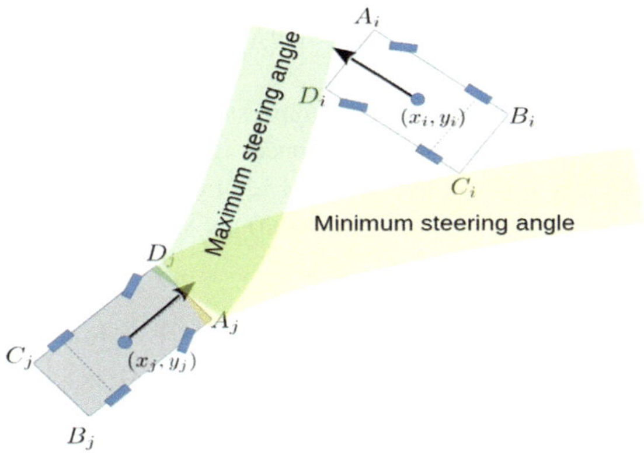

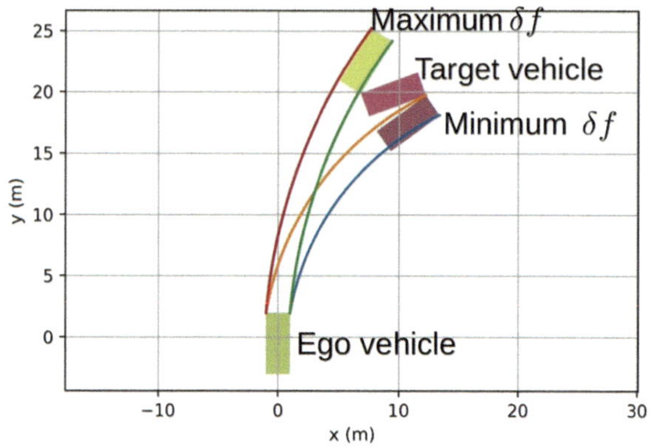

Fig. 13.5 Blockage interval example

The computation of the CG trajectory can be achieved through the utilization of Eqs. (13.4) and (13.5). When presented with a coordinate $\left(x^{D_i}, y^{D_i}\right)$ within the trajectory of corner D_i, it becomes imperative to ascertain the appropriate steering angle δ^f. Consequently, the problem of determining δ^f can be transformed into a minimization problem denoted as follows:

$$t^*, \delta^{f*} = \underset{t,\delta^f}{\operatorname{argmin}}\left(y^{A_j}\left(t|\delta^f\right) - y^{D_i}\right)^2 + \left(x^{A_j}\left(t|\delta^f\right) - x^{D_i}\right)^2 \tag{13.9}$$

The implication of Eq. (13.9) is grounded in the understanding that, given $\left(x^{D_i}, y^{D_i}\right)$, the trajectory of corner A is required to traverse through the designated point by employing δ^{f*}, and the time of arrival at said point is t^*. This minimization procedure is repeated for all eight corners (four corners per vehicle), as depicted below:

$$t^*, \delta^{f*} = \underset{t,\#,t,\delta^f}{\operatorname{argmin}}\left(\underset{t,\delta^f}{\operatorname{argmin}}\left(y^{\#}\left(t|\delta^f\right) - y^{\dagger}\right)^2 + \left(x^{\#}\left(t|\delta^f\right) - x^{\dagger}\right)^2\right), \tag{13.10}$$
$$\# \in \left\{A_j, B_jC_j, D_j\right\}, \ \dagger \in \{A_i, B_i, C_i, D_i\}$$

The maximum and minimum steer angles emanate from the highest and lowest angles among the aforementioned eight optimization outcomes. A specific numerical outcome is illustrated in Fig. 13.5. Furthermore, the ultimate states (corresponding to the precise collision condition) are visualized as well.

13.5.1.3 Permitted Travel Distance, Evasion Condition and Evasion Distance

In a deadlock situation, vehicles are obstructed by each other. The obstruction can occur when one vehicle is blocked by another vehicle in a one-by-one case or when one vehicle is blocked by several other vehicles in a one-by-many case. In the latter case, the blocked vehicle can only be released if all the blocking vehicles move simultaneously. Initially, we define the one-by-one blockage scenario and then extend it to encompass one-by-many and many-by-many blockage scenarios. Figure 13.6a illustrates a basic one-by-one blockage scenario where vehicle 0 is blocked by vehicle 1. In this blockage setting, the predetermined paths of the vehicles are assumed to remain unchanged, with fixed steering angles (variable steering angles will be discussed later). When vehicle 1 is stationary, vehicle 0 can only move forward by a certain distance, represented by the red dashed rectangle, which we define as the permitted traveling distance. Conversely, as vehicle 1 advances by a certain distance, vehicle 0 is able to move an additional distance. We can express the maximum permitted traveling distance of vehicle 0 as a function of the distance that vehicle 1 has traveled, denoted as $\mathrm{l}_0 = \mathcal{L}_{0\rightarrow 1}(\mathrm{l}_1)$. The subscript $0 \rightarrow 1$ indicates that vehicle 0 is obstructed by vehicle 1. This relationship between l_0 and $\mathcal{L}_{0\rightarrow 1}(\mathrm{l}_1)$ is referred to as the restriction function and will be further developed later. As vehicle 1 moves forward to location 2 in Fig. 13.6, it moves out of vehicle 0's projected trajectory and therefore no longer blocks vehicle 0. In this particular situation, vehicle 0 is able to "escape" from the blockage caused by vehicle 1. The distance that vehicle 1 needs to travel in order to release vehicle 0 is defined as the escaping condition of vehicle 0, while the distance traveled by vehicle 0 just before being released is referred to as the escaping distance.

These three distances are indicated by the following symbols: the permitted traveling distance $\widehat{\mathrm{l}_{0\rightarrow 1}}$, the evasion condition $\underline{\mathrm{l}_{1\leftarrow 0}}$, and the evasion distance $\overline{\mathrm{l}_{0\rightarrow 1}}$. The underline signifies the minimum distance, while the overline denotes the maximum distance. The first quantity $\widehat{\mathrm{l}_{0\rightarrow 1}}$ represents the distance traveled by vehicle 0 to reach location 1 in Fig. 13.6-a. The second distance $\underline{\mathrm{l}_{1\leftarrow 0}}$ corresponds to the distance

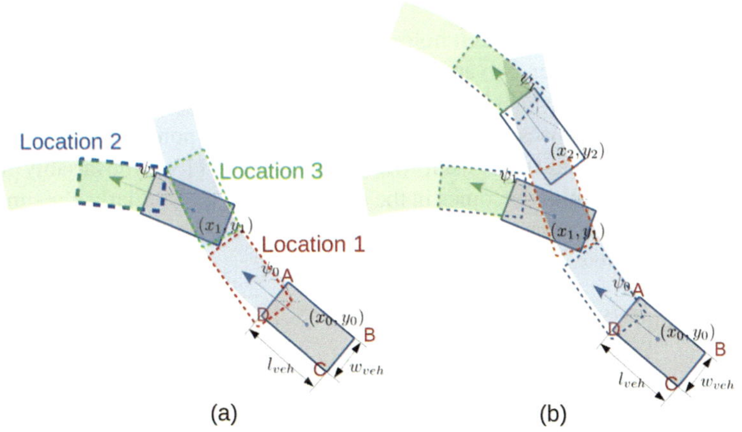

Fig. 13.6 Lock state definition for fixed path

traveled by vehicle 1 to reach location 2 in Fig. 13.6-a, and the third distance $\overline{l_{0\to 1}}$ represents the distance traveled by vehicle 0 to reach location 3 in Fig. 13.6.

In order to delineate the blockage, an initial assumption is made where vehicle 1 is considered stationary. Subsequently, an evaluation is conducted to ascertain whether vehicle 1 resides along the projected trajectory of vehicle 0 (i.e., potential collision between the two vehicles would occur if vehicle 0 continues forward in motion). If a potential collision is identified, it can be inferred that either one vertex of vehicle 1 lies within the encompassed region of vehicle 0, or vice versa, as specified in Eq. (13.7). The coordinates of the center of gravity of vehicle 0 can be derived from Eqs. (13.4) and (13.5), thereby enabling the real-time determination of the coordinates of all four vertices using Eq. (13.6). Consequently, the obstruction of vehicle 0 by vehicle 1 can be numerically assessed utilizing Eq. (13.7), although this calculation has been omitted within this context.

Hereafter, the focal point of our study centers on the development of algorithms necessary for computing the aforementioned three distances. More specifically, the permissible traveling distance is determined by the point at which vehicle 0 will intersect with vehicle 1 during its journey by using the critical-collision condition. In this case, the corner of vehicle 0 will either fall within the confines of vehicle 1's area or one of vehicle 1's corners will occupy a position within the body of vehicle 0. Given that the corners of the associated vehicles are in a constant state of flux, we represent the four corners of vehicle 0 as $A_0(t)$, $B_0(t)$, $C_0(t)$ and $D_0(t)$, respectively. To derive the allowable travel distance, it is necessary to determine the timestamp at which a specific corner of vehicle 0 will lie within the body of vehicle 1 by employing the following formula for optimization:

$$t^{*0} = \arg\min_{\#,t}\big(S_{\Delta\#A_1B_1} + S_{\Delta\#B_1C_1} + S_{\Delta\#C_1D_1} + S_{\Delta\#D_1A_1} - w_{veh}(l_F + l_R) - d_{buffer}\big)^2 \# \in \{A_0(t), B_0(t), C_0(t), D_0(t)\} \tag{13.11}$$

The aforementioned equation holds a significant physical interpretation, stating that the disparity between two areas (one being the combined area of four triangles, and the other being $w_{veh}(l_F + l_R)$) must not deviate from d_{buffer}, a minute positive value (for optimization purposes, a value of 0.1 is employed). Upon collision between the two vehicles, the equal relationship (referred to as Eq. (13.7)) invariably remains intact, leading to infinite outcomes in the absence of d_{buffer}. Similarly, the timestamp at which a specific corner of vehicle 1 aligns with the body of vehicle 0 can be determined as follows:

$$t^{*1} = \operatorname{argmin}_{\#,t}\big(S_{\Delta\#A_0(t)B_0(t)} + S_{\Delta\#B_0(t)C_0(t)} + S_{\Delta\#C_0(t)D_0(t)} + S_{\Delta\#D_0(t)A_0(t)} - w_{veh}(l_F + l_R) - d_{buffer}\big)^2; \# \in \{A_1, B_1, C_1, D_1\} \tag{13.12}$$

The minimum moment at which the two vehicles collide can be denoted as $t^* = \min(t^{*0}, t^{*1})$. At this moment, the coordinate of the center of gravity (CG) of vehicle 0 can be represented as $(x_0(t^*), y_0(t^*))$. The calculation of the allowable travel distance is given by the following equation:

$$\widehat{l_{0\to 1}} = \int_{0}^{t^*} \sqrt{x'(t)^2 + y'(t)^2 dt} = v_0 \cdot t^* \tag{13.13}$$

Regarding the evasion condition $\underline{l_{1\leftarrow 0}}$, it refers to the distance traveled by vehicle 1 in order to avoid colliding with vehicle 0. Once this distance has been traveled, the trajectory of vehicle 0 will no longer overlap with the boundary of vehicle 1. To determine this distance, we first optimize the moment when vehicle 1 releases vehicle 0 precisely, using the following optimization process:

$$\begin{aligned} t_0^*, t_1^* = &\operatorname*{argmin}_{\#,t_0,t_1}\Big(\big(S_{\Delta \# A_1(t_1)B_1(t_1)} + S_{\Delta \# B_1(t_1)C_1(t_1)} + S_{\Delta \# C_1(t_1)D_1(t_1)}\big) \\ &+ S_{\Delta \# D_1(t_1)A_1(t_1)}\Big)(-w_{veh}(l_F + l_R) - d_{buffer})^2, \\ &\# \in \{A_0(t_0), B_0(t_0), C_0(t_0), D_0(t_0)\} \end{aligned} \tag{13.14}$$

The aforementioned optimization objective reflects the scenario in which both vehicles move simultaneously, and the moment at which they diverge (without collision) is obtained by minimizing the difference in areas. Consequently, the evasion condition $\underline{l_{1\leftarrow 0}}$ and the evasion distance $\overline{l_{0\to 1}}$) can be computed as $v_0 \cdot t_0^*$ and $v_1 \cdot t_1^*$, respectively.

13.5.1.4 Restriction Function and Blockage Graph

There exists a specific correlation between the permissible travel distance of vehicle 0 and the travel distance of vehicle 1 when vehicle 0 is obstructed by vehicle 1. This correlation, denoted as $l_0 = \mathcal{L}_{0\to 1}(l_1)$ and referred to as the restriction function, illustrates a dependence between the blocked vehicle's travel distance and the blocking vehicle's distance. Notably, the restriction function is only applicable within the range $\left[0, \underline{l_{1\leftarrow 0}}\right]$ as the blockage relationship ceases when the blocking vehicle's travel distance surpasses the evasion condition of the blocked vehicle. Additionally, the value co-domain interval is $\left[\widehat{l_{0\to 1}}, \overline{l_{0\to 1}}\right]$.

The process of deriving the restriction shares similarity with that of deriving the permitted travel distance, but the nonlinearity of the vehicle dynamics results in a nonlinear restriction function that can only be obtained through numerical methods. To determine the restriction function, the interval $$\left[{0,{\text{~}}\underline{{{\text{l}}_{{1 \leftarrow 0}} }} } \right]$$ is discretized into a sequence of travel distances using a predefined discretization step, after which the corresponding permitted travel distance for each travel distance is determined. The result of this approach is a constructed restriction function. An example of such a function is shown in Fig. 13.7. The restriction function 13.7 , which depicts the function $\mathcal{L}_{i\to j}(\cdot)$ with an appropriate domain. In this example, vehicle 1 assumes an initial state of [0,0,π/2], while vehicle 2 assumes an initial state of [20,35,π/2], with the steering angles of the two vehicles being $5^{\circ} and -5^{\circ}$, respectively. It is apparent that the initial position of vehicle 2 obstructs the path of vehicle 1. However, if vehicle 2 were to travel a distance of 5.519 m, vehicle 1 would be free from vehicle 2's blockage.

The aforementioned definitions, namely the permitted distance of travel, the condition of evasion, and the evasive distance, are predicated on the assumption of fixed steering angle and speed. They can be applied more broadly to situations involving varying speed and steering angle. Additionally, their applicability extends effortlessly to the one-to-many scenario depicted in Fig. 13.6b. Let us suppose that the collection of vehicles obstructing vehicle i is denoted by $\mathbb{C}_i$. Then, the authorized travel distance can be determined as $\widehat{l_{i \to \mathbb{C}_i}} = \min_{j \in \mathbb{C}_i} \widehat{l_{i \to j}}$; the evasive distance can be defined as $\overline{l_{i \to \mathbb{C}_i}} = \max_{j \in \mathbb{C}_i} \overline{l_{i \to j}}$. The evasion condition varies from the aforementioned two measures and must be established for each individual vehicle in $\mathbb{C}_i$, which remains true in the one-to-one case.

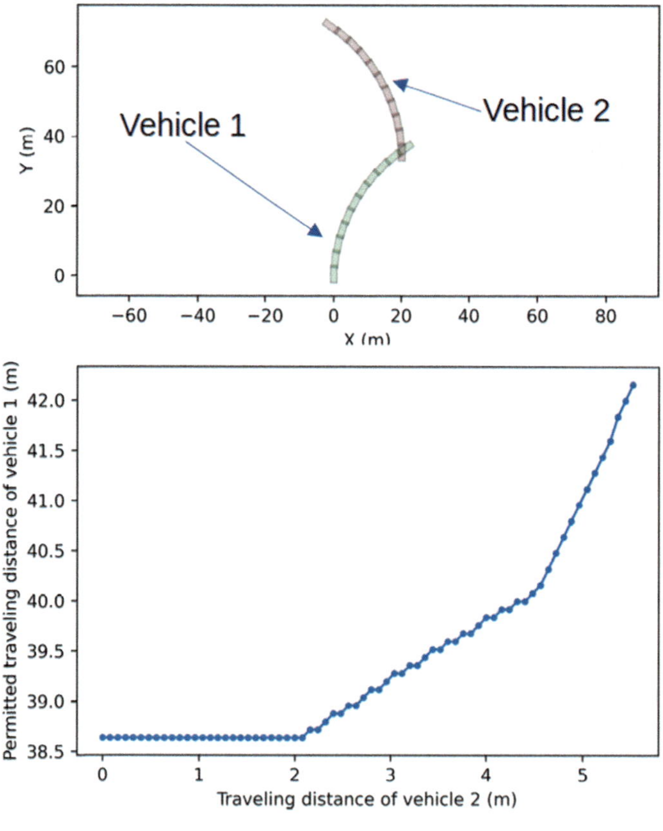

Fig. 13.7 The restriction function

13.5.1.5 Blockage Graph Construction

Given the provided information, when the steering angle and velocity of a vehicle are known, the resulting trajectory can be determined. It is possible to establish a relationship of inter-blockage among vehicles. Between any two vehicles, A and B, either vehicle A acts as an obstruction to vehicle B, or vice versa, or they have no obstructive effect on each other. By utilizing the notion of blockage, a blockage graph (referred to as BG) can be constructed. Each vehicle involved is represented as a node within this graph. Specifically, if vehicle A obstructs vehicle B, an edge is created from B to A. Consequently, the blockage graph is denoted as BG($\mathfrak{N}$, $\mathfrak{E}$). $\mathfrak{N} = \{1, 2 \ldots\}$ representing the set of nodes and each node corresponding to a distinct vehicle. $\mathfrak{E}$ denotes the set of edges. The value of $|\mathfrak{N}|$ signifies the number of vehicles present within the intersection in question. Importantly, it should be noted that as vehicles enter or exit the intersection in a dynamic manner, the blockage graph BG($\mathfrak{N}$, $\mathfrak{E}$) evolves over time. In the context of deadlock detection, BG($\mathfrak{N}$, $\mathfrak{E}$) serves as the fundamental data structure. Moreover, it is important to acknowledge that each edge within BG($\mathfrak{N}$, $\mathfrak{E}$) inherently possesses a restriction function and three distances.

13.5.2 *Properties of Weak Deadlock*

In light of the prevailing conditions of obstruction, constraint, and circumvention exhibited by vehicles, it is now incumbent upon us to expound upon the definitions of deadlocks. To begin with, we shall consider the simplified scenario of one-to-one blockage, wherein each vehicle is obstructed by a maximum of one other vehicle. It should be noted that the subsequent definition can readily be extended to encompass the one-to-many case.

In the absence of any loops within the graph BG($\cdot, \cdot$), deadlock is precluded. Conversely, the advent of a loop in the aforementioned graph entails the potential occurrence of a deadlock. Let us scrutinize the k-th loop denoted as $\mathbb{L}_k$ within BG($\mathfrak{N}$, $\mathfrak{E}$). Without loss of generality, the vehicles within this loop are assigned indices from the set $\mathbb{L}_k = \{1, 2, 3, ..\mathbb{L}_k\}$. Notably, vehicle 1 assumes precedence within the loop, with vehicle 2 obstructing vehicle 1, and so forth. Concomitant with the configuration of the blockage loop, vehicle 1 proceeds to obstruct vehicle $\mathbb{L}_k$. By virtue of the restriction function, it follows that $\mathcal{L}_{m \to m+1}(\cdot)$, $\forall m = 1 \sim |\mathbb{L}_k| - 1$ and $\mathcal{L}_{\mathbb{L}_k \to 1}(\cdot)$. Inductively, for any given vehicle m situated within the loop, the following relationship holds true:

$$\begin{aligned} l_{m-1} &= \mathcal{L}_{m-1\to m}(l_m); \\ l_{m-2} &= \mathcal{L}_{m-2\to m-1}(l_{m-1}) \\ &\cdots \\ l_3 &= \mathcal{L}_{3\to 4}(l_4) \\ l_2 &= \mathcal{L}_{2\to 3}(l_3) \end{aligned} \tag{13.15}$$

The aforementioned recurrent equation implies that in order for vehicle m to cover distance l_m, vehicle 2 must advance by distance l_2. We consolidate this relationship into a single formula denoted as $l_2 = \Im_{2\rightsquigarrow m,\mathbb{L}_k}(l_m)$, which effectively conveys the connection between l_2 and l_m. This formulation can be expanded to encompass a broader scenario involving two arbitrary vehicles, denoted as vehicle i and vehicle j, within the loop $\mathbb{L}_k$. Specifically, we define the route restriction function as $l_i = \Im_{i\rightsquigarrow j,\mathbb{L}_k}(l_j)$, assuming a consistent directionality along the loop, whereby $\Im_{i\rightsquigarrow j,\mathbb{L}_k}(\cdot)$ differs from $\Im_{i\rightsquigarrow j,\mathbb{L}_k}(\cdot)$. Additionally, for any general vehicle j within the loop, the expression $\Im_{j\rightsquigarrow j,\mathbb{L}_k}(l)$represents the maximum distance that vehicle j can eventually travel, after which it is assumed to have traveled distance l and sequentially releases the space for subsequent follower vehicles.

When expressing the permissible travel distance of vehicle 1, denoted as $\widehat{l_{1\to 2}}$, using the route restriction function, we obtain:

$$l_1 = \Im_{1\rightsquigarrow 1,\mathbb{L}_k}\left(\widehat{l_{1\to 2}}\right) \tag{13.16}$$

In the context described, "vehicular 1" travels a prescribed distance $\widehat{l_{1\to 2}}$, known as the permitted travel distance, thereby creating space for subsequent vehicles within the same cycle. Subsequent vehicles, except for "vehicular 1", are then able to sequentially cover a determined distance. Ultimately, "vehicular 1" is able to cover a distance denoted as l_1. In the event that $l_1 = \Im_{1\rightsquigarrow 1,\mathbb{L}_k}\left(\widehat{l_{1\to 2}}\right) < \overline{l_{1\to 2}}$, it indicates that "vehicular 1" has reached its maximum travel capacity while generating a suitable space for the following vehicles to proceed, rendering its own evasive distance (l_1) smaller. Consequently, "vehicular 1" experiences a state of self-constraint, resulting in a complete standstill for all vehicles within loop $\mathbb{L}_k$. This study adopts the subsequent deadlock definition:

Definition 1 weak deadlock (one-to-one case deadlock): Within an ordinary loop $\mathbb{L}_k$, if a specific vehicle j in the loop satisfies the subsequent condition, we denote the existence of a weak deadlock in loop $\mathbb{L}_k$.

$$l_j = \Im_{j\rightsquigarrow j,\mathbb{L}_k}\left(\widehat{l_{j\to j+1}}\right) < \overline{l_{j\to j+1}} \tag{13.17}$$

The aforementioned criterion is established to apply to a singular vehicle, denoted as j, within the loop. If this criterion is satisfied by one vehicle, all other vehicles within the same loop will also comply with this criterion, provided that the restriction functions exhibit monotonicity (the formal demonstration is not included in this context). Conversely, once a vehicle fails to meet this criterion, none of the other vehicles in the loop are bound by this condition either. Hence, the detection of deadlock can be assessed by selecting a vehicle at random.

Based on the mentioned condition, in order to identify the presence of a weak deadlock, it becomes necessary to recursively compute the distance that each vehicle must travel in order to create sufficient space for vehicle j. The proposed algorithm for this purpose is known as the escape distance propagation (EDP) algorithm, which will be expounded upon subsequently.

Definition 1 solely pertains to instances of one-to-one blockage between vehicles. The generalization of said definition to encompass one-by-many blockage is anticipated to be straightforward, given the similarity in underlying principles. With regards to the generalization, it becomes imperative to explicate the structure of loops applicable to the one-to-many scenario. Given the existence of one-to-many blockages amidst vehicles, the loops become intertwined or overlapping in nature. Several loops share common nodes, referring to the vehicles themselves. Figure 13.8 presents an exemplification of a deadlock in a one-to-many setting. For the sake of brevity, only a section of the loop is depicted. The diagram illustrates two distinct loops: $0 \rightarrow 2 \rightarrow 4 \rightarrow 0$ and $0 \rightarrow 1 \rightarrow 3 \rightarrow 4 \rightarrow 0$, denoted as $c\mathbb{L}_1$ and $\mathbb{L}_2$ respectively. It is apparent that vehicle 0 encounters simultaneous blockage from both vehicle 1 and vehicle 2. As a consequence, the evasion distance for vehicle 0 is defined as $\overline{\mathrm{l}_{0\rightarrow\{1,2\}}} = \max\limits_{\mathrm{j}\in\{1,2\}} \overline{\mathrm{l}_{0\rightarrow\mathrm{j}}}$, representing the maximum evasion distance within the context of all blockage relationships. Meanwhile, the permissible travelling distance is computed as $\widehat{\mathrm{l}_{0\rightarrow\{1,2\}}} = \min\limits_{\mathrm{j}\in\{1,2\}} \widehat{\mathrm{l}_{0\rightarrow\mathrm{j}}}$. By applying the deadlock condition outlined by Eq. (13.17) to both loops $\mathbb{L}_1$ and $\mathbb{L}_2$ individually, the resulting deadlock conditions can be determined.

$$\begin{cases} \Im_{0\rightsquigarrow 0,\mathbb{L}_1}\left(\widehat{\mathrm{l}_{0\rightarrow\{1,2\}}}\right) \geq \overline{\mathrm{l}_{0\rightarrow\{1,2\}}} \ (a) \\ \Im_{0\rightsquigarrow 0,\mathbb{L}_2}\left(\widehat{\mathrm{l}_{0\rightarrow\{1,2\}}}\right) \geq \overline{\mathrm{l}_{0\rightarrow\{1,2\}}} \ (b) \end{cases} \tag{13.18}$$

In instances where either Eq. (13.18)-a or b is satisfied, the vehicles become immobilized, resulting in a state of deadlock at the intersection. By generalizing the scenario depicted in Fig. 13.8 to encompass multiple interconnected loops, the following definition of weak deadlock applies to situations where one-to-many blockage occurs.

Definition 2 (weak deadlock in one-to-many case): Let there be multiple loops within the graph BG($\mathfrak{N}$, $\mathfrak{E}$). These loops share common nodes, forming a connected undirected graph when considering their respective edges. If the following condition is met, a state of weak deadlock arises and all vehicles within the loops become immobilized:

$$\Im_{\mathrm{j}\rightsquigarrow\mathrm{j},\mathbb{L}_k}\left(\widehat{\mathrm{l}_{\mathrm{j}\rightarrow\mathrm{C_j}}}\right) \geq \overline{\mathrm{l}_{\mathrm{j}\rightarrow\mathrm{C_j}}}, \ \exists \mathrm{j}, \mathrm{k} \tag{13.19}$$

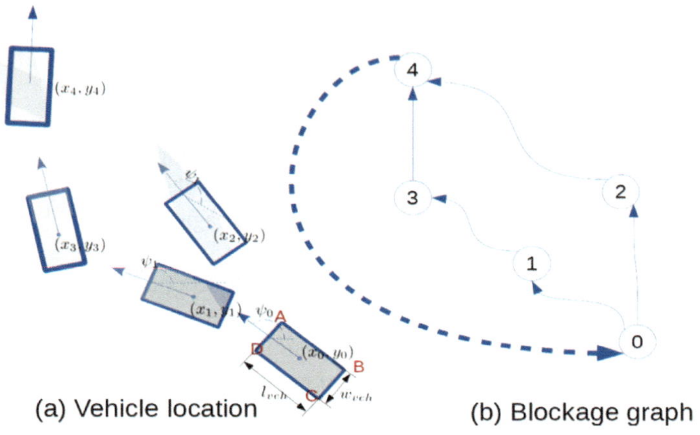

Fig. 13.8 One-to-many blockage in the loop

Similar to the aforementioned definitions, C_j denotes the set of vehicles that are blocking vehicle j; $\widehat{l_{j\to C_j}}$ represents the minimum among the arrival times to each vehicle $k \in C_j$; $\overline{l_{j\to C_j}}$ is defined as the maximum among the departure times from each vehicle $s \in C_j$.

13.5.3 Properties of Strong Deadlock

In the context of the weak deadlock definition (Definition 1 and Definition 2), it is postulated that the steering angle (referred to as ψ in the vehicle's state space model) of the vehicles remains constant, thereby resulting in fixed expected trajectories. However, it should be acknowledged that in practice, the vehicles have the ability to deviate from the congestion by altering their steering. Hence, in order to account for this flexibility in steering angle, we extend the applicability of Definition 1 and Definition 2. In this scenario, the steering angle of connected and automated vehicles (CAVs) is constrained solely by their physical capabilities.

$$-\delta^{f,MAX} \leq \delta^f \leq \delta^{f,MAX} \tag{13.20}$$

In line with the weak deadlock scenario, the initial examination focuses on the individual deadlock situation, utilizing the iterative process denoted as $\mathbb{L}_k$. For the sake of brevity, we assume all vehicles within the loop to be connected and autonomous (CAV). In the event that a particular vehicle, denoted as j, is a human-driven vehicle (HDV), its steering angle δ_j^f is predetermined and beyond control. In order to establish the state of deadlock, similar to the case of fixed steering angle, it is imperative to define the constraint function between any two vehicles that block each other as well as the restriction function alongside loop $\mathbb{L}_k$. To clarify, the constraint function corresponding to the two blocking vehicles, vehicle i and vehicle j, under specific steering angle configurations can be expressed as $l_i = \mathcal{K}_{i\to j}\left(l_j, \delta_i^f, \delta_j^f\right)$. In this scenario, vehicle j obstructs vehicle i. The maximum distance vehicle i can travel is contingent upon vehicle j, and consequently, in order to cover a distance of l_i, vehicle j must advance by a distance of l_j.

Figure 13.9 illustrates an instance where $l_i = \mathcal{K}_{i\to j}\left(l_j, \delta_i^f, \delta_j^f\right)$. In this scenario, vehicle j obstructs vehicle i. Both vehicles possess two steering angles, namely δ_i^f and $\delta_i^{f'}$ for vehicle i, and δ_j^f and $\delta_j^{f'}$ for vehicle j. Supposing that vehicle j adheres to the trajectory dictated by $\delta_j^{f'}$, and vehicle i adheres to the trajectory dictated by δ_i^f, vehicle j must traverse a distance of l_j (depicted by the black dashed box in Fig. 13.9) for vehicle i to advance by a distance of l_i (depicted by the red dashed box in Fig. 13.9). It is crucial to note that the blockage relationship between the two vehicles is a prerequisite for the formula $\mathcal{K}_{i\to j}(\cdot,\cdot,\cdot)$. Should a specific steering angle disrupt the blockage relationship (i.e., vehicle i evades vehicle j's obstruction by steering away), the function $\mathcal{K}_{i\to j}(\cdot,\cdot,\cdot)$ becomes invalid. Similar to the fixed steering angle scenario, with slight abuse of notation, we express the allowed travel distance, evasion condition, evasion distance as $\widehat{l_{i\to j}}|_{\delta_i^f}$, $\underline{l_{j\leftarrow i}}|_{\delta_i^f,\delta_j^f}$ and $\overline{l_{i\to j}}|_{\delta_i^f,\delta_j^f}$, respectively. The presence of vertical bars denotes that these distances are contingent upon certain conditions.

In resemblance to the characterization of the function $\Im_{\cdot\rightsquigarrow\cdot,\mathbb{L}_k}(\cdot)$ along a closed trajectory, we establish the propagation of distance along said trajectory for the manipulation of variable steering angle as follows:

$$l_i = \mathfrak{K}_{i\rightsquigarrow j,\mathbb{L}_k}\left(l_j, \delta_i^f, \delta^f\right) \tag{13.21}$$

The inherent connotation of $\mathfrak{K}_{i\rightsquigarrow j,\mathbb{L}_k}\left(l_j, \delta_i^f, \delta^f\right)$ is readily perceivable: vehicle i assumes the steering angle δ_i^f, while vehicle i + 1 through vehicle j employ the corresponding steering angles denoted by the vector δ^f, resulting in vehicle j obstructing vehicle j–1, vehicle j–1 impeding vehicle j–2, and so forth, with eventual inhibition of vehicle i + 1 by vehicle i. In order for vehicle i to traverse a distance l_i, vehicle j

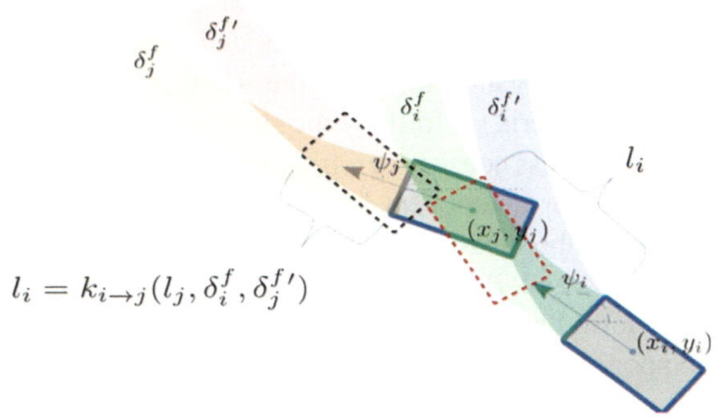

Fig. 13.9 Blockage definition between two vehicles for variable steering angle

must advance a length of l_j. Employing $\mathfrak{K}_{i\rightsquigarrow j,\mathbb{L}_k}(\cdot,\cdot,\cdot)$ facilitates the construction of the deadlock definition, an extension of Definition 1, with relative ease.

Definition 3 (strong deadlock): The occurrence of strong deadlock can be defined as follows: When considering the states of all vehicles (both CAVs and HDVs) at the intersection, if there is any possible steering angle δ_i^f $\forall i$, such that a resulting $BG(\mathfrak{N}, \mathfrak{E})$ contains a loop $\mathbb{L}_k$ satisfying the deadlock condition specified in either Definition 1 or Definition 2, then it can be concluded that strong deadlock is present. This definition is derived from either Definition 1 or Definition 2, and can be expressed mathematically as the following deadlock condition:

$$l_j = \mathfrak{K}_{j\rightsquigarrow j,\mathbb{L}_k}\left(\widehat{l_{j,j+1}}|_{\delta_j^f}, \delta_j^f, \delta^f\right) > \overline{l_{j\rightarrow j+1}}|_{\delta_j^f,\delta_{j+1}^f}, \forall\delta_j^f, \delta^f \tag{13.22}$$

13.6 Deadlock Detection

In this section, we introduce a technique for recognizing intersection deadlocks in mixed autonomous vehicles settings with input obtained from both CAVs and HDVs. We initially outline the deadlock detection approach. Subsequently, an augmented Kalman filtering model is formulated to deduce the steering angles of HDVs which are otherwise unknown but essential to the proposed model. Finally, we present the two deadlock detection models for weak and strong deadlocks, respectively.

13.6.1 Deadlock Detection Procedure

Given the existence of two distinct forms of deadlocks, it is important to note that the stringent criteria set forth by the strong deadlock definition exceeds that of the weak deadlock definition. Consequently, in the subsequent outlined procedure, the initial examination focuses on the weak deadlock scenario. In the event that the conditions outlined by the weak definition fail to be met, the identification process is terminated without the discovery of any deadlocks. However, in the event that the prerequisites outlined by the weak definition are met, the algorithm proceeds with further investigation to determine whether the identified deadlock aligns with the stringent criteria set forth by the strong deadlock definition.

The process consists of a tripartite system that has been thoroughly documented. BLOCK1, or Component 1, serves the purpose of modelling HDV behaviour, which will be elaborated on in the ensuing section. Although Components 2 and 3, or BLOCK2 and BLOCK3, are akin, they possess slight discrepancies, and have been formulated to address weak and strong deadlocks, respectively. These two components will be fully expatiated upon subsequently.

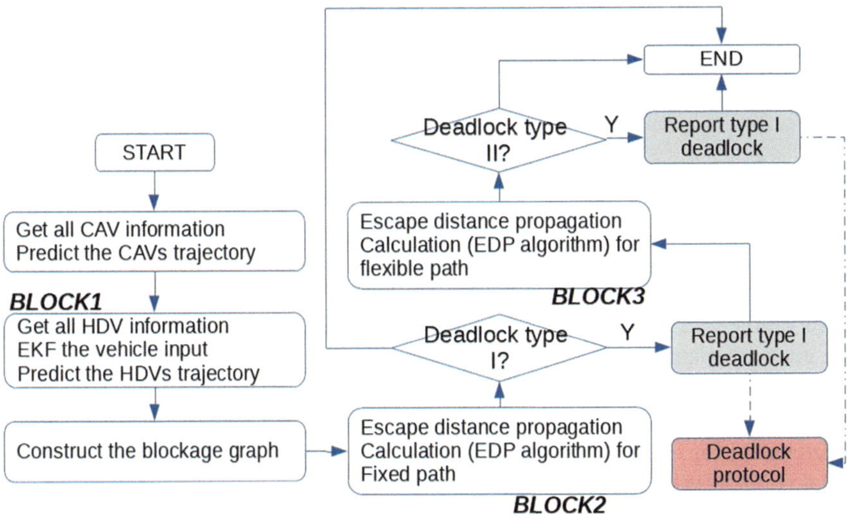

Fig. 13.10 Deadlock detection process

The output of the deadlock detection procedure provides evidence for the existence of two distinct types of deadlock. Specifically, there are three conceivable outcomes, namely: (1) zero deadlocks; (2) the existence of a weak deadlock, without a concomitant strong deadlock; (3) existence of a strong deadlock (in addition to weak deadlock). The findings from this procedure will be implemented in the development of the deadlock protocol design (the pink block depicted in Fig. 13.10).

13.6.2 HDV Model by Extended Kalman Filtering

The real-time states of CAVs can be reported, whereas HDVs do not transmit such data, making their states difficult to fully observe. Nonetheless, it is assumed that the HDVs' real-time locations can be obtained through roadside units or other sensors. Despite this, the steering angle of HDVs must be inferred to initiate the deadlock detection process. To this end, the real-time coordinates of the HDV are provided, and the steering angle is one of the outputs. In the vehicle model described by Eq. (13.1), the controller input is the steering angle (δ^f), which must be estimated from the trajectory by transforming the state equation. Hence, the following state vehicle model and observation model are proposed for HDVs:

$$\begin{cases} \dot{\mathbf{z}} = \begin{bmatrix} \dot{x} \\ \dot{z} \\ \dot{\psi} \end{bmatrix} = f(\mathbf{z}, \mathbf{u}) + \Upsilon = f\left(\begin{bmatrix} x \\ y \\ \psi \end{bmatrix}, \begin{bmatrix} v \\ \delta_f \end{bmatrix} \right) + \Upsilon \ (a) \\ w = \begin{bmatrix} x \\ y \end{bmatrix} = \begin{bmatrix} 1\,0\,0 \\ 0\,1\,0 \end{bmatrix} \begin{bmatrix} x \\ y \\ \psi \end{bmatrix} + \Gamma \qquad (b) \end{cases} \tag{13.23}$$

The second equation represents the act of observing HDV, in which the real-time coordinates of HDVs can be determined. The system error, denoted as Υ and observation error, denoted as Γ, are assumed to be independent and their covariance are given by matrices Q and R, respectively.

As the aforementioned system is nonlinear, it must be discretized about a reference point, such as u_R and u_R, in order to be estimated using the Kalman method. Assuming a reference trajectory of $\dot{z_R} = f(z_R, u_R)$, which is represented by Eq. (13.1), and employing the vehicle model found in Eq. (13.23), the system must first be linearized at a specific point by means of the Jacobi matrix.

$$\dot{\mathbf{z}} = f(\mathbf{z}, \mathbf{u}) = f\left(\mathbf{z}^{\mathbf{R}}, \mathbf{u}^{\mathbf{R}}\right) + \left.\frac{\partial f}{\partial z}\right|_{\mathbf{z}^{\mathbf{R}}} (z - z_R) + \left.\frac{\partial f}{\partial u}\right|_{\mathbf{u}^{\mathbf{R}}} (u - u_R) + \Upsilon \\ + \text{high order term} \tag{13.24}$$

where the two Jacobi matrixes are defined as:

$$\left.\frac{\partial f}{\partial z}\right|_{z_R} = \begin{bmatrix} 0\ 0 & -v \sin\left(\psi + \arctan\left(\frac{\tan(\delta_f)}{2}\right)\right) \\ 0\ 0 & v \cos(\psi + \arctan\left(\frac{\tan(\delta_f)}{2}\right) \\ 0\ 0 & 0 \end{bmatrix};$$

$$\left.\frac{\partial f}{\partial u}\right|_{u_R} = \begin{bmatrix} \cos\left(\psi + \arctan\left(\frac{\tan(\delta_f)}{2}\right)\right) & -\frac{v\left(\frac{1+\tan(\delta_f)^2}{2}\right)\sin\left(\psi + \arctan\left(\frac{\tan(\delta_f)}{2}\right)\right)}{1+\frac{\tan(\delta_f)^2}{1}} \\ \sin\left(\psi + \arctan\left(\frac{\tan(\delta_f)}{2}\right)\right) & \frac{v\left(\frac{1+\tan(\delta_f)^2}{2}\right)\cos\left(\psi + \arctan\left(\frac{\tan(\delta_f)}{2}\right)\right)}{1+\frac{\tan(\delta_f)^2}{4}} \\ \sqrt[4]{1 + \frac{\tan(\delta_f)^2}{4}} & \frac{v\left(1+\tan(\delta_f)^2\right)}{4\sqrt{1+\frac{\tan(\delta_f)^2}{4}}} - \frac{2v\left(1+\tan(\delta_f)^2\right)\tan(\delta_f)^2}{32\left(1+\frac{\tan(\delta_f)^2}{4}\right)} \end{bmatrix} \tag{13.25}$$

Then we use the discretized state space model to approximate the original non-linear dynamic system:

$$
\begin{aligned}
\dot{\mathbf{z}} &= \boldsymbol{f}(\mathbf{z}^{\mathbf{R}}, \mathbf{u}^{\mathbf{R}}) + \left.\frac{\partial f}{\partial z}\right|_{\mathbf{z}^{\mathbf{R}}} (z - z_R) + \left.\frac{\partial f}{\partial u}\right|_{\mathbf{u}^{\mathbf{R}}} (u - u_{\mathrm{R}}) + \Upsilon \Rightarrow \\
\mathbf{z}_{\mathbf{k}+1} - \mathbf{z}^{\mathbf{R}} &= \Delta t \cdot \boldsymbol{f}(\mathbf{z}^{\mathbf{R}}, \mathbf{u}^{\mathbf{R}}) + \Delta t \cdot E(z_{k+1} - z_R) + \Delta t \cdot F(\mathrm{u}_{k+1} - \mathbf{u}^{\mathbf{R}}) \\
&\quad + \Delta t \cdot \Upsilon \Rightarrow \\
(\boldsymbol{I} - \boldsymbol{E})\mathbf{z}_{\mathbf{k}+1} - \mathbf{F}\mathrm{u}_{\mathrm{k}+1} &= \boldsymbol{f}(\mathbf{z}_{\mathbf{k}}, \mathbf{u}_{\mathbf{k}}) - \Delta t \cdot E\mathbf{z}_{\mathbf{k}} - \Delta t \cdot F\mathbf{u}_{\mathbf{k}} + \Delta t \cdot \Upsilon \Rightarrow \\
[I - \Delta t \cdot E \quad & - \Delta t \cdot F]\begin{bmatrix} \mathbf{z}_{\mathbf{k}+1} \\ \hline \mathrm{u}_{\mathrm{k}+1} \end{bmatrix} \\
= [I - \Delta t \cdot E \quad & - \Delta t \cdot F]\begin{bmatrix} \mathbf{z}_{\mathbf{k}} \\ \hline \mathrm{u}_{\mathrm{k}} \end{bmatrix} + \Delta t \cdot \boldsymbol{f}(\mathbf{z}_{\mathbf{k}}, \mathbf{u}_{\mathbf{k}}) + \Delta t \cdot \Upsilon
\end{aligned} \tag{13.26}
$$

In the above transformed state space model, the system state is $\begin{bmatrix} \mathbf{z}_{k+1} \\ \mathrm{u}_{k+1} \end{bmatrix}$, which include the control input. For easy reference, we express the block matrix using the following symbols:

$$
G = \begin{bmatrix} I - \Delta t \cdot E & -\Delta t \cdot F \end{bmatrix}; H = \begin{bmatrix} -\Delta t \cdot E & -\Delta t \cdot F \end{bmatrix} \tag{13.27}
$$

Then we have the following linearized system:

$$
G\begin{bmatrix} \mathbf{z}_{\mathbf{k}+1} \\ \hline \mathrm{u}_{\mathrm{k}+1} \end{bmatrix} = H\begin{bmatrix} \mathbf{z}_{\mathbf{k}} \\ \hline \mathrm{u}_{\mathrm{k}} \end{bmatrix} + \Delta t \cdot \boldsymbol{f}(\mathbf{z}_{\mathbf{k}}, \mathbf{u}_{\mathbf{k}}) + \Delta t \cdot \Upsilon \tag{13.28}
$$

In the above equation, $\boldsymbol{f}(\mathbf{z}_{\mathrm{k}}, \mathbf{u}_{\mathrm{k}})$ is a constant which is the evaluation of $\boldsymbol{f}(\cdot)$ at point $(\mathbf{z}_{\mathrm{k}}, \mathbf{u}_{\mathrm{k}})$. On the other hand, the observation equation in Eq. (13.23)b can be reorganized as:

$$
\begin{aligned}
w &= \begin{bmatrix} x \\ y \end{bmatrix} = \begin{bmatrix} 1\,0\,0 \\ 0\,1\,0 \end{bmatrix}\begin{bmatrix} x \\ y \\ \psi \end{bmatrix} + \Gamma \Rightarrow \\
w &= \begin{bmatrix} x \\ y \end{bmatrix} = \begin{bmatrix} 1\,0\,0\;\;0\,0 \\ 0\,1\,0\;\;0\,0 \end{bmatrix}\begin{bmatrix} x \\ y \\ \psi \\ v \\ \delta_f \end{bmatrix} = \mathrm{M}\begin{bmatrix} x \\ y \\ \psi \\ v \\ \delta_f \end{bmatrix} + \Gamma
\end{aligned} \tag{13.29}
$$

For simplicity we put the transformed state-space model and the observation model together, which yields:

$$
\begin{cases} G\begin{bmatrix} \mathbf{z}_{\mathbf{k}+1} \\ \hline \mathrm{u}_{\mathrm{k}+1} \end{bmatrix} = H\begin{bmatrix} \mathbf{z}_{\mathbf{k}} \\ \hline \mathrm{u}_{\mathrm{k}} \end{bmatrix} + \boldsymbol{f}(\mathbf{z}_{\mathbf{k}}, \mathbf{u}_{\mathbf{k}}) + \Delta t \cdot \Upsilon & \text{(a)} \\ w = \mathrm{M}\begin{bmatrix} \mathbf{z}_{\mathbf{k}} \\ \hline \mathrm{u}_{\mathrm{k}} \end{bmatrix} & \text{(b)} \end{cases} \tag{13.30}
$$

The objective of this task is to deduce $\begin{bmatrix} \mathrm{z}_{k+1} \\ \mathrm{u}_{k+1} \end{bmatrix}$ for every time interval by utilizing observation w, according to the model presented in Eq. (13.31). Assuming that the estimation error covariance matrix at step k is denoted as $\mathrm{P}_{k|k}$, the optimal estimator can be obtained as derived in [16, 23].

$$\begin{bmatrix} \mathbf{z}^*_{k+1} \\ \mathrm{u}^*_{k+1} \end{bmatrix} = P_{k+1|k+1}\left(G^{\mathrm{T}}\left(\Delta t \cdot Q + H\mathrm{P}_{k|k}H^T\right)^{-1} H \begin{bmatrix} \mathbf{z}^*_k \\ \mathrm{u}^*_k \end{bmatrix} + M^T R^{-1} w_{k+1}\right) \tag{13.31}$$

Consequently, the control input can be inferred. In the aforementioned equation, $\begin{bmatrix} \mathbf{z}^*_k \\ \mathrm{u}^*_k \end{bmatrix}$ represents the estimation of $\begin{bmatrix} \mathbf{z}_k \\ \mathrm{u}_k \end{bmatrix}$ based on the measurements up until time step k, while $\mathrm{P}_{k|k}$ complies with the subsequent generalized Riccati difference equation.

$$\mathrm{P}_{k+1|k+1} = \left(\mathrm{G}^{\mathrm{T}}\left(\Delta t \cdot Q + H\mathrm{P}_{k|k}H^T\right)^{-1} G + M^T R^{-1} M\right)^{-1} \tag{13.32}$$

To summarize, in the process of estimating the steering angle through the observation of HDV, we initially compute the Jacobian matrix along the current state during each time step. Subsequently, the control input (speed and steering angle) is estimated employing Eq. (13.32), followed by the update of the estimation error covariance using Eq. (13.32).

13.6.3 Evasion Distance Propagation Algorithm-Based Weak Deadlock Detection

In the case where no cycle exists within the blockage graph, the identification of deadlock becomes unnecessary, as the presence of a loop is a prerequisite for deadlock occurrence. Deadlock may be present if a loop is detected. An evasive distance propagation algorithm (EDP)-based model for weak deadlock detection is devised. By definition, the detection process can be transformed into the verification of the condition $\mathrm{l}_{j^{\mathbb{L}_k}} = \Im_{j\rightsquigarrow j,\mathbb{L}_k}\left(\widehat{\mathrm{l}_{j,j+1}}\right) \geq \overline{\mathrm{l}_{j\to j+1}}$ for any arbitrary vehicle j along loop $\mathbb{L}_k$. To initiate the process, a random vehicle in the loop is chosen as the starting point. For the sake of simplicity, let us assume that the starting vehicle has an index of 1, and the total number of vehicles in loop $\mathbb{L}_k$ is denoted as $|\mathbb{L}_k|$. The calculation of $\Im_{1\rightsquigarrow 1,\mathbb{L}_k}\left(\widehat{\mathrm{l}_{1\to 2}}\right)$ and $\overline{\mathrm{l}_{1\to 2}}$ can be fragmented throughout the entire loop:

$$\Im_{1\rightsquigarrow 1,\mathbb{L}_k}\left(\widehat{\mathrm{l}_{1\to 2}}\right) = \mathcal{L}_{1\to 2}\left(\ldots \mathcal{L}_{|\mathbb{L}_k|-3\to|\mathbb{L}_k|-2}\left(\mathcal{L}_{|\mathbb{L}_k|-2\to|\mathbb{L}_k|-1}\left(\mathcal{L}_{|\mathbb{L}_k|\to 1}\left(\widehat{\mathrm{l}_{1\to 2}}\right)\right)\right)\ldots\right) \tag{13.33}$$

Henceforth, the iterative computation of $s\mathcal{L}_{j\to j+1}(\cdot)$, $\forall j$, for each j in backward progression throughout the loop, commencing from vehicle 1, is imperative, whilst also considering the provided value of $\widehat{l_{1\to 2}}$. The aforementioned calculation is to be iterated $|\mathbb{L}_k|$ times. The denotation employed to represent the process of evaluating $\Im_{1\rightsquigarrow 1,\mathbb{L}_k}(\cdot)$ is evasion distance propagation algorithm, commonly referred to as EDP. In the event that, during the execution of EDP, an awareness arises that the evasion distance for a specific vehicle can be achieved within the confines of its authorized travel distance, EDP may be terminated, assuring the absence of any deadlock, despite the presence of a loop. The pseudo code for EDP is presented in Table 13.1 for reference.

The above procedures are designed for one-by-one blockage case. For one-by-many case, the extension is easy, which is shown in Table 13.2.

Table 13.1 Pseudo code of EDP for one-by-one blockage

Input	• The information of all vehicles. It includes the information of all CAVs and HDVs. The information of CAV are speed and steering angle, while the information of the HDVs are real-time coordinates
Output	• Weak deadlock set $\mathcal{D}$ • The vehicles within deadlock
	Initialize deadlocks set $\mathcal{D} = \emptyset$
	Estimate the steering angle of all HDVs;
	Construct the BG($\mathfrak{N}$, $\mathfrak{E}$) using method described in Sect. 0;
	IF loops exist in BG($\mathfrak{N}$, $\mathfrak{E}$):
	Identify all loops in BG($\mathfrak{N}$, $\mathfrak{E}$) sas set $\mathcal{L} = \{\mathbb{L}_k\}$;
	FOR each loop $\mathbb{L}_k$ in $\mathcal{L}$:
	Randomly select one vehicle (say vehicle 1 without loss of generality) to check deadlock existence;
	Calculate $\overline{l_{1\to 2}}$ and $\widehat{l_{1\to 2}}$;
	Calculate $\Im_{1\rightsquigarrow 1,\mathbb{L}_k}\left(\widehat{l_{1\to 2}}\right)$;
	IF $\Im_{1\rightsquigarrow 1,\mathbb{L}_k}\left(\widehat{l_{1\to 2}}\right) < \overline{l_{1\to 2}}$:
	$\mathcal{D} = \mathcal{D} \cup \{\mathbb{L}_k\}$;
	ELSE
	No deadlock is found
	RETURN $\mathcal{D}$;

Table 13.2 Pseudo code of EDP for one-by-many blockage

Input	• The same as Table 13.1
Output	• The same as Table 13.1
	Initialize deadlocks set $\mathcal{D} = \emptyset$ Estimate the speed and the steering angle of all HDVs; Construct the BG($\mathfrak{N}$, $\mathfrak{E}$) using method described in Sect. 0; IF loops exist in BG($\mathfrak{N}$, $\mathfrak{E}$): Identify all loops in BG($\mathfrak{N}$, $\mathfrak{E}$) as set $\mathcal{L} = \{\mathbb{L}_k\}$; FOR each vehicle j, determine blocking vehicles set C_j; Randomly select one vehicle (say vehicle 1 without loss of generality) to check deadlock existence; Calculate $\overline{l_{1\to C_1}}$ and $\widehat{l_{1\to C_1}}$; IF exist loop $\mathbb{L}_k$ satisfies $\Im_{1\rightsquigarrow 1,\mathbb{L}_k}\left(\widehat{l_{1\to 2}}\right) < \overline{l_{1\to 2}}$: $\mathcal{D} = \cup\{\mathbb{L}_m\}$ ELSE No deadlock is found RETURN $\mathcal{D}$;

13.6.4 *Evasion Distance Propagation Algorithm-Based Strong Deadlock Detection*

13.6.4.1 Steering Angle Dependent Blockage Graph Construction

The issue of strong deadlock presents a greater degree of complexity compared to the fixed steering angle scenario. This is due to the fact that changes in the steering angle directly impact the blockage graph, rendering the existing EDP designed for weak deadlock (where the blockage graph remains static) unsuitable for direct application. An illustrative example of the blockage graph construction under variable steering angles is depicted in Fig. 13.11a. In this example, Vehicle 0 has a range of steering angle options, spanning from δ_0^{f} to $\delta_0^{f''}$ (with positive values indicating left turns and negative values indicating right turns). If the steering angle falls within the range of $\left(\delta_0^{f''}, \delta_0^{f'}\right)$, Vehicle 0 will encounter a blockage from Vehicle 2. On the other hand, if the steering angle falls within $\left[\delta_0^{f''}, \delta_0^{f}\right]$, Vehicle 0 will be blocked by Vehicle 1. Consequently, a steering-angle-dependent blockage graph is formulated, as shown in Fig. 13.11b, where each edge represents a specific interval denoting the blockage requirement. The blockage graph is then decomposed based on the intervals assigned to each edge, resulting in the formation of distinct sub-blockage-graphs. The outcomes of this decomposition process are presented in Fig. 13.11c and d. Figure 13.11c depicts the scenario where Vehicle 0's steering angle resides within the interval $\left(\delta_0^{f}{}', \delta_0^{f}\right)$, leading to the occurrence of a blockage solely between Vehicle 1 and Vehicle 0. In Fig. 13.11d, Vehicle 0's steering angle is defined within the interval $\left(\delta_0^{f''}, \delta_0^{f'}\right)$, resulting in blockages from both Vehicle 1 and Vehicle 2. In

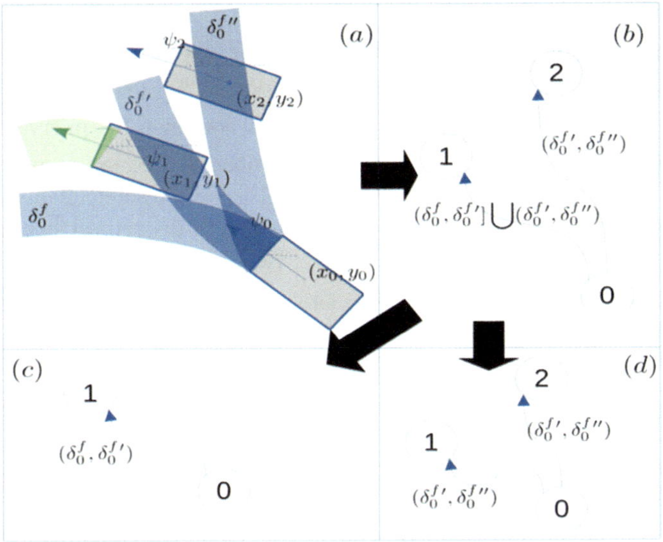

Fig. 13.11 The decomposition of the blockage graph topology

summary, Vehicle 0 exhibits two distinct sub-blockage-graphs, as demonstrated in Fig. 13.11c and d.

The aforementioned division procedure of a graph obstructed by obstacles is universalized. Let it be assumed that the directional angles of a standard vehicle "j" are subdivided into consecutive intervals, denoted as $\left(\delta^{f}_{j,m}, \overline{\delta^{f}_{j,m}}\right)$ for the m-th interval. The quantity of intervals for vehicle "j" is denoted as $\left]\delta^{f}_{j}\right[$. Consequently, the overall number of resulting subgraphs obstructed by obstacles is equal to the product of all quantities $\prod_{j}\left]\delta^{f}_{j}\right[$. The collection of all these obstructed graphs is represented as $\overleftrightarrow{BG}\left(\mathfrak{N}, \mathfrak{E}|\delta^{f}\right) = \left\{BG(\mathfrak{N}, \mathfrak{E})\middle|\left\{\left(\underline{\delta^{f}_{j,m'}}, \overline{\delta^{f}_{j,m'}}\right)\right\}, \forall j, m\right\}$ Each particular obstructed graph $BG(\mathfrak{N}, \mathfrak{E})$ is distinct from another obstructed graph $BG(\mathfrak{N}, \mathfrak{E})$ in that the edges within the former are assigned intervals in a precise manner. Furthermore, the intervals assigned to edges originating from the same node (or equivalently vehicle) are identical.

13.6.4.2 Strong Deadlock Detection

To ascertain the existence of a strong deadlock, it is imperative to evaluate the deadlock condition for each graph $BG(\mathfrak{N}, \mathfrak{E})$ within the collection $\overleftrightarrow{BG}\left(\mathfrak{N}, \mathfrak{E}|\delta^{f}\right)$. The determination of a strong deadlock is contingent upon all blockage graphs meeting the stipulated condition in Eq. (13.22). Consequently, should there be a blockage graph devoid of any cyclic paths, unlocking the deadlock becomes feasible.

In order to describe the process of detecting strong deadlocks, we will describe a typical sub-blockage-graph BG($\mathfrak{N}$, $\mathfrak{E}$). Specifically, we refer to the steering angles interval for a vehicle labelled "j" in a given loop within the sub-blockage-graph as the closed interval $\left(\underline{\delta_j^f}, \overline{\delta_j^f}\right)$. It is important to note that this interval is continuous, and as such, there exists an infinite number of steering angles within it. To address this issue, we discretize the interval using a discrete step size of $\Delta\delta^f$. Consequentially, for each vehicle j, we obtain a set of discretized steering angles, which we label as.$\delta_{j,1}^f$, $\delta_{j,2}^f$, $\delta_{j,3}^f$ $\ldots\delta_{j,|m_j|}^f$ Here,$|m_j|$ represents the total number of discretized steering angles. To identify cases of strong deadlock, we need to check the relationship between the following two quantities for all δ^f (a vector containing all steering angles of vehicles in the loop), index term (i.e., $\alpha \in \{1, 2, \ldots |m_j|\}$, $u \in \{1, 2, \ldots |m_j|\}$, and $\beta \in \{1, 2, \ldots |m_{j+1}|\}$ in the following Eq. (13.34)a, and the definition presented in Definition 3.

$$\begin{cases} \mathfrak{K}_{j\rightsquigarrow j,\mathbb{L}_k}\left(\widehat{l_{j\to j+1}}|_{\delta_{j,\alpha}^f}, \delta_{j,u}^f, \boldsymbol{\delta^f}\right) & (a) \\ \overline{l_{j\to j+1}}|_{\delta_{j,\alpha}^f,\delta_{j+1,\beta}^f} & (b) \end{cases} \tag{13.34}$$

In cases where Eq. (13.34)a < Eq. (13.34)b holds true for all α, u, β, and δ^f, there exists a strong deadlock. The degree of computational effort required increases with the number of vehicles and the number of discretized steering angles. While the calculation of Eq. (13.34)b is straightforward, the calculation of Eq. (13.34)a is more difficult, as it must be performed cyclically along the loop and traverse the steering angles of each vehicle, beginning with a selected vehicle j.

Drawing inspiration from the dynamic programming methodology, we propose an EDP method for the detection of deadlock with variable steering angles. The approach involves checking the maximum distance propagated by each vehicle in the loop and determining whether it leads to a deadlock, i.e., if the cyclically propagated distance is greater than the permitted travel distance. If the maximum distances result in deadlock, all other distances, which are necessarily shorter and therefore release less space for the upstream vehicles in the same loop, would also result in deadlock. Figure 13.12 provides an illustration, where the arrow indicates the direction of vehicle movement. For simplicity, only a portion of the loop is shown, with each column representing the discretized steering angle values (note that the number of discretized steering angle values may differ for each vehicle). Beginning from vehicle j, the strong deadlock detection process involves propagating the permitted travel distance $\widehat{l_{j\to j+1}}|_{\delta_{j,\alpha}^f}$ for a specified index term $\alpha \in \{1, 2, \ldots |m_j|\}$ backward along the loop. The propagated travel distance for vehicle j-1, the upstream vehicle for vehicle j, if it drives using the steer angle of $\delta_{j-1,\beta}^f$ is expressed as follows:

$$l_{j-1}|_{\beta,\alpha} = \mathcal{K}_{j-1\to j}\left(\widehat{l_{j\to j+1}}|_{\delta_{j,\alpha}^f}, \delta_{j-1,\beta}^f, \delta_{j,\alpha}^f\right) \tag{13.35}$$

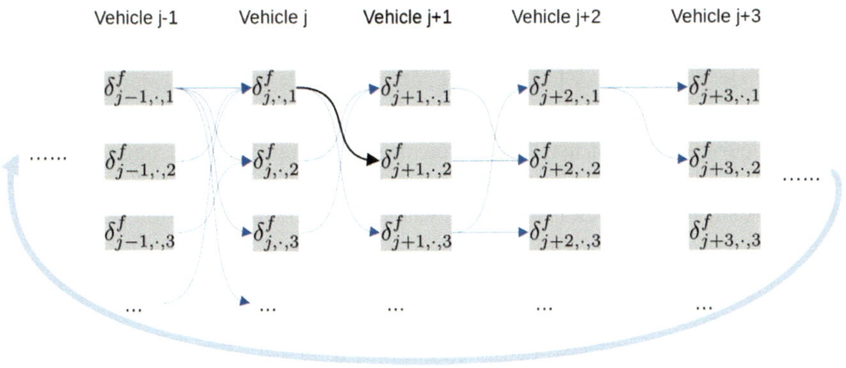

Fig. 13.12 EDP for type II deadlock detection

The propagated travel distance of vehicle j–1, denoted as $l_{j-1}|_{\beta,\alpha}$, represents the distance traveled by vehicle j–1 under the condition of β, assuming vehicle j travels using steering angle $\delta^f_{j,\alpha}$. In a similar manner, the propagated travel distance of vehicle j–2 can be expressed as follows when it is driven using steering angle $\delta^f_{j-2,\gamma}$:

$$l_{j-2}|_{\gamma,\beta} = \mathcal{K}_{j-2\to j-1}\left(l_{j-1}|_{\beta}, \delta^f_{j-2,\gamma}, \delta^f_{j-1,\beta}\right) \tag{13.36}$$

It is evident that different β values will result in varied $l_{j-2}|_{\gamma}$ distances. Thus, we define the maximal propagated travel distance as follows:

$$l_{j-2}|_{\gamma} = \max_{\beta} l_{j-2}|_{\gamma,\beta} = \max_{\beta} \mathcal{K}_{j-2\to j-1}\left(l_{j-1}|_{\beta}, \delta^f_{j-2,\gamma}, \delta^f_{j-1,\beta}\right) \tag{13.37}$$

If this distance leads to a deadlock, it implies that all other β values will also result in a deadlock. This can be attributed to the fact that other β values would cause vehicle j–2 to travel a shorter distance, consequently releasing less space, ultimately leading to a deadlock. We can express the aforementioned maximal propagated travel distance as follows:

$$l_{j-2}|_{\gamma} = \mathcal{W}_{j-2\rightsquigarrow j,\mathbb{L}_k}\left(l_j, \delta^f_{j-2,\gamma}, \delta^f_{j,\alpha}\right) \tag{13.38}$$

The physical interpretation of $\mathcal{W}_{j-2\rightsquigarrow j,\mathbb{L}_k}\left(l_j, \delta^f_{j-2,\gamma}, \delta^f_{j,\alpha}\right)$ is the maximum travel distance that vehicle j–2 can cover when employing the steering angle $\delta^f_{j-2,\gamma}$, considering that the terminal vehicle j travels a distance of l_j using the steering angle $\delta^f_{j,\alpha}$, regardless of the angle used by vehicle j–1. Based on the definition of $\mathcal{W}(\cdot,\cdot,\cdot)$, we can establish the following one-step recursive equation for a general vehicle

$$\begin{aligned} &\mathcal{W}_{\Gamma-1\rightsquigarrow \mathrm{j},\mathbb{L}_k}\left(\mathrm{l_j}, \delta^{\mathrm{f}}_{\Gamma-1,\gamma}, \delta^{\mathrm{f}}_{\mathrm{j},\alpha}\right) \\ &= \max_{\omega} \mathcal{K}_{\Gamma-1\rightarrow\Gamma}\left(\mathcal{W}_{\Gamma\rightsquigarrow \mathrm{j},\mathbb{L}_k}\left(\mathrm{l_j}, \delta^{\mathrm{f}}_{\Gamma,\omega}, \delta^{\mathrm{f}}_{\mathrm{j},\alpha}\right), \delta^{\mathrm{f}}_{\Gamma-1,\gamma}, \delta^{\mathrm{f}}_{\Gamma,\omega}\right) \end{aligned} \tag{13.39}$$

The recursive equation used in dynamic programming shares a strong resemblance with Eq. (13.39), serving as the basis for designing a strong deadlock detection system. A strong deadlock is identified when the following condition is met:

$$\mathcal{W}_{\mathrm{j+1}\rightsquigarrow \mathrm{j},\mathbb{L}_k}\left(\widehat{\mathrm{l_{j\rightarrow j+1}}}|_{\delta^{\mathrm{f}}_{\mathrm{j},\alpha}}, \delta^{\mathrm{f}}_{\mathrm{j+1},\omega}, \delta^{\mathrm{f}}_{\mathrm{j},\alpha}\right) \le \overline{\mathrm{l_{j\rightarrow j+1}}}|_{\delta^{\mathrm{f}}_{\mathrm{j},\alpha}, \delta^{\mathrm{f}}_{\mathrm{j+1},\omega},}, \forall \alpha, \omega \tag{13.40}$$

The left-hand side (LHS) of the condition Eq. (13.40) indicates the maximum distance vehicle $\mathrm{j}+1$, the downstream vehicle of vehicle j, can travel when vehicle j has reached its maximal permitted distance and released as much space as possible for upstream vehicles. (Given the loop structure, vehicle $\mathrm{j}+1$ is also considered the upstream vehicle of vehicle j.) Vehicle j becomes locked by itself once Eq. (13.40) is valid, rendering it unable to release space along the loop for self-evasion.

The pseudo code of the EDP for strong deadlock is given in Table 13.3. It is important to note that the execution of the last statement in the procedure implies that none of the "No strong deadlock" conditions were satisfied, indicating the existence of a strong deadlock.

13.7 Cooperative Deadlock Avoidance and Recovery Protocol

The primary objective of the deadlock protocol is to prevent the occurrence of a deadlock by utilizing information obtained from sensors. Conversely, if a deadlock has already transpired, the protocol should aim to restore the flow of traffic to its normal state using cooperative CAV movement. Hence, the design of the deadlock protocol necessitates the consideration of two distinct scenarios: one pertaining to the imminent possibility of a deadlock, and the other concerning an existing deadlock. The figure below showcases the protocols devised for both avoidance and recovery of dead (Fig. 13.13).

13.7.1 Deadlock Avoidance Protocol

The phenomenon of deadlock in vehicular traffic is identified by the characteristic loop-style waiting behavior. To circumvent this problem, it has been proposed that one effective approach is to counteract the blocking behavior between vehicles caused by the loop structure. To this end, two different strategies have been devised: chain spillover free and loop free strategies.

Table 13.3 Pseudo code for type II deadlock

Input	• The information of all vehicles. It include the information of all CAVs and HDVs. The information of CAV are speed and front steering angle, while the information of the HDVs are real-time coordinates
Output	• Strong deadlock set $\mathcal{D}$ • The vehicles within deadlock
	Initialize deadlocks set $\mathcal{D} = \emptyset$
	Estimate the speed and the steering angle of all HDVs;
	Construct the extend blockage graph and decompose it into sub-blockage graphs, i.e.:
	$\overleftrightarrow{\mathbf{BG}}\left(\mathfrak{N}, \mathfrak{E} \vert \delta^f\right) = \left\{\mathbf{BG}(\mathfrak{N}, \mathfrak{E}) \middle\vert \left\{\left(\underline{\delta^f_{j,m}}, \overline{\delta^f_{j,m}},\right)\right\}, \forall j, m\right\}$;
	IF there exist certain BG($\mathfrak{N}$, $\mathfrak{E}$) without loop:
	RETURN "**No strong deadlock**"
	FOR BG($\mathfrak{N}$, $\mathfrak{E}$) in $\overleftrightarrow{\mathbf{BG}}\left(\mathfrak{N}, \mathfrak{E} \vert \delta^f\right)$ if there is loop in BG($\mathfrak{N}$, $\mathfrak{E}$):
	FOR $\mathbb{L}_k$ in BG($\mathfrak{N}$, $\mathfrak{E}$)
	Discretize the steering angles interval $\left(\underline{\delta^f_j}, \overline{\delta^f_j},\right)$ for each vehicle into $\delta^f_{j,1}, \delta^f_{j,2}, \ldots$;
	Random select one vehicle j (say j = 1) to start the deadlock detection process;
	FOR α in $\{1, 2, 3 \ldots \vert m_1 \vert\}$ and β in $\{1, 2, 3 \ldots \vert m_2 \vert\}$:
	SET $\mathcal{W}_{1 \rightsquigarrow 2, \mathbb{L}_k}\left(\overline{l_{1\to 2}}\vert_{\delta^f_{1,\alpha}, \delta^f_{2,\beta}}, \vert_{\delta^f_{1,\alpha}, \delta^f_{2,\beta}}\right) = \mathcal{K}_{1\to 2}\left(\overline{l_{1\to 2}}\vert_{\delta^f_{1,m_1,\alpha}, \delta^f_{2,m_2,\beta}}, \vert_{\delta^f_{1,m_1,\alpha}, \delta^f_{2,m_2,\beta}}\right)$
	FOR j in $\{3, 4, 5, \ldots.\} \in \mathbb{L}_k$:
	FOR α in $\{1, 2, 3 \ldots \vert m_1 \vert\}$,$\beta$ in $\{1, 2, 3 \ldots \vert m_2 \vert\}$, $\in$ in $\left\{1, 2, 3 \ldots \left\vert m_{j-1} \right\vert\right\}$, γ in $\left\{1, 2, 3 \ldots \left\vert m_j \right\vert\right\}$:
	Calculate $\mathcal{W}$ iteratively using:
	$\mathcal{W}_{1 \rightsquigarrow j, \mathbb{L}_k}\left(\overline{l_{1\to 2}}\vert_{\delta^f_{1, m_1, \alpha'}, \delta^f_{2, m_2, \beta}}, \delta^f_{1, m_1, \alpha}, \delta^f_{j, m_j, \gamma}\right) =$
	$\min_{\in} \mathcal{K}_{j-1\to j}\left(\begin{matrix} \mathcal{W}_{1 \rightsquigarrow j, \mathbb{L}_k}\left(\overline{l_{1\to 2}}\vert_{\delta^f_{1,m_1,\alpha}, \delta^f_{2,m_2,\beta}}, \delta^f_{1,m_1,\alpha}, \delta^f_{j-1,m_{j-1},\in}\right), \\ \delta^f_{j-1,m_{j-1},\in}, \delta^f_{j,m_j,\gamma} \end{matrix}\right)$
	Calculate permitted traveling distance of vehicle j, i.e. $\widehat{l_{j\to j+1}}\vert_{\delta^f_{j,m_j,\gamma}}$
	IF $\exists \alpha$, β, γ, that the following holds:
	$\mathcal{W}_{1 \rightsquigarrow j, \mathbb{L}_k}\left(\overline{l_{1\to 2}}\vert_{\delta^f_{1,m_1,\alpha}, \delta^f_{2,m_2,\beta}}, \delta^f_{1,m_1,\alpha}, \delta^f_{j,m_j,\gamma}\right) \le \widehat{l_{j\to j+1}}\vert_{\delta^f_{j,m_j,\gamma}}$
	RETURN "**No strong deadlock**"
	RETURN "Strong deadlock"

The chain spillover free strategy is illustrated in Figs. 13.14a and 13.15, where each movement represents one or multiple vehicles. In Fig. 13.14a, Movement 2 blocks Movement 1 and thereby incurs a spillover of Movement 2. If Movement 3 blocks Movement 2, a chain spillover transpires, resulting in the spillover of Movements 3 and 2. As a chain-style spillover can lead to a deadlock, it is necessary to avoid such

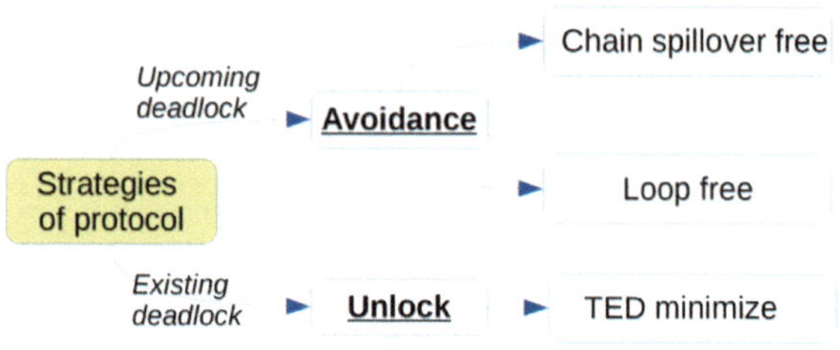

Fig. 13.13 Strategies of deadlock protocol

a scenario. A fundamental strategy to evade the chain-style spillover is displayed in Fig. 13.15. Vehicle 3 and Vehicle 4 constitute one movement, while Vehicle 2 and Vehicle 1 belong to another movement. The third movement only involves Vehicle 0. Vehicle 1 has yet to join the queue and thus conflicts with Vehicle 0 for priority. If preference is given to Vehicle 1, then the vehicle arrangement might evolve into the situation in Fig. 13.15b, which is an instance of chain spillover. Therefore, to circumvent this problem, Vehicle 1 must cede the right of way to Vehicle 0, as demonstrated in Fig. 13.15c, to promote an uninterrupted flow of traffic.

Both the chain spillover free strategy and the loop free strategy are developed based on two key elements: spillover identification and CAV (Connected and Automated Vehicle) control. Figure 13.16 illustrates the process involved. The initial steps align with those outlined in Fig. 13.10. Once the blockage graph is constructed, the identification of spillovers occurs. Subsequently, any conflicts arising between CAVs or between CAVs and HDVs (Human-Driven Vehicles) are identified. The prediction of spillover evolution follows, with the assumption that priority is assigned to each conflicting vehicle. In the event that the spillover develops into either a chain-spillover or a loop-spillover, a give-way command is issued to the CAVs.

To prevent the occurrence of chain spillovers or loop spillovers in CAVs, adjustments to their deceleration are necessary. If the distance of the CAV in question to the confliction area is represented as $l_{\#}$ and its current speed as $v_{\#}$, the deceleration required to bring it to a stop in front of the confliction area can be accurately calculated using Eq. (13.41). The CAV's motion can then be regulated using the vehicle dynamic model defined by Eq. (13.2), which utilizes the aforementioned result as an input. In the case of connected but non-automated vehicles, where direct control over acceleration may be unavailable, the vehicle can be stopped until the vehicle

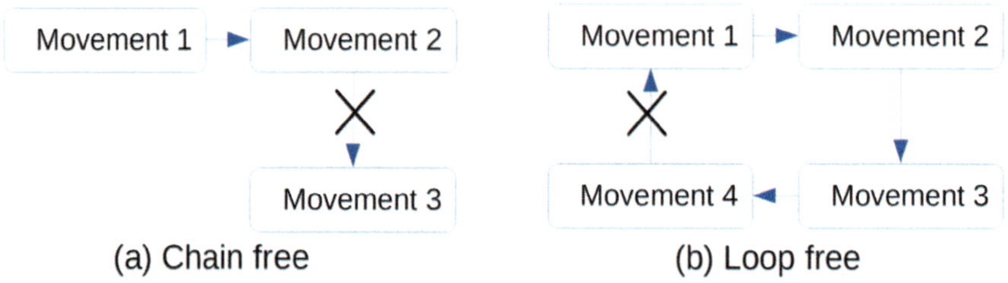

Fig. 13.14 Illustration of deadlock avoidance strategy

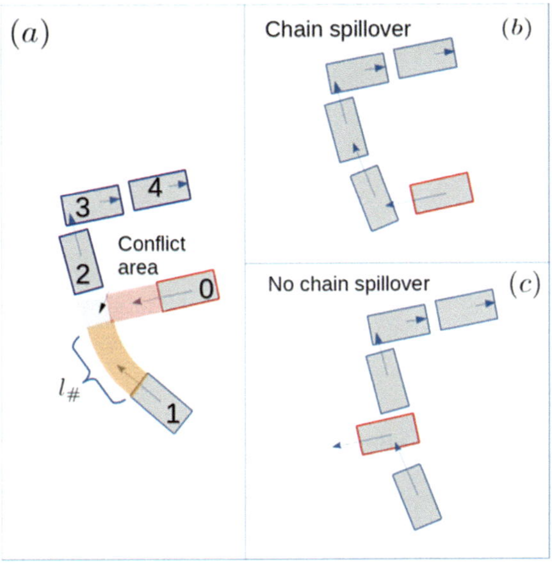

Fig. 13.15 Chain spillover free strategy

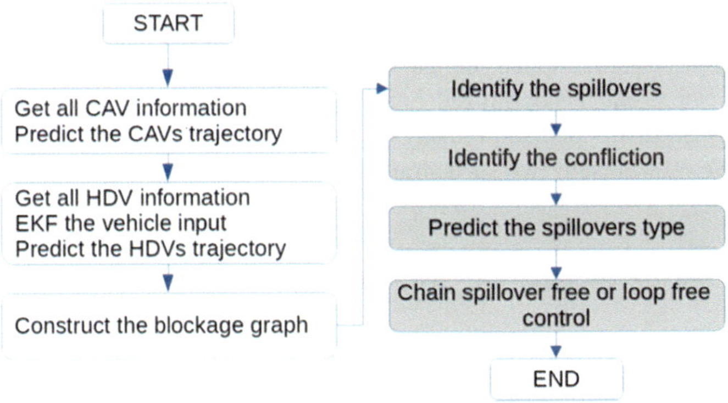

Fig. 13.16 Chain spillover free and loop free strategy process

with priority has successfully passed the confliction point.

$$a_\# = \frac{v_\#^2}{2l_\#} \tag{13.41}$$

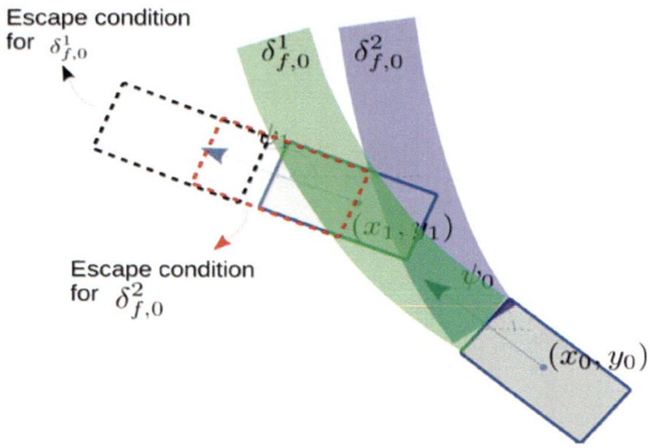

Fig. 13.17 Reroute strategy for type I deadlock

13.7.2 Cooperative Recovery Protocol

When faced with an existing deadlock, it becomes necessary to resolve it. The resolution of this deadlock requires coordinated movement of the involved vehicles. In this context, an unlock plan is defined as a composite of three distinct sequences: (1) a sequence of vehicles that must move sequentially to create enough space for the unlocking process, (2) a sequence of the corresponding steering angles of these vehicles, and (3) a sequence specifying the travel distances of these vehicles. These three sequences represent the optimized outcomes derived from specific unlock strategies. An illustrative video showcasing an actual unlocking sequence is available on github.[1] We now present two such strategies: reroute and backward-forward.

13.7.2.1 Unlock Strategy for Weak Deadlock: Reroute

In cases where a strong deadlock is not present, there exists a specific combination of steering angles that can guide the vehicles within the loop out of deadlock. Consequently, it becomes necessary for certain vehicles within the loop to adhere to corresponding steering angles and travel predetermined distances in a prescribed sequence, thus eliminating the occurrence of weak deadlock. The fundamental principle can be seen in Fig. 13.17, where Vehicle 0 (represented by the grey box) has two options for steering angles: $\delta^1_{f,0}$ and $\delta^2_{f,0}$. When $\delta^1_{f,0}$ is employed, Vehicle 1 (indicated by the green box) obstructing Vehicle 0 must advance to the location denoted by the black dashed box. Conversely, if $\delta^2_{f,0}$ is chosen, Vehicle 1 only needs to move to the red dashed box, requiring a shorter driving distance.

Theoretical considerations entail a multitude of steering options that necessitate the selection or optimization of final angles for each CAV. Subsequently, resolving the issue of CAV rerouting to alleviate deadlocks requires the simultaneous search for optimal steering angles and optimal driving distances. While a straightforward

[1] Unlock_realworldcase.mp4, https://github.com/QHS2020/deadlocksubmissionmaterials.

approach involves random selection of steering angles using the EDP algorithm, the resulting outcome may not be optimal in terms of the unlocking process. An alternative method involves employing Distance Escaping from Deadlock (DED) as a performance metric for determining the optimal rerouting plan. The DED is obtained by summing the travel distances of vehicles within the loop, where the corresponding steering angles enable deadlock resolution. For a loop denoted as $\mathbb{L}_k$ with a set of vehicles {1,2,3,...}, assuming the combinations of steering angles capable of unlocking the deadlock are represented by δ_j^f, $j = 1, 2, 3...$, the escaping distance of vehicle j (the initiating vehicle for the unlocking process) is denoted as $\overline{l_{j\to j+1}}$. Subsequently, the DED can be calculated as follows:

$$\mathrm{PI}_{\mathrm{DED}}^{1}\left(\left\{\delta_j^f, \delta^f\right\}\right) = \sum_{\substack{m \in \mathbb{L}_k \\ m \neq j}} \overline{l_{j\to j+1}} + \mathfrak{K}_{j\rightsquigarrow m,\mathbb{L}_k}\left(\widehat{l_{j\to j+1}}, \delta_j^f, \delta^f\right) \tag{13.42}$$

Minimize $\mathrm{PI}_{\mathrm{DED}}^{1}$ will result the optimal steering angles, i.e. $\left\{\delta_j^{f*}, \delta^{f*}\right\}$

$$\left\{\delta_j^{f*}, \delta^{f*}\right\} = \underset{\left\{\delta_j^f, \delta^f\right\}}{\mathrm{argmin}}\, \mathrm{PI}_{\mathrm{DED}}^{1}\left(\left\{\delta_j^f, \delta^f\right\}\right) \tag{13.43}$$

By discretizing the decision variable, δ_j^f, for each vehicle within the loop $\mathbb{L}_k$, problem Eq. (13.43) can be transformed into a multi-stage decision problem, with the number of stages coinciding with the amount of vehicles present within the loop. The permissible actions available at each given stage correspond to discretized steering angles. A standard dynamic programming approach can be employed to solve Eq. (13.43), however, a detailed step-by-step solution will not be provided at this time.

13.7.2.2 Unlock Strategy for Strong Deadlock: Backward-Forward Movement

According to the definition of a strong deadlock, it is impossible to unlock the deadlock by any combination of steering angles. The sole option available is to employ a backward-forward movement technique, whereby certain vehicles first move backward and subsequently move forward. The fundamental concept is demonstrated in Fig. 13.18. In the deadlock scenario, vehicle 0 is obstructed by vehicle 1. To escape from the deadlock, vehicle 0 must initially execute a backward movement utilizing the steering angle $\delta_{0,2}^f$ and subsequently perform a forward movement using the steering angle $\delta_{0,3}^f$. Following the successful implementation of the backward-forward strategy by the chosen vehicle, the remaining vehicles can then proceed sequentially.

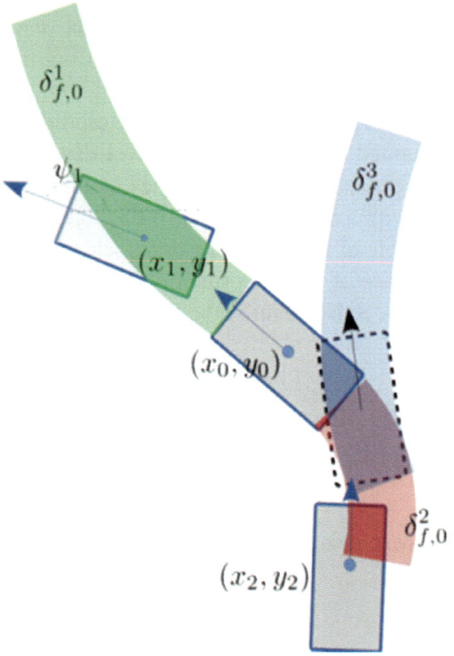

Fig. 13.18 The protocol of backward-and-forward

The intricate steps involved in solving this issue consist of identifying the appropriate vehicle for executing the backward movement, determining the four decision variables associated with this vehicle, specifically the angle and distance of the backward and forward movements, and computing the steering angles of the remaining vehicles. Likewise, akin to the reroute strategy, the DED serves as the selected performance indicator. If a deadlock occurs within the loop $\mathbb{L}_k$, the decision variables can be itemized as follows: (a) the vehicle (j) responsible for executing the backward movement and its four parameters, namely, the steering angle and distance for the backward movement, as well as the steering angle and distance for the forward movement $\delta_j^{f-}, l_j^-, \delta_j^{f+}, l_j^+$; and (b) the steering angles of the remaining vehicles that form the loop $\mathbb{L}_k$, denoted as $\delta^f = \{\delta_m^f\}$, $m \neq j$. Accordingly, we can represent the resolution of the deadlock.

Consequently, we can express the unlock problem by:

$$\mathrm{PI}^2_{\mathrm{DED}}\left(\left\{j, \delta_j^{f-}, l_j^-, \delta_j^{f+}, \delta^f\right\}\right) = \sum_{\substack{m \in \mathbb{L}_k \\ m \neq j}} l_j^- + l_j^+ + \mathfrak{K}_{j \rightsquigarrow m, \mathbb{L}_k}\left(l_j^+, \delta_j^f, \delta^f\right) \tag{13.44}$$

Minimizing $\mathrm{PI}^2_{\mathrm{DED}}$ will result the optimal steering angles and driving distance for deadlock unlock, i.e. $\left\{\delta_j^{f^*}, \delta^{f^*}\right\}$

Fig. 13.19 The test intersection in Carla

$$\left\{\delta_j^{f-^*}, 1_j^{-*}, \delta_j^{f+^*}, 1_j^{+*}, \delta^{f*}\right\} = \underset{\left\{j, \delta_j^{f-}, 1_j^{-}, \delta_j^{f+}, \delta^{f}\right\}}{\operatorname{argmin}} \mathrm{PI}_{\mathrm{DED}}^{2}\left(\left\{j, \delta_j^{f-}, 1_j^{-}, \delta_j^{f+}, \delta^{f}\right\}\right) \tag{13.45}$$

13.8 Numeric Analysis

13.8.1 Scenario Setup

Numeric analysis is conducted within the context of the Carla simulation platform, which is an opensource software tool employed for the evaluation of autonomous vehicles. Whilst empirical experimentation is desirable, identifying an intersection that features repeated vehicular gridlocks in the real world is implausible. To garner sufficient data points for analysis, a standard four-leg intersection, as depicted in Fig. 13.19, has been constructed for the purposes of experimentation. Each intersection approach is comprised of three turning lanes, and the incoming and outgoing roads are identical across all four approaches. The width of each lane has been set at 3.5 m. To create the deadlock condition, vehicle demand on the roadway segment is set beyond capacity, thereby producing the oversaturation scenario. Information related to the vehicles is then collected to infer the presence of any deadlocks, and if detected, eliminated using a proposed approach. Subsequently, visual examples of the deadlock detection and recovery process are presented, followed by an analysis highlighting the benefits of the suggested model.

13.8.2 Illustrative Examples of the Deadlock Detection

In this section, we present an illustration exemplifying the detection of deadlock. Figure 13.20 visually depicts a specific instance of weak deadlock occurring within an intersection. Figure 13.20 a provides a snapshot capturing the states of various vehicles when the deadlock manifests. Each vehicle is represented by a box, with directional arrows indicating their intended movements. Figure 13.20b displays the planned trajectories of the vehicles, while Fig. 13.20c showcases a screenshot from the Carla simulation. It is worth noting that the scenario involves a total of 14 vehicles, with only one vehicle being unblocked, as depicted in Fig. 13.20a. The remaining vehicles become immobilized, with three specific vehicles playing pivotal roles in the deadlock loop. These three vehicles are visually identified as the 0th, 1st, and 2nd vehicles by the deadlock detection protocol, as shown in Fig. 13.20a. Notably, this loop also constitutes a weak deadlock. The inter-vehicle restriction functions within the loop are detailed in Fig. 13.20d. Vehicle 0 is permitted to travel a distance of 1.84 m, while vehicles 1 and 2 can propagate distances of 2.31 m and 2.5 m, respectively. Moreover, the evasion condition for vehicle 0 (i.e., the distance vehicle 1 needs to travel) is determined to be 5.52 m, surpassing the 2.31 m threshold. Consequently, unless the three vehicles adjust their steering angles, a deadlock will inevitably ensue.

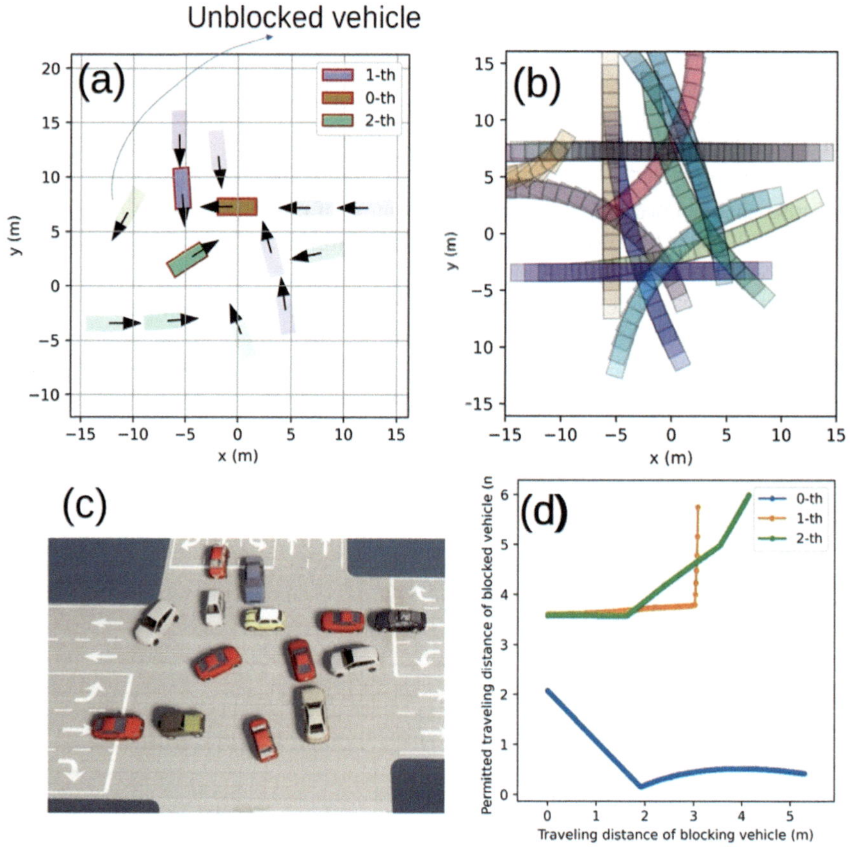

Fig. 13.20 The weak deadlock example

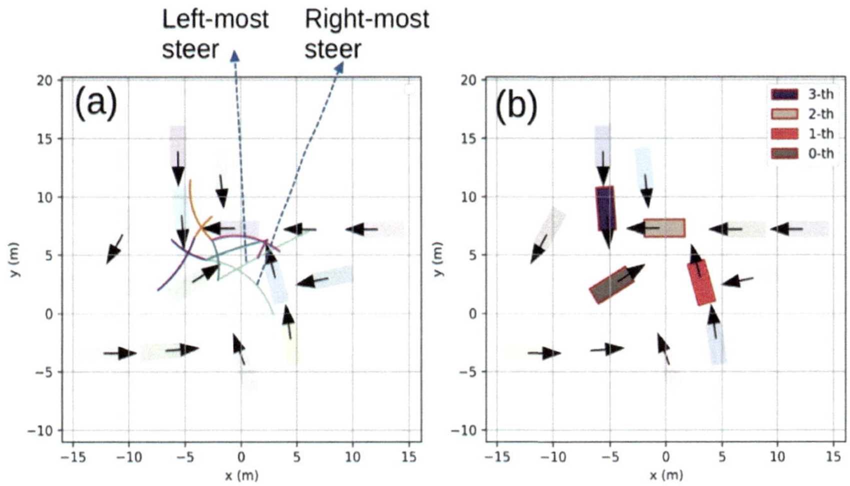

Fig. 13.21 Another loop for variable steers

We proceed to assess the status of the resulting deadlock in terms of its strength. The intervals for the steering angles are computed for each pair of obstructing vehicles according to Eqs. (13.9) and (13.10). Subsequently, the sub-graphs are extracted. It is worth noting that there is an additional loop consisting of four vehicles, as depicted in Fig. 13.21b. The identification of the vehicles within the loop follows the direction of the loop. The steer intervals responsible for causing the blockage are presented in Fig. 13.21a. For instance, in the case of the pair of vehicles denoted as 0 and 1 in Fig. 13.21b, the intervals of the extreme left and right steering directions are indicated in Fig. 13.21a. These intervals are visually represented by the trajectories of the corresponding vehicle corners. Specifically, the trajectory of the right-front corner represents the extreme left steering direction, while the trajectory of the left-front corner represents the extreme right steering direction. If vehicle 0 adopts any steering angle within this interval, it will encounter an obstruction from vehicle 1. The strong deadlock condition is examined for the loops presented in Figs. 13.20a and 13.21b. Both loops do not fulfill the criteria for a strong deadlock. Consequently, we conclude that only a weak deadlock exists and there is no presence of a strong deadlock.

The link[2] showcases a video showcasing the outcome of deadlock detection. Four representative snapshots have been extracted and depicted in Fig. 13.22. The trajectories have been computed (based on the vehicle model) and visually represented in real-time. The green hue denotes a normal state without any obstructions; the blue hue signifies a loop scenario (in other words, vehicles with a blue trajectory are part of a loop, as illustrated in Fig. 13.22b; the red hue indicates a weak deadlock scenario, and lastly, the black hue indicates a strong deadlock scenario. From Fig. 13.22a–d, a

[2] https://github.com/QHS2020/deadlocksubmissionmaterials, deadlockawareness.mp4.

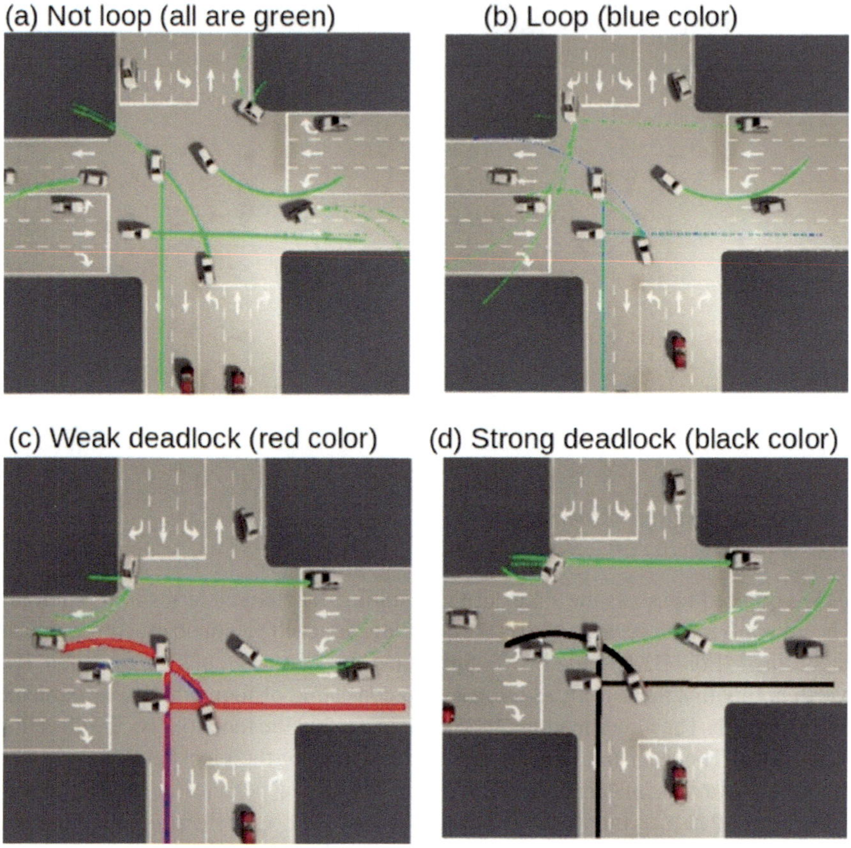

Fig. 13.22 The formation of a strong deadlock

strong deadlock materializes rapidly. Notably, in Fig. 13.22d, one can visually infer that the vehicles can maneuver away and there should be a rerouting plan to unlock the deadlock.

13.8.3 Benefits of the Proposed Model

To examine the advantages of the proposed model, we log the total number of vehicles traversing the intersection, also known as the intersection throughput. It is worth mentioning that in numerous studies, the intersection capacity is typically assessed at the stop line, beyond which vehicles are excluded from the traffic state analysis. However, in the event of a deadlock occurrence, vehicles may remain trapped within the intersection. Consequently, we document the count of vehicles exiting the intersection as a performance metric. We proceed to compare two scenarios: implementation of the deadlock avoidance protocol followed by the deployment of the unlock protocol.

The contrast between scenarios wherein a deadlock avoidance protocol is present and those devoid of said protocol is illustrated in Fig. 13.23. The bold orange line represents the scenario in which the avoidance protocol is implemented, while the blue line portrays the benchmark scenario without the incorporation of any avoidance protocol. The temporal scope spans approximately 10 min or 600 s. Initially, during the first 2 min, the two curves exhibit almost complete overlap. However, subsequent to this period, the orange curve experiences a rapid and substantial ascent that surpasses the trajectory of the alternative curve. This entails that the proposed deadlock avoidance protocol significantly augments the overall capacity of the depicted intersection. In terms of the ultimate number of vehicles throughout this 10-min simulation, the benchmark scenario yields 140 vehicles, whereas the orange curve employing the deadlock avoidance model demonstrates an escalation to 190 vehicles, thus producing a substantial 35.7% increase in intersection capacity.

Subsequently, a comparison was made between the situation wherein a deadlock recovery protocol was implemented (indicated by the thick orange line), and the situation wherein only a deadlock avoidance model was employed, but with no recovery capability (represented by the thin blue line). The findings were displayed in Fig. 13.24, which yielded a similar conclusion. Evidently, during the initial minute of analysis, the two curves were almost indistinguishable, but thereafter, the orange curve exhibited a rapid and substantial growth rate far exceeding that of the other

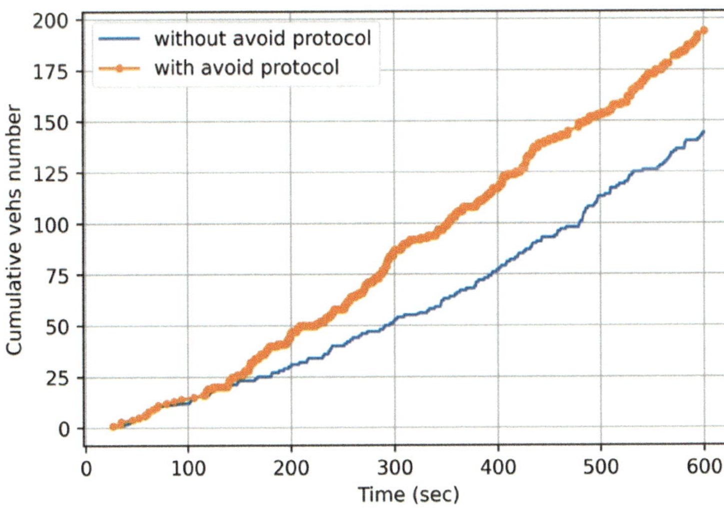

Fig. 13.24 The comparison between with and without avoid protocol

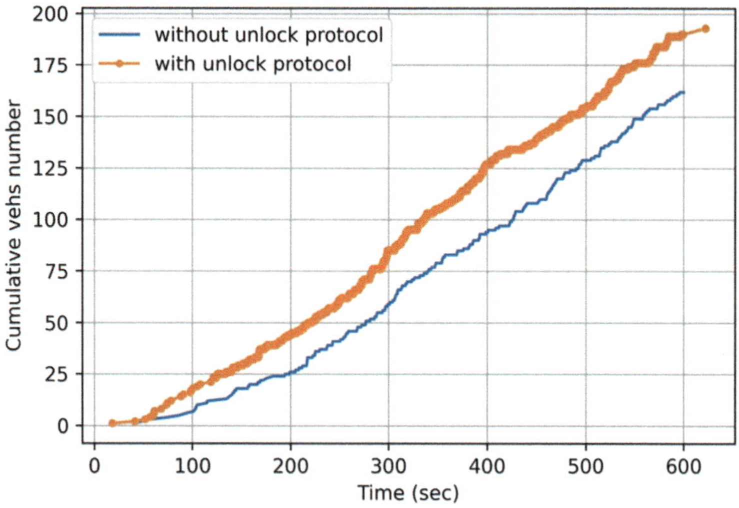

Fig. 13.24 The comparison between with and without unlock protocol

curve. This signifies that the proposed deadlock recovery protocol was able to significantly boost the overall intersection capacity. From the 10-min experiment, it can be inferred that the intersection capacity advanced by an additional 18%.

13.8.4 Leading Time for Deadlock Detection

In this section, an analysis is conducted on the duration required for detecting deadlocks, which is a crucial metric in traffic management systems. Swift detection of deadlocks allows authorities to take prompt action in recovering normal traffic flow. The EDP algorithms introduced earlier are employed to pinpoint the initiation of deadlocks. Subsequently, the speed of vehicles within the intersection is continuously monitored. If the average speed decreases to zero and remains at that level, it is determined that a deadlock has indeed occurred. Figure 13.25 visually portrays the progression of average vehicle speeds (measured in meters per second) during a mild deadlock scenario. At approximately 213 s, all vehicles in the loop access the intersection. Generally, the average speed exhibits a downward trend, although minor fluctuations are observed. By the 226th second, the EDP algorithm successfully detects the presence of a mild deadlock, with the corresponding average speed recording 2 m/s. At the 231st second, the average speed falls to zero, indicating a standard deadlock scenario where no vehicle is able to move. The time interval between detection of the deadlock and the point where average speed reaches zero can be interpreted as the lead time of deadlock detection.

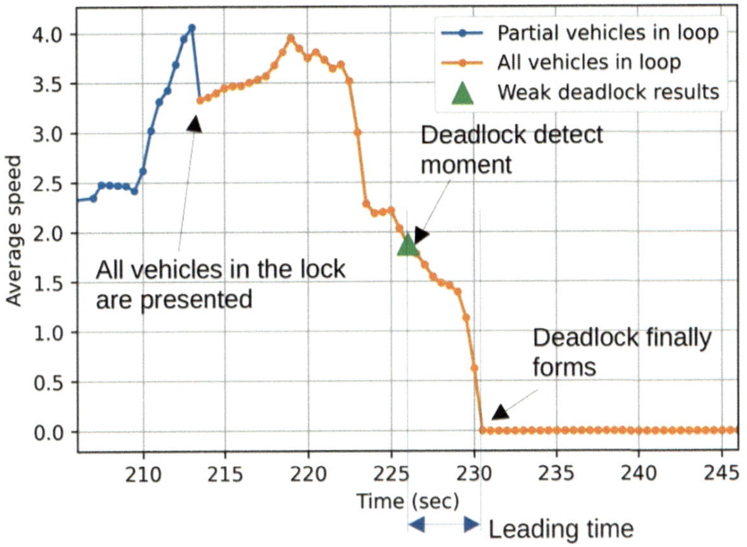

Fig. 13.25 The leading time of the deadlock detect

We conducted 10 simulations of intersection deadlock detection, each with a time horizon of 10 min. The resulting distribution of the time interval preceding deadlock detection is presented in the form of a histogram in Fig. 13.26, where the X-axis corresponds to the lead time for detection of deadlock, and the Y-axis denotes the frequency of such detections. Based on the results, it was observed that the average lead time for deadlock detection was approximately 13 s. The maximum frequency of occurrences was encountered with a 6-s lead time, which constituted more than half of the total sample. It was further noted that an increasing lead time was associated with a decrease in the probability of deadlock detection occurring, with a steep decline in frequency noted at longer detection times.

13.8.5 The Time of Recovery Time

When a deadlock situation arises, a specific duration is necessary to restore normal traffic conditions. Minimizing the time required to resolve the impasse would result in a larger capacity available for the intersection. Consequently, the moment at which the deadlock is successfully resolved becomes crucial for the operation of the intersection. An illustration of the re-routing unlock strategy is depicted in Fig. 13.27. The blue curve denotes the average velocity of vehicles within the deadlock region, while the orange curve symbolizes the average velocity of all vehicles encompassed by the loop. The green markers delineate the instances at which deadlock occurrence is detected. At approximately the 505th second, all vehicles have congregated

Fig. 13.26 The leading time of the deadlock detect and the average speed

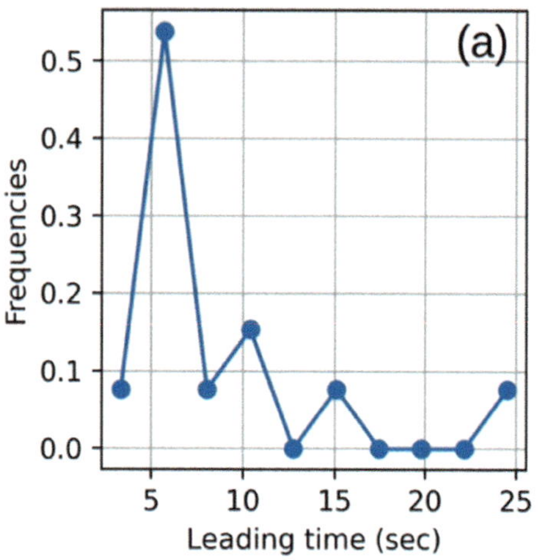

at the intersection, resulting in a consistent decline in speed commencing from the 511th second. Around the 512th second, identification of the deadlock is ascertained, subsequently prompting the implementation of the re-routing strategy. Due to the requirements imposed by this strategy, the speed of the vehicles no longer experiences a decline. Following the 518th second, the deadlock is successfully resolved and the velocities of these vehicles progressively recover. Consequently, the duration necessary for deadlock resolution spans approximately 6 s (518–512 = 6 s).

The recovery process was subjected to ten simulations, each with a ten-minute duration, and the resulting recovery time distribution was plotted as displayed in Fig. 13.28. Notably, the proposed methodology successfully resolved over 80% of

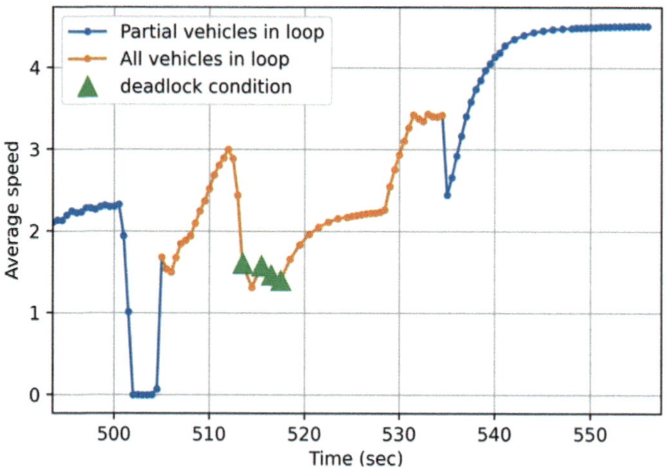

Fig. 13.27 The time used in the unlocking

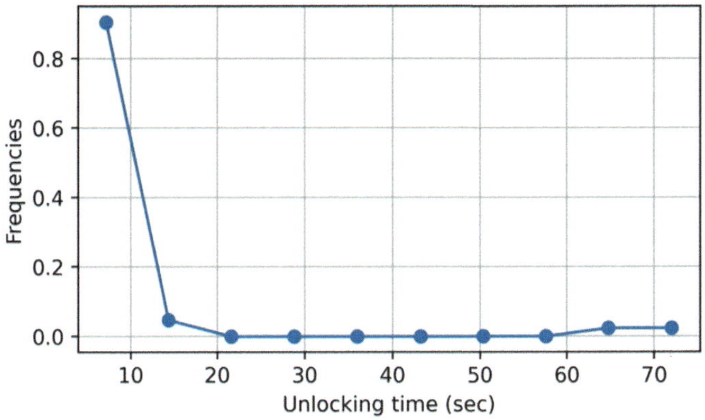

Fig. 13.28 Unlocking time distribution

deadlocks within a time frame of less than 10 s. A rapid decrease in frequency was observed as the recovery time increased. While the highest recorded recovery time was 70 s, it was associated with a negligible probability. On average, the unlocking process took approximately 6.01 s.

13.9 Notes

This chapter is a revised version of [19]. In the context of rush hour traffic, intersection deadlock represents an extreme state wherein traffic from all directions comes to a complete halt. Although prior literature extensively covers autonomous vehicle trajectory planning and joint optimization with signal control, the explicit modeling of intersection deadlock and cooperative control protocols for autonomous vehicles remains understudied. Consequently, deadlocks can persist and traffic flow cannot autonomously return to normal. This chapter aims to rectify this gap by presenting a framework for the detection, avoidance, and recovery of traffic deadlocks using connected and autonomous vehicle technology. We propose a comprehensive set of models and protocols that leverage connected and autonomous vehicle technology to effectively detect, avoid, and recover from traffic deadlocks. To test the efficacy of our proposed model, we utilize the Carla simulation platform, which yields satisfactory outcomes. Our findings indicate that, on average, our algorithm can detect impending deadlocks 13 s prior to their occurrence. Furthermore, upon activation of the recovery protocol, it takes approximately 6 s to restore traffic flow. Moreover, our results demonstrate that the proposed deadlock avoidance algorithm enhances traffic throughput by 35.7%, while the proposed deadlock recovery protocol further improves throughput by an additional 18%.

The proposed model is rigorously evaluated using the Carla simulation platform. Our results demonstrate that our algorithm consistently detects impending deadlocks an average of 13 s in advance, and once the recovery protocol is triggered, it takes approximately 6 s to restore traffic flow. Moreover, our findings indicate that the proposed deadlock avoidance algorithm enhances traffic throughput by 35.7%. Furthermore, the proposed deadlock recovery protocol further increases throughput by an additional 18%.

References

1. Arvin R, Khattak AJ, Kamrani M, Rio-Torres J (2020) Safety evaluation of connected and automated vehicles in mixed traffic with conventional vehicles at intersections. J Intell Transp Syst 25(2):170–187. https://doi.org/10.1080/15472450.2020.1834392
2. Bauza, R., & Gozálvez, J. (2013). Traffic congestion detection in large-scale scenarios using vehicle-to-vehicle communications. J. Netw. Comput. Appl. 36(5): 1295–1307. https://doi.org/10.1016/j.jnca.2012.02.007.
3. Cabri G, Gherardini L, Montangero M, Muzzini F (2021) About auction strategies for intersection management when human-driven and autonomous vehicles coexist. Multimed. Tools Appl. 80:15921–15936. https://doi.org/10.1007/s11042-020-10222-y
4. Coffman EG, Elphick M, Shoshani A (1971) System deadlocks. ACM Comput Surv 3(2):67–78. https://doi.org/10.1145/356586.356588
5. Dresner, K., Stone, P. (2004). Multiagent traffic management: a reservation-based intersection control mechanism. In: Proc. Third Int. Jt. Conf. Auton. Agents Multiagent Syst, 2004. Vol. 3 AAMAS, pp. 530–537.
6. Ghaderi Z, Alqahtani A, Bagherzadeh N (2018) AROMa: aging-aware deadlock-free adaptive routing algorithm and online monitoring in 3D NoCs. IEEE Trans Parallel Distrib Syst 29(4):772–788
7. Gorrieri A, Martalò M, Busanelli S, Ferrari G (2015) Clustering and sensing with decentralized detection in vehicular ad hoc networks. Ad Hoc Netw 36:450–464. https://doi.org/10.1016/j.adhoc.2015.05.019
8. Khayatian M, Mehrabian M, Andert E, Dedinsky R, Choudhary S, Lou Y, Shirvastava A (2020) A survey on intersection management of connected autonomous vehicles. ACM Trans. Cyber-Phys. Syst. 4(4):1–27. https://doi.org/10.1145/3364182
9. Liberzon D (2011) Calculus of variations and optimal control theory: a concise introduction. Princeton University Press, Princeton
10. Luo J, Wan Y, Wu W, Li Z (2020) Optimal Petri-net controller for avoiding collisions in a class of automated guided vehicle systems. IEEE Trans Intell Transp Syst 21(11):4526–4537
11. Luo J, Liu Z, Zhou M (2019) A Petri net based deadlock avoidance policy for flexible manufacturing systems with assembly operations and multiple resource acquisition. IEEE Trans. Ind. Inform. 15(6):3379–3387
12. Feng, Y., Xing, K., Zhou, M., Wang, X., and Liu, H. (2020). Robust deadlock prevention for automated manufacturing systems with unreliable resources by using general Petri nets. IEEE Trans. Syst. Man Cybern.: Syst. 50(10): 3515–3527.
13. Miculescu D, Karaman S (2020) Polling-systems-based autonomous vehicle coordination in traffic intersections with no traffic signals. IEEE Trans Autom Control 65(2):680–694. https://doi.org/10.1109/TAC.2019.2921659
14. Mishra KN, Kumar N (2020) Voting–priority-based deadlock prevention in multi-server multi-cs distributed systems. Natl Acad Sci Lett 43(7):625–630
15. Mirheli A, Tajalli M, Hajibabai L, Hajbabaie A (2019) A consensus-based distributed trajectory control in a signal-free intersection. Transp. Res. Part C Emerg. Technol. 100:161–176. https://doi.org/10.1016/j.trc.2019.01.004

16. Nikoukhah, R., Willsky, A. S., and Levy, B. C. (1992). Kalman filtering and Riccati equations for descriptor systems. In 29th IEEE Conference on Decision and Control, Honolulu, HI, USA, December 1990. IEEE, pp. 2886–2891.
17. Palmer GI, Harper PR, Knight VA (2018) Modelling deadlock in open restricted queueing networks. Eur J Oper Res 266(2):609–621
18. Rajamani, Rajesh (2011). Vehicle dynamics and control. Springer Science & Business Media, German.
19. Qi H, Song Y, Huang Z, Hu X (2023) Deadlock detection, cooperative avoidance and recovery protocol for mixed autonomous vehicles in unstructured environment. IET Intelligent Trans Sys itr2.12338. https://doi.org/10.1049/itr2.12338
20. Silberschatz A, Peterson G (1991) Operating Systems Concepts. Addison-Wesley, Reading, MA
21. Wu B, Qian L, Lu M, Qiu D, Liang H (2019) Optimal control problem of multi-vehicle cooperative autonomous parking trajectory planning in a connected vehicle environment. IET Intell Transp Syst 13(11):1677–1685
22. Yang K, Guler SI, Menendez M (2016) Isolated intersection control for various levels of vehicle technology: Conventional, connected, and automated vehicles. Transp. Res. Part C Emerg. Technol. 72:109–129. https://doi.org/10.1016/j.trc.2016.08.009
23. Zasadzinski, M., Mehdi, D., ans Darouach, M. (1991). Recursive state estimation for singular systems. In 1991 American Control Conference, Boston, June 1991. IEEE, pp. 2850–2851.
24. Zhou Y, Hu H, Liu Y, Lin SW, Ding Z (2020) A distributed method to avoid higher-order deadlocks in multi-robot systems. Automatica 112:108706

Chapter 14
Closing Thoughts: Future Microscopic Modeling

14.1 Systematic Modeling of Mobility Behavior

The first consideration is the multi level modeling of the driving behavior. Figure 14.1 presents the basic framework for such consideration. It should be noted that the meaning of the term 'behavior' in transportation or traffic engineering are manifold. Figure 14.1 distinguishes four levels of behaviors regarding the automobile:

- Travel behavior model. This level contains the highest travel behavior, which include the departure time choice and route choice (suppose the traveler select the automobile already). The two determines the moment when the vehicle enter the road network and which route (consecutive list of road segments) the vehicle would traverse. In this stage, the detailed movement of this vehicle is not known yet.
- In the second level, the microscopic traffic behavior emerges. Traditionally, the microscopic traffic behavior has been classified into three categories, car following and lane changing in roads, as well as gap acceptance within intersection. These microscopic traffic behaviors can also be decomposed into two dimensions, i.e., longitudinal and lateral movement. Various models are developed to cope with the above movements. The main theme of this book is devoted to this level of behavioral description.
- The third level concerns the driving behavior. The difference between traffic behavior (the second level) and driving behavior is that, driving behavior need to consider the vehicle kinetics and dynamics. In traffic behavior modeling, the vehicle is often represented by a mass point, where its influence can be encapsulated by Newtonian mechanics in deterministic mode or stochastic mode. However, the vehicle body is with certain size. In contrast, in driving behavior modeling, the vehicle model is explicitly considered. The control input of microscopic traffic behavior is acceleration, while the control input in the driving

© The Author(s), under exclusive license to Springer Nature Singapore Pte Ltd. 2024

H. Qi, *Stochastic Two-Dimensional Microscopic Traffic Model*, Lecture Notes in Intelligent Transportation and Infrastructure,
https://doi.org/10.1007/978-981-97-3597-6_14

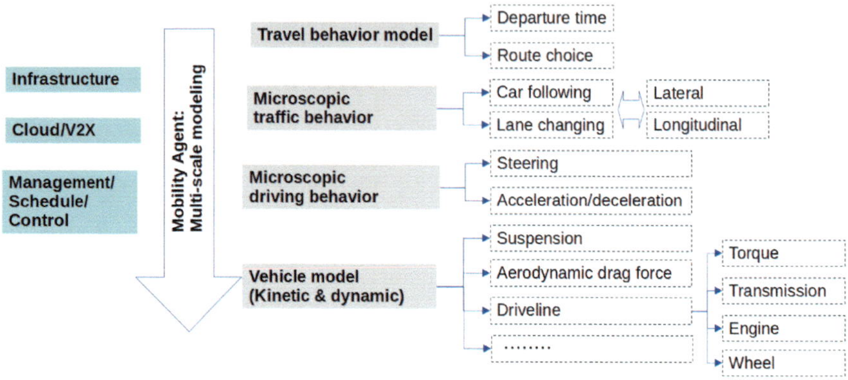

Fig. 14.1 Framework of future microscopic traffic modeling

behavior should include the steering and acceleration. In this level, the vehicular movement can be described in kinetic or dynamic perspective.

- In the lowest level, the detailed vehicle model plays indispensible role. To make the vehicle move along the trajectories specified by driving behavior model or microscopic traffic model, the components of vehicles need to be work properly, including suspension component, driveline component, pavement-tire interaction etc. Also the aerodynamic forces come into play.

The above four levels together describes how vehicle move in the road network, under the influence of surround vehicles. This is summarized as "mobility agent: multi level modeling" in Fig. 14.1. However, there are also other stakeholders that need to be considered, which is listed at the left of Fig. 14.1. These include the following:

- Infrastructure. The infrastructure is manifold. They include the pavement, traffic signs, other facilities, etc. These infrastructure influences the behavior of the drivers and as a result, the vehicle movement subject to change. The concept of virtual boundary and virtual potential are typical effort toward the modeling of the influence of infrastructure on the driving behavior. However, there is also other more complex infrastructure such as the roundabout. Thus more efforts are required to understand the behaves of the multi scale behavioral modeling.
- Cloud/V2X. One of the promising technologies that can promote the efficiency and enhance the safety of the transportation system is the connected technology. Through this, various components of the system are connected and can exchange information. The vehicle can also access this type of infrastructure and the resulting capability is termed as V2X. With V2X, the vehicle can access information either near field or far field. As a result, the capability of the vehicles can be greatly enhanced in several levels. As an example, the vehicle can make long term trajectory planning based on the information provided by its own sensors and those by V2X.

– Management and control of the transportation system. This refers to the control or management on the traffic flow, including speed limit, signal control, etc.

If the models considering the above attributes are developed, the vehicular movement then can capsulate the following characteristics: (1) driver behaviors. The drivers differ from person to person. Their behavioral attributes are not the same. For instance, the reaction time, risk preference can influence the vehicular moment; (2) different vehicular dynamics. Different types of vehicles (passenger, truck, etc.) have different vehicular attributes; (3) different locations/countries. As demonstrated in Chap. 2, the difference of the traffic flow between different countries can be quantitatively measured. It implies that current methodologies, where the difference between some unknown attributes are neglected. With the above consideration, the resulting microscopic traffic behavioral model would be more realistic and useful.

14.2 Incorporating Advanced Technology: LLM and AIGC

As of 2023, there is a hot topic of technology that transforms the transportation sector, as well as auto industry. That is AI, artificial intelligence. Notably driven by the LLM (large language model), the AI industry begins its booming. After its successful debut at the end of 2022 by openAI company, ChatGPT, which is an instance of LLM has been applied to many fields and it is envisioned that various fields would be reshaped. The microscopic modeling and application of traffic flow, which has a sub-field of data driven approach, has been influenced by this trend either. The influences of AI on the topic discussed in this book include but not limited to the following aspects:

(1) New approach toward data-driven microscopic modeling. Conventional data driven modeling approach, such as MLP, clustering etc. only can accommodate simple scenarios. Complex behavioral instance, such as inference chain of the driver and the resulting complex vehicular dynamics represented by the trajectories may not be captured by machine learning techniques developed in previous decades. Incorporating new powerful methods from AI field is a possible solution. Actually, from the beginning of the booming of ChatGPT, various AI techniques such as transformers are introduced into vehicular modeling. These developments include the re-construction of data driving microscopic traffic modeling. However, there are also concerns of the suitability of the data-driven approach toward microscopic modeling.
(2) Trajectory generation from behavioral perspective. In AVs test, the background traffic flow can be provided by traffic simulation software. Thus the fidelity of the background traffic flow is crucial for the algorithms, as if these background traffic flow cannot represent the real world traffic flow operation, the developed AV algorithm may not work perfectly when they are deployed to open road. AIGC, AI-Generated Content is a popular technology to generate desired contents based on user specification. Actually, the method of AIC has been employed to the trajectory generation. However, there are still challenges: the

generated trajectory may not subject to the physical constraints. For instance, how to generate trajectory for different vehicle types?

14.3 Modeling Irregular Behavior: Safety and Distraction Considerations

Most microscopic traffic flow model assume that the driver would keep certain safe distance from either leader or neighbors. This assumption is reasonable most of the time. However, there are certain moments when the driver deviate from normal behavior of driving. These moments include distraction, or safety critical condition. As an example of the latter type, suppose two neighboring vehicles drive at parallel lanes. If one vehicle change lane without considering the risk, another vehicle need to steer away from the previous vehicle. When crash is about to occur, driver behavior is driven by instinct rather than assumptions used in micro-traffic models. Under distraction scenario such as using phone, the driving behavior may deviate from the conventional microscopic models. Hence traditional model may not be suitable for such scenario.

This book investigates the irregular driving behavior from jerk perspective. The view point is that, jerk represents the abrupt change of the acceleration, and it closely relate to the traffic flow operational risk. Thus a jump-diffusion approach is used. The jerk is modeled by the compound Poisson process. However, this approach may have drawbacks regarding modeling the safety critical scenario and the distraction scenario.

To model the above irregular scenarios and other types such as drunk driving, drug driving, etc, firstly the behavioral characteristics should be analyzed based on real world dataset. However, this type of data is rare, which bring challenge to the modeling. This problem can be addressed by using the simulator. After that, new methods to model the irregular behaviors can be developed.

14.4 Developing Platforms Support Simulation, Generation, Inference and More

Main commercial and open source microscopic traffic have incorporated the microscopic traffic models. The drawbacks in current practice of simulations are manifold, and gradually solved by the academic groups or industrial practitioners. Simulation packages become more and more important to the AVs industry, and play an important role in the prototype design and test of AV algorithms.

The books enhance current microscopic traffic flow models with stochasticity consideration and two-dimensional extension. Besides, the vehicle body is also considered both in deterministic and stochastic configurations. This makes the resulting vehicular movement more realistic. It would be use full to embed the proposed model into the well developed simulation packages. However, this is not an easy task. As an example, in current simulation settings, the lateral dynamics in SUMO are not activated if the lateral resolution setting is not switched on. After the lateral resolution is activated, the lateral dynamics is discrete determined. Therefore, the real world vehicular movement regarding the lateral dimension deviates from the real world nature. When the simulation package is applied to investigate the AV's safety when HDV's lateral movement is of concern, the resulting outputs may far from realistic.

14.5 Multi Scale Modeling

As the transportation system transitions from HDVs to connected and CAVs, transportation agencies face the challenge of effectively managing traffic flow. Unlike the automotive industry, these agencies primarily focus on network-level or link-level traffic flow as a whole, rather than individual vehicle-level characteristics. While significant progress has been made in developing micro/macroscopic traffic flow models, traffic control models, and capacity evaluation models for HDVs in transportation regulation, there is still a need for redesigned theoretical and technical tools to address the mixed flow of CAVs. Some studies have examined capacity adjustment factors specifically for CAV flow in certain facilities, but further research is required in this area.

Despite these efforts, challenges persist in merging two branches of theory: autonomous vehicle research and traffic flow research, in order to develop analytical tools applicable to both fields. The key to this merger lies in understanding the behavioral disparities between HDVs and CAVs. CAVs exhibit reduced reaction delays, lower stochasticity, accurate goal pursuit, and cooperative driving abilities. In contrast, HDVs have longer reaction delays, higher uncertainty, greater volatility, and at times, exhibit competitive behavior. Moreover, CAVs can acquire information beyond their sensing radius and pursue objectives at multiple levels. These behavioral differences suggest that conventional network and link-level macroscopic traffic models, which overlook behavioral heterogeneity, may not be adequate. For instance, the behavior of a single CAV locally can impact the global traffic flow characteristics, as the gap-regulating behavior of an individual CAV during lane changes might not be captured by link-level macroscopic models. Additionally, conventional models fail to incorporate heterogeneous information from different types of vehicles, which is valuable for multi-scale modeling. Apart from the challenges arising from behavioral disparities, other difficulties exist in multi-scale modeling for mixed AV flow, particularly in congested traffic conditions.

(1) The increasing number of vehicles in mixed AV flow introduces a curse of dimensionality, as AVs need to infer or predict the behavior of HDVs across different timescales.

(2) The complex game behavior interactions among HDVs and AVs, as well as among AVs themselves, make microscopic/macroscopic state prediction more challenging. With numerous opportunities for interaction among CAVs and HDVs, a single game behavior can lead to a chain influence, causing decision trees to expand exponentially.
(3) Under congested scenarios, the decision space of AVs may significantly reduce, limiting their ability to execute optimal driving behaviors.
(4) The issues of information radius and decision-influence radius are interconnected, presenting both opportunities for global optimization and analytical challenges.

The aforementioned challenges underscore the need for new theories, models, and methods. One possible approach is to adapt the concepts of mesoscopic and macroscopic models, which handle mixed autonomous vehicles in a pseudo-aggregated manner. This would enable the derivation of traffic flow dynamics at the lane/link/network levels while preserving the outcomes of individual CAVs. In essence, new models and method stacks are required to capture the behavioral differences between CAVs and HDVs while ensuring consistency at both the individual CAV level and aggregate level.

GPSR Compliance

The European Union's (EU) General Product Safety Regulation (GPSR) is a set of rules that requires consumer products to be safe and our obligations to ensure this.

If you have any concerns about our products, you can contact us on ProductSafety@springernature.com

In case Publisher is established outside the EU, the EU authorized representative is:

Springer Nature Customer Service Center GmbH
Europaplatz 3
69115 Heidelberg, Germany

Batch number: 10703867

Printed by Printforce, the Netherlands